THERMOMECHANICAL PROCESSING OF MICROALLOYED AUSTENITE

THERMOMECHANICAL PROCESSING OF MICROALLOYED AUSTENITE

Proceedings of the International Conference on the Thermomechanical Processing of Microalloyed Austenite sponsored by the Ferrous Metallurgy Committee of The Metallurgical Society of AIME held in Pittsburgh, Pennsylvania, August 17-19, 1981.

Edited by

A.J. DeARDO
University of Pittsburgh
Pittsburgh, Pennsylvania

G.A. RATZ
Molycorp , Inc.
Pittsburgh, Pennsylvania

P.J. WRAY
United States Steel Corporation
Monroeville, Pennsylvania

A Publication of   The Metallurgical Society of AIME

A Publication of The Metallurgical Society of AIME
420 Commonwealth Drive
Warrendale, Pennsylvania 15086
(412) 776-9000

The Metallurgical Society and American Institute of
Mining, Metallurgical, and Petroleum Engineers are
not responsible for statements or opinions in this publication.

Printed in the United States of America.
Library of Congress Card Catalogue Number 82-61571
ISBN Number 0-89520-398-7

ORGANIZING COMMITTEE

A.J. DeArdo (General Chairman)
University of Pittsburgh

John B. Ballance
Journal of Metals

A.T. Davenport
Republic Steel Corporation

P.D. Deeley
Shieldalloy Corporation

M. Korchynsky
Union Carbide Corporation

P.L. Mangonon
Foote Mineral Company

J.W. Morrow
Climax Molybdenum Company

G.A. Ratz
Molycorp, Inc.

W.G. Wilson
Niobium Products Company

P.J. Wray
United States Steel Corporation

FOREWORD

One of the most important developments in the field of ferrous metallurgy during the last two decades has been the thermomechanical processing of microalloyed steels. The evolution of this processing, which produces such important improvements in the strength and toughness of steels, can be traced back to at least the BISRA Conference "Metallurgical Developments in Carbon Steels" at Harrogate, England, in 1963. At a subsequent conference, "Strong, Tough Structural Steels" at Scarborough, England, in 1967, the potential benefits of thermo-mechanical processing in the austenite range were emphasized, but it was only in the 1970s that interest in the commercial benefits of thermomechanical processing really took hold. This commercial interest was highlighted by the Low-Alloy High-Strength Steel Conference in Nuremberg, Germany, in 1970, and matured with the Microalloying '75 Conference in Washington, D. C.

As the thermomechanical processing of microalloyed steels was being adopted in production, it became necessary to predict ever more precisely the influences of rolling schedule and microalloy content on the austenite microstructure. Spurred by the commercial need to develop this predictive capability, the thermomechanical pro-cessing of microalloyed austenite became a fertile area of steel re-search, and accounts of some of the fairly basic studies were presented at the symposium "Hot Deformation of Austenite" sponsored by The Metallurgical Society of AIME in 1975. In the following years, separate accounts of additional research were published, but as the decade ended there was a general realization that our basic understanding of the phenomena associated with the thermomechanical processing of microalloyed austenite was still inadequate.

To help improve the scientific bases of thermomechanical processing, a series of informal discussions were begun in Pittsburgh in 1979. Out of the Pittsburgh discussions came the proposal to hold the present conference. An organizing committee was formed, the first task of which was to define the scope of the program. After some deliberation, the committee decided to limit the range of interest to the evolution of the microstructure of austenite and ferrite during processing. Accordingly, the arrangement of the thirty-three papers presented at the Conference generally follows the evolution of austenite microstructure, beginning with the coarse grains at high temperatures where microalloying additions are dissolved, through the region of complex interactions between recrystallization and precipitation, to below the recrystallization temperature where grain elongation occurs, and finishing at the austenite transformation. We hope the reader will agree that emphasis on the evolution of microstructure provides a better understanding of thermomechanical processing. We also hope that the better understanding will lead to more rapid improvement in thermomechanical processing as practiced in commericial production.

A. J. DeArdo
University of Pittsburgh
Pittsburgh, Pennsylvania

G. A. Ratz
Molycorp, Inc.
Pittsburgh, Pennsylvania

P. J. Wray
United States Steel Corporation
Monroeville, Pennsylvania

August   1982

ACKNOWLEDGMENTS

The Organizing Committee wishes to express their thanks to the following:

.....Ferrous Committee of The Metallurgical Society of AIME for its support.

.....J.M. Gray, S.S. Hansen, M. Korchynsky, J.W. Morrow, G.R. Speich, H. Stuart, and P.J. Wray for serving as Session Chairmen.

.....K. Mielityinen-Tiitto for her invaluable help with editing chores.

.....D. Davis (Union Carbide Corporation), K. Horney and N. McDermott (Niobium Products Company), S. Jones and P. Sawyer (University of Pittsburgh), and L.Smith (Molycorp, Inc.) for their secretarial assistance.

.....And for their generous financial contributions
Climax Molybdenum
Foote Mineral Company
Molycorp, Inc.
Niobium Products Company
Shieldalloy Corporation
Union Carbide Corporation

GRAIN SIZE CONVERSION TABLE

Equivalent Measures of an
Equiaxed Grain Structure

ASTM No. G	Mean Intercept Length $\bar{\ell}$, μm	Surface to Volume Ratio S_V, mm^{-1}
00	453	4.41
0	320	6.25
1	226	8.84
2	160	12.5
3	113	17.7
4	80	25
5	56.6	35.3
6	40	50
7	28.3	70.7
8	20	100
9	14.1	142
10	10	200
11	7.07	283
12	5	400

Note: $G = 10 - 2 \log_2 (\bar{\ell}/10)$

$\quad\quad = 10 - 6.6439 \log_{10} (\bar{\ell}/10)$

and $S_V = 2000/(\bar{\ell})$.

TABLE OF CONTENTS

AUSTENITE GRAIN COARSENING AND THE EFFECT OF THERMO-

MECHANICAL PROCESSING ON AUSTENITE RECRYSTALLISATION

R K Amin

British Steel Corp'n
Sheffield Laboratories
Rotherham, England

F B Pickering

Department of Metallurgy
Sheffield City Polytechnic
Sheffield, England

A study of the grain coarsening characteristics of Nb and V steels has shown that Nb steels have higher grain coarsening temperatures than V steels, and increasing N content has little effect. A maximum grain coarsening temperature occurs at the stoichiometric ratio in Nb steels, but this was less marked in V steels. The results have been tested against a model for predicting the grain coarsening temperature with encouraging results, but it has been shown that in the V steels which are Al treated, the grain coarsening temperature is controlled more by AlN than by VN. During thermo-mechanical working, Nb was shown to retard recrystallisation at high temperatures by a solute drag effect, and at lower temperatures by strain induced Nb(CN) precipitation. These effects have been discussed in terms of the nucleation and growth of recrystallising grains. V in solution has much less effect than Nb in retarding recrystallisation, but at ~950°C VN can be strain induced to precipitate and so retard recrystallisation. The complex interactions between composition, microstructure and processing parameters in controlling recrystallisation have been described, and the various nucleation sites for recrystallisation have been identified.

1. Introduction

The purpose of thermo-mechanical processing is to control the structure, morphology and grain size of the austenite in H.S.L.A. steels so that the eventual transformation of the austenite produces the optimum ferritic structure. Microalloying additions are often used to help realise this. The important features are:-

(i) the reheating temperature which controls the initial austenite grain size and the solution and precipitation of the micro-alloy carbides and nitrides. Undissolved precipitates can control the grain size, finer grains recrystallising more rapidly than coarser grins, whilst dissolved solute influences the recrystallisaton by virtue of strain induced precipitation, as also do large undissolved particles which may serve as nuclei for recrystallised grains,

(ii) the rolling treatment which influences the degree of deformation of the austenite grains, the extent of recrystallisation and also the intensity of strain induced precipitation,

(iii) the composition, particularly with respect to the microalloy: carbon/nitrogen stoichiometric ratio, which controls the temperature dependence of the carbide/nitride solubility and thereby the recrystallisation,

(iv) inter-pass times or holding times which influence the degree of recrystallisation and grain growth between passes.

The evolution of the austenite recrystallisation has been studied by many workers using various methods of deformaton such as rolling(1-10), compression(11-13), tension(14,15) and torsion(16). The mechanism by which microalloying retards recrystallisation is not entirely understood, both solute drag(1, 9, 15, 17) and strain induced precipitation in Nb (1, 14, 15, 18, 19) and V(6, 13, 20) steels being cited as the controlling process. On many occasions, both mechanisms have been suggested to be operative, but their relative importance has not been clearly established, particularly with respect to the nucleation and growth stages of recrystallisation.

The present work describes the effects of Nb and V on austenite grain growth and recrystallisation, and attempts to define some of the mechanisms involved.

2. Experimental Methods

Three series of steels were made to a base composition 0.08%C, 0.9%Mn, 0.3%Si, 0.02/0.05%Al with additions of up to 1.03%Nb and up to 0.98%V, the V steels being at both low N (0.006/0.009%) and high N (0.015/0.020%) levels. In addition a steel containing very low interstitial contents of 0.005%C and 0.003%N with 0.29%Nb was also made in order to try to distinguish between solute drag and precipitation effects on the retardation of austenite recrystallisation. The analysis of these steels are given in Table I.

TABLE I - Analyses of Experimental Steels

Steel Series	Steel No.	Composition (mass %)							
		C	Mn	Si	Al	Nb	V	N	M/C
Base Steel	1	0.08	0.76	0.24	0.04	–	–	0.008	–
Nb Steels	2	0.09	0.89	0.25	0.05	0.07	–	0.008	0.85
	3	0.09	0.93	0.27	0.07	0.11	–	0.009	1.16
	4	0.09	0.79	0.18	0.10	0.16	–	0.007	1.81
	5	0.09	0.82	0.23	0.04	0.46	–	0.006	5.22
	6	0.09	0.96	0.35	0.06	0.73	–	0.007	8.11
	7	0.09	0.94	0.28	0.06	1.03	–	0.007	11.98
Low N-V Steels	8	0.09	0.84	0.25	0.02	–	0.06	0.007	0.70
	9	0.09	0.87	0.28	0.02	–	0.11	0.007	1.21
	10	0.09	0.93	0.29	0.02	–	0.14	0.007	1.59
	11	0.08	0.86	0.26	0.03	–	0.22	0.008	2.65
	12	0.09	0.85	0.26	0.05	–	0.55	0.008	6.40
	13	0.08	0.74	0.22	0.02	–	0.98	0.008	11.51
High N-V Steels	14	0.08	0.85	0.25	0.03	–	0.14	0.020	1.80
	15	0.07	0.91	0.24	0.03	–	0.25	0.015	3.31
	16	0.08	0.82	0.21	0.02	–	0.46	0.018	6.13
	17	0.08	0.96	0.27	0.03	–	0.98	0.018	12.40
Low C-N Nb	18	0.005	1.13	0.30	0.05	0.29	–	0.003	58.0

   All the steels were solution treated at 1300°C to produce a uniform initial structure and condition. Grain coarsening studies were carried out at temperatures between 950°C and 1300°C, using a soaking period of 0.5h, followed by rapid cooling to ∿900°C and holding for 5 minutes, which was then followed by quenching in iced brine. This facilitated the etching of the prior austenite grain boundaries for grain size measurements, and did not affect the austenite grain size.

   Thermo-mechanical processing was carried out by hot rolling, Fig.1 showing a flow chart for the various conditions used. Specimens were reheated at 1150°C or 1300°C with rolling at 900/950°C or 1200/1250°C to 20% or 50% reduction in one pass, followed by quenching in iced brine. Other specimens were also held at 950°C and 1250°C for times up to 1000s, and then iced brine quenched.

   The metallographic examination of the austenite grain structure was carried out on specimens etched in saturated picric acid solutions in either water or alcohol to which had been added a small amount of wetting

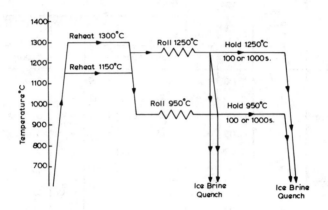

Fig.1.   Flow Sheet of Experimental Rolling Programme

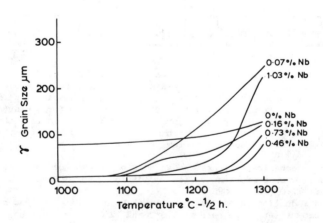

Fig.2.   Austenite Grain Coarsening Curves for Nb Steels

agent either with or without 2% ammonium persulphate. The austenite grain sizes and the volume fraction of recrystallised austenite were determined by standard quantitative metallographic techniques, using the mean linear intercept method and point counting respectively. Relative errors of these measurements were <5% of the measured values.

3. Experimental Results

3.1 Austenite Grain Coarsening Characteristics

(a) Nb Steels. Austenite grain coarsening curves for the Nb steels are shown in Fig.2. Whilst all the Nb steels were consistently finer grained than the base steel below 1200°C, at 1300°C abnormal grain growth in the 0.07% and 1.03%Nb steels produced coarser grain sizes than the base steel. Abnormal grain growth decreased towards, and was virtually absent at the Nb:C stoichiometric ratio. Increasing Nb content refined the austenite grain size at all temperatures up to 1200°C, Fig.3(a). Up to 1050°C, Nb contents as low as 0.07% produced an austenite grain size of ∿10μm, but increasing Nb had no further effect. At the higher temperatures of 1150/1200°C increasing Nb progressively refined the grain size up to about the stoichiometric Nb:C ratio, but higher Nb contents slightly coarsened the grain size. These effects were also reflected in the grain coarsening temperature, Fig.4, which increased to >1200°C at the stoichiometric Nb:C ratio, but decreased sharply at the higher Nb contents.

(b) Low N-V steels. All the low N-V steels had finer austenite grain sizes than the base steel at temperatures up to 1000°C, Fig.3(b), but at 1050°C this effect was only observed for V contents above 0.14%. Generally there was little effect of V on the grain size at temperatures above 1150°C due to the fact that the grain coarsening temperature never exceeded 1100°C. The effect of V on the grain coarsening temperature, Fig.4, was an increasing temperature up to about the stoichiometric V content, and only a slight decrease in grain coarsening temperature at higher V contents. The grain coarsening temperatures of the low N-V steels were consistently lower than those of the Nb steels, and did not exceed 1050°C.

(c) High N-V Steels. The behaviour was not dissimilar to that of the low N-V steels, Fig.5, but it can be seen that at 1150°C, in contrast to the low N-V steels, increasing V progressively refined the austenite grain size. At 1000°C all the V steels had finer grain sizes than the base steel, but at 1050°C this was only observed for V contents above 0.14%. The effect of V on the grain coarsening temperature was similar in high N steels to that in the low N steels, Fig.4, except that above 0.46%V the grain coarsening temperature was slightly higher in the high N steels.

3.2 Effect of Thermo-mechanical Treatment

3.2.1 Recrystallisation Effects

(a) Nb steels. To avoid strain induced Nb(CN) precipitation during rolling, and so to study the effect of Nb in solution, the steels were reheated at 1300°C and rolled at 1250°C. After both 20% and 50% reduction there was complete recrystallisation in the base steel, but the 0.07% and 0.16%Nb steels were only partially recrystallised ater 20% reduction. After 50% reduction the 0.07%Nb steel was fully recrystallised but increasing Nb produced less recrystallisation, and holding at 1250°C only resulted in a slow increase in the amount of recrystallisation. Typical partially recrystallised structures are shown in Fig.6, for the 0.46% and

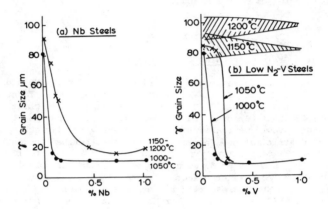

Fig.3.   Effect of Nb and V on Austenite Grain Size
          in low N Steels

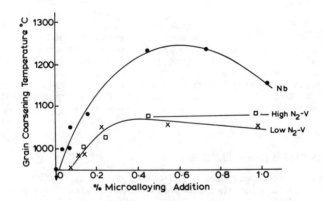

Fig.4.   Effect of Nb and V on Austenite Grain
          Coarsening Temperature

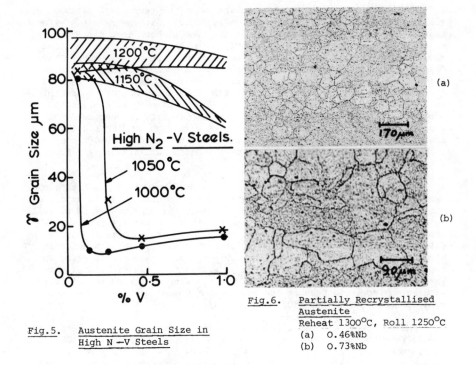

Fig.5. Austenite Grain Size in High N —V Steels

Fig.6. Partially Recrystallised Austenite
Reheat 1300°C, Roll 1250°C
(a) 0.46%Nb
(b) 0.73%Nb

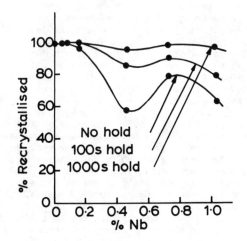

Fig.7. Effect of Nb on Recrystallisation of Austenite at 1250°C
(Reheat 1300°C, Roll 1250°C

0.73%Nb steels. The effect of Nb on the extent of recrystallisation during rolling and holding at 1250°C is shown in Fig.7, which reveals the effectiveness of Nb in retarding recrystallisation even at such high temperatures. At this rolling temperature of 1250°C, more than ∿0.10%Nb was required to retard recrystallisation and the effect increased up to ∿0.46%Nb. Above 0.46%Nb the extent of recrystallisation increased due to the undissolved Nb(CN) acting as nuclei for recrystallised grains. Decreasing the holding temperature to 950°C after rolling at 1250°C, allowed Nb(CN) to precipitate in the unrecrystallised austenite. The effects of Nb were similar to those observed during holding at 1250°C, in that increasing Nb decreased the amount of recrystallisation. However, the very high 1.03%Nb steel showed an increased amount of recrystallisation during holding at 950°C compared with 1250°C, and the increased recrystallisation rate also gave a finer recrystallised grain size.

To investigate the effect of strain induced Nb(CN) precipitation on recrystallisation, the rolling temperature was lowered to 950°C whilst still using the 1300°C reheating temperature. The progress of recrystallisation on holding at 950°C is shown in Fig.8(a). After 20% reduction, recrystallisation did not start even after 1000s at 950°C in the 0.07% and 0.16%Nb steels, but after 50% reduction recrystallisation started after ∿100s and increased slowly with increasing holding time. Increasing Nb tended to decrease the extent of recrystallisation, Fig.8(a), due to a larger amount of strain induced NbC precipitation as the degree of supersaturation increased because the composition moved nearer towards the stoichiometric ratio for NbC.

The effect of a finer initial austenite grain size and less dissolved Nb was investigated by decreasing the reheating temperature to 1150°C, and rolling and holding at 950°C. The effects are shown in Fig.8(b), the extent of recrystallisation being considerably greater than after reheating at 1300°C. Reductions of 20% and 50% produced very similar results.

As can be seen in Fig.9, Nb contents up to 0.16% markedly refined the recrystallised austenite grain size. Higher Nb contents had a decreasing effect on the grain size. The effect of holding at 1250°C and 950°C on the recrystallised grain size is shown in Figs.10(a) and (b). Rolling and holding at 1250°C gave little or no grain growth during the first 100s, but thereafter with an increased holding time of 1000s, growth occurred at 0.07% but not at 0.16%Nb Fig.10(a). Holding at 950°C resulted in much less grain growth in the 0.07%Nb steel, and again grain growth was inhibited with more than 0.16%Nb, Fig.10(b).

(b) V Steels. With both 20% and 50% reduction at 1250°C, after reheating at 1300°C, both the high and low N-V steels were completely recrystallised. Holding at 1250°C simply produced growth of the recrystallised grains. Reducing the rolling temperature to 950°C, and holding at 950°C, showed that whilst the base steel was partially recrystallised immediately after rolling, Fig.11, the V steels were completely unrecrystallised, Fig.12. Progressive recrystallisation occurred during the holding period, Fig.11. The amount of recrystallisation decreased with increasing V content in both the high and low N steels, Figs.11(a) and (b), and increasing the rolling reduction from 20% to 50% increased the amount of recrystallisation in all the V steels. Compared with the low N-V steels, the high N-V steels showed more recrystallisation after holding for 100s at 950°C, but the high N content slowed down the rate of recrystallisation so that after 1000s holding at 950°C the high N-V steels were less recrystallised than the low N-V steels. As can be seen from Fig.13, increasing the V content decreased the amount of recrystallisation in both high and low N steels, but the effect of V was more pro-

8

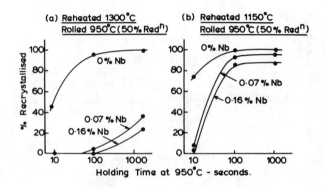

(a) Reheated 1300°C
Rolled 950°C (50% Red^n)

(b) Reheated 1150°C
Rolled 950°C (50% Red^n)

% Recrystallised

0% Nb

0.07% Nb
0.16% Nb

0% Nb

0.07% Nb

0.16% Nb

Holding Time at 950°C - seconds.

Fig.8.   Effect of Holding at 950°C on Recrystallisaton
of Austenite in Nb Steels

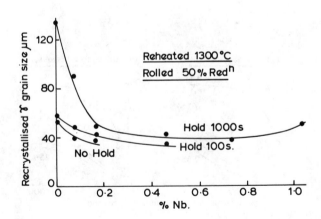

Recrystallised γ grain size μm

Reheated 1300°C
Rolled 50% Red^n

Hold 1000s

Hold 100s.

No Hold

% Nb.

Fig.9.   Effect of Nb on Recrystallised Austenite Grain
Size after Rolling and Holding at 1250°C

9

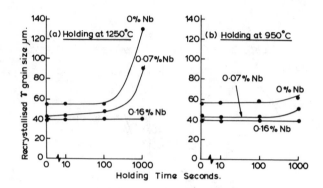

Fig.10.   Effect of Holding Nb Steels at 1250°C and 950°C
           after Reheating at 1300°C and Rolling 50%
           reduction at 1250°C on Recrystallised Grain Size

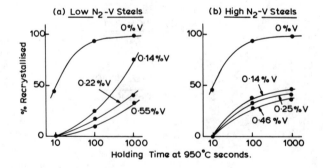

Fig.11.   Effect of Holding at 950°C in V Steels on
           recrystallisaton after Reheating at 1300°C
           and Rolling 50% Reduction at 950°C

(a)   (b)

Fig.12.   Unrecrystallised Austenite Structures in V Steels
           Reheat 1300°C, Roll 950°C.
           (a) 0.14%V
           (b) 0.22%V

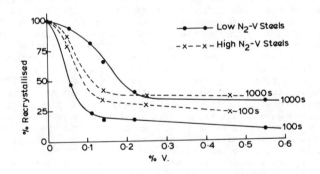

Fig.13.   Effect of V and Holding Time at 950°C on
           recrystallisation after Reheating at 1300°C
           and Rolling 50% Reduction at 950°C

nounced up to ∿0.2% and thereafter had relatively little further effect.

The effect of decreasing the reheating temperature to 1150°C, and rolling and holding at 950°C, is showing in Fig.14. Increasing the rolling reduction increased the amount of recrystallisation. Whilst the base steel was heavily recrystallised immediately after rolling, the V steels showed only a small amount of recrystallisation. Holding at 950°C, however, resulted in marked recrystallisation in the V steels, and the high N steels showed less recrystallisation than the low N steels. The rate of recrystallistion decreased with increasing holding time at 950°C in all the V steels, Fig.14, but recrystallisation was still incomplete even after holding for 1000s.

The effect of V on the recrystallised austenite grain size in the high and low N steels is shown in Fig.15, after reheating at 1300°C, rolling at 1250°C and holding at 950°C. Increasing the V content at both N contents refined the recrystallised grain size, which was smaller in the higher N steels containing more than 0.14%V. Growth of the recrystallised grains in the base steel occurred on holding at both 950°C and 1250°C, but increasing the V content tended to inhibit grain growth. Grain growth was completely inhibited at 0.55%V. An increase in V content slowed down the growth of recrystallised grains in both the high and low N steels at both 950°C and 1250°C.

4. Discussion of Results

4.1 Austenite Grain Coarsening

(a) Nb Steels. Increasing the Nb content refined the austenite grain size at all temperatures up to 1200°C, Figs.2 and 3, presumably due to the increased volume fraction of undissolved Nb(CN). However, Fig.3 shows that there seems to be a limiting austenite grain size below which even the highest Nb contents cannot produce further refinement. This limiting grain size increases with increasing temperature because of both solution of the Nb(CN) and particle coarsening. The undissolved precipitates at the highest reheating temperature are coarse whilst those precipitated during cooling or during reheating to the austenitising temperature are much finer. It has been shown many times that it is the finer pinning precipitates which are most effective in inhibiting grain growth, and it is possible from the published solubility data to calculate the amount of Nb present as these fine precipitates after solution treatment at 1300°C and then re-heating at any particular austenitising temperature. The results shown in Fig.16 clearly confirm the efficiency of such fine precipitates in refining the austenite grain size at any particular austenitising temperature. They also show that for a constant volume fraction of Nb(CN), the increasing particle size with increasing temperature results in a coarser austenite grain size. The effect of volume fraction and particle size of precipitates on the grain size of the resulting austenite have been depicted graphically (21,22), and results from the present work are shown on such a graph in Fig.17. The effect of increasing temperature in decreasing the volume fraction and increasing the particle size of the precipitates on the ensuing austenite grain size is clearly seen.

The grain coarsening temperature was a maximum at approximately the stoichiometric Nb:C ratio , Fig.4, and this can be seen to be due to the fact that the estimated volume fraction of fine Nb(CN) precipitates was a maximum at this ratio, Fig.18. The reason for this is that the temperature

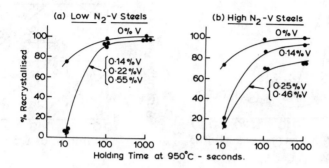

Fig.14. Effect of Holding Time at 950°C on
Recrystallisation in V Steels Reheated at
1150°C and Rolled 50% Reduction at 950°C

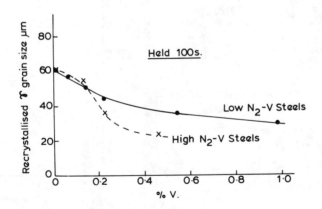

Fig.15. Effect of V on Recrystallised Grain Size after
Reheating at 1300°C, Rolling 50% Reduction at
1250°C and Holding at 950°C

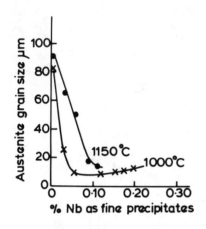

Fig.16. Effect of Nb as Fine
Precipitates on the
Austenite Grain Size

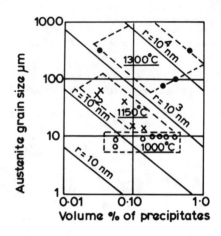

Fig.17. Effect of Particle Size
and Amount of Precipitate
on Austenite Grain Size

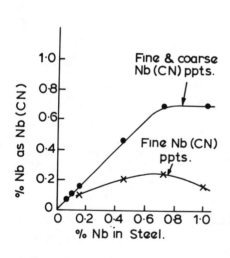

Fig.18. Effect of Nb in Steel on
Fine and Coarse Nb(CN)
Precipitates

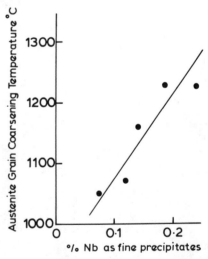

Fig.19. Effect of Fine Nb(CN)
Precipitates on Austenite
Grain Coarsening Temperature

dependence of solubility of NbC is a maximum at the stoichiometric ratio. This effect of fine precipitates in controlling the grain coarsening temperature is further confirmed by the relationship shown in Fig.19, the grain coarsening temperature increasing with the amount of fine Nb(CN). A similar effect has been reported for grain coarsening in steels containing AlN (23).

A model has been proposed for the calculation of the grain coarsening temperature (23), which involves:-

(i) the calculation of the critical precipitate particle size r_c which can just pin grains of radius R, using the equation (21):-

$$r_c = \frac{6Rf}{\pi} \cdot \left\{ \frac{3}{2} - \frac{2}{Z} \right\}^{-1} \quad \dots\dots\dots\dots\dots\dots\dots\dots\dots\dots\dots \quad (1)$$

where f = volume fraction of precipitates

and Z = a grain size heterogeneity factor defined as the ratio of the radii of growing grains to matrix grains.

The volume fraction of precipitates varies with temperature, as defined by the solubility equation, and hence r_c varies with temperature, and can be calculated. The heterogeneity factor may vary between 1.5 and 3.0 (24).

(ii) the calculation of the particle size due to Ostwald ripening of the precipitates, r, as a function of temperature using the equation obtained for NbC and AlN (23):-

$$\log_{10} r = - \frac{5167}{T} + 2.593 + \frac{\log_{10} t}{3} \quad \dots\dots\dots\dots\dots\dots \quad (2)$$

where T = temperature in Kelvin

and t = time in hours

This equation is only valid for specimens quenched prior to austenitising so that it applies to precipitates forming and growing at the austenitising temperature, and also assumes that precipitate growth is not dependent on supersaturation.

Values of r_c and r were calculated, using Z = 1.5, for the Nb steels and plotted aginst temperature. The value for R in equation (1) was taken as 10μm, which is the average value for the austenite grain size below the grain coarsening temperature. The predicted grain coarsening temperature is that at which $r_c = r$. Justification for using equation (2) to calculate r was that the rate of cooling from the pre-austenitising treatment of 1300°C was sufficiently rapid to inhibit precipitation of Nb(CN) during cooling. The calculated and observed grain coarsening temperatures are given in Table II. It can be seen that the calculated values show a maximum grain coarsening temperature at about the stoichiometric ratio, as has been shown by the observed values, Fig.4, but the agreement between calculated and observed values was only good for the 0.73% and 1.03% Nb steels. For the lower Nb steels, the calculated value was higher than the observed value. A possible reason for this may be the value of Z used to calculate r_c, and the actual value of Z to produce agreement between the calculated and observed grain coarsening temperatures is shown in Table II.

15

TABLE II - Calculated and Observed Grain Coarsening
Temperatures for Nb Steels

| %Nb | Grain Coarsening Temperature °C | | Z value required for agreement |
	Calculated	Observed	
0.07	1110	1060	1.62
0.16	1160	1080	1.70
0.46	1190	1250	1.43
0.73	1210	1230	1.47
1.03	1180	1180	1.50

It can be seen that this value of Z was close to 1.5, as has been
suggested (23), lying in the range 1.4-1.7, but there was a clear trend
for Z to be a minimum at the stoichiometric ratio, at which there was
also the maximum volume fraction of fine Nb(CN), Fig.18. A large volume
fraction of fine Nb(CN) particles will decrease the interparticle spacing,
giving a more uniformly pinned grain size and thus less grain size
heterogeneity, or low Z value. This interpretation agrees with the
observation that in the low Nb steels the austenite grain sizes were
distinctly more heterogeneous (or mixed) at temperatures below the grain
coarsening temperature than in the high Nb steels.

Even at the pre-austenitising temperature of 1300°C there were
undissolved Nb(CN) particles in all the steels, which increased in amount
with increasing Nb content. The particle sizes however, were very large
compared with the fine Nb(CN) precipitates and had little influence on the
grain coarsening temperature. It is suggested, however, that the precipitates do not determine the grain size established on heating through the
critical range because the driving force for the $\alpha \rightarrow \gamma$ transformation is too
large for the pinning action of the precipitates to be operative. The
grain size established immediately upon austenitisation is therefore
controlled by the austenite nucleation and growth rates, and the Z value
established will be dependent initially on these parameters. Once the
$\alpha \rightarrow \gamma$ transformation is complete, if the nucleation rate has been high the
austenite grains will grow to the equilibrium value established by the
pinning particles and also by the value of Z.

(b) V Steels. Because of the increased stability and lower solubility of
VN compared with VC, it is the VN which is usually believed to be
responsible for pinning the austenite grain boundaries and thereby
inhibiting grain growth. It seemed probable therefore, that the grain
coarsening temperature would be controlled by the solubility and coarsening rate of VN in austenite. It has been shown by Erasmus (25) that the
grain coarsening temperature is always less than the VN solvus temperature,
due to the temperature dependent growth of VN. The results from the
present investigation confirm this effect, Fig.20, but two features of
interest may be seen:-

(i) The grain coarsening temperature varies, as might be
expected, with V content but the high N steels have
virtually identical grain coarsening temperatures to
the low N steels.

16

(ii) up to about 0.2%V, the grain coarsening temperature
      is almost identical with the VN solvus, which is
      unexpected as it would seem that VN solubility rather
      than growth is the controlling factor.

A possible reason for these effects is the complex precipitation
behaviour in V steels which have been Al treated. AlN is more stable and
has a lower solubility in austenite than VN. Hence AlN will tend to
occur in preference to VN. It has been reported (26) that in Al-V-N
steels, Al does not affect the solubility of VN and V does not affect
the solubility of AlN. In the present work the Al contents of steels
which are close to the stoichiometric V:C ratio tended to be high, and
it is this effect which is responsible for the apparent trend for the
grain coarsening temperature to increase to a maximum at ∿0.5%V. The
calculated AlN solvus for the appropriate Al contents shows that the
observed grain coarsening temperature is always well below the AlN solvus,
as would be expected when coarsening of AlN is the more important factor
compared with the solubility. At the higher Al contents, the observed
grain coarsening temperature departs more markedly from the AlN solvus
curve, probably due to the more rapid coarsening of AlN at high supersaturations.

The grain coarsening temperature of the V steels was calculated,
using equations (1) and (2), on the basis that AlN was the grain
boundary pinning precipitate, the the value of Z was 1.5 and that the
grain size below the grain coarsening temperature was the observed value
of ∿10μm, Figs.3 and 5. The results are shown in Table III, from which
it can be seen that there was excellent agreement for all the low N-V
steels and for the high N- higher V steels. For the high N -lower V
steels the calculated grain coarsening temperature was higher than that
observed, and this was a tendency for steels which contained high Al
contents. It is tentatively suggested that this reflects an increased
coarsening rate of AlN at higher supersaturations, which is not allowed
for in equation (2). It seems that in V steels which have been Al
treated, the grain coarsening behaviour is mainly controlled by AlN
rather than by VN. However, if thermal treatments are used which form
both VN and AlN, the precipitation of AlN and its growth will be
dependent on the kinetics of solution of the less stable VN. There was
no apparent correlation between the calculated and observed grain
coarsening temperatures when the solubility of VN was used in
equation (1). This may be because equation (2) was determined for AlN
and Nb(CN) (23) and may not be applicable to VN.

The grain size was measured at temperature intervals of 50°C in
order to determine the grain coarsening temperature, which is therefore
no more accurate than ±25°C. The correlation between the observed and
calculated grain coarsening temperatures showed that 65% of the results
were within ±25°C and over 80% were within ±50°C. Bearing in mind the
simplifying assumptions made in the model used, this agreement is
encouraging.

From the present analysis, it is clear that the effects of microalloying additions on austenite grain growth depend on the solubility
and growth rate of the micro-alloy carbides/nitrides. As Nb(CN) is more
stable and less soluble than VC or VN it will be more effective in
inhibiting grain growth, and consequently result in a higher grain
coarsening temperature. This is shown in Fig.21 and is consistent also
with the data in Fig.4. It is suggested that the mutual effects of Al,

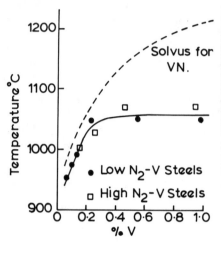

Fig.20.   Relationship between VN
          Solvus and Grain Coarsening
          Temperature

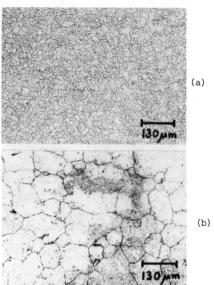

Fig.21.   Comparison of grain size
          of Nb and V steels
          reheated at 1200°C
          (a)   0.46%Nb
          (b)   0.55%V

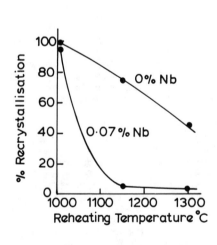

Fig.22.   Effect of Reheating
          Temperature and Nb on
          Recrystallisation at 950°C

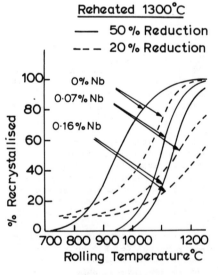

Fig.23.   Effect of Rolling
          Temperature, Reduction
          and Nb on Recrystallisation

V, Nb and N on the particle coarsening of the micro-alloy nitrides is worth further examination.

TABLE III - Calculated and Observed Grain Coarsening
            Temperatures for V Steels, Using Solubility of AlN

%V	%N	%Al	Grain Coarsening Temperature °C	
			Calculated	Observed
0.06	0.007	0.02	1000	960
0.11	0.007	0.02	1000	1000
0.14	0.007	0.02	1010	1000
0.02	0.008	0.03	1040	1050
0.55	0.008	0.05	1070	1050
0.14	0.020	0.03	1115	1020
0.25	0.015	0.03	1110	1040
0.46	0.018	0.02	1080	1070
0.98	0.018	0.02	1110	1075

4.2   Recrystallisation of Austenite

The recrystallisation of a deformed alloy may be retarded by dissolved solute, strain induced precipitates, a coarse grain size and the absence of second phase particles capable of nucleating recrystallisation. In addition, lower temperatures and smaller deformations also tend to retard recrystallisation. Many of these effects are illustrated by the present results, but the interactions between the processing parameters and the microstructural features can be very complex.

(a) Nb Steels. Varying the reheating temperature varies the amount of Nb in solution, the amount of undissolved coarse NbC and the initial grain size. In the base steel, only the initial grain size will be affected, and Fig.22 shows how the coarser initial grain size with increasing reheating tempera- ture retards recrystallisation at 950°C. This effect will also be present in the Nb steels, but less so with increasing Nb due to the refinement of the initial austenite grain size. It can be seen from Fig.22 however, that 0.07%Nb markedly reduced recrystallisation as the reheating temperature increased because of the increased solution of Nb and the decreased amount of undissolved Nb(CN). Whether this effect is due to Nb in solution or to strain induced Nb(CN) precipitation is not clear, but during rolling at 950°C, Nb(CN) readily precipitates and it is believed that this is the major effect. It will be shown later that Nb in solution can inhibit recrystallisation. The increased amount of recrystallisation in the Nb steel at the lowest reheating temperature is due to the fine grain size, the small amount of dissolved Nb and the increased amount of undissolved Nb(CN).

Decreasing the rolling temperature and reduction decreased the extent of recrystallisaton, as expected, Fig.23 and this was more pronounced with increasing Nb content due to the greater amount of Nb(CN) in solution as the stoichiometric ratio was approached. This led to more strain induced precipitation which was more effective in retarding recrystallisation, particularly as the rolling temperature decreased to 950°C and below. Strain induced precipitation retards both the nucleation (8, 27) and the growth (8, 28) of recrystallised grains.

The effect of Nb was to decrease the amount of recrystallisation, and this effect has been attributed either to solute drag and/or precipitation effects. For rolling at 1250°C, increasing Nb clearly decreases the amount of recrystallisation Fig.7 and at this temperature the effect was most probably due to solute drag because precipitation is very slow (29). However this interpretation is not unequivocal as it can be seen from Fig.24 that the amount of recrystallisation decreased with increasing potential for Nb(CN) precipitation at 1250°C, but further evidence for solute drag effects will be presented later. At the higher Nb contents, Fig.7, the amount of recrystallisation increased due to the nucleating effect of undissolved Nb(CN) (30, 31) and also due to the finer initial austenite grain sizes. A similar effect of Nb retarding recrystallisation was observed for rolling at 950°C, but at this temperature there would be copious strain induced Nb(CN) precipitation (29) which could be the reason for the slower recrystallisation. It was also observed that the effect of Nb on the retardation of recrystallisation was more effective as the stoichiometric composition was approached, and the temperature required for complete recrystallisation increased similarly as shown in Fig.25 which is constructed from diagrams such as Fig.23. This is due to the temperature dependence of the solubility and the potential for precipitation being a maximum at the stoichiometric ratio. Fig.25 also shows that the rolling temperature for complete recrystallisation increased with increasing Nb content, with decreasing rolling reduction and with increasing reheating temperature prior to rolling, as would be expected.

Increasing the Nb content up to 0.16% refined the recrystallised austenite grain size immediately after rolling at 1250°C. Higher Nb contents gave partial recrystallisation which became total recrystallisation after holding for 1000s at 1250°C, Fig.7. During the holding period, grain growth occurred particularly at the lower Nb contents. The refinement of the recrystallised grain size by Nb has been well documented (32, 33) and seems to be at variance with Nb decreasing the rate of recrystallisation. However, by increasing the extent of strain induced precipitation in the unrecrystallised austenite due to increased supersaturation,an increase in Nb probably retards the growth of the recrystallising grains and thereby refines the recrystallised grain size. Moreover, above the stoichiometric Nb:C ratio, the smaller amount of strain induced precipitation allows more rapid growth of the recrystallising grains, and thus a coarser grain size, Fig.9. This suggestion is supported by the fact that as the Nb content approached stoichiometry, the growth of the recrystallised grains with increasing holding time at 1250°C was completely inhibited, Fig.10, and this effect was more pronounced at the lower holding temperatures of 950°C where strain induced precipitation was greater. It was also, observed, in agreement with other work (34), that small rolling reductions resulted in coarser recrystallised austenite grains, and inhomogeneous grain sizes were observed. An analysis of the grain size as a function of Nb in solution, showed that it did not inhibit growth of the recrystallised grains, whereas 0.07%Nb precipitated as Nb(CN) did.

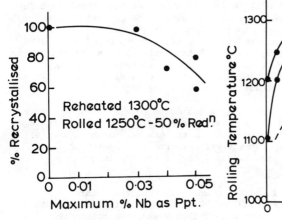

Fig.24. Effect of Precipitated Nb(CN) on Recrystallisaton

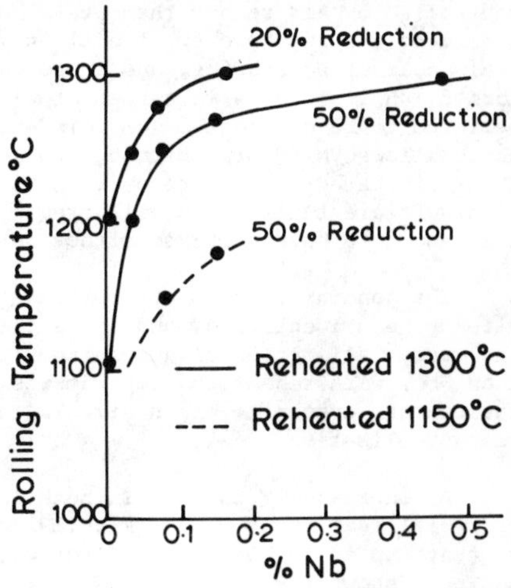

Fig.25. Effect of Nb on the Rolling Temperature for Complete Recrystallisation

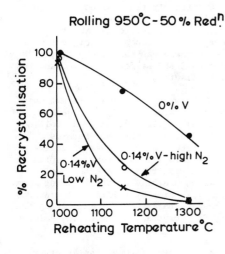

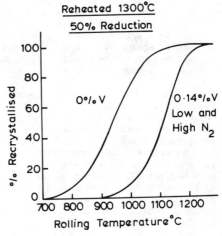

Fig.26. Effect of Reheating Temperature and V on Recrystallisation

Fig.27. Effect of Rolling Temperature on Recrystallisation in V Steels

(b) **V Steels**. It has been suggested that in the steels treated with Al, the grain coarsening behaviour during heating in the austenite region was controlled by AlN rather than by VN. In the case of the thermo-mechanical treatment studies however, the times were much shorter than those used for grain coarsening studies, and there was definite evidence that VN was formed (35). This is in agreement with other published work (36), and is believed to be due to the very different crystal structures of AlN and VN which allows VN to precipitate with less lattice misfit than can AlN. Hence VN precipitates in preference to AlN due to more favourable kinetics, but with sufficient time, AlN will precipitate and VN will dissolve. This has been shown to have a marked effect on strengthening (35).

The general factors controlling the austenite recrystallisation have already been identified, and it is clear from Figs.26 and 27 that V delayed the recrystallisation of austenite. Due to the greater solubility of VC than NbC, with reheating temperatures prior to rolling greater than 1150°C the coarser austenite grain size in the V steels results in less recrystallisation.

An increased V content in both low and high N steels retarded recrystallisation at 950°C, Figs.13 and 14, the major effect being with V contents up to ∿0.20%. The effect with higher V was less pronounced. It has been shown (29, 35) that VN can be strain induced to precipitate at 950°C, and the retardation of recrystallisation during rolling at 950°C is due to these precipitates, which are shown on the sub-boundaries in the austenite, and were identified as V(CN) by electron diffraction, Fig.28. Precipitates seen on sub-boundaries are probably VN, whilst those forming as interphase precipitates probably are VC. In addition, VN precipitates were observed on the austenite grain boundaries. All the strain induced VN precipitates are able to delay the growth of the recrystallising austenite grains, and thereby recrystallisation generally. It is suggested that with about 0.15/0.20%V, sufficient VN precipitation has occurred to produce the major retardation of recrystallisation, and thereafter any further VN precipitation has less effect. In contrast to the Nb steels, there was no evidence of accelerated recrystallisation at the highest V contents, because even in the high N steels there was relatively little VN undissolved at the reheating temperatures used and so few large undissolved precipitates to act as nuclei for recrystallisation.

In both the low and high N steels, there was no obvious retardation of recrystallisation at the highest rolling temperature of 1250°C, again because of the insufficient VN precipitation, Fig.27, but at lower rolling temperatures where VN precipitates more liberally, V markedly retards recrystallisation (15, 36). It may be inferred that V is less effective than Nb in retarding recrystallisation by a solute drag effect. Despite reports to the contraray (8, 15) the present results indicate that solute does not retard the growth of recrystallising grains. As precipitates seem to be most important in retarding recrystallisation at rolling temperatures of 950°C, it is perhaps not surprising that increasing N, by increasing the volume fraction of precipitated VN, retards recrystallistion during holding at 950°C, Fig.11.

With regard to the grain size of the recrystallised austenite after rolling at 1250°C, increasing V refined the grain size, Fig.15. Because 1250°C is above the VC and VN solvus, this effect cannot be due to the inhibition of growth of the recrystallising grains by VN precipitates. At this rolling temperature, V did not retard recrystallisation, and it seems probable that in order to refine the recrystallised grain size, more nucleation sites must be activated. It is possible that by retarding

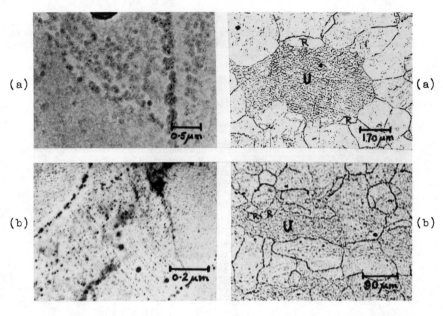

(a)

(a)

(b)

(b)

Fig.28. V(CN) precipitates at
        sub-boundaries in
        controlled rolled
        austenite.
        Reheat 1300°C
        Roll 950°C
    (a) 0.55%V
    (b) 0.22%V

Fig.29. Recrystallisation
        nuclei at interface
        between recrystallised
        and unrecrystallised
        austenite.
        Reheat 1300°C
        Roll 1250°C
    (a) 0.46%Nb
    (b) 0.73%Nb

Fig.30. Recrystallisation at deformation band.

recovery processes, V can increase the stored energy and so allow more recrystallisation nuclei to be activated. Increasing N also tended to refine the recrystallised grain size for V contents >0.14%. As there were no strain induced precipitates, V may further increase the stored energy by V-N complexes locking defects and further inhibiting recovery. It was observed, however, that the absence of VN precipitates at 1250°C allowed marked growth of the recrystallised grains, but at 950°C there had been sufficient VN precipitation in the austenite to inhibit grain growth.

4.3 Mechanisms of Recrystallisation

The main sites for nucleation of recrystallisation were the deformed austenite grain boundaries, which were often observed to be serrated, particularly at low deformations. This suggests that nucleation occurred by strain induced boundary migration induced by local strain energy gradients, which were more pronounced due to the local strain inhomogeneity at the lower deformations.

Recrystallisation also nucleated at the recrystallised-unrecrystallised interface, Fig.29, and this type of nucleation has occasionally been reported (8, 37). It is interesting to consider why this should occur in preference to the growth of recrystallised grains into the unrecrystallised matrix. Such nucleation can only occur at stationary pinned interfaces, and it is significant that it was only observed in the Nb steels in which there was heavy and uniform precipitation of strain induced Nb(CN) capable of effective pinning the recrystallised-unrecrystallised interface.

Recrystallisation nucleation on deformation bands in heavily deformed austenite have also been reported (38, 39), and such a form of nucleation has been observed in the present work, Fig.30. However, not all deformation bands acted as nuclei, possibly due to variations in the strain energy associated with them, as has been suggested (39). This form of nucleation occurred at 50% reduction during rolling, and it cannot be regarded as a major type of nucleation.

It has been suggested that undissolved particles of Nb(CN) caused acceleration of recrystallisation in the higher Nb steels, Fig.7, and such coarse precipitates act as recrystallisation nuclei, providing their size exceeds the critical nucleus size. As the critical nucleus size decreases with increasing rolling strain, smaller undissolved particles can act as nuclei at greater reductions (40). The smallest particle size for nucleating recrystallised grains is reported to be ∿0.5μm (41) and it has been suggested that the more usual particle size for this form of nucleation is >1-2μm (24). Both the strain gradient developed around the particle, and its interfacial energy favour nucleation of the recrystallised grains. An anlysis of the size distribution of Nb(CN) particles in the present work showed that steels with more than 0.46%Nb contained particles >0.5μm, and with increasing Nb content the proportion of particles of larger sizes became greater. It is not surprising therefore, that steels with more than 0.46%Nb showed increased recrystallisation at high temperatures.

4.4   The Retardation of Recrystallisation by Nb

Whilst the evidence showed conclusively that strain induced Nb(CN) precipitation, and also that of VN, retarded recrystallisation of austenite at the lower rolling temperature of 950°C, there were also indications that Nb in solution could retard recrystallisation by a solute drag effect, Fig.7 and Fig.31(a). In the 0.46%Nb steel, in which there was least

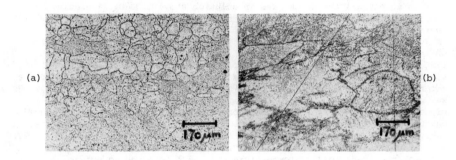

(a)

(b)

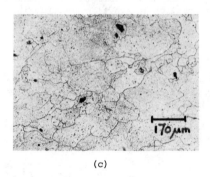

(c)

Fig.31.   <u>Effect of Solute drag on Recrystallisation</u>
(a)   0.73%Nb. Reheat 1300°C, Roll 1250°C
(b)   Low interstitial - 0.29%Nb, Reheat 1300°C,
      Roll 1250°C, no hold.
(c)   Low interstitial - 0.29%Nb, Reheat 1300°C,
      Roll 950°C, hold 950°C.

recrystallisation, 0.14%Nb was in solution but 0.05%Nb was present as Nb(CN) in a particle size range which might have inhibited nucleation. The evidence was therefore, not unequivocal.

To try to confirm that Nb solute drag can inhibit recrystallisation, a very low interstitial (C + N) steel containing 0.29%Nb was made, Table I. In this steel, all the Nb(CN) was in solution above 1100°C. Reheating at 1300°C and rolling at 1250°C produced no sign of recrystallisation, Fig.31(b), whereas the base steel and the 0.07%Nb steel were fully recrystallised immediately after rolling. This is a very clear demonstration of dissolved Nb inhibiting nucleation for recrystallisation. The steel was then held at 1250°C, and recrystallisation was complete after 100s whilst further holding simply resulted in grain growth. This suggested that Nb in solution did not retard grain growth. Rolling at 1250°C but holding at 1150°, 1050° and 950°C was carried out to investigate the temperature dependence of the effect of dissolved Nb on recrystallisation. Complete recrystallisation was produced on holding at 1150° and 1050°C, but holding at 950°C only produced partial recrystallisation, Fig.31(c). At 950°C, strain induced Nb(CN) is formed (29) and it appears to be this that inhibits recrystallisation during holding, although the lower holding temperature will also have had an effect. Whilst some strain induced Nb(CN) precipitation might be expected at 1050°C because it is below the solvus, the amount was apparently too small to inhibit recrystallisation to any marked extent. Two mechanisms by which dissolved Nb may inhibit recrystallisation are either solute-vacancy interaction which inhibits dislocation climb and the formation of the sub-grains necessary for nuclei, or solute-dislocation interaction possibly involving Nb-interstitial complexes, which retards the movement of dislocations necessary for sub-grain formation.

4.5   The Retardation of Recrystallisation by V

The evidence suggests that dissolved V is relatively ineffective in retarding recrystallisation, but that the precipitation of VN at temperatures of 950°C and below can have an inhibiting effect on recrystallisation. Increasing either V, or more particularly N, can enhance this effect by increasing the supersaturation,which increases the volume fraction of VN precipitates. However the stoichiometric ratio should not be exceeded.

5. Concluding Remarks

The study of the grain coarsening characteristics of Nb and V steels has shown that the Nb steels have considerably higher grain coarsening temperatures than the V steels, and the amount of N in Al treated V steels has virtually little or no effect. Nb refined the austenite grain size at temperatures up to 1200°C, the major effect being produced by up to 0.07%Nb whilst higher Nb contents produced little further refinement. A maximum grain coarsening temperature was observed at approximately the stoichiometry ratio. Increasing V also refined the austenite grain size at temperatures up to ∿1050°C, but had little effect above 1100°C. Whilst V increased the grain coarsening temperature, a maximum at the stoichiometric ratio was not very apparent. A model proposed for estimating the grain coarsening temperature, based on the decrease in volume fraction and increase in particle size of precipitates with increasing temperature was generally applicable for Nb steels providing the effect of stoichiometry on the heterogeneity of grain size was taken into account. For V steels the grain coarsening temperatures could not be predicted by the model, probably because in these Al-treated steeels AlN tended to form instead of VN. When the model was applied to the

V steels on the assumption that AlN controlled the grain coarsening tempera-ture, good agreement was obtained for the low N steels and for the high N-higher V steels. Any remaining discrepancies can be explained by the super-saturation dependence of the growth of AlN at high Al contents.

The study of the recrystallisation of austenite during thermo-mechanical treatment has shown that Nb can retard recrystallisation at high temperatures by a solute drag effect, which inhibited nucleation rather than the growth of recrystallising grains. At lower temperatures where strain induced Nb(CN) precipitation occurs, the retardation of recrystallisation is even more pronounced because precipitation inhibited both nucleation and growth of the recrystallising grains. In the V steels, due to the shorter times at tempera-ture and the kinetic advantage of VN precipitation over AlN precipitation, the recrystallistion is largely controlled by VN precipitation. V in solution has no observable effect on recrystallisation at high temperatures, i.e. by solute drag, but at lower temperatures of $\sim 950^{\circ}C$, VN precipitation causes retardation of recrystallistion, and N accentuates this effect by increasing the volume fraction of precipitated VN.

The complex interactions between micro-alloy addition dissolution in the austenite, undissolved micro-alloy carbides-nitrides, initial austenite grain size, strain induced precipitation, stoichiometry, rolling/holding temperature, and the reduction during rolling have been discussed in an attempt to explain the different observed recrystallisation and grain size effects. The various nucleation sites for recrystallisation nuclei have also been identified. It is suggested that increasing the N content of V steels tends to make them behave in a manner more similar to Nb steels, particularly at the lower rolling and holding temperatures around $950^{\circ}C$.

Acknowledgements

Acknowledgement is made to Union Carbide Corporation for providing support for this work, and to Mr M Korchynsky for many stimulating discussions. Our thanks are also extended to Dr A W D Hills, Head of the Department of Metallurgy, Sheffield City Polytechnic, for providing facilities for the work to be carried out.

References

1. J D Jones and A B Rothwell, "Controlled Rolling of Low Carbon-Niobium Treated Mild Steels", pp 78-82 in Deformation under Hot Working Conditions, Report No 108, The Iron and Steel Institute, London, 1967.

2. J J Irani, D Burton and D J Latham, "Observations on the Recovery and Recrystallisation Processes in Steels", Report No MG/C/31/68, British Iron and Steel Research Association, 1968.

3. R Priestner, C C Early and J H Randall, "Observations on the Behaviour of Austenite during the Hot Working of some Low Carbon Steels". J.I.S.I., 206 (1968) pp 1252-1262.

4. K J Irvine, T Gladman, J Orr and F B Pickering, "Controlled Rolling of Structural Steels", J.I.S.I., 208 (1970) pp 717-726.

5. (a) T Tanaka, T Funakoshi, M Ueda, J Tsuboi, T Yasuda and
         C Utahashi, "Development of High Strength Steel with Good
         Toughness at Arctic Temperatures for Large Diameter Line Pipe",
         pp 399-409 in Microalloying '75, Union Carbide Corporation,
         New York, 1977.

    (b) I Kozasu, C Ouchi, T Sampei and T Okita, "Hot Rolling as a High
         Temperature Thermo-Mechanical Process", pp 120-135 in
         Microalloying '75, Union Carbide Corporation, New York, 1977.

6. A J DeArdo and E L Brown, "Hot Rolling Behaviour of Austenite Micro-
    alloyed with Vandadium and Nitrogen", Journals of Metals, 29 (2)
    (1977) pp 26-29.

7. H Sekine and T Maruyama, "Retardation of Recrystallisation of
    Austenite during Hot Rolling in Nb-containing Low Carbon Steels",
    Trans I.S.I. Japan, 1 (1976) pp 427-436.

8. W Roberts, "Studies pertaining to Austenite Recrystallisation during
    Controlled Rolling of Nb-microalloyed H.S.L.A. Steels", Report No I.M.-
    1211, Swedish Institute for Metals Research, Stockholm, 1977.

9. T M Hoogendoorn and M H Spanraft, "Quantifying the Effect of Micro-
    alloying Elements on Structures during Processing", pp 75-84 in
    Microalloying '75, Union Carbide Corporation, New York, 1977.

10. L J Cuddy, "Laboratory Examination of Recrystallisation and Trans-
    formation in Deformed Austenite", pp 169-185 in The Hot Deformation
    of Austenite, John B Ballance, ed; A.I.M.E., New York, 1977.

11. M J Stewart, "The Effects of Niobium and Vanadium on the Softening of
    Austenite", pp 233-249 in The Hot Deformation of Austenite,
    John B Ballance, ed; A.I.M.E., New York, 1977.

12. W Roberts, "Hot Deformation Studies on a Vanadium Microalloyed Steel",
    Report No I.M.-1333, Swedish Institute for Metals Research,
    Stockholm, 1978.

13. M J White and W S Owen, "Effects of Vanadium and Nitrogen on Recovery
    and Recrystallisation during and after Hot Working some H.S.L.A.
    Steels", Met Trans A, 11A (1980) pp 597-604.

14. M Lamberigts and T Greday, "Precipitation and Recrystallisation in
    Dispersion Hardemed Steels", pp 31-38 in Report No 38, Centre National
    de Recherches Metallurgiques, Liege, 1974.

15. J N Cordea and R E Hook, "Recrystallisation Behaviour in Deformed
    Austenite of High Strength Low Alloy Steels", Met Trans A, 1A (1970)
    pp 111-120.

16. A LeBon, J Rofes-Vernis and C Rossard, "Recrystallisation and
    Precipitation during Hot Working of a Nb-bearing H.S.L.A. Steel",
    Metal Science Journal, 9 (1975) pp 36-40.

17. A T Davenport, R E Miner and R A Kot, "The Recrystallistion of
    Austenite during the Hot Rolling of a Cb-bearing H.S.L.A. Steel",
    pp 186-203 in The Hot Deformation of Austenite, John B Ballance, ed;
    A.I.M.E., New York, 1977.

18. I Kozasu, T Shimizu and H Kubata, "Recrystallisation of Austenite in Si-Mn Steels with Minor Alloying Elements after Hot Rolling", Trans. I.S.I. Japan, 11 (1971) pp 367-375.

19. P L Magnonon and W E Heitmann, "Sub-grain and Precipitation Strengthening Effects in Hot Rolled Columbium bearing Steels" pp 59-70 in Microalloying '75, Union Carbide Corporation, New York, 1977.

20. V K Heikkinen and H D Boyd, "The Effect of Processing Variables on the Mechanical Properties and Strain Ageing of H.S.L.A. Vanadium and Vanadium-Nitrogen Steels", Canadian Metallurgical Quarterly, 15 (3) (1976) pp 219-226.

21. T Gladman, "On the Theory of the Effect of Precipitate Particles on Grain Growth in Metals", Proc. Roy. Soc., 266(A) (1966), pp 298-309.

22. T Gladman, "The Effects of Inclusions on Recrystallisation and Grain Growth", pp 172-181 in Inclusions, F B Pickering, ed; The Institution of Metallurgists, London, 1979.

23. T Gladman and F B Pickering, "Grain Coarsening of Austenite", J.I.S.I. 205 (1967) pp 653-664.

24. T Gladman, "The Effect of Second Phase Particles on Grain Growth", pp 183-192 in Recrystallisation of Multiphase and Particle Containing Materials, N Hansen, A R Jones and T Leffers, eds; RISO National Laboratory, Roskilde, Denmark, 1980.

25. L A Erasmus, "Effect of Small Additions of Vanadium on the Austenite Grain Size, Forgeability and Impact Properties of Steels" J.I.S.I., 202 (1964) pp.128-134.

26. T Gladman and I D McIvor, "The Distribution of Interstitials in Carbon-Manganese and Grain Refined Steels". Scand. J. of Met. 1 (1972) pp 247-253.

27. J J Jonas, "Recovery, Recrystallisation and Precipitation under Hot Working Conditions", pp.976-1002 in Proc. of Fourth Int. Conf. on Strength of Metals and Alloys, (3), Nancy, 1976.

28. T L Capelitti, L A Jackman and W J Childs, "Recrystallisation following the Hot Working of a High Strength Low Alloy Steel and a 304 Stainless Steel at the Temperature of Deformation", Met Trans A, 3A (1972) pp 789-796.

29. R K Amin, G Butterworth and F B Pickering, "Effects of Rolling Variables and Stoichiometry on Strain Induced Precipitation of Nb(CN) in C-Mn-Nb Steels", pp 27-31 in Hot Working and Forming Processes, The Metals Society, London, 1979.

30. T Gladman, B Holmes and F B Pickering, "Work Hardening of Low Carbon Steels", J.I.S.I., 209 (1970) pp 172-183.

31. U Koster, "Recrystallisation Involving a Second Phase", Metal Science Journal, 8 (1974) pp 151-160.

32. I L Dillamore, R F Dewsnap and M G Frost, "Metallurgical Aspects of Steel Rolling Technology" Report No CDL/MT/2, British Steel Corporation, 1974.

33. H Sekine and T Marayama, "Controlled Rolling for obtaining a Fine and Uniform Structure in High Strength Steels", pp 85-88 in The Microstructural Design of Metals and Alloys, The Metals Society, London, 1974.

34. C M Sellars and J A Whiteman, "Recrystallisation and Grain Growth in Hot Rolling", Metal Science, 3 (1974) pp 187-194.

35. R K Amin, M Korchynsky and F B Pickering, "Effect of Rolling Variables on Precipitation Strengthening in High Strength Low Alloy Steels Containing Vanadium and Nitrogen" Metals Technology 8 (7) (1981) pp 250-261.

36. E L Brown, A J DeArdo and J H Bucher, "The Microstructure of Hot Rolled High Strength Low Alloy Steel Austenite", pp 250-285 in The Hot Deformation of Austenite, John B Ballance, ed., A.I.M.E. New York 1977.

37. B Ahlblom. Tekn. Doctoral Thesis, Royal Institute of Technology, Stockholm, 1977.

38. A P Moreno. Ph.D Thesis, University of Sheffield, 1975.

39. I L Dillamore, "Recrystallisation in Heavily Deformed Metals". J Australian Inst. Met. 1 (3) (1978) pp 136-145.

40. F J Humphreys, "The Nucleation of Recrystallisation at Second Phase Particles in Deformed Aluminium", Acta Metallurgia, 25 (1977) pp.1325-1344.

41. T Gladman, I D McIvor and F B Pickering, "Effect of Carbide and Nitride Particles on the Recrystallisation of Ferrite", J.I.S.I. 209 (1971) pp 380-390.

Q: Because initial grain size controls recrystallization behavior, it is important to know whether the initial grain size of the steels was controlled during the deformation experiments. It appears that only one or two reheating temperatures were used for the variety of alloys.

A: Each material was reheated at one temperature. The authors feel that by reheating a material at only one temperature the grain size is constant. It must be remembered that the niobium-containing steel, the vanadium-containing steel and the base steel were originally homogenized at 1350°C. They were cooled to room temperature before reheating. If the question is whether the austenite grain size of the vanadium and niobium steels was the same, the answer is they probably were not.

Q: Several of the micrographs of the prior austenite microstructures seem to have an unusual etching appearance. For example, the last one (Figure 31), showed an unrecrystallized grain with an apparent substructure in the center. Was any special etching technique employed?

A: Yes, the chemical etching technique is given in the paper.

Q: This question has to do with the measurement of the recrystallized grain size in different steels which have different fractions of recrystallization. It is normally observed that as a material recrystallizes, the size of the recrystallizing grains also increases. Therefore, with increasing alloy content, when the fraction of recrystallized material under any condition decreases, one would expect to see that the grain size has decreased. What effect on grain size was due to the different fractions of recrystallization, and what was directly associated with the effect that the solutes would have on the more fully recrystallized grain size?

A: If the recrystallization is less, you will expect the grain size to be smaller. Is this essentially what your asking?

Q. Yes. Normally during static recrystallization, most nucleation occurs early. These new grains grow progressively as recrystallization proceeds so that if you measure the size of grains in a partially recrystallized structure, they will be smaller than in a fully recrystallized structure. Therefore, when you are comparing a base steel, with 100% recrystallization, with say a niobium steel with about 50% or 30% recrystallization, then I wonder how the recrystallized grain size of the austenite can be defined.

THE INFLUENCE OF NIOBIUM, VANADIUM AND NITROGEN ON THE RESPONSE

OF AUSTENITE TO REHEATING AND HOT DEFORMATION IN MICROALLOYED STEELS

I. Weiss,* G.L. Fitzsimons,** K. Mielityinen-Tiitto** and A.J. DeArdo**

*Dept. of Engineering, Wright State Univ., Dayton, OH   45435
**Dept. of Metallurgical/Materials Engineering
Univ. of Pittsburgh, Pittsburgh, PA   15261

The influence of Nb, V and N on the reheating and subsequent hot deformation behavior of austenite has been studied. Grain coarsening studies were performed at temperatures between 950 and 1300°C under iso- thermal conditions. The grain coarsening temperature was observed to increase with microalloying elements in the order V, Nb and V+Nb. An increase in the N level also led to an increase in this temperature. In the vicinity of 1200°C, unexpectedly small grains ($\sim$60µm) resulted from high N levels in the V+Nb steels. Axisymmetric hot compression tests were conducted at constant true strain rates of 0.006, 2 and 13 s^{-1} and at tem- peratures in the range of 900 to 1225°C. At the two higher strain rates all the steels displayed a retardation of dynamic softening which was only weakly dependent on temperature and strain rate, and appears to be a solid solution effect. At the low strain rate the retardation increased sharply and was found to be associated with precipitation effects. High nitrogen levels were necessary for postponement of dynamic softening in vanadium steels at low temperature and strain rate.

Introduction

Properly rolled microalloyed steels provide a balanced package of properties at strength levels between those of as-rolled C-Mn and quenched and tempered grades of low alloy steels. One of the principal objectives of high strength, low alloy (HSLA) steel processing has been to produce extremely fine ferrite grain sizes in the as-rolled condition. Ferrite grain size is important since it is the only microstructural parameter that can simultaneoulsy increase the strength and toughness levels of the steel.(1) Fine ferrite grains can be achieved through control of either the transformation temperature (2) or the metallurgical condition of the prior-austenite by hot rolling.(3)

It has been recognized for some time that hot rolling can exert a strong influence on final microstructure by changing the densities and types of defects in the austenite. Several studies have shown the benefit that can result from hot rolling schedules in which a large portion of the deformation takes place at temperatures below the "recrystallization" temperature of the austenite.(4,5) Unfortunately, low temperature rolling often leads to roll separating forces and torques that exceed the capacity of commercial strip and plate mills. The large forces encountered during low temperature rolling have prompted investigations into high temperature thermomechanical treatment as a possible way of refining the austenite and the ferrite grain size in as-rolled low alloy steel.

A frequent observation during hot rolling experiments of low alloy steel is that when the deformation takes place at temperatures above about 1000°C, the prior-austenite grains that are observed upon quenching immediately after rolling exhibit an equiaxed morphology. The equiaxed prior-austenite grains that are found after rolling are often somewhat smaller than those present prior to rolling, i.e. the hot rolling refined the austenite grain size.(6-8) It is generally assumed that this new set of equiaxed prior-austenite grains results from the recrystallization of the original grains, and that this recrystallization is a consequence of the hot rolling treatment.

Depending on deformation conditions, there are four mechanisms that could individually cuase an original set of equiaxed grains to be replaced by a second set during a thermomechanical process: a) static recrystallization, b) dynamic recrystallization, c) metadynamic recrystallization, and d) continuous recrystallization. Static recrystallization occurs when both the nucleation and growth of the new grains take place after the deformation.(9) Dynamic recrystallization results when both the nucleation and growth take place during the deformation.(10) Metadynamic recrystallization obtains when the nucleation of the new grains occurs during deformation but the growth takes place after deformation.(11) Continuous recrystallization is somewhat different from the others in that it is not strictly a recrystallization process but rather an advanced form of recovery in which there is no gross movement of high angle grain boundaries.(12-13) Since there are several ways that hot rolling can lead to a final, equiaxed, prior-austenite grain morphology, it would appear quite tenuous to try to assign a mechanism to this process merely on the basis of observations of grain morphology made before and after rolling.

An alternative method of assessing which mechanism may be operating under a given set of processing conditions is to study the hot flow curves as a function of temperature and strain rate. The shape of the flow curve reflects the balance between hardening and softening mechanisms. As a result of this balance, the flow curve may show a peak at some level of strain, ε_p. Previous studies using single-phase f.c.c. materials have shown that, in this particular case, the strengthening results from work hardening and solid solution effects while the softening results from dynamic recovery and recrystallization.(14,15) Precipitation influences this balance and strongly affects ε_p.(16-18) Previous work has also indicated that ε_p can be related to the onset of dynamic recrystallization. (14,16-21)

One of the objectives of the present study was to investigate high temperature recrystallization through the use of the hot compression test. Hot flow curves were generated for several microalloyed steels under a wide range of testing conditions. The resulting data were analyzed in terms of changes in ε_p as influenced by solid solution and precipitation effects.

Experimental Procedure

Materials and Processing

The compositions of the steels used in this study are shown in Table I. These materials were produced as air-melted, silicon-killed laboratory ingots (22 kg). These ingots were homogenized at 1200°C, hot rolled in several passes to plates 12.7mm thick with a finishing temperature of 1000°C. The plates were air cooled to room temperature after hot rolling. This practice resulted in ferrite-pearlite microstructures air cooled to room temperatures.

TABLE I - Chemical Composition of the Materials (wt.%)

C	Mn	Si	V	Nb	N
.08	1.24	.42	–	–	.006
.08	1.24	.38	–	–	.025
.08	1.23	.42	.14	.073	.025
.08	1.25	.39	.14	.073	.015
.08	1.23	.42	.14	.073	.006
.08	1.24	.40	–	.073	.025
.08	1.23	.40	.14	–	.025
.08	1.24	.39	.14	–	.006
.08	1.22	.39	.14	.038	.025
.08	1.25	.41	.14	.037	.006

P $\leq$ .005; S $\leq$ .005 for all materials

Reheating Studies

The specimens for the reheating study were small cubes, one cm on an edge, which had been removed from the hot rolled plates. These specimens were heated in a box furnace to different temperatures ranging between 950 to 1300°C. Following austenitization for 30 min., specimens were quenched in ice water.

Metallography

The linear intercept technique was used to determine the grain size of the prior-austenite in the reheated specimens. A modified picric acid etch was used to reveal the prior-austenite grain boundaries.

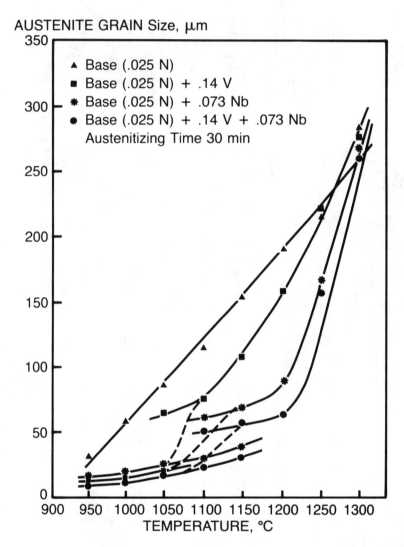

Figure 1 - The influence of microalloying elements on the grain coarsening behavior of austenite in high nitrogen steels.

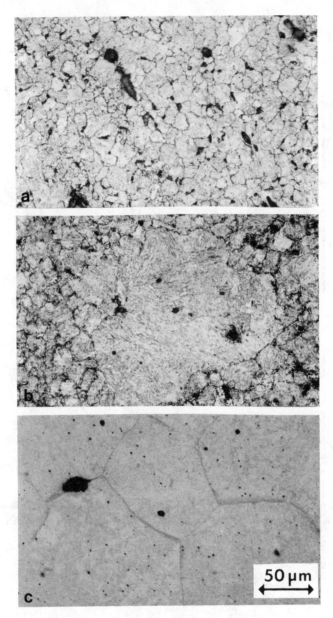

Figure 2 – Optical micrographs illustrating grain coarsening of austenite in the base (.025 N) + .14 V + .073 Nb steel at temperatures: (a) 1050°C, (b) 1100°C and (c) 1200°C.

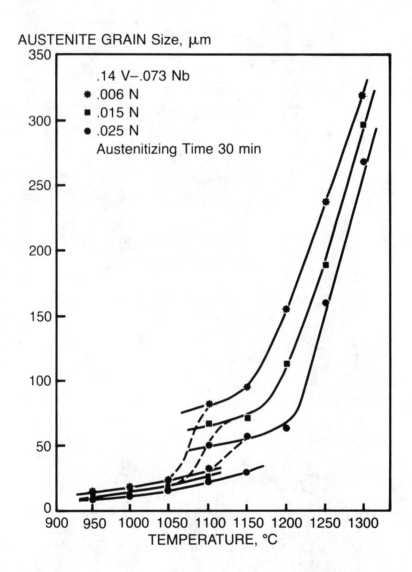

Figure 3 - Effect of nitrogen on the grain coarsening behavior of austenite in the .14 V + .073 Nb steel.

Thin foil and replica electron microscopy was conducted on as-quenched specimens in order to assess the state of precipitation in both reheated and deformed specimens.

Hot Compression Testing

Compression tests were carried out using a MTS machine modified for constant strain rate compression testing.(22) Following solution treatment at 1200°C for 30 min., specimens were air cooled to test temperatures ranging between 900 and 1225°C and deformed at constant strain rates of 0.006 and 13s^{-1}. After deformation to various strains up to 1.25 the samples were quenched within 1 sec. in ice water and the as-quenched structures were analyzed. Glass lubricant was applied to the compression specimens to reduce friction and barreling. The use of constant strain rate uniaxial compression testing permitted to flow curves to be determined as functions of temperature and strain rate. From these flow curves, the strains to the peak stress (ε_p) were measured. The value of ε_p obtained in a uniaxial compression test run at constant strain rate $\dot{\varepsilon}$ can be converted to t_p, the time of deformation required to reach the peak stress at a given $\dot{\varepsilon}$, since

$$t_p = \varepsilon_p / \dot{\varepsilon} \ .$$

A useful way of analyzing this type of data is to plot t_p against temperature for various strain rates.(16,23) The usual format plots T along the ordinate and t_p along the abscissa. The resulting plots closely resemble the classical time-temperature-transformation diagrams and are presented in the following analysis.

Results and Discussion

Reheating Studies

The grain coarsening behavior of austenite for some of the high nitrogen (0.025 N) steels is summarized in Figure 1. The carbon steel exhibits normal grain growth over the entire temperature range (950-1300°C). On the other hand, the microalloyed steels showed three stages of grain coarsening, as previously reported.(24) The microstructures corresponding to each of the three stages of coarsening are displayed in Figure 2.

The nitrogen content of the steels influenced the grain coarsening behavior of microalloyed austenite as shown in Figure 3. Since precipitates can strongly influence the grain coarsening behavior of austenite, precipitate studies were conducted. The relative distribution of precipitate sizes was determined by electron microscopy using both thin foil and replica techniques. These results are given in Table II. Representative precipitate distributions for the V-Nb steels are shown in Figure 4 for two levels of nitrogen and for reheating temperatures of 1050 and 2100°C.

As noted in connection with Figure 1, microalloying elements strongly modified the grain coarsening behavior of austenite. The carbon steel showed continuous grain growth with increasing temperature, whereas the Nb and V steels exhibited three stages of grain coarsening. In all microalloyed steels studied, normal but sluggish grain growth occurred at low temperatures leading to small austenite grains. Heterogeneous grain structures with abnormally large grains were found for intermediate reheating temperatures. At sufficiently high temperatures, normal grain growth resumed once again.

TABLE II - Relative Frequency of Precipitate Sizes
in the .14 V + .073 Nb Steels

N (wt.%)	.006		.025		
T(°C)	1050	1200	1050	1100	1200
Particle Size (nm)					
20- 50	M	–	H	M	L
50- 70	M	L	H	H	M
70-100	–	L	–	H	M
100-150	–	–	–	–	M
150-200	M	–	H	H	–
200-250	L	M	L	L	H
>250	L	L	L	L	M

H = High, M = Medium, L = Low

The extent to which grains coarsen at each of the three stages can be
controlled by the microalloying elements. In steels with a high nitrogen
content niobium seems to be more effective than vanadium in reducing the
grain size of austenite. However, the results for the Nb-V steel indicate
that the grain size of the Nb steel can be even further reduced by vanadium
additions. Three interesting features for the high N steel can be observed
in Figure 1.

i) The first temperature at which abnormal grains occurred was
higher for the Nb and Nb-V steels (1100°C) than for the V steel (1050°C).

ii) The range of grain sizes at intermediate temperatures, where
abnormal grain growth occurred, was smaller in the Nb-V steel than in the
Nb or V steel.

iii) Increasing the nitrogen content from .006 to .025% in the Nb-V
steel reduced the austenite grain size, especially above 1050°C, as seen
in Figure 3. At 1200°C, a surprisingly small grain size of 60 µm was
obtained.

The results of thin foil and replica studies, given in Table II and
figure 4, clearly demonstrate that precipitates of different sizes ranging
from 20 to more than 250 nm were present at reheating temperatures. The
multimodal distribution of particle sizes can be understood on the basis of
prior treatment of the steels. The air-cooled slabs, being the starting
material in these studies, already contained particles from the soaking and
rolling treatments as well as from the cooling of the slabs. Some precipitation may also have taken place during reheating. Special attention was
paid to the precipitates which were present after reheating at 1050°C,
which is just below the grain coarsening temperature* in these steels.
These particles were also considered to be important since the subsequent
coarsening and dissolution rates at higher temperatures would be expected
to be related to their composition and stability.

*Following Gladman, et al., (25) the grain coarsening temperature corre-
sponds to the development of 20-30% of the coarse grains.

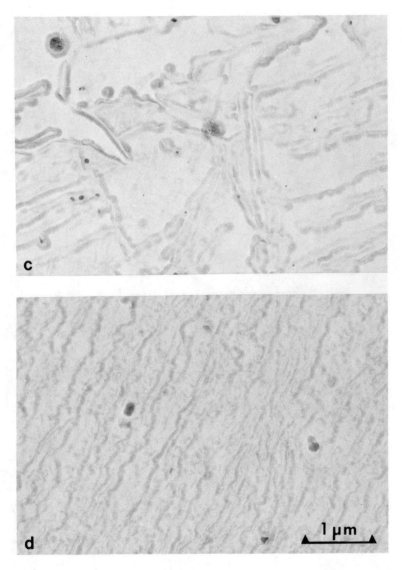

Figure 4 – Replica electron micrographs illustrating precipitates in the .14 V + .073 steel at two N levels after reheating for 30 min.: (a) 1050°C; .025 N, (b) 1050°C; .006 N, (c) 1200°C; .025 N, (d) 1200°C; .006 N.

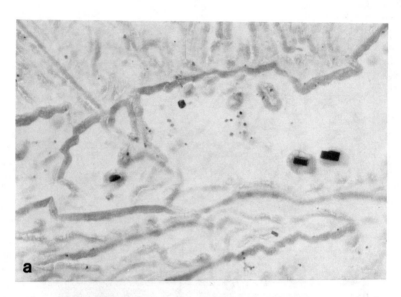

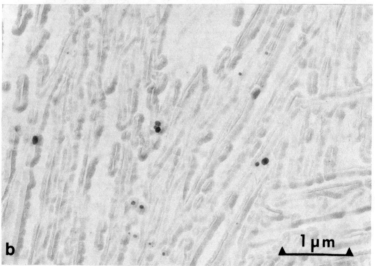

Figure 4(a)(b)   Cont'd.

According the the grain coarsening theory of Gladman, et al.,(25) the critical particle size required to pin grain boundaries increases with increasing matrix grain size and volume fraction of precipitates. It can be observed in Table II that a considerable number of small precipitates are present at 1050°C in the steels which contain either .006 or .025 N. The grain sizes of austenite at that temperature were fine in both steels. At 1200°C, only a few, small precipiates were found in the .006 N steel, and the corresponding austenite grains were large (160 μm). In the .025 N steel, small precipitates were numerous even at 1200°C, and the austenite grain size was about 60 μm.

The results discussed above indicate that grain growth in these steels is restricted whenever a considerable amount of fine precipitate is present. To be more precise, the fine precipitates are not only effective in suppressing the normal grain growth of the primary austenite grains, in accordance with the Gladman theory,(25) but also act to hinder the rate of coarsening of the grains which are growing above the grain coarsening temperature. This latter effect is especially pronounced at 1200°C where there is a very large difference in grain sizes of the .006 and .025 N steels.

The difference in the size and volume fraction of precipitates in the .006 N and.025 N steels is related to the thermodynamic stability of the particles. Since increasing the nitrogen content of a steel increases the nitrogen content of the precipitates,(26) the particles in the .025 N steel have a higher nitrogen content than those in the .006 N steel. Nordberg, et al.,(27) have shown that the stability of Nb(C,N) precipitates increases with increasing nitrogen content. Thus, the precipitates in the .025 N steel should be more stable than those in the .006 N steel. These precipitates in the high-nitrogen steel, therefore, can persist to temperatures far above the grain coarsening temperature, and are responsible for the grain refinement observed at those temperatures.

Hot Compression Studies

Typical flow curves for the different steels tested are displayed in Figure 5. These flow curves are characteristic of materials undergoing dynamic recrystallization. The strain to the peak stress (ε_p) has been shown (16-20,23) to be determined by recrystallization and influenced by precipitation.

The hot flow curves of the base steel were observed to vary systematically with T and $\dot{\varepsilon}$ as incorporated in the Zener-Holloman parameter Z in a manner consistent with previous observations.(15) That is, the flow stress was observed to increase and the peaks in the flow curves became broader and flatter as Z increased. In addition, the flow curves of the microalloyed steels investigated showed some deviation from this behavior. At high strain rates ($\dot{\varepsilon} \gtrsim 2s^{-1}$), the trend was the same; a systematic increase in ε_p with microalloying content was observed, Figure 5. However, at low $\dot{\varepsilon}(\sim .006s^{-1})$, dramatic deviations from this general trend were observed. At sufficiently low temperatures, large increases in ε_p were observed for the microalloyed steels, but not for the base steels. Behavior of this type suggested that the most convenient way of discussing the data is in terms of high and low strain rate experiments.

High Strain Rate Experiments. Results of tests at $13s^{-1}$ are given in Figures 6 and 7. Specimens had been given a constant reheating treatment at 1200°C for 30 min. This practice resulted in different as-reheated austenite grain sizes for the various steels, Figures 1 and 3. It also

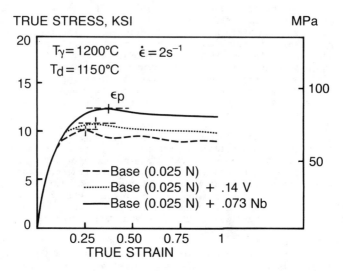

Figure 5 - Typical high temperature flow curves for a base steel, a V-bearing steel and a Nb-bearing steel.

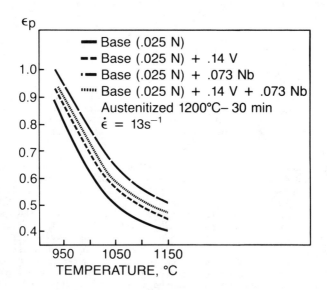

Figure 6 - Effect of Nb and V additions on the dependence of εp with temperature in the high strain rate experiments.

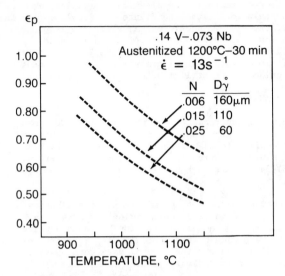

Figure 7 - Effect of N level on the
dependence of εp with initial austenite
grain size for the base steel (.025N).

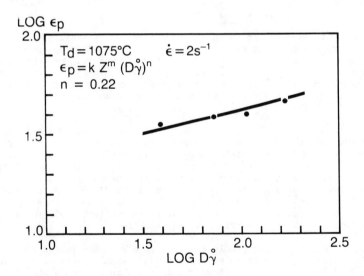

Figure 8 - Dependence of εp with initial austenite grain
size for the base steel (.025N).

resulted in different amounts of undissolved particles, especially in the steels which contain .025 nitrogen, Figure 4. Previous work has shown that the presence of undissolved particles can alter the response of austenite to hot deformation.(20, 28,29) In addition, an increase in the reheated grain size will shift ε_p to larger values.(30-32) It has been suggested that this shift in ε_p with initial grain size can be accounted for by the use of a correction factor utilizing the following equation.

$$\varepsilon_p = kZ^m(D_\gamma^o)^n ,$$

where k, m and n are constants.(33) The value of n determined in our experiments is 0.22 for the base steel, Figure 8. It was assumed that this value could be applied to all of the steels used in the high $\dot{\varepsilon}$ experiments. The value of n = 0.22 compares favorably with recent determinations ($.17 \leq n \leq .33$)(34,35) although higher values have been reported in earlier work. (30-33).

This correction was used to eliminate the contribution of the initial grain size to the shift in ε_p by normalizing the data to a constant initial grain size of 195 μm. Once this correction has been applied, any remaining shift in ε_p must be caused by the microalloying elements, nitrogen or Z. Furthermore, as discussed earlier, the ε_p data were converted to t_p by dividing by the appropriate strain rate.

When the data of Figure 6 are corrected for the grain size effect, this can be replotted in terms of t_p, Figure 9. A comparison of these two figures reveals a change in the relative positions of the curves. In the corrected figure, the V-Nb-N steel shows a larger value of t_p than does the Nb-N steel, whereas the opposite was true in Figure 6. A similar correction can also be used to assess the effect of nitrogen.

Results of the high strain rate experiments were summarized in Figures 10 and 11. In Figure 10 it can be seen that all microalloying elements, either alone or in combination, produce an increase in t_p. Of particular interest in these diagrams is the relative change in t_p with $\dot{\varepsilon}$ and T for a given microalloyed steel with respect to that of the base steel. Variations in these relative changes can be taken as a measure of the retardation of dynamic softening.(16,23)

The data of Figures 10 and 11 reveal that the smallest shift in t_p is obtained with the vanadium addition whereas the largest shift is due to the vanadium-niobium addition. The addition of .073 Nb leads to t_p values which fall between these two, as shown in Figure 9. The magnitude of the relative increases in t_p are independent of the temperature and strain rate, provided the strain rates are above about 1 sec^{-1}. These observations suggest that the delay of dynamic softening is a solid solution effect. A conclusion of this type would be expected, given the very small values of t_p encountered in the high strain rate experiments.

The influences of nitrogen level and strain rate are presented in Figure 11. Variations in nitrogen content had a small effect on t_p compared to the effect of microalloying elements, as shown in Figures 9 and 10. Again, this influence of nitrogen variations was found to be essentially independent of temperature and strain rate, provided the strain rate is above about 1 sec^{-1}.

Low Strain Rate Experiments. The results for a strain rate of .006 sec^{-1} are given in Figure 12 for a variety of steels. Within the range of deformation temperature used ($900 \leq T \leq 1225°C$), two general patterns of

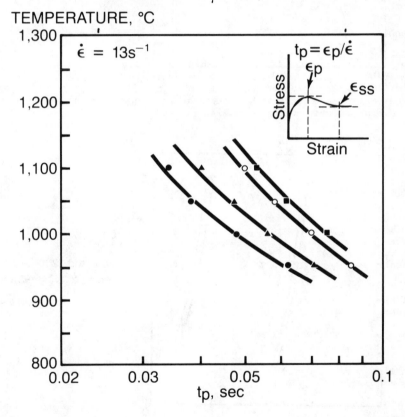

- Base (.025 N)
- ▲ Base (.025 N) + .14 V
- ○ Base (.025 N) + .073 Nb
- ■ Base (.025 N) + .14 V + .073 Nb
  Austenitized 1200°C– 30 min
  Corrected to $D_\gamma^\circ = 195\mu m$

Figure 9 – Dependence of tp on deformation temperature, corrected to the same initial austenite grain size (195μm). Data taken from Figure 6.

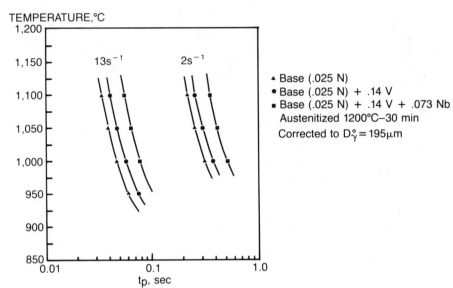

Figure 10 - Dependence of t_p on deformation temperature and strain rate for some high nitrogen steels.

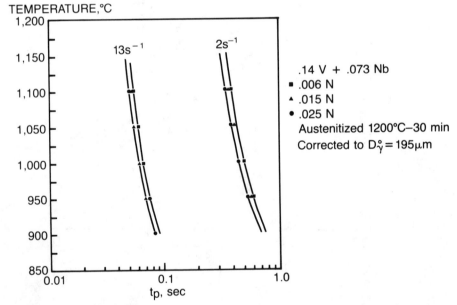

Figure 11 - Dependence of t_p on deformation temperature and strain rate for a V+Nb steel at three levels of N.

48

behavior were observed. The vanadium steel containing low nitrogen (.006 N) and the base steel displayed behavior that resembled that which was found in the high strain rate experiments, Figures 10 and 11. That is, the relative shift in t_p was observed to be essentially independent of temperature. In the case of the other three microalloyed steels the shift in t_p was found to be strongly temperature dependent; below a well-defined temperature, T*, a further reduction in deformation temperature led to a drastic increase in t_p. These temperatures, T*, at which the first deviation from the high temperature trend in t_p is observed, are given in Figure 12.

Among the three steels which showed the drastic increase in t_p with decreasing deformation temperature, it can be seen that both the high and low nitrogen niobium-bearing steels exhibit similar values T* (1075-1090°C). This temperature did not appear to be very sensitive to the nitrogen content. The vanadium-high nitrogen steel had a T* in the vicinity of 970°C. In contrast to the V-Nb steels, the nitrogen level had a profound effect on the response of the vanadium steel to hot deformation; unlike the high nitrogen steel, the vanadium steel with .006 N showed no drastic increase in t_p with decreasing temperature.

In addition to the mechanical testing just described, a systematic investigation of precipitation in the deformed and quenched specimens was conducted. An example of the change in shape of a flow curve with precipitation is shown in Figure 13. The absence of dynamic precipitation is associated with well-defined peaks and sharp drops in flow stress beyond the peak, Figure 13, curve 1. A small amount of precipitation caused a slight increase in ε_p or t_p, Figure 13, curve 2. A larger amount of precipitation led to broad, flat peaks with very large ε_p and t_p, Figure 13, curve 3. Examples of the precipitation found in the latter two conditions are shown in Figure 14.

Further precipitation studies were conducted on the deformed and quenched specimens used to construct Figure 12. Each of the microalloyed steels shown in Figure 12 were analyzed for precipitation at temperatures both above and below the T* for that particular steel. Under no circumstances were fine precipitates found in specimens quenched from above T*. Conversely, fine precipitates were always found in specimens deformed at temperatures somewhat below T*. Of particular interest is the comparison of the precipitates that were responsible for similar, large increase in t_p for the microalloyed steels used in Figure 12. To this end, specimens of the V-N-.025N steel [T*=980°C] were deformed at 915°C. These precipitation observations are shown in Figure 14. It should also be noted that the V-Nb steel with .006N likewise exhibited a distribution of fine precipitates when deformed below 980°C [T*=1075°C].

It is well known that the peak in the flow stress curve is a result of the dynamic balance between hardening and softening events which are simultaneously occurring during the deformation. In single phase materials, these events have been thoroughly studied and identified as work hardening, dynamic recovery and dynamic recrystallization.(14,15) More precisely, the specimen becomes strengthened as a result of an increase in dislocation density but softened as a result of dynamic recrystallization. However, an additional event such as precipitation may occur. If precipitation occurs, it might be expected to increase the rate of work hardening and also retard recrystallization kinetics. This interaction with recrystallization kinetics would lead to values of ε_p much larger than those expected for single phase alloys. In the present study, very large values of t_p were shown to be associated with precipitates. Moreover, the testing conditions,

49

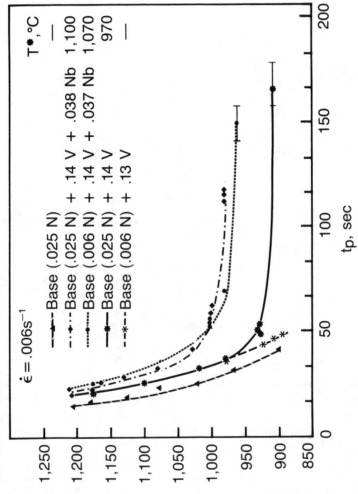

Figure 12 – Dependence of t_p on deformation temperature at a low strain rate for some microalloyed steel containing two levels of N.

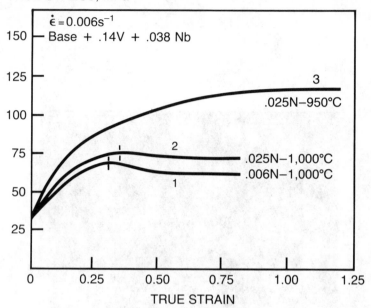

TRUE STRESS, MPa

$\dot{\epsilon} = 0.006s^{-1}$
Base + .14V + .038 Nb

3
.025N–950°C

2
.025N–1,000°C
.006N–1,000°C

1

TRUE STRAIN

Figure 13 - Influence of dynamic precipitation on the flow curves of a
V + Nb microalloyed steel: (1) no precipitation, (2) slight precipitation
and (3) copious precipitation.

(T, $\dot{\epsilon}$) which led to large t_p were those that represented both a large solute
supersaturation (low T) and sufficient time (low $\dot{\epsilon}$) for this precipitation
to occur. Similarly, those compositional variables which would be expected
to alter the precipitation, would also be expected to affect t_p. Hence,
an increase in the nitrogen content increases the solubility temperature
of precipitates leading to higher values of T*. The different microalloying elements in the steels were also shown to affect T*'s in a manner
consistent with this interpretation.

Conclusions

The influence of Nb, V and N on the reheating and subsequent hot
deformation behavior of austenite has been studied. The results of this
research lead to the following conclusions.

1. The grain coarsening temperature is controlled by the stability of
the precipitates which are pinning the austenite grains. The stability of
these precipitates is controlled by their composition which, in turn, is
related to the overall composition of the steel. The precipitates in the
Nb + V steel containing the highest N content are most stable and are
responsible for the fine austenite grains which were observed after reheating to high temperatures in the vicinity of 1200°C.

2. Compression tests revealed that at high strain rate the delay of
dynamic recrystallization as measured by relative increases in t_p were only
weakly dependent on temperature and strain rate. This delay is most likely

Figure 14 – Dark-field electron micrographs of precipitates formed dynamically during deformation. Strain rate $\dot{\varepsilon} = .006s^{-1}$.
(a)   .14 V + .038 Nb + .025 N steel; $T_{deform} = 1000°C$.
(b)   Steel as above in (a); $T_{deform} = 950°C$.
(c)   .14 V + .025 N steel; $T_{deform} = 915°C$.

Figure 14(b)

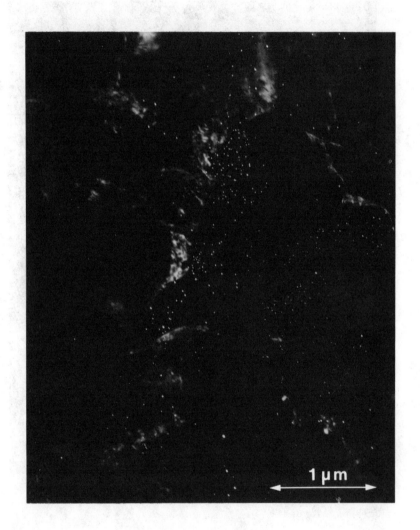

Figure 14(c)

caused by solute drag effects. In this regard, Nb was found to be somewhat more effective than V; however, the combination of Nb+V was the most effective. Interstitial N had no significant retarding effect.

3. At low strain rates the delay of dynamic recrystallization became strongly temperature dependent. Retardation at high temperatures could again be attributed to solute effects. At low temperatures, however, the delay was found to be much more intense and was shown to be caused by precipitation. In the V+Nb steels, this strong, low temperature delay was only weakly dependent on the N level. On the other hand, a high N level was shown to be essential in producing a strong, low temperature retardation. In this context, V steels with high N exhibited a pattern of behavior similar to that of V+Nb steels.

Acknowledgment

The authors want to thank the Metals Division of Union Carbide Corporation and Niobium Products, a subsidiary of CBMM for supporting this research. Thanks are also due to the Graham Research Laboratory of J & L Steel Co. for providing the steels.

References

1. K.J. Irvine, Symposium on Low-Alloy High-Strength Steels, The Metallurgy Companies, Nuremburg, BRD, 1970.

2. J.M. Gray, Met. Trans., 3 (1972), p. 1495.

3. T. Gladman, D. Dulieu and I.D. McIvor, Microalloying '75, M. Korchynsky, ed., Union Carbide Corporation, New York, p. 32 (1977).

4. T. Tanaka, N. Tabata, T. Hatomura and C. Shiga, Reference 3, p. 107.

5. T. Tanaka, International Metals Review, No. 4 (1981), p. 185.

6. I. Kozasu, C. Ouchi, T. Sampei and T. Okita, Microalloying '75, M. Korchynsky, ed., Union Carbide Corporation, New York, p. 120 (1976).

7. T.G. Oakwood, W.E. Heitmann and E.S. Madrzyk, The Hot Deofrmation of Austenite, J. Ballance, ed., TMS-AIME, New York, p. 204 (1977).

8. E.L. Brown, A.J. DeArdo and J.H. Bucher, ibid., p. 250.

9. J.G. Byrne, Recovery, Recrystallization and Grain Growth, MacMillan, New York, p. 60 (1965).

10. J.J. JOnas, C.M. Sellars and W.J. McG. Tegart, Met Rev., 14 (1969), p.1.

11. R.A.P. Djaic and J.J. Jonas, Met. Trans., 4 (1973), p. 621.

12. U. Koster, Metal Sci., 8 (1974), p. 151.

13. H. Ahlborn, E. Hornbogen and U. Koster, J. Mater. Sci., 4 (1969), p. 944.

14. M.J. Luton and C.M. Sellars, Acta Met., 17 (1969), p. 1033.

15. H.J. McQueen and J.J. Jonas, Plastic Deformation of Metals, vol. 6, R.J. Arsenault, ed., Academic Press, New York, p. 393 (1975).

16. M.G. Akben, Ph.D. Thesis, McGill University, 1981, Montreal, Canada.

17. I. Weiss and J.J. Jonas, Met. Trans A, 10A (1979), p. 831.

18. M.J. Luton, R. Dorvel and R.A. Petkovic, Met. Trans. A, 11A (1980), p. 411.

19. A. LeBon, J. Rofes-Vernis and C. Rossardi, Met. Sci. J., 9 (1975), p. 36.

20. C.M. Sellars, Recrystallization and Grain Growth of Multi-Phase and Particle Containing Materials; N. Hansen, A.R. Jones and T. Leffers, eds., Risø National Laboratory, Roskilde, Denmark, p. 291.

21. H.J. McQueen and S. Bergerson, Metal Sci. J., 6 (1972), p. 25.

22. G. Fitzsimons, H.A. Kuhn and R. Venkateshwar, J. of Metals, 33, No. 5, (1981), p. 11.

23. M.G. Akben, I. Weiss and J.J. Jonas, Acta Met., 29 (1981), p. 111.

24. T. Gladman and F.B. Pickering, J. Iron Steel Inst., 205 (1967), p. 653.

25. T. Gladman, Proc. Roy. Soc., 294A (1966), p. 298.

26. L. Meyer, H. Buhler and F. Heisterkamp, Thyssem Forschung, August 1971, Thyssen-Hitte AG, Duisburg-Hamborn, p. 8.

27. H. Nordberg and B. Aronsson, J. Iron Steel Inst., 206 (1968), p. 1263.

28. M. Santella, Ph.D. Thesis, Department of Metallurgical and Materials Engineering, University of Pittsburgh, 1981.

29. M. Santella and A.J. DeArdo, this Proceedings.

30. J.P. Sah, G.J. Richardson and C.M. Sellars, Metal Sci., 8 (1974), p. 325.

31. W. Roberts, Metal Sci., 13 (1979), p. 195.

32. S. Sakui, T. Sakai and K. Takeishi, Trans. Iron and Steel Institute of Japan, 17 (1975), p. 718.

33. C.M. Sellars, "proceedings of the Conference on Hot Working and Forming Processes", C.M. Sellars and G.J. Davies, eds., The Metals Society, London, England, 1980, p. 3.

34. T. Sakai, University of Electro-communications, Tokyo, Japan, private communication.

35. P. Alvarado, M.S. Thesis, Department of Metallurgical and Materials Engineering, University of Pittsburgh, 1981.

DISCUSSION

Q: Were your tests conducted at constant velocity or at constant strain rate?

A: Constant true strain rate.

Q: I noticed from your compression and flow curves, that you went up to a strain rate of 1.25. Did you get any folding over, or was the flow even on the surfaces of the samples?

A: There was no folding what so ever.

Q: How about flowing over. Was the spread even on the surfaces?

A: Yes.

AUSTENITE RECRYSTALLIZATION AND GRAIN GROWTH DURING THE

HOT ROLLING OF MICROALLOYED STEELS

D. R. DiMicco and A. T. Davenport
Republic Steel Corporation
Research Center
Independence, Ohio

The ability to obtain homogeneous austenite grain refinement and avoid mixed-grain sizes is a key factor to obtaining high levels of toughness in controlled-rolled HSLA steels. In this paper, the results of multipass laboratory rolling experiments will be reported in which the hot deformation process in low-carbon microalloyed steels has been studied in bainitic steels alloyed to facilitate the retention of austenite structure in heavy sections. This paper is concerned with the effects of both compositional and processing variations on the development of recrystallized austenite structure and the growth, both normal and abnormal, of the austenite structure. Three main areas of study will be reported: (1) grain refinement by recrystallization, (2) grain growth between passes and during prolonged intermediate holding, and (3) strain induced grain growth. The relevance of these areas in achieving the optimum refinement of structure and the avoidance of mixed microstructural effects will be discussed.

Introduction

It is now well accepted that the structural refinement of austenite during hot-rolling plays a major role in determining the mechanical properties of contolled-rolled microalloyed HSLA steels. At high temperatures where the austenite rapidly recrystallizes, refinement is produced by successive recrystallization events during repeated rolling reductions. At low temperatures below the temperature range where austenite recrystallizes, the austenite grains become deformed (or pancaked) during rolling and, in effect, become refined by the increase in grain boundary area per unit volume and the introduction of shear bands. There is, therefore, much interest in the effect of hot-rolling in the austenite recrystallization and non-recrystallization regimes, particularly in steels containing strong recrystallization inhibitors, notably, alloy elements such as columbium.

The significance of columbium's role as a potent inhibitor of austenite recrystallization at low rolling temperatures of $< 954°$ C ($1750°$ F) is well recognized. However, it is also recognized that unless there is careful control over the hot working process the addition of columbium can have a deleterious effect on mechanical properties, in particular impact toughness (1-4).

Much of the research into hot-rolled austenite has focused on low-carbon manganese-type steels where considerable experimental problems exist in directly observing austenite because of their low hardenability. In short, as the ability to retain prior austenite structure is dependent upon the amount of martensite or bainite formed on quenching, the low hardenability of the HSLA steel compositions of common interest presents a major handicap. Largely because of this problem, many of the studies into the direct observation of hot-rolled austenite structure in these steels have been confined to thin sections usually subjected to simple one or two rolling reductions. The changes occurring during the more complex rolling schedules of commercial production is an area which has yet to receive much attention.

The research in the present paper concerns a laboratory study in which an attempt has been made to examine the austenite structural changes in complicated (multipass) commercial-type rolling schedules in low-carbon bainitic steels of substantially more hardenability than typically found in the usual HSLA steel variety. These steels are of a hardenability level sufficient to allow the preservation of prior austenite structure in quenched sections in excess of 63.5 mm (2-1/2 inches) thick, this dimension being the maximum thickness of the initial slabs examined in this study.

Many investigators (4-11) have observed the formation of mixed or duplex ferrite-pearlite microstructures in steels, in particular controlled rolled steels. This observation can be related back to a mixed or duplex austenite structure. Several explanations have been proposed for this observation including:

- Partial recrystallization of the austenite (4-6).

- Normal grain growth between rolling passes and during delay periods (7-10).

- Inhomogeneous formation of shear bands during pancaking (5, 6, 11).

- Abnormal grain growth due to strain induced grain boundary migration (6, 11-13).

The multipass rolling studies reported in this paper examine the factors which govern the formation of very fine, uniform austenite grain structures. Experimentation was also directed at an investigation of the formation of mixed and duplex austenite structures through a phenomenon known as Strain Induced Abnormal Grain Growth.

Experimental

Compositions

All steels were produced from Al-killed, laboratory air-induction melted heats which fall into the following two categories of nominal base composition (wt. %):

- Bainitic Steels containing
  1.4 Mn - 1.4 Cr - 1.4 Ni - 0.40 Mo - 0.25 Si - 0.70 Cu, and

- Carbon-Maganese Steels containing
  1.4 Mn - 0.5 Si.

In both categories, C and Cb variations were emphasized with a limited variation in V included in the bainitic steels. Compositional details of the bainitic and C-Mn steels are listed in Tables I and II, respectively.

TABLE I- CHEMISTRY: BAINITIC STEELS

STEEL I.D.	C	Mn	P	S	Si	Al	V	Cb	N
A	0.05	1.35	<0.01	<0.01	0.28	0.042	—	0.04	0.008
B	0.055	1.40			0.30	0.026	—	0.05	0.007
C	0.063	1.35			0.27	0.04	—	0.10	0.006
D	0.11	1.35			0.28	0.05	—	—	0.007
E	0.095	1.35			0.25	0.03	—	0.037	0.006
F	0.12	1.40			0.32	0.03	—	0.051	0.007
G	0.11	1.35			0.29	0.03	—	0.11	0.006
H	0.11	1.40			0.30	0.02	0.10	—	0.006
I	0.24	1.7	0.012	0.010	0.33	0.033	0.15	—	0.019
J	0.23	1.7	0.012	0.01	0.32	0.032	0.15	0.017	0.020
K	0.24	1.7			0.31	0.04	0.14	0.035	0.021
L	0.23	1.7			0.31	0.04	0.14	0.067	0.021

ALL STEELS CONTAIN $\frac{Cr}{1.4}$ $\frac{Ni}{1.4}$ $\frac{Mo}{0.40}$ $\frac{Cu}{0.70}$

TABLE II- CHEMISTRY C-Mn STEELS

STEEL ID	C	Mn	P	S	Si	Al	V	Cb	N
AI	0.065	1.38	0.016	0.014	0.50	0.035	—	—	0.006
BI	0.058	1.40	0.013	0.013	0.50	0.05	—	0.04	0.007
CI	0.06	1.40	0.013	0.013	0.50	0.05	—	0.06	0.006
DI	0.06	1.35	0.013	0.013	0.49	0.05	—	0.11	0.007
EI	0.11	1.40	0.016	0.014	0.51	0.045	—	—	0.006
FI	0.11	1.40	0.014	0.013	0.53	0.05	—	0.04	0.006
GI	0.10	1.40	0.014	0.013	0.53	0.04	—	0.06	0.006
HI	0.11	1.40	0.014	0.013	0.51	0.04	—	0.11	0.006

Processing

After preliminary hot-forging into slabs, rolling was performed on a laboratory 2 high, 35.5 cm (14-inch) roll diameter mill after initial reheating for 1 to 1-1/2 hours at 1288° C (2350° F). Rolling temperatures were monitored by an imbedded thermocouple, inserted into the center of each slab prior to the first pass. Samples were water quenched from various stages of rolling in order to assess the austenite structure.

A total of twelve different rolling schedules were examined, details of which can be found in Figures 1-4. These experiments were designed to provide information on the following aspects of hot-rolling:

- Grain refinement during continuous rolling
  - Schedules I-III, Figure 1;

- Grain growth between passes and during delay periods
  - Schedules IV-V, Figure 2;

- Grain growth during isothermal holding at intermediate rolling temperatures
  - Schedules VI-IX, Figure 3;

- Strain induced abnormal grain growth
  - Schedules X-XII, Figure 4.

The starting slab thickness for all schedules was 63.5 mm (2.5 inches) with exception of Schedules II and III where the thickness was 50.8 mm (2.0 inches).

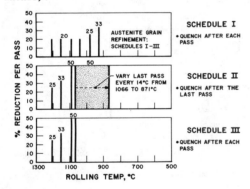

Fig. 1 - The Rolling Schedules Used Primarily for the Accumulation of Austenite Grain Refinement Data, Schedules I-III.

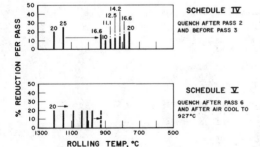

Fig. 2 - The Rolling Schedules Used Primarily for the Accumulation of Austenite Grain Growth Data During Rolling, Schedules IV and V.

62

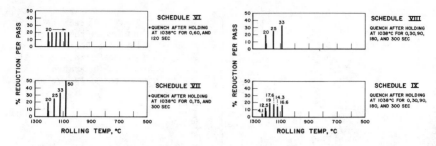

Fig. 3 - The Rolling Schedules Used Primarily for the Accumulation of
Austenite Grain Growth Data During Isothermal Holding Following Rolling
at 1038° C (1900° F), Schedules VI-IX.

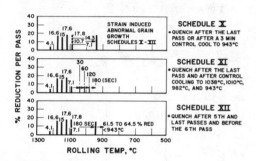

Fig. 4 - The Rolling Schedules
Used for the Accumulation of
Strain Induced Abnormal Grain
Growth Data, Schedules X-XII.

Metallography

Metallographic analysis was conducted on through-thickness sections
(taken from near the vicinity of the thermocouple tip) which were oriented
normal to the rolling plane and containing the rolling direction. The principal etchant used was a saturated picric acid solution modified with a wetting agent (14). Assessment of the austenite structure was confined to sections, away from the rolling surface, which were representative of the bulk
of the sample. In general, the extent of the surface effects was small.
Grain size measurements were made using a circular intercept procedure in
which the average grain diameter was obtained from five measurements made
throughout the thickness.

Results

Austenite Grain Refinement

In all of the steels examined, the austenite was progressively refined
by rolling above 1050° C (1922° F). Below this temperature, nonrecrystallization effects (pancaking) set in at various temperatures depending upon
the Cb level. Analysis of the various factors controlling grain refinement
above the nonrecrystallization range follows.

Total Reduction. As shown in Figure 5, the most important factor controlling grain refinement is the total reduction. The data in this
figure are largely taken from bainitic steels processed with Schedule I,
and these data are tabulated in Table III. However, supplementary data

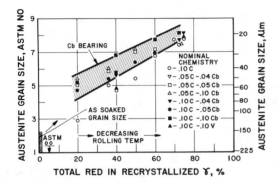

Fig. 5 – The Effect of Total Reduction in the Recrystallization Region on the Austenite Grain Size. The Data Were Generated From Schedules I, IV, V, and VI.

TABLE III - SCHEDULE I GRAIN SIZE RESULTS

STL ID	NOMINAL CHEMISTRY		PASS NO	1	2	3	4	5	6
			% RED / PASS	20	20(25)[+]	20(33)[+]	20(-)[+]	25	33
			% TOTAL RED	20	36(40)[+]	50(60)[+]	60(60)[+]	70	80
			PASS TEMP	1204°C	1149°C	1093°C	1038°C	982°C	927°C
D	10C-			2.9	5.0	5.4	6.8	7.7	P*
F	.10C-.05Cb			4.7	5.4	6.4	6.9	P	P
G	.10C-.10Cb[+]	AUSTENITE GRAIN SIZE (ASTM NO)		5.2	5.7	7.7	—	P	P
B	.05C-.05 Cb			5.1	5.9	7.0	7.1	P	P
C	.05C-.10 Cb			5.4	6.0	7.1	P	P	P

+ NOTE DIFFERENT PER PASS AND TOTAL REDUCTIONS FOR THE .10 C -.10 Cb STEEL

* P - PANCAKED OR PARTIALLY RECRYSTALLIZED

were also included from other schedules where appropriate. Notwithstanding the overriding effect of total reduction in Figure, 5 there is enough scatter in the data to suggest the influence of other variables.

As-Soaked Grain Size. Although not illustrated in Figure 5, an effect of Cb content on the starting (as-soaked) austenite grain size was observed. This grain size varied from a uniform ASTM No. 00 for the base and vanadium bearing steels to a relatively uniform ASTM No. 2 for the 0.10 wt. % Cb steel. At lower columbium contents the structure consisted of a mixture of grains from 00 to 2, with the structure being progressively less mixed and finer as the columbium content approached 0.10 wt. %. This data suggests that there was incomplete solution of the Cb at the 1288° C (2350° F) soaking temperature. Chemical extraction data indicated that for a one hour soak at 1288° C (2350° F) the 0.03-0.04, 0.05, and 0.10 wt. % Cb steels contained 0.008, 0.011, and 0.015 wt. % Cb remaining undissolved, respectively. It is believed that these grain refinement effects on the as-soaked structure may be associated with the presence of Cb N in small amounts, some of which may be beyond the limits of detection of the extraction technique used here.

There are indications in Figure 5 that the finer starting grain size of the Cb steels is inherited during the first few rolling reductions and that this effect diminishes as the percent total reduction increases.

64

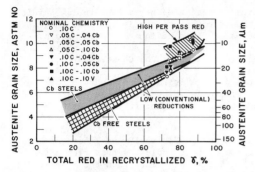

Fig. 6 - The Effect of Increased Percent Reduction per Pass on the Austenite Grain Size. These Data Were Generated From Schedules II, III, VI and VII.

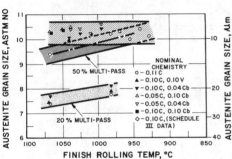

Fig. 7 - The Effect of Finish Rolling Temperature on the Austenite Grain Size. These Data Were Taken From Schedule II (50% MP) and Schedules V and VI (20% MP).

Percent Reduction/Pass. Figure 6 illustrates the additional austenite grain refinement that can be realized by increasing the percent reduction taken on each rolling pass. Attention is drawn to several sets of data obtained from Schedules II, III, VI, and VII in which the last two passes of each schedule were increased from 20% to 33 and 50%, to 50 and 50% reduction/pass. Clearly, another factor leading to greater grain refinement above and beyond that of total reduction is that of the reduction/pass. All things being equal, increasing the reduction per pass increases the level of grain refinement.

Temperature. It will be appreciated that another variable influencing the trend in Figure 5 is temperature, since temperature decreases with increasing total reduction (and number of passes). In order to distinguish the effect of rolling temperature from other variables, Steels D, E, and H were given identical 20% multipass schedules starting at 1204° C (2200° F) and finishing at 1066° C (1950° F) and 982° C (1800° F), Schedules V and VI, respectively. As shown in Figure 7, lowering the temperature over this range produced finer grains but the effect was small. Data obtained from Schedules II and III, in which the final pass temperature was the only rolling variable, are also included in Figure 7 and these data reinforce the conclusion that rolling temperature is of minor importance at least at temperatures between 1066 and 927° C.

The base 0.11 wt. % C steel processed to Schedules II and III does indicate a slight temperature dependency at temperatures between 982° C (1800° F) and 1066° C (1950° F). This is consistent with the results obtained from Schedules V and VI for the Cb-free steels.

These experiments tend to reinforce the contention that total reduction is primarily responsible for determining grain size.

Composition. Figure 5 demonstrates that the Cb-bearing steels exhibited a finer austenite grain size than the Cb-free steels and that increasing the Cb content resulted in increased grain refinement. This was particularly true during the initial rolling passes and this may have been related to the finer as-soaked grain sizes of the Cb steels.

This result is reinforced by the data illustrated in Figure 8 for the base and 0.10 wt. % Cb (Bainitic) steels which were processed to Schedule III. This grain refinement occurs throughout the processing sequence.

Comparison of the data in Figure 9 and in Tables IV and V for the bainitic and C-Mn steels of similar Cb content indicates that even after identical processing the bainitic steels are finer than the C-Mn steels. As will be discussed later, this appears to be an effect of the additional alloying elements (Cr, Ni, Mo, and Cu) in the bainitic steels. A related effect dealing with the recrystallization behavior of austenite during rolling has been discussed in an earlier work (15).

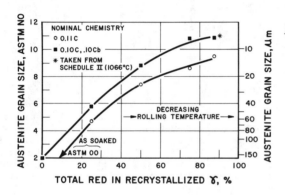

Fig. 8 – A Comparison Between the Austenite Grain Sizes Achieved for the 0.11 wt. % C and 0.10 C-0.10 Cb (wt. %) Steels Demonstrating the Grain Refining Effects of Cb During Hot Rolling. These Data Were Generated From Schedule III.

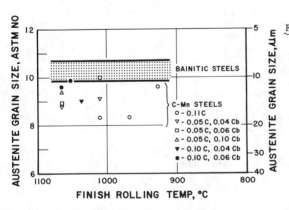

Fig. 9 – This Figure Illustrates the Additional Grain Refinement Achieved for the Bainitic Steels Over That Achieved for the C-Mn Steels.

66

TABLE IV - SCHEDULE II GRAIN SIZE RESULTS
50% MULTI-PASS SCHEDULE WITH A 87.5% TOTAL REDUCTION

STL I.D.	NOMINAL CHEMISTRY	ROLLING TEMPERATURE OF LAST PASS,°C										
		1066	1052	1038	1024	1010	996	982	968	954	940	927
	C-Mn											
A I	.05C-	—	—	—	—	8.3	—	—	8.3	—	—	9.6
B I	.05C-.04Cb	8.8	—	—	—	9.1	—	—	—	—	—	—
C I	.05C-.06Cb	9.2	—	—	—	10.0	—	—	—	—	—	—
D I	.05C-.10Cb	9.4	—	—	—	—	—	—	—	—	—	—
F I	.10C-.04Cb	—	—	9.0	—	—	—	—	—	—	—	—
G I	.10C-.06Cb	9.6	9.9	—	—	—	—	—	—	—	—	—
	BAINITE											
D	.11C-	—	—	—	—	—	—	10.2	10.3	10.0	10.4	10.1
A	.05C-.04Cb	—	—	—	10.4	10.4	10.2	10.6	—	—	—	—
C	.05C-.10Cb	10.3	10.4	10.5	—	—	—	—	—	—	—	—
E	.10C-.04Cb	—	—	10.0	10.2	10.2	10.4	—	—	—	—	—
G	.10C-.10Cb	10.3	10.4	10.0	—	—	—	—	—	—	—	—

TABLE V - SCHEDULE III GRAIN SIZE RESULTS

STEEL I.D.	NOMINAL CHEMISTRY		PASS NO	1	2	3	4
			% RED/PASS	25	33	50	50
			% TOTAL RED	25	50	75	87.5
			PASS TEMP,°C	1204	1149	1093	1066
D	.10 C			4.7	7.4	8.6	9.4
G	.10C-.10Cb (BAINITE)	AUSTENITE ASTM NO GRAIN SIZE		5.8	8.8	10.4	*
H I	.10C-.10Cb C-Mn			—	8.6	9.8	*

* PARTIALLY RECRYSTALLIZED

The addition of 0.10 wt. % V had little, if any, grain refining effect over that of the base 0.11 wt. % C steels as indicated in Figures 5 and 7.

Austenite Grain Growth During Rolling

Continuous Cooling. Schedules IV and V were schedules designed to investigate the effect of interpass time on grain growth. The data shown in Table VI indicates that little if any normal grain growth (and no abnormal grain growth) was occurring during either the interpass time of 120 seconds of Schedule IV which involved an air cool from 1149° C to 927° C (2100 to 1700° F) or the much shorter interpass time of approximately 25 seconds of Schedule V from 982° C to 927° C (1800 to 1700° F). This was true regardless of the chemistry variations examined.

Isothermal Holding. Figures 10 and 11 and Table VII demonstrate the low susceptibility of the bainitic steels to exhibit grain growth during isothermal holds at intermediate temperatures, i.e., 1038° C (1900° F). There is little, if any, observed grain growth for steels containing Cb levels of ≥ 0.05 wt. % or combinations of Cb, V, and N (Steel J), even for hold times of 300 seconds in duration.

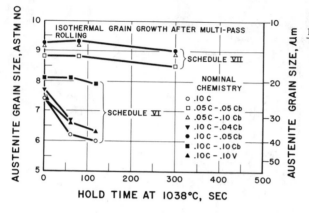

Fig. 10 – The Effect of Isothermal Hold Time at 1038° C (1900° F) on the Austenite Grain Growth Observed in Several Steels Processed to Schedules VI and VII.

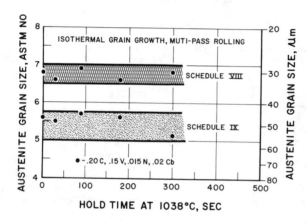

Fig. 11 – The Effect of Isothermal Hold Time at 1038° C (1900° F) on the Austenite Grain Growth Observed in Steel J Processed to Schedules VIII and IX.

TABLE VI – AUSTENITE GRAIN GROWTH DURING ROLLING[1]

		RECRYSTALLIZED AUSTENITE GRAIN SIZE (ASTM NO)			
		SCHEDULE IV		SCHEDULE V	
STEEL I.D.	NOMINAL CHEMISTRY	AFTER PASS 2[2] (1149°C)	BEFORE PASS 3[3] (927°C)	AFTER PASS 6[4] (982°C)	AFTER AIR COOLING TO 927°C (25 SEC)
D	.10 C	4.7	4.8	7.8	7.2
E	.10C-.04Cb	5.6	4.9	8.1	p[5]
G	.10C-.10Cb	6.8	6.8	P	P
H	.10C-.10V	4.8	5.0	7.9	7.6

1 - 20% PER PASS, MULTI-PASS ROLLING
2 - SLAB THICKNESS 38.1mm; TOTAL REDUCTION 40%
3 - AIR COOLING TO 927°C TOOK 2 MINUTES
4 - SLAB THICKNESS 15.8 mm; TOTAL REDUCTION OF 75%
5 - P = PANCAKED OR PARTIALLY RECRYSTALLIZED

TABLE VII- AUSTENITE GRAIN GROWTH FOLLOWING
ISOTHERMAL HOLDING AT 1038°C

SCHEDULE NO	STL ID	NOMINAL CHEMISTRY	FINISH ROLLING TEMP°C	HOLDING TIME AT 1038°C, SEC							
				0	30	60	75	90	120	180	300
VI 73%(1)	D	.10C	1066	7.4(3)	—	6.2	—	—	6.0	—	—
	E	.10C-.04Cb	"	7.7	—	6.7	—	—	—	—	—
	G	.10C-.10Cb	"	8.1	—	8.1	—	—	7.9	—	—
	H	.10C-.10V	"	7.4	—	6.6	—	—	6.3	—	—
VII 80%(1)	F	.10C-.05Cb	1079	9.3	—	—	9.3	—	—	—	9.0
	B	.05C-.05Cb	"	8.8	—	—	8.8	—	—	—	8.5
	C	.05C-.10Cb	"	9.2	—	—	9.2	—	—	—	8.9
VIII 60%(1)	J(2)	.20C-.15V-.02Cb-.015N	1093	6.8	6.6	—	—	6.9	—	6.6	6.8
IX 60%(1)	J(2)	.20C-.15V-.02Cb-.015N	1093	5.6	5.5	—	—	5.7	—	5.6	5.1

(1)- TOTAL REDUCTION
(2)- SCHEDULE X HAD HEAVIER PER PASS REDUCTIONS
(3)- ASTM GRAIN SIZE NUMBER

However, grain growth was observed for Steels D, E, and H after 60 seconds at 1038° C (1900° F), with little additional growth occurring after 120 seconds. The grain growth observed here was normal and not abnormal in nature, i.e., there was a uniform and progressive coarsening of the entire structure.

Strain Induced Abnormal Grain Growth. Schedules X-XII were designed to investigate the phenomenon of Strain Induced Abnormal Grain Growth during hot working. The experiments conducted examined five points:

1) the existence of this phenomenon during the hot working of steel,

2) the effect of percent reduction on the last pass prior to a delay period,

3) the effect of delay time following the last pass,

4) the effect of heavy low temperature reductions (pancaking) on the mixed structures produced by the phenomenon, and

5) the effect of Cb additions from 0 to 0.67 wt. % on the grain growth behavior (Steels I-L).

The results of this experimentation, as presented in Figures 12-14, Tables VII and IX, and summarized in Figure 15, are as follows:

1) Light rolling reductions ($\leq$ 10%) on the pass prior to a delay period can stimulate the growth of abnormally coarse austenite grains of ASTM No. 0-2. This can occur in a matrix of fine grains of ASTM 5-6 as is illustrated in Figure 12. These coarse grains can make-up more than 50% of the structure.

2) The length of the delay period, while accentuating the growth of these abnormal grains, is not as critical as the percent reduction. A significant amount of these abnormal grains was observed after a short 30 second cool to 1038° C (1900° F), as is illustrated in Figure 12c.

69

TABLE VIII - AUSTENITE GRAIN SIZE DATA STRAIN INDUCED ABNORMAL GRAIN GROWTH

SCHEDULE I D	STEEL I D	NOMINAL CHEMISTRY	TIME (SEC) TO COOL TO QUENCH TEMP	QUENCH TEMP,°C	% RED ON FINAL PASS 7.1 (48)	10.7 (50)	14.3 (52)	17.8 (54)
	I	.20 C, .15 V, .015 N	0	1093	4.8*	6.0	6.0	6.1
			180	943	**	5.4	5.3	5.8
	J	.20 C, .15 V, .015 N, .017 Cb	0	1093	4.6	5.5	5.9	6.0
X			180	943	**	5.0	5.0	5.8
	K	.20 C, .15 V .015 N, .035 Cb	0	1093	5.7	6.1	6.1	5.9
			180	943	**	**	5.9	6.0
	L	.20 C, .15 V, .015 N, .067 Cb	0	1093	6.6	6.1	6.3	6.6
			180	943	6.6	**	5.4	6.8

*-AUSTENITE GRAIN SIZE (ASTM NO), **-ABNORMAL GRAIN GROWTH

3) Reductions of 14.3 and 17.8% on the last pass prevented the formation of these abnormal grains. The length of the delay period had little, if any, coarsening effect on the structures thus formed, as illustrated in Figure 13.

4) Once these mixed structures had formed, they could not be removed through the use of heavy reductions (> 60%) below 943° C (1730° F), as is illustrated in Figure 14.

5) This phenomenon was demonstated in steels containing high levels of V (0.15 wt. %), N (0.02 wt. %), and Cb (0 to 0.67 wt. %). However, increase in Cb content to levels ≥ 0.035 wt. % did appear to minimize the severity (i.e., the number and the size of the coarse grains) with which this phenomenon occurred.

Other interesting observations were made including the fact that, as is illustrated in Figure 16(b), the austenite grains present immediately following the light last pass reduction (≤ 10%) for steels I-L possessed wavy or ragged grain boundaries. These wavy boundaries were not present in the materials quenched out prior to the last pass, as is illustrated in Figure 16(a), or those processed with ≥ 14.3% reduction, Figure 16(c), or for Steel I after cooling 30 seconds to 1038° C (1900° F), as is illustrated in Figure 16(d). These observations are the subject of future research aimed at a better understanding of the phenomenon.

It was also observed, as indicated in Table VIII, that Steel L (0.067 wt. % Cb) did not show evidence of the abnormal grain growth following a 7.1% reduction and a 180 second delay. It did exhibit the phenomenon after the 10.7% reduction and delay. This apparent anomaly was not anticipated and it is the subject of further research.

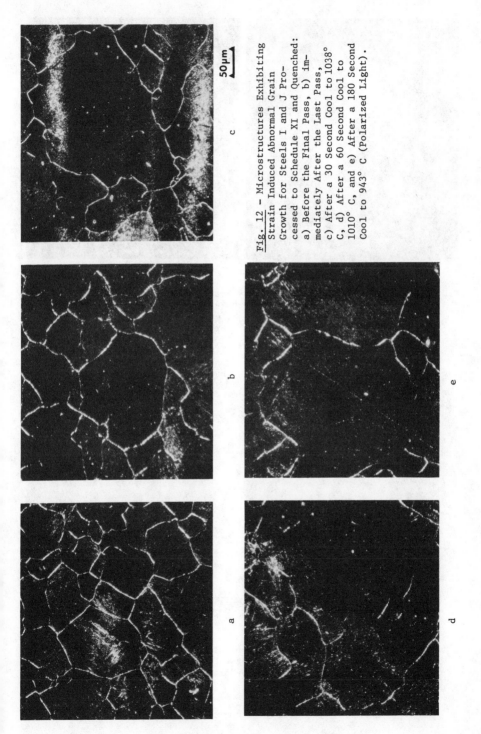

50μm

Fig. 12 - Microstructures Exhibiting
Strain Induced Abnormal Grain
Growth for Steels I and J Processed to Schedule XI and Quenched:
a) Before the Final Pass, b) immediately After the Last Pass,
c) After a 30 Second Cool to 1038°
C, d) After a 60 Second Cool to
1010° C, and e) After a 180 Second
Cool to 943° C (Polarized Light).

71

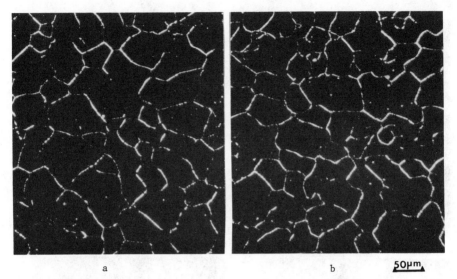

a   b   50μm

Fig. 13 - Microstructures for Steels I-L Given a 17.8% Reduction on
the Last Pass of Schedule X and Held for a) 0, and b) 180 Seconds
to 943° C (Polarized Light).

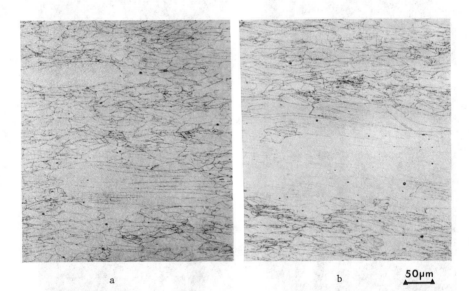

a   b   50μm

Fig. 14 - Microstructures for Steels I and J Processed to Schedule XII
Exhibiting the Existence of Coarse Flattened Austenite Grains Even
After the Mixed Structure was Deformed > 60% Below 927° C.
(Polarized Light)

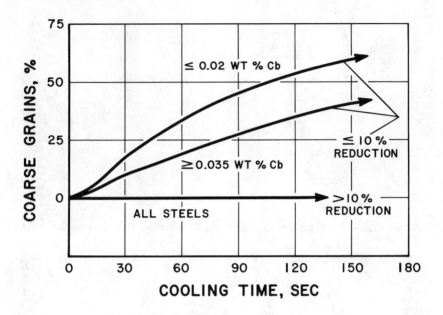

Fig. 15. The Effects of Percent Reduction and Cb Content on the Susceptability of Steels to Strain Induced Abnormal Grain Growth During Rolling.

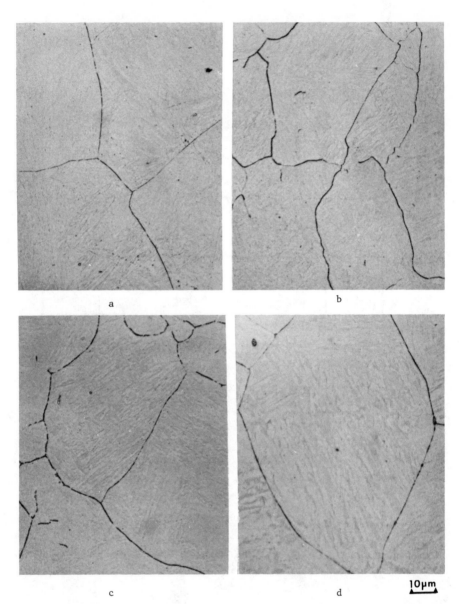

a

b

c

d

10μm

Fig. 16 – Microstructures for Steels I-L Processed to Schedule XI
Demonstrating the Nature of the Austenite Grain Boundaries a) Before
the Last Pass Reduction of 7.1%, b) Immediately After the Last Pass
Reduction of 7.1%, c) Immediately After a Last Pass Reduction of
≥ 14.3%, and d) After Cooling 30 Seconds to 1038° C (Steel I Only).

74

TABLE IX - AUSTENITE GRAIN SIZE DATA STRAIN INDUCED ABNORMAL GRAIN GROWTH

SCHEDULE I D	STEEL I D	NOMINAL CHEMISTRY	TIME (SEC) TO COOL TO QUENCH TEMP	QUENCH TEMP,°C	% RED ON FINAL PASS 7.1 (48)	10.7 (50)	14.3 (52)	17.8 (54)
XI	I	.20C, .15V, .015 N	0	1093	4.5	—	—	—
			30	1038	**	—	—	—
			60	1010	**	—	—	—
			120	982	**	—	—	—
			180	943	**	—	—	—
	J	.20C, .15V .015 N, .017 Cb	0	1093	4.6	—	—	—
			30	1038	**	—	—	—
			60	1010	**	—	—	—
			120	982	**	—	—	—
			180	943	**	—	—	—

*-AUSTENITE GRAIN SIZE (ASTM NO), **-ABNORMAL GRAIN GROWTH

Discussion

The research presented in this paper concerns a laboratory study in which an attempt was made to examine the austenite structural changes occurring as a result of complicated commercial-type rolling schedules. These experiments were conducted on both low-carbon bainitic and C-Mn type steels. In general, the results have indicated the following key points:

- The austenite grain refinement data presented agrees well with the concepts of static recrystallization since they illustrate the importance of such factors as; starting grain size, total reduction, percent reduction/pass, temperature and composition.

- Austenite grain growth during rolling (i.e., continuous cooling) does not appear to be a problem in the bainitic steels when reduction/pass is sufficient to promote complete recrystallization.

- Abnormal austenite grain growth can occur during rolling when the reduction/pass is not sufficient to promote recrystallization but does provide sufficient energy to the system to allow for grain boundary migration.

Austenite Grain Refinement

The results illustrated in Figures 5-9 indicate that the factors contributing to the refinement of the austenite structure during hot-rolling include, in decreasing order of importance, the following:

- Total Reduction

- Percent Reduction/Pass

- Composition

- Temperature, and

- Starting Grain Size.

In general, these factors do not act independently of each other but all contribute concurrently to the development of the final austenite structure.

It was observed in Figure 5 that, as the total reduction increased to the limiting value of 75%, the difference in grain size between the Cb-bearing and Cb-free steels diminished. This observation might be interpreted to be solely the result of a saturation effect of total reduction. However, as the total reduction increases, the rolling temperature decreases; hence the contribution of temperature must also be considered. As illustrated in Figure 7 there is a small but measurable effect of temperature that results in a slightly finer structure with decreasing rolling temperature. It is important to observe here that the Cb-free steels, when processed to lower rolling temperatures, possess austenite grain size virtually identical to the Cb-bearing steels processed to higher rolling temperatures. [Note: Cb's strong retarding effect on recrystallization prevents the processing of the Cb steels to the same low temperatures.] Thus, the convergence in the austenite grain size for the Cb and Cb-free steels as observed in Figure 5 is more likely due to a combined effect of both total reduction and temperature. The ability of the Cb steels to have a finer recrystallized austenite grain size at the higher processing temperatures (for a given amount of total reduction) relates to its ability to negate the effects of temperature that are present for the Cb-free steels, i.e., grain growth after recrystallization. This may be due to either solute drag effects, precipitation effects, or a combination of both.

As illustrated in Figure 9, the presence of Cr, Ni, Mo, and Cu in the bainitic steels appears to contribute to the total level of grain refinement achieved in these steels over that of the C-Mn steels. This could be attributed to a solute drag mechanism which affects the growth of the austenite grains after recrystallization. In related experimentation, as reported in an earlier work (15), the presence of these elements in a Cb-free steel was found to affect the austenite recrystallization process and to promote austenite grain flattening (pancaking) at higher temperatures than a comparable C-Mn steel.

It is interesting to speculate from these results on the similarities between the effects on austenite grain refinement of the additional alloying in the bainitic steels and the effects of Cb in both the bainitic and C-Mn steels. Whereas the mechanism for the enhanced grain refinement achieved for the Cb-free bainitic steels (vs. Cb-free C-Mn steels) could only be explained using a solute drag argument, that of the Cb steels could be explained on the basis of either solute drag, precipitation, or a combination of both. This work may be indicating that the influence of Cb on grain size is through a solute drag effect on grain growth after recrystallization. However, it is also of interest here to note that the Cb effect occurs in both the bainitic and C-Mn steels and imparts additional grain refinement above that attributable to the Cr, Mo, etc., present in the bainitic steels. Research work is continuing in this area in an effort to clarify the relative effects of solute drag and precipitation, in Cb-bearing steels, on austenite grain refinement, grain growth, and grain flattening.

The effect of Cb on grain size is evidenced not only during rolling but also prior to rolling by the as-soaked grain structure. Cb acts to refine the as-soaked austenite structure as discussed earlier. There remains some question as to whether the effects of Cb on the refinement of the structure during the initial rolling passes results from the finer as-soaked structure or simply to temperature effects during rolling, i.e., grain growth after recrystallization. This point will be the subject of future research efforts.

Austenite Grain Growth During Rolling

It has been discussed (7-10) that austenite grain growth during rolling could be responsible for the presence of the duplex microstructures often observed in many controlled rolled steels. Common to most controlled rolling practices, there is a designed delay period to allow the slabs (or bars) to reach the low temperatures desired for finish rolling. Typically this delay period can occur over the temperature range of 1093-927° C (2000-1700° F). It has been the presence of this delay period and also a concern for what happens between rolling passes at high temperatures that has drawn attention to the phenomenon of grain growth during rolling.

The results of the grain growth studies reported here demonstrate that under processing conditions such that the austenite is recrystallized with each reduction (or at lower temperatures flattened with each reduction in Cb steels) there is no significant grain growth taking place during rolling, over this temperature range, that might result in mixed or duplex structures (austenite or ferrite-pearlite). However, as discussed by Tanaka (11, 12) and as demonstrated here, if at high and intermediate temperatures the reductions are such that recrystallization does not take place (i.e., <10-14%), a phenomenon known as Strain Induced Grain Boundary Migration can and will occur, resulting in the formation of abnormally coarse grains. The results reported earlier demonstrate that abnormally coarse grains can be present within very short times following the critical reduction. These times would be compatible with the interpass times present during commercial rolling, with longer delay times promoting coarse, more mixed austenite structures. The resulting mixed austenite structures persist despite the use of heavy reductions at low rolling temperatures. These structures then transform to mixed ferrite-pearlite structures which can severely decrease the impact toughness of the final product.

The Formation of Mixed or Duplex Structures

The formation of mixed or duplex ferrite-pearlite microstructures relates directly to the state of the austenite prior to the transformation. There have been many viable explanations put forth in attempting to explain how the austenite structure obtains its duplex nature. Some of the more frequently discussed include:

- Insufficient or partial recrystallization during high and intermediate temperature rolling (4-6).

- Normal grain growth during rolling between passes and during rolling delay periods (7-10).

- Strain Induced Abnormal Grain Growth during rolling due to very light per pass reductions which are insufficient to promote recrystallization (11-13).

- Insufficient refinement of the austenite grains at high and intermediate temperatures.

- The inhomogeneous formation of deformation or shear bands during low temperature grain flattening (5, 6, 11).

The presence of elements like Cb which retard austenite recrystallization during rolling can act as a potential solution to, aggravation of, or cause of the problem. For instance Cb additions can minimize or prevent grain growth during rolling. Cb additions also appear to lessen the severity of the structural effects produced by Strain Induced Abnormal Grain Growth. However, Cb additions do not appear to be able to eliminate the occurrence of this phenomenon. The presence of Cb can also promote the formation of partially recrystallized structures. The presence of coarse grains, prior to the reductions which result in a partially recrystallized austenite, will aggravate the mixed nature of the austenite grains.

The fact that at low rolling temperatures Cb will act to prevent recrystallization complicates all of the above factors. The heavy low temperature finishing operations common to controlled rolled steels will only flatten and not recrystallize the incoming austenite structure. The mixed nature of the incoming austenite structure will persist. Finally, the inhomogeneous formation of shear bands during the grain flattening or pancaking stages will only act to further aggravate the mixed nature of the austenite structure. If the incoming austenite is not mixed but consists of a uniformly coarse austenite grain structure, the inhomogenous formation of shear bands could still act to promote a mixed or duplex transformed structure.

It is obvious that Cb is of tremendous benefit in the production of stronger and tougher steels. It does this primarily because of its strong influence on austenite recrystallization and grain growth. It is also obvious that this very advantage makes the processing of Cb steels a complex procedure. A strong understanding of the interaction effects between processing conditions and Cb-content is critical to achievement of the desired benefits.

Of the many possible explanations for the formation of mixed or duplex structures the phenomenon of strain-induced-abnormal-grain-growth appears to be not only one of the most likely explanations but also the most sinister. In fact, some of the others can be reduced in significance. For example, the concept of normal grain growth during rolling in practice appears to be of small concern. The incidence of partial recrystallization during rolling, while a reality, does not always explain why an austenite structure, which is significantly and uniformly refined by high and intermediate temperature reductions coupled with a low soaking practice, results in the formation of a mixed structure. In general, the partial recrystallization explanation will be limited by the incoming state of the austenite. If it consists of coarse or mixed (coarse + fine) austenite grains, then partial recrystallization will aggravate the mixed nature of the austenite. The phenomenom of strain-induced-abnormal-grain growth on the other hand does not appear to depend on the existing state of the austenite and it can produce a grossly mixed structure from an extremely fine austenite structure.

This is where this phenomenon picks up its sinister tone. The presence of one light reduction at a critical stage in the rolling sequence can negate all the processing controls which were applied for the purpose of producing an extremely fine grained and heavily flattened austenite structure. Processing controls such as low soaking, heavy high and low temperature reductions and even the working of the steel in the two-phase region can be rendered ineffective by strain induced grain growth.

Although most of the work contained herein was done on alloyed (bainitic) steels with additions of Cr, Ni, Mo, and Cu, the results obtained provide insight into the processing of austenite in C-Mn and C-Mn-Cb steels and the eventual achievement of the desired mechanical properties.

Conclusions

1. The processing parameters of total reduction, percent reduction/pass and temperature contribute to the achievement of fine, uniform recrystallized austenite grain structures. The refinement of the austenite grain size increases with:

   • Increasing Total Reduction,

   • Increasing Percent Reduction/Pass, and, to a lesser extent,

   • Decreasing Rolling Temperature.

2. Cb additions of up to 0.10 wt. % can have a strong grain refining effect both on the as-soaked grain size and recrystallized austenite grain size achieved during rolling.

3. The bainitic steels processed in this study, which contained significant levels of Cr, Ni, Mo, and Cu, demonstrated a finer austenite grain size than C-Mn steels of similar Cb content. This can be rationalized on the basis of a solute drag effect on the recrystallized grain boundaries.

4. The grain refining effect of lower rolling temperatures appeared to have more significance for the Cb-free steels than for the Cb-bearing steels over the ranges of temperature where the Cb steels do not pancake. This point was discussed in terms of possible Cb solute drag and/or precipitation effects on grain growth after recrystallization.

5. It was observed that, under the conditions that the austenite is completely recrystallized with each rolling pass, grain growth during rolling or during extended delay periods was not found to be a problem.

6. The existence of Strain Induced Abnormal Grain Growth as discussed by Tanaka and others (11-13) was confirmed and can be a significant factor causing the formation of duplex microstructure at intermediate rolling temperatures.

7. It was found that reductions of <10-14% and temperatures in the range of 1093-943° C (2000-1730° F) can provide the driving force for this phenomenon and that delay periods following such reductions can increase the severity to which this phenomenon affects the austenite structure (i.e., coarser grains occurring with a greater frequency). It is entirely possible that this phenomenon could occur during interpass times consistent with commercial rolling practices.

8. While Cb is known to be an effective inhibitor of grain growth the work conducted herein indicates that, at best, Cb additions of up to 0.07 wt. % will do no more than decrease the severity to which the austenite structure is affected by abnormal grain growth; it does not prevent the phenomenon from occurring.

9. The only effective solution to abnormal grain growth as indicated by the results of this study, is to insure that the reduction/pass is >10 to 14%, particularly when delay periods are included in the schedules.

References

1. G. Glover and J. Hauranek, "Niobium Bearing Hot Rolled Steel Strip", BHP Technical Bulletin, 16, No. 1, May 1972.

2. R. Rousser, "The Effect of Niobium on the Microstructure of Hot-Rolled Low-Carbon-Manganese Steels", Trans. ISIJ, Vol. 13, 1973.

3. D. R. DiMicco and A. T. Davenport, "The Effect of Finishing Temperature on the Structure and Properties of a 0.10 C-0.10 Cb Ferrite-Pearlite Plate Steel", Republic Steel Research Center Report, PR-12,045-77-3, November, 1977.

4. J. J. Irani, D. Burton, J. D. Jones, and A. B. Rothwell, "Beneficial Effects of Controlled Rolling in the Processing of Structural Steels", pp. 110-122, Strong Tough Structural Steels, ISI Publication 104, 1967.

5. I. Kozasu, C. Ouchi, T. Sampei, and T. Okita, "Hot Rolling as a High-Temperature Thermo-Mechanical Process", pp. 126-128, Microalloying 75, New York, N.Y., 1977.

6. T. Tanaka, T. Enami, M. Kirmura, Y. Saito, and T. Hatomura, "Formation Mechanism of Mixed Austenite Grain Structure Accompanying Controlled-Rolling of Niobium-Bearing Steel", To Be Published in Proceeding of "Thermomechanical Processing of Microalloyed Austenite", Conference Held August 17-19, 1981, TMS-AIME.

7. A. B. LeBon and L. N. de Saint-Martin, "Using Laboratory Simulations to Improve Rolling Schedules and Equipment", p. 93, Microalloying 75, New York, N.Y., 1977.

8. K. J. Irvine, T. Gladman, J. Orr and F. B. Pickering, "Controlled Rolled Structural Steels", JISI, pp. 717-726, August, 1970.

9. F. B. Pickering, "High-Strength, Low-Alloy Steels – A Decade of Progress", p. 17, Microalloying 75, New York, N.Y., 1977.

10. C. Ouchi, T. Sampei, T. Okita, and I. Kozasu, "Microstructural Changes of Austenite During Hot Rolling and Their Effects on Transformation Kinetics", pp. 316-340, Hot Deformation of Austenite, J. B. Ballance, ed., AIME, New York, N.Y., 1977.

11. Tanaka, N. Tabata, T. Hatomura, and C. Shiga, "Three Stages of the Controlled-Rolling Process", p. 115, Microalloying 75, New York, N.Y., 1977

12. T. Tanaka, T. Funakoshi, M. Ueda, J. Tsuboi, T. Yasuda, and C. Utahashi, "Development of High-Strength Steel With Good Toughness at Arctic Temperatures for Large Diameter Line Pipe", pp. 399-409, Microalloying 75, New York, N.Y., 1977.

13. L. Cuddy, "Microstructures Developed During Thermomechanical Treatment of HSLA Steels", Met. Trans. A, Vol. 12A, pp. 1313-1320, July, 1981.

14. A. T. Davenport, R. E. Miner, and R. A. Kot, "The Recrystallization of Austenite During the Hot-Rolling of a Cb-Bearing HSLA Steel", p. 186, Hot Deformation of Austenite, John B. Ballance, ed, AIME, New York, N.Y., 1977.

15. A. T. Davenport and D. R. DiMicco, "The Effect of Columbium on the Austenite Structural Changes During the Hot-Rolling of Low-Carbon Bainitic and Ferrite-Pearlite Steels", pp. 1237-1248, Proceedings of the International Conference on Steel Rolling, ISIJ, Tokyo, Japan, 1980.

DISCUSSION

Q: When you dealt with constant reductions and then switched to a higher reduction per pass, how did you deal with the time-temperature element? Did you end up with the same temperature by holding the bar between passes?

A: Yes, we tried to roll at constant temperature and not at constant time.

Q: Did you feel there was any effect of holding time between the passes when you dealt with the higher reductions? Was there any recrystallization there that might have affected the work at all?

A: My interpretation is that recrystallization was coincident with the reduction shortly after the reduction pass and that there was little granular reduction between the passes.

Q: In other words, you might have gone from a nine pass schedule down to five passes in dealing with the two situations.

A: It was something like six passes down to a four or five pass reduction schedule.

Q: It wasn't quite clear whether the strain-induced grain coarsening is simply an effect of small strains on statically recrystallized grain size, or whether it's some different phenomenon.

A: I'm not sure I attempted to explain the cause of the phenomenon as much as I demonstrated its existence. You probably also noticed that in quenching our steels immediately after the processing of the light pass reduction, or just prior to it, the structures went from nice recrystallized grains with uniform boundaries to deformed grains with straight boundaries to grains of similar sizes but with boundaries wavy or ragged in appearance.

81

THE HOT ROLLING BEHAVIOR OF AUSTENITE IN Nb-BEARING STEELS:

THE INFLUENCE OF REHEATED MICROSTRUCTURE

M. L. Santella and A. J. DeArdo

University of Pittsburgh
Pittsburgh, Pennsylvania

The appealing package of properties exhibited by HSLA steels is a direct result of fine ferrite grain size (d_α). Previous work has shown that controlled rolling can promote small d_α; the goal of this work was to try to achieve similar small d_α through the use of high-temperature rolling. Specifically, the recrystallization of HSLA steel austenite which accompanies high-temperature hot rolling was studied, with the influence of the reheated microstructure on recrystallization behavior being of primary concern. The alloys investigated were a C-Mn, a C-Mn-Nb, and a C-Mn-Nb-N steel. The results indicated that the Nb-steels were more resistant to recrystallization than the C-Mn steel only when large Nb supersaturations were developed prior to rolling. Initial austenite grain size effects were found to have no influence on recrystallization behavior. NbCN particles which remained undissolved at the rolling temperature were observed to enhance austenite grain refinement.

Introduction

Two of the most desirable properties of hot rolled low-alloy steels are high yield strength and low impact-transition temperature. There are various mechanisms available for obtaining each of these properties, but control of ferrite grain size is perhaps the most important. It is the only method by which simultaneous improvements in both strength and toughness can be produced. Ferrite grain refinement in hot rolled steels results from increasing the number of nucleation sites for ferrite in the austenite, or increasing the undercooling at which the austenite-to-ferrite transformation occurs. The first of these two factors is the more important one, and is the reason that the subject of controlled rolling, i.e. heavily deforming austenite below its recrystallization temperature, has received so much attention over the last two decades.

Ideally, during controlled rolling, the austenite grains are flattened, or "pancaked", prior to transformation to ferrite. This austenite grain shape coupled with the high defect concentration remaining in the unrecrystallized austenite produces a large effective interfacial area (S_V) in the austenite grains, and therefore a relatively large number of nucleation sites for ferrite.(1) While controlled rolling is capable of producing considerable ferrite grain refinement, it frequently suffers from two major drawbacks.(2) Namely, the requirement of large rolling reductions below the austenite recrystallization temperature tends to overload most existing rolling mills. In addition, the need to allow the material to cool to sufficiently low temperatures prior to final rolling in the finishing train leads to decreases in mill productivity.

With this in mind, the overall goal of the present work was to investigate alternate means of producing large effective interfacial areas in austenite by high-temperature (T > 1000°C) hot rolling. In general, hot rolling or thermomechanical treatment may be divided into three broad areas: reheating, hot deformation, and cooling. This investigation was undertaken specifically to characterize the austenite behavior during reheating and hot rolling, as well as to examine the relationships which may exist between the two; that is, to examine the influence of the microstructure of reheated austenite on its subsequent response to high-temperature hot rolling.

Experimental Details

Three alloys were chosen for study: a C-Mn, a C-Mn-Nb, and a C-Mn-Nb-N steel, and their compositions are given in Table 1. All three were prepared as silicon-killed, high-purity, vacuum-melted, laboratory heats. Following homogenization at 1300°C, the 113 Kg ingots were air-cooled to 1175°C and rolled immediately to 25mm thick slabs with a finishing temperature of about 1000°C in each case. From this point the slabs were air cooled to ambient temperature. These slabs provided directly the starting material condition for all subsequent reheating and rolling experiments.

Table I -- Alloy Compositions in weight percent

	C	Mn	Si	Nb	N
A1 (C-Mn)	0.09	1.02	0.20	--	0.005
A3 (C-Mn-Nb)	0.09	1.01	0.22	0.07	0.006
A7 (C-Mn-Nb-N)	0.08	0.98	0.23	0.07	0.023

84

The austenite reheating behavior was determined via standard isothermal grain growth experiments. Small pieces of the slab materials were
austenitized from 900 to 1250°C for times up to 1000 min. Precautions were
taken to prevent excessive oxidation of the specimens. Following the
required annealing treatment, all specimens were immediately quenched.

The rolling experiments were performed on a laboratory-scale rolling
mill at an $\bar{\varepsilon} \simeq 3.8s^{-1}$. The dimensions of the rolling specimens were
64 x 50 x 12.5 mm. These specimens were prepared by milling equal amounts
of material from each original rolling plane surface to arrive at the final
shape. The 64mm length was parallel to the original rolling direction of
the slab, and was retained as the rolling direction in subsequent hot rolling. Thermocouples, inserted into the center of each specimen, were used
to monitor their temperature before and during rolling. A single 50%
reduction in thickness was imparted to each specimen at roll entrance temperatures of 1050 to 1250°C. After austenitizing to a predetermined condition, specimens were either rolled straightaway at the reheating temperature, or were air-cooled to a lower rolling temperature. Specimens were
quenched immediately (less than 3s) upon exiting the rolls.

All austenite grain structures were analyzed by standard metallographic
techniques. An etchant based on a saturated aqueous solution of picric
acid was used to reveal the prior austenite grain boundaries. In all
cases, the plane of the specimen surface used for metallographic purposes
coincided with a through-thickness section parallel to the rolling direction. For the reheating experiment the grain size distribution was qualitatively assessed as being either uniform or duplexed, and the average
austenite grain size was determined for the equiaxed grain structures.
Similarly, for the rolling experiment, the extent of recrystallization was
assessed as being complete, partial, or totally unrecrystallized, and the
average austenite grain size was determined for the fully recrystallized
grain structures.

The presence and nature of NbCN precipitate particles in the austenite
were determined by transmission electron microscopy (TEM), X-ray diffraction, and microchemical analysis.

Results and Discussion

Since the slab material provided the starting material condition for
all subsequent treatments, the initial step in this study was to characterize its microstructural state. All three of the slab materials contained similar ferrite-pearlite microstructures with ferrite grain sizes
ranging from 25 to 30 μm. TEM was used to establish the differences among
these materials. Close examination of A1 (C-Mn steel) revealed that this
material consisted of pearlite in a ferrite matrix which contained no
appreciable level of any other type of precipitation.

The microstructures of A3 (C-Mn-Nb steel) and A7 (C-Mn-Nb-N steel) at
low magnification are shown in Fig. 1. Small particles are evident in
both steels, but a general dispersion of much larger particles (up to about
500Å in size) was additionally present only in A7. At higher magnification
some of the small particles were found to be aligned in rows characteristic
of the interphase precipitation of alloy carbonitrides in ferrite.(3)
These are shown in Fig. 2(a). Other particles, as shown in Fig. 2(b),
formed a more general dispersion in the ferrite. These particles were often
associated with dislocations. Most of the small particles were observed
to have the Baker-Nutting orientation relationship with the ferrite.(4)
Some of these small particles possessed no orientation relationship with

the ferrite, presumably having precipitated in the austenite during the slabbing operation.(5) All of the particles, both large and small, were identified as cubic niobium carbonitride (NbCN) by electron diffraction.

It was assumed at this point that most of the niobium supersaturation in the A3 and A7 slab material had been relieved by a combination of NbCN precipitation in both the austenite and the ferrite. Thus when the slab specimens were reaustenitized for the final reheating and rolling treatments, a certain volume fraction of the initial NbCN distribution should have dissolved to satisfy solubility considerations. The remaining volume fraction of NbCN should have coarsened.

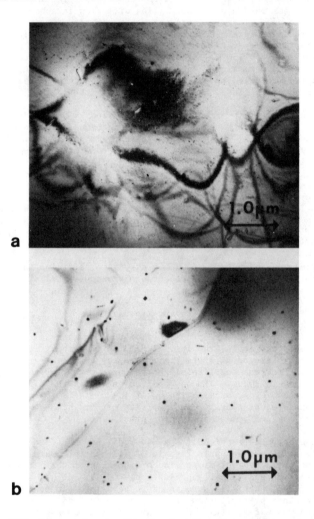

Figure 1   TEM micrographs showing the ferrite microstructures of (a) A3 and (b) A7.

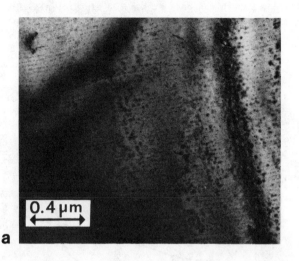

Figure 2a   Bright field TEM micrograph of NbCN interphase precipitation
typically found in the slab material of both Nb steels.

Figure 2b   Centered dark field TEM micrograph of general NbCN precipitation
in ferrite typically found in the slab material of both Nb
steels.

The reheating study provided the information necessary for selecting various austenite microstructures for subsequent hot rolling. The results of this experiment indicated that an austenitizing time of 100 min. would yield a wide range of grain structures prior to rolling. The isochronal austenite grain growth response at t = 100 min. for all three steels is summarized in Fig. 3.

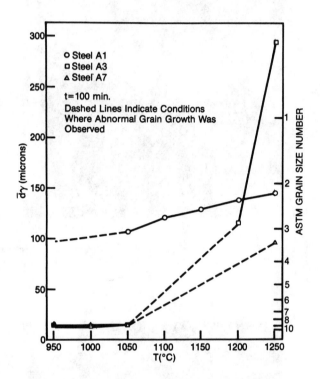

Figure 3   Isochronal austenite grain growth behavior
for A1, A3, and A7 at t = 100 min.

At temperatures above 1050°C, A1 coarsened very slowly by normal grain growth. Only a gradual increase in average austenite grain size accompanied increasing temperature in this range. This agrees with other observations on the austenite grain growth of plain carbon steels.(6) Furthermore, no significant level of precipitation of any kind was observed in the austenitized specimens of A1.

In steels A3 and A7 both normal and abnormal austenite grain growth occurred in the temperature range examined. Duplexed grain structures were observed in A3 and A7 at 1100 and 1150°C, as well as in A7 at 1200°C. Uniform, equiaxed grain structures were observed in A3 at 1200 and 1250°C, but only at 1250°C for A7. A large increase in grain size in A3 occurred between 1200 and 1250°C, where the austenite grains were extremely large.

The more complicated grain coarsening behavior for steels A3 and A7 stems from the presence of NbCN in the reheated austenite. Furthermore, the difference observed in the coarsening behavior of these two alloys was due to a difference in the particles present in each. Specimens identical to those used for determining the grain coarsening behavior were subjected to TEM analysis. Carbon replicas and thin foils were used to examine and identify the particles in both alloys. The experimentally observed variation of NbCN particle size with temperature for A3 and A7 is given in Fig. 4, along with similar data taken from the literature.(7,8) The results indicate that, in general, the average particle size, $\bar{r}$, in A7 is greater than that in A3*; however, the particle coarsening rate is somewhat lower. They also indicate that the solution temperature of NbCN in A7 is above 1250°C, while that in A3 lay between 1150 and 1200°C. The typical NbCN particle dispersions found in A3 and A7 at 1100°C are shown in Fig. 5. At 1250°C, NbCN particles could only be found in A7, and these are shown in Fig. 6.

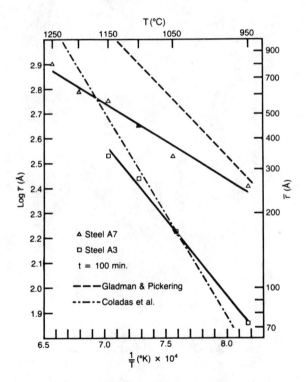

Figure 4   Variation of NbCN precipitate particle size
with austenitizing temperature for A3 and
A7.

*It should be noted that both large ($\bar{r} \cong 200\text{-}500\text{Å}$) and small ($\bar{r} \cong 60\text{-}150\text{Å}$) NbCN particles were found in A7 after low temperature austenitizing ($T \leq 1050°C$). These small particles, however, were not quite as prevalent as the larger ones. Thus the particle sizes measured in A7 at these temperatures are weighted toward the larger particles sizes.

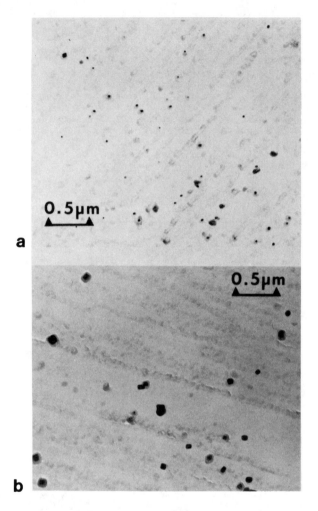

Figure 5   Carbon extraction replica TEM micrographs of NbCN particles in:
(a) A3, and (b) A7. T = 1100°C, t = 100 min.

The TEM results prompted further study of the NbCN particles present
in A3 and A7 by X-ray diffraction and microchemical analysis. Both of
these analyses were performed on chemically isolated residues from specimens of A3 and A7 reheated to 950, 1100, and 1250°C for 100 min. The
isolation technique and the microchemical method are described elsewhere.(9,10) The X-ray diffraction technique used was the Debye-Scherrer
powder method. This permitted a determination of a crystal structure and
a lattice parameter for each residue. The only crystalline phase detected
by X-ray was cubic NbCN, which was present in every residue except that
for A3 at 1250°C. No crystalline phase was detected in this specimen,
which agrees directly with the TEM analysis. The results of the lattice

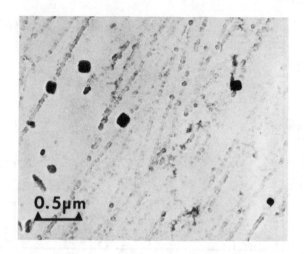

Figure 6   Carbon extraction replica TEM micrograph of NbCN particles in
            A7. T = 1250°C, t = 100 min.

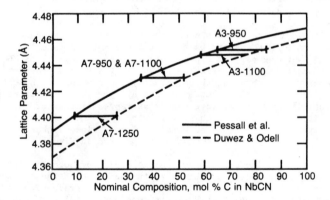

Figure 7   Variation of lattice parameter with carbon content for NbCN.
            The lattice parameters of the residues isolated from A3 and
            A7 are shown.

parameter measurements indicated that the lattice parameters of the A7
residues were always smaller than those of the A3 residues. The experimentally determined lattice parameters are plotted against the known
variation of NbCN lattice parameter with composition in Fig. 7.(11,12)
This indicates that the NbCN precipitate particles in A7 are more nitrogen
rich than those in A3. The microchemical analysis of the residues confirmed that the compositions predicted by the X-ray analysis were
accurate.

    Thus, the combination of the TEM, X-ray diffraction, and microchemical

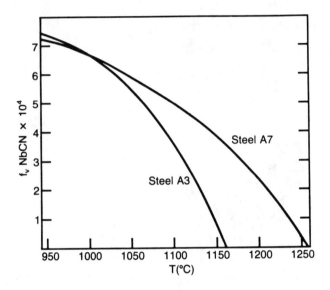

Figure 8   Calculated variation of volume fraction of NbCN with austenitizing temperature for A3 and A7.

analyses allowed the relationship between austentizing temperature and volume fraction of NbCN particles to be estimated for A3 and A7. This was accomplished by using various expressions for the solubility of NbCN in austenite to calculate the variation in volume fraction of these particles with temperature. Afterward, the solubility expressions which best appeared to describe the behavior of NbCN in A3 and A7, particularly with regard to the solution temperature, were chosen as representative of these alloys. The solubility expressions which best fit the experimental observations of A3 and A7 were those of Chino and Wada,(13) and Irvine, Pickering, and Gladman,(14) respectively. These data are plotted in Fig. 8.

Thus, the reheating study provided the basis for determining or estimating the following factors with regard to the state of the austenite prior to hot rolling: the austenite grain size, and grain size distribution; the size and distribution of NbCN precipitate particles; the level of niobium in equilibrium solid solution; and, the level of niobium in supersaturated solid solution.

Hot Rolling

The rolling schedule used for this experiment is given in Fig. 9. All of the rolled specimens for Al were completely recrystallized, and these recrystallized grain sizes, $\bar{d}_\gamma$, are plotted as a function of the Zener-Holloman parameter, Z, in Fig. 10. On the other hand, not all of the specimens of steels A3 and A7 were completely recrystallized during rolling. However, for those which did, $\bar{d}_\gamma^{-1}$ vs. lnZ is plotted in Fig. 11. The data for the niobium steels fit two separate straight lines with the lower dashed line being that previously determined for Al. Those points which fit the upper line correspond to two-phase mixtures of austenite plus NbCN prior to rolling; the points for A3 on the lower line were approximately single phase prior to rolling.

Reheating Temperature	Rolling Temperature (°C)				
	1250	1200	1150	1100	1050
1250°C	1	2	3	4	5
1200°C	—	6	7	8	9
1150°C	—	—	10	11	12
1100°C	—	—	—	13	14

Figure 9   Austenite hot rolling schedule. Each number designated a different rolling schedule.

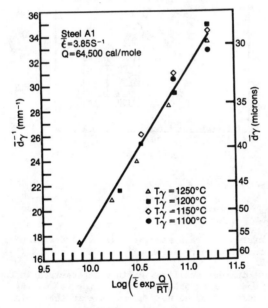

Figure 10   Variation of recrystallized austenite grain size with the Zener-Holloman parameter for Al.

The good correlation between $\bar{d}_\gamma^{-1}$ and lnZ is taken as an indication that dynamic recrystallization occurred for these deformation conditions.[15] Any dynamic recrystallization, however, may have been rapidly followed by static recrystallization due to the unstable nature of dynamically recrystallized microstructures [16] and the very high static recrystallization rates observed at these temperatures.[17,18] Also adding support to the contention that dynamic recrystallization has occurred is the observation

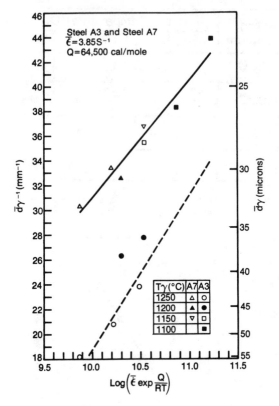

Figure 11 Variation of recrystallized austenite
grain size with the Zener-Holloman
parameter for A3 and A7. Dashed line
for A1.

that the initial austenite grain size or grain size distribution had no
significant impact on the correlation between $\bar{d}_\gamma^{-1}$ and lnZ.(19)

The refinement in recrystallized grain size observed in the niobium
steels containing undissolved NbCN particles prior to rolling can be considered to have resulted from either an increase in the nucleation rate of
recrystallization, or a decrease in the growth rate of the recrystallized
grains. Employing the Gladman theory of particle-inhibited growth,(20)
it can be shown that the distributions of undissolved NbCN particles for
the reheating conditions of the top line in Fig. 11 would be unable to
impede the growth of grains as small as the observed recrystallized grains.
Thus, it must be concluded that these pre-existing NbCN particles had no
effect on the growth stage of recrystallization.

In contrast, there is a possibility that these particles did influence
the nucleation of recrystallization. Humphreys (21) has discussed this
effect of second-phase particles in some detail. He concluded that even
clusters of closely spaced small particles (where the interparticle spacing
within the clusters, as well as the particle size, is less than 1 μm) may

94

act cooperatively to enhance the nucleation of static recrystallization. While a considerable amount of effort has been focused on the interaction of second-phase particles with static recrystallization, the same is not true for the case of dynamic recrystallization. Some investigators (22,23) have shown that relatively large NbCN particles can influence the hot flow behavior of austenite causing it to approach that of plain carbon steels. However, no metallographic evidence presently exists that would suggest that such particles could not enhance the nucleation of dynamic recrystallization and refine recrystallized grain sizes. Given the overall similarities between dynamic and static recrystallization, it is reasonable to expect that pre-existing dispersions of insoluble particles could have similar effects on the evolution of microstructure in both cases.

The metallographic techniques of this experiment are not sufficient for determining whether dynamically recrystallized grains, or grains which may have subsequently statically recrystallized were being observed. However, the fact remains that specimens of the Nb-steels containing undissolved NbCN particles at the rolling temperature showed a marked refinement of recrystallized grain size compared to those which did not. Apparently this grain size refinement resulted from enhanced nucleation of recrystallization, whether dynamic or static, in those specimens containing such particles.

The lower line in Fig. 11 indicates that for certain conditions under which A3 completely recrystallized, its recrystallized grain size is not significantly different from that of A1. These particular conditions for complete recrystallization in A3 correspond to combinations of reheating and rolling temperatures for which all of the niobium (0.07 wt. %) in this alloy remains substantially in equilibrium solid solution during processing. For these same reheating temperatures but lower rolling temperatures, recrystallization in A3 is severely inhibited. Calculations based on the solubility relation for NbCN in austenite established for A3 indicate that these lower rolling temperatures coincide with the development of large niobium supersaturations.

The results of the calculations of niobium supersaturation at the rolling temperature for A3 and A7 are given in Fig. 12. On this figure is shown the condition of the austenite grains following hot rolling, where R, P and U mean, respectively, that the austenite is fully recrystallized, partially recrystallized, or unrecrystallized. The number in the upper right hand corner of each block in Fig. 12 denotes the calculated niobium supersaturation for that particular combination of reheating and rolling temperatures. The quantity is expressed as a percent of the total niobium in the alloy.

Fig. 12 shows that for a given rolling temperature, e.g. 1100°C in A3, as the initial reheating temperature is increased, the niobium supersaturation also increases. Simultaneously, the extent of recrystallization accompanying rolling decreases. This same trend also exists in A3 rolled at 1050°C and in A7 rolled at 1150, 1100, and 1050°C. Even in the light of the errors which may be involved with the use of the solubility expressions chosen, Fig. 12 shows a clear correlation between the level of niobium supersaturation at the rolling temperature, and the tendency of the austenite to recrystallize.

Conclusions

The major findings of this study on the high temperature hot rolling of low-alloy steels using a single 50% reduction in the temperature range

Reheating Temperature	Rolling Temperature (°C)				
	1250	1200	1150	1100	1050
1250°C	R[0]	R[0]	R[8]	U[44]	U[67]
1200°C	—	R[0]	R[8]	P[44]	U[67]
1150°C	—	—	R[0]	P[35]	U[59]
1100°C	—	—	—	R[0]	R[23]

a)

Reheating Temperature	Rolling Temperature (°C)				
	1250	1200	1150	1100	1050
1250°C	R[0]	R[28]	P[49]	P[65]	U[77]
1200°C	—	R[0]	P[21]	P[37]	U[49]
1150°C	—	—	R[0]	P[16]	P[28]
1100°C	—	—	—	P[0]	P[12]

b)

Figure 12   The microstructural state of the austenite following hot rolling. Included in the corner boxes are the calculated Nb supersaturations given as the percentage of the total Nb.   a)   Steel A3;   b) Steel A7.

of 1050 to 1250°C are:

1. In all cases for which fully recrystallized grain structures were observed for steels A1, A3, and A7, the recrystallized grain size was determined to be a function of the hot rolling temperature only.

2. NbCN particles which remained undissolved prior to rolling were observed to result in significantly finer recrystallized austenite grain sizes compared to the plain-C steel. Presumably, these particles enhanced the nucleation of recrystallization accompanying rolling.

3. Niobium which remained substantially in equilibrium solid solution during rolling was observed to have no significant effect on the extent of recrystallization or the recrystallized grain size in A3, as compared to the plain-C steel. This held true as long as A3 was austenitized above the solution temperature for NbCN.

4. Whenever a substantial proportion of niobium was supersaturated

in austenite, with respect to the precipitation of NbCN, recrystallization was severely inhibited in both niobium steels.

Acknowledgements

The authors would like to thank: The American Iron and Steel Institute, and Niobium Products, Ltd., a subsidiary of CBMM, for providing financial support for this research work; National Steel Corp., for providing financial support for MLS; U.S. Steel Corp. (Dr. L.J. Cuddy and G. Krapf), for providing materials and certain analytical facilities; and the Jones and Laughlin Steel Corp. (Dr. J. Butler, G. Staib, and L. Stirling), for providing access to their pilot facilities and expert supervision of the hot rolling experiments.

References

1. I. Kozasu, C. Ouchi, T. Sampei, and T. Okita, "Hot Rolling As a High-Temperature Thermo-Mechanical Process" in Microalloying '75, an International Symposium on High-Strength Low-Alloy Steels, Washington, D.C., 1975 (New York: Union Carbide Corporation, 1975), Session 1, pp. 106-114.

2. R.B.G. Yeo, A.G. Melville, P.E. Repas, and J.M. Gray, "Properties and Control of Hot-Rolled Steels", J. of Metals, Vol. 20 (June, 1968), pp. 33-43.

3. A.T. Davenport and R.W.K. Honeycombe, "Precipitation of Carbides at γ/α Phase Boundaries in Alloy Steels", Proc. Roy. Soc., Vol. 322A (1971), pp. 191-205.

4. R.G. Baker and J. Nutting, "The Tempering of a Cr-Mo-V-W and a Mo-V Steel", in Precipitation Processes in Steels (London: The Iron and Steel Institute, 1959), pp. 1-22.

5. A.T. Davenport, L.C. Brossard, and R.E. Miner, "Precipitation in Microalloyed High-strength Low-alloy Steels",, J. of Metals, Vol. 27 (June 1975), pp. 21-27.

6. O.O. Miller, "Influence of Austenitizing Time and Temperature on Austenite Grain Size of Steel", Trans. ASM, Vol. 43 (1951), pp. 260-289.

7. T. Gladman and F.B. Pickering, "Grain-Coarsening of Austenite", J.I.S.I., Vol. 205 (June, 1967), pp. 653-664.

8. R. Coladas, J. Masounave, G. Guérin, and J.P. Baïlon, "Austenite Grain Growth in Medium and High-Carbon Steels Microalloyed with Niobium", Met. Sci., Vol. 11 (1977), pp. 509-517.

9. G. Krapf, J.L. Lutz, L.M. Melnick, and W.R. Bandi, "A DTA-EGA Study of the Chemical Isolation of Fe_3C, Amorphous Carbon, and Graphite from Steel and Cast Iron", Thermochim. Acta, Vol. 4 (1972), pp. 257-271.

10. G. Krapf and W.R. Bandi, "Identification and Determination of Titanium Sulphide and Carbosulphide Compounds in Steel", Analyst, Vol. 104 (September 1979), pp. 812-821.

11. P. Duwez and F. Odell, "Phase Relationships in the Binary Systems of Nitrides and Carbides of Zirconium, Columbium, Titanium, and Vanadium", J. Electrochem. Soc., Vol. 97 (October, 1950), pp. 299-304.

12. N. Pessall, R.E. Gold, and H.A. Johansen, "A Study of Superconductivity in Interstitial Compounds", J. Phys. Chem. Solids, Vol 29 (1968), pp. 19-38.

13. H. Chino and K. Wada, Thermodynamic Study of the Deoxidation and Precipitation of Carbides and Nitrides, Yahat Technical Report, No. 251 (1965), pp. 75-100.

14. K.J. Irvine, F.B. Pickering, and T. Gladman, "Grain-Refined C-Mn Steels", J.I.S.I., Vol. 205 (February, 1967), pp. 161-182.

15. H.J. McQueen and J.J. Jonas, "Recovery and Recrystallization during High Temperature Deformation", in Treatise on Materials Science and Technology, Volume 6, ed. by H. Herman (New York: Academic Press, Inc., 1975), pp. 621-624.

16. S. Sakui, T. Sakai, and K. Takeishi, "Hot Deformation of Austenite in a Plain Carbon Steel", Trans. I.S.I.J., Vol. 17 (1977), pp. 719-725.

17. I. Kozasu, T. Shimizu, and H. Kubota, "Recrystallization of Austenite of Si-Mn Steels with Minor Alloying Elements after Hot Rolling", Trans. I.S.I.J., Vol. 11 (1971), pp. 367-375.

18. R.A.P. Djaic and J.J. Jonas, "Static Recrystallization of Austenite Between Intervals of Hot Working", J.I.S.I., Vol. 210 (1972), pp. 256-261.

19. J.P. Sah, G.J. Richardson, and C.M. Sellars, "Grain-Size Effects during Dynamic Recrystallization of Nickel", Met. Sci., Vol. 8 (1974), pp. 325-331.

20. T. Gladman, "On the Theory of the Effect of Precipitate Particles on Grain Growth in Metals", Proc. Roy. Soc., Vol 294A (1966), pp. 298-309.

21. F.J. Humphreys, "Recrystallization Mechanisms in Two-phase Alloys", Met. Sci., Vol. 13 (1979), pp. 136-145.

22. A. leBon, J. Rofes-Vernis, and C. Rossard, "Recrystallization and Precipitation during Hot Working of a Nb-Bearing HSLA Steel", Met. Sci., Vol. 9 (1975), pp. 36-40.

23. I. Weiss and J.J. Jonas, "Dynamic Precipitation and Coarsening of Niobium Carbonitrides During the Hot Compression of HSLA Steels", Met. Trans., Vol. 11A (March, 1980), pp. 403-410.

DISCUSSION

Q: In your second conclusion, you say that recrystallization is controlled by the niobium supersaturation. I think from your results, that this could be seen as a correlation and that the one thing correlates with the other. For example, it would also correlate with the equilibrium concentration of vacancies, or any other thing that depends on temperature. But it's not clear that there is a cause-and-effect relationship between the two. You did mention that, in the TEM work, there were inconclusive results as far as looking for precipitates with what is concerned. For example, do you find that in some cases you had unrecrystallized austenite with no precipitates present?

A: We feel that the relationship between niobium supersaturation and recrystallization temperature is more than just a fortuitous correlation. The solubility equations used in the calculations were chosen based on the goodness of fit between the theoretical and observed carbonitride dissolution temperatures. Once the appropriate solubility equation was chosen for each alloy, the supersaturations for various conditions were calculated for each alloy. The fact that the same basic trend was observed in two different steels using solubility equations appropriate for each steel indicates, at least to us, that the relationship is one that is more meaningful and fundamental than fortuitous.

Our TEM studies of unrecrystallized austenite formed at high niobium supersaturations revealed that very small particles of NbCN could often be found. However, the particles were never uniformly distributed, but rather occurred in clusters. This observation appears to suggest that the niobium was segregated in the austenite. The significance of these observations regarding the distribution of NbCN has yet to be determined.

Comment: The fact that particles that are undissolved prior to rolling lead to extra grain refinement is an important one. I think the same type of result was reported in the paper by Prof. Pickering earlier in this session. Pickering showed that lower reheating temperatures led to undissolved particles which led to finer as-rolled austenite grains.

Q: I'd like to know why the value of Q = 645 cal/mole was chosen for use in the parameter Z? Also, why did you choose to normalize your 1/D data with the Z parameter?

A: The value of Q used in this study is a standard one which has been determined previously and is available in the literature. This value of Q is very close to the activation energy for the self-diffusion coefficient of iron in austenite.

Previous published work has shown that steady state grain sizes and stress levels vary with the parameter Z. Our plots of 1/D-vs-ln Z simply follow this accepted format.

RECRYSTALLIZATION OF AUSTENITE IN HIGH-TEMPERATURE
HOT-ROLLING OF NIOBIUM BEARING STEELS

M. Katsumata*, M. Machida*, and H. Kaji**

* Central Research Laboratory,
  Kobe Steel, Ltd. Kobe, Japan

** Kakogawa Works, Kobe Steel, Ltd.
   Kakogawa, Japan

The recrystallization behavior of austenite at high temperatures was
investigated in niobium bearing steels. In the range of practical rolling
reductions, the recrystallized austenite grain size had a good correlation
with the effective interfacial area, which was defined as the sum of the
elongated austenite grain boundary area and the interfacial area of deformation bands. This result indicates that austenite recrystallization can
be attributed to the static process. On the other hand, single pass rolling with low reduction of about 5 to 10% formed a mixed austenite grain
structure containing coarse austenite grains. The fraction of coarse
grains markedly increased with the rise in rolling temperatures. Using
the range of practical rolling reductions, it was difficult to produce a
uniform structure from the mixed austenite grain structure. Coarse
austenite grains were transformed into coarse upper bainite, a constituent
which causes the toughness to deteriorate.

Introduction

The controlled rolling process has developed rapidly and is used to manufacture as-rolled high-strength low-alloy (HSLA) steels in line pipes for better strength and toughness combinations. In recent years, the controlled rolling process has been applied to both high strength and low temperature service steels for ships, because it can save energy and mineral resources and improve weldability. In the near future, it may be applied to low temperature service steels for pressure vessels.

The controlled rolling process as a thermomechanical treatment consists mainly of two steps. The first is the refinement of austenite grains at temperatures above the "recrystallization temperature" and the second is the significant deformation of austenite grains below this temperature. "Recrystallization temperature" means the lowest temperature at which austenite completely recrystallizes immediately after hot-rolling. The two steps, however, have the common purpose of increasing the area of the interfaces such as grain boundaries. It is known that the increase in the interfacial area in the austenite grain structure results in the refinement of the transformed structure, which causes strength and toughness to improve simultaneously (1-6). Thus, a great deal of effort has been made to refine the recrystallized austenite grain structure and to deform it heavily in the controlled rolling process of HSLA steels. To optimize the control of the hot rolling process above the recrystallization temperature, a clear understanding of the recrystallization behavior of austenite is required.

This paper is concerned with the following five topics: (1) An investigation of the change in grain size of austenite recrystallized immediately after hot-rolling, as a function of rolling reduction, temperature, and initial grain size. (2) A discussion of the recrystallization mechanism based on the relationship between recrystallized austenite grain size and the interfacial area in the austenite structure during hot-rolling. (3) An observation of the formation of the mixed austenite grain structure containing coarse grains. (4) An investigation of the subsequent recrystallization behavior of the mixed structure. (5) A discussion of the effect of recrystallized austenite grain size on the formation of unwanted coarse bainite.

Experimental Procedure

Material

The two niobium bearing low alloy steels for API X 65 and X 70 line pipes were used in this study. The chemical compositions of the steels are shown in Table I. These steels were obtained from commercially rolled slabs and cut into specimens with a thickness of 50 to 27 mm.

Table I   Chemical Compositions of The Steels

	C	Si	Mn	P	S	Cu	Ni	Cr	Mo	V	Nb	Al
X65 Steel	.11	.26	1.35	.011	.008	—	—	—	—	.046	.027	.016
X70 Steel	.06	.14	2.04	.012	.007	.04	.05	.11	.34	—	.049	.020

Hot rolling and evaluation of austenite grain size

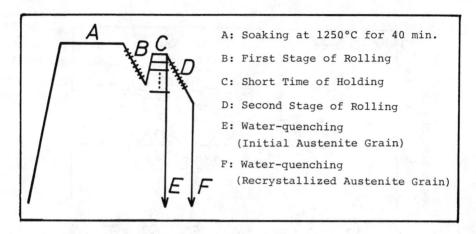

A: Soaking at 1250°C for 40 min.

B: First Stage of Rolling

C: Short Time of Holding

D: Second Stage of Rolling

E: Water-quenching
   (Initial Austenite Grain)

F: Water-quenching
   (Recrystallized Austenite Grain)

Fig. 1   Laboratory two-stage hot-rolling operation

The two-stage hot rolling operation indicated in Fig. 1 was performed using a laboratory two-high mill whose roll radius and roll speed are 250 mm and 13 r.p.m., respectively. Following soaking at 1250°C for 40 min , the first stage of rolling was carried out with several rolling temperatures and reductions in order to provide various initial austenite grain sizes in the range of grain size number -2 to 5.6. The rolling temperature was measured using a contact-type thermometer. The specimen thickness after the first stage of rolling was kept constant at 20 mm. The specimens were then trans- ferred to a furnace which was held at the temperature of the second stage of rolling. After a short time of holding not exceeding five minutes, the second stage of rolling was performed with a rolling temperature between 1,250 and 1,000°C and rolling reduction up to 80%. Within one second after rolling, the rolled specimens were water-quenched. Initial austenite grain size was taken as the grain size of the specimen obtained prior to the second stage of rolling. Metallographic specimens were etched in saturated solution of picric acid containing a surface active reagent to reveal the grain boundaries. In these specimens, the mean grain size was determined by comparing it with the JIS austenite grain size using an optical microscope.

Results and Discussion

Recrystallization behavior of austenite

Figure 2 shows typical examples of the change in the grain size of aus- tenite recrystallized immediately after hot-rolling with rolling reduction for a number of rolling temperatures. The X 65 and X 70 steels had initial grain size numbers of 2.5 and 0.9, respectively. The broken line parallel to the vertical axis indicates the critical rolling reduction necessary for austenite to recrystallize completely. As is well-known (3,4), the critical rolling reduction for recrystallization increases with decreasing rolling temperature. It was found that austenite grains were refined with an in- crease in rolling reduction and/or a decrease in rolling temperature above the critical rolling reduction for recrystallization. Grain coarsening occurred by light rolling with reductions below 10%. This phenomenon will be described in detail in a later section.

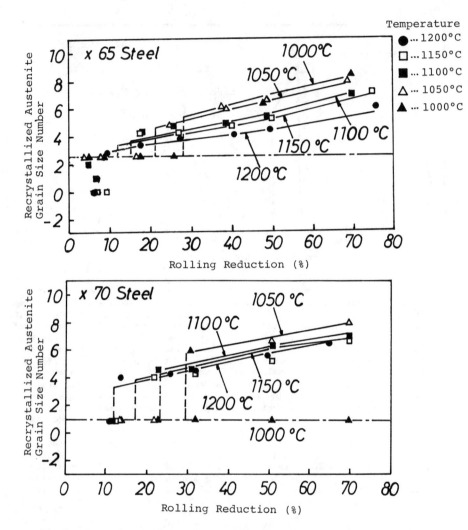

Fig.2　Typical examples of the change in the grain size of
austenite with rolling reduction

　　　The influence of initial austenite grain size and rolling temperature
on the critical rolling reduction for recrystallization is summarized in
Fig. 3 . It can be seen that a refinement in initial grain size and an
increase in rolling temperature produced a decrease in the critical rolling
reduction and reduced its temperature and initial grain size dependences,
respectively. The critical rolling reduction was hardly influenced by the
initial grain size when the grain size was finer than number 2. This result
reveals that a greater rolling reduction per pass is needed to recrystallize
austenite by lowering rolling temperatures at the step where the finer grain
is obtained in practical multiple-pass rolling operation. Figure 3 also
suggests that the critical rolling reduction is the same for X 65 and X 70
steels.

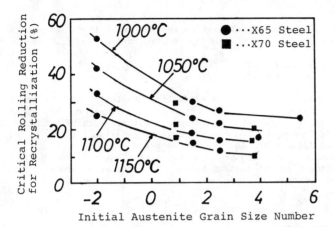

Fig.3 Influence of initial austenite grain size and rolling temperature on the critical rolling reduction for recrystallization

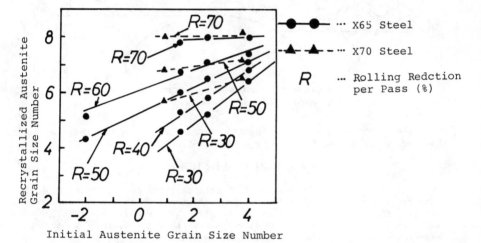

Fig.4 Effect of initial austenite grain sizes and rolling reductions on recrystallized austenite grain size at 1050°C

Figure 4 demonstrates the effect of initial grain sizes and rolling reductions on recrystallized austenite grain size at a rolling temperature of 1,050°C. Recrystallized austenite grain size numbers are linearly related to initial grain size numbers for a fixed rolling reduction. The dependence of recrystallized austenite grain size on initial grain size decreased with rolling reduction, until at a rolling reduction of 70%, recrystallized grain sizes was independent of the initial grain size. The initial grain size dependence was smaller in X 70 steel than in X 65 steel. This is attributed to the easier refinement of X 70 steel.

Mechanism of recrystallization

The recrystallization of austenite may be dynamic or static (4,7). It is reported that in dynamic recrystallization the recrystallized austenite grain size is not affected by the initial grain size or the amount of deformation (4,7-9). As suggested in Fig. 4, although dynamic recrystallization appears to occur at rolling reductions above 70%, static recrystallization is predominant in the extended rolling condition.

In static recrystallization, the recrystallized austenite grain size is assumed to be determined by the number of nucleation sites, which depends on the interfacial area per unit volume. Austenite grains become elongated with rolling reduction and deformation bands may be induced within the grains, as illustrated in Fig. 5. Subsequently, nuclei of recrystallization should take place preferentially at elongated grain boundaries and interfaces of deformation band. The interfacial areas of elongated grains and deformation bands were estimated as follows.

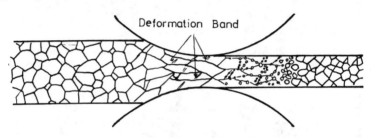

Fig.5 Schematic illustration of static recrystallizing process during hot rolling

The interfacial area of elongated grains per unit volume, Svg (mm^2/mm^3), has been given by the following equation (10):

$$Svg = 0.429\ Mh + 1.571\ Mv \qquad (1)$$

where Mh and Mv are the mean number of intersections per unit length of the directed secant parallel to the rolling and the thickness directions, respectively. Mh and Mv are postulated to be given by the mean number M of intersections per unit length of the direct secant with the equiaxed grain boundary traces and the rolling reduction R, so that

$$Mh = M \times (1 - R/100)$$
$$Mv = M/(1 - R/100) \qquad (2)$$

The relationship between M and austenite grain size number N is described by the following equation:

$$M = 1.12 \times 2^{(N+3)/2} \qquad (3)$$

By combining equations (2) and (3) with equation (1), we obtain:

$$Svg = 1.12 \times 2^{(N+3)/2}\{\,0.429 \times (1 - R/100) + 1.571/(1 - R/100)\} \qquad (4)$$

It is well-known that deformation below the austenite recrystallization temperature introduces deformation bands into the grains. It was rare to observe them above this temperature in hot-rolling. Sellars et al.(8), however, observed that deformation bands were introduced into the grains and became nucleation sites of recrystallization, when nickel was used in the torsion test above this temperature. In high temperature hot-rolling, deformation bands are expected to be induced in austenite grains. We previously investigated the relationship between the surface density of the deformation bands, Svd, and rolling temperature, reduction, and initial austenite grain size at non-recrystallization temperatures below 1000°C, as shown in Fig 6 (11). It was found that the density of the deformation bands depended mainly on rolling reduction above 30% and was hardly affected by rolling temperature and initial grain size. This relation is assumed to hold at higher temperatures such as 1,000 to 1,250°C. From Fig 6, the density of the deformation bands per unit volume, Svd (mm^2/mm^3), can be estimated using

$$Svd = 0 \qquad (R < 30\%)$$
$$Svd = 0.8R - 24 \qquad (R \geq 30\%) . \qquad (5)$$

Interfacial area per unit volume, Sv (mm^2/mm^3), is defined as the sum of Svg and Svd:

$$Sv = 2^{(N+3)/2} \left\{ 0.48(1 - R/100) + 1.76/(1 - R/100) \right\}$$
$$(R < 30\%)$$
$$Sv = 2^{(N+3)/2} \left\{ 0.48(1 - R/100) + 1.76/(1 - R/100) \right\}$$
$$+ 0.8R - 24 \qquad (R \geq 30\%) \qquad (6)$$

The number of nucleation sites per unit of Svg or Svd was considered to be equivalent.

Figure 7 shows the variation of recrystallized austenite grain size number with Sv. The recrystallized grain size correlates with Sv at a fixed temperature and is refined by initially increasing Sv. It subsequently tends to a certain value which depends on the rolling temperature. This

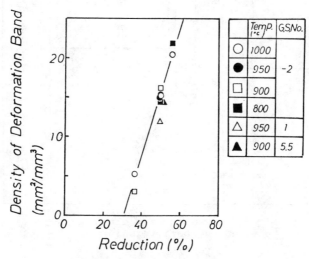

	Temp. $(°C)$	G.S.No.
○	1000	
●	950	-2
□	900	
■	800	
△	950	1
▲	900	5.5

Fig.6 Relationship between the surface density of the deformation
       bands and rolling temperature, reduction and initial
       austenite grain size

result suggests that the refinement in austenite grain structure due to initially increasing Sv is attributable to static recrystallization. Since a large rolling reduction is not practical in industrial rolling, only static recrystallization is likely to occur.

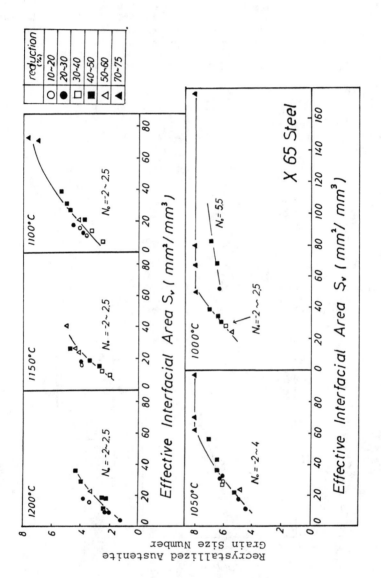

Fig.7   Variation in recrystallized austenite grain size number with interfacial area Sv

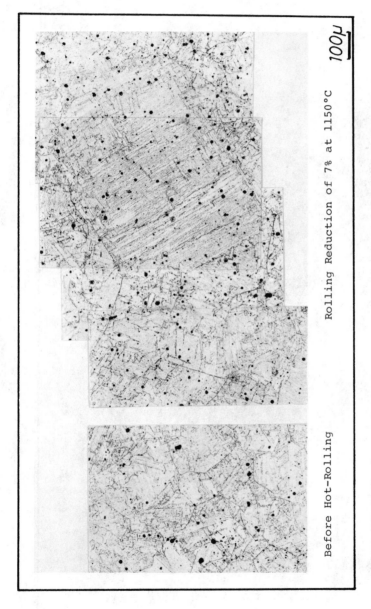

Before Hot-Rolling   Rolling Reduction of 7% at 1150°C

100μ

Fig.8 Mixed austenite grain structure caused by light reduction (<10%)

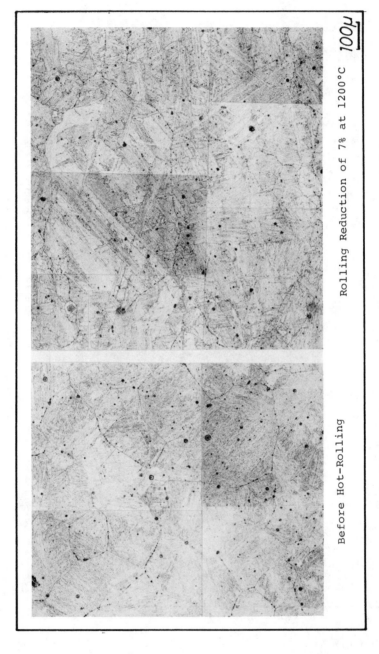

Before Hot-Rolling   Rolling Reduction of 7% at 1200°C   100μ

Fig.9 Mixed austenite grain structure caused by light reduction (<10%)

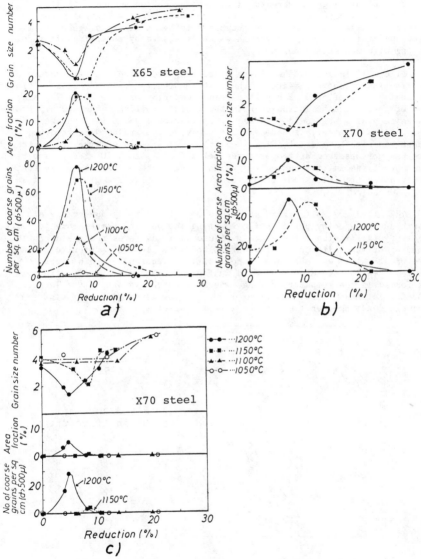

Fig.10   Formation of coarse grains in the mixed austenite
          grain structure by rolling with light reduction
          at high rolling temperatures. a) X65 steel,
          initial grain size number 2.5, b) X70 steel,
          initial grain size number 1.0, c) X70 steel,
          initial grain size number 3.5-4.0

Formation of the mixed austenite grain structure

Single pass light rolling with reductions below 10% resulted in the formation of mixed grain structures containing coarse austenite grains (6,12) as was indicated in Fig. 2. Figures 8 and 9 show typical mixed grain structures. This phenomenon appears to be similar to the well-known abnormal ferrite grain growth which is caused by annealing after cold working with low strain below 10%.

Figure 10 indicates the effect of rolling reduction and temperature on the fraction of coarse grains in the mixed austenite grain structure. Grain coarsening took place markedly at rolling reduction of 5 to 10%. The fraction of coarse grains increased with the rise in rolling temperature. Coarse austenite grains were seldom formed by rolling at 1,050°C in X 65 steel (Fig 10-a). In X 70 steel, it appears that fewer coarse grains were produced in comparison with X 65 steel (Fig. 10-b, c). The mixed austenite grain structure was not observed at temperatures below 1,100°C in X 70 steel. The refinement of the initial austenite grains reduced the number of coarse austenite grains but diminished the uniformity of the austenite grain structure (Fig. 10-b,c).

The mixed austenite grain structure may also form by abnormal grain growth during a delay period while lowering the rolling temperature to below the recrystallization temperature. Figure 11 illustrates the grain growth behavior of austenite due to isothermal holding. Initial austenite grain size of JIS No. 7 was obtained by hot-rolling. Austenite grain sizes due to reheating are also indicated in Fig. 11. The mixed grain structure was formed in the temperature range where reheated grain size was larger than initial grain size, except at 1,200°C. The mixed grain structure was produced for a shorter time at the intermediate temperature.

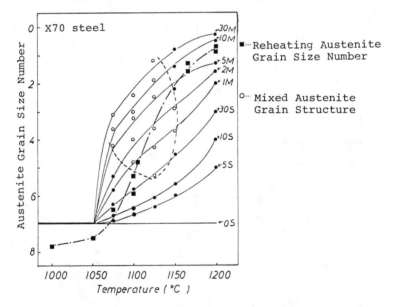

Fig.11 Grain growth during isothermal holding up to 30 min at high temperatures (X70 steel, initial grain size number:7)

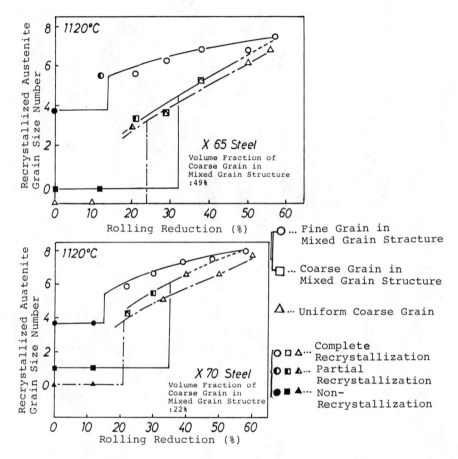

Fig.12   Recrystallization behavior of coarse and fine grains
          in mixed austenite grain structure

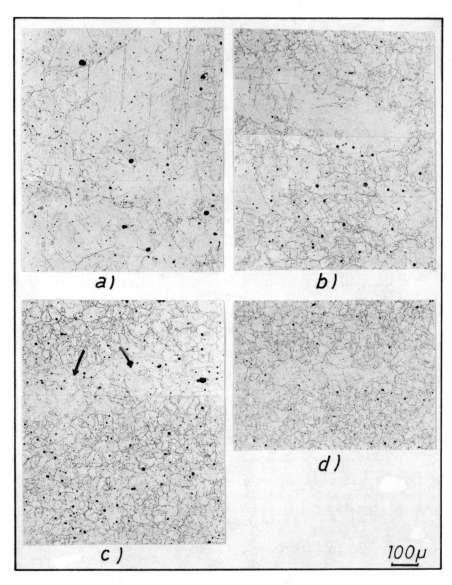

Fig.13   Change in mixed austenite grain structure with rolling
          reduction at 1120°C in X70 steel.   a) before rolling,
          b) 21%, c) 38%, d) 57%

Recrystallization behavior of mixed austenite grain structure

Figures 12 and 13 show the recrystallization behavior of the coarse
and fine grains in the mixed austenite grain structure. Both grain structures were refined with rolling reductions above the critical reduction for
recrystallization. The difference in size between both grain structures
gradually decreased with increasing rolling reduction. The critical rolling
reduction for recrystallization was higher in the coarse grains than in the
fine grains. This promoted a non-uniformity of austenite grain structure at
rolling reductions between the critical reduction of the coarse grains and
that of the fine grains, as shown in Fig. 14. Decreasing rolling temperature
tended to increase this non-uniformity and to extend the rolling reductions
where the non-uniformity took place. Figure 12 also indicates the recrystallization behavior of the uniform coarse austenite with the same grain
size as the coarse grains in the mixed grain structure. It was found that
the coarse grains in the mixed grain structure were more difficult to recrystallize than the uniform coarse grains. This may be attributed to the
predominant deformation of the fine grains in the mixed grain structure.
Figure 15 explains the critical rolling reduction for recrystallization of
both coarse grains. The critical reduction was about 10% higher in the
coarse grains of the mixed grain structure than in the uniform coarse grains.
The results suggest that a large reduction per pass is needed to recrystallize
and refine the coarse austenite grains in the mixed grain structure. Such
a rolling reduction is seldom possible in industrial rolling so that it
appears to be difficult to completely eliminate the coarse grains once formed,
with subsequent hot-rolling process.

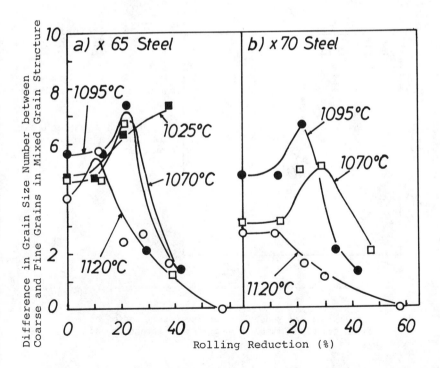

Fig.14   Effect of rolling reduction and temperature on non-uniformity
           of austenite grain

115

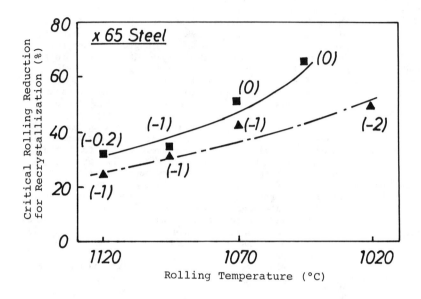

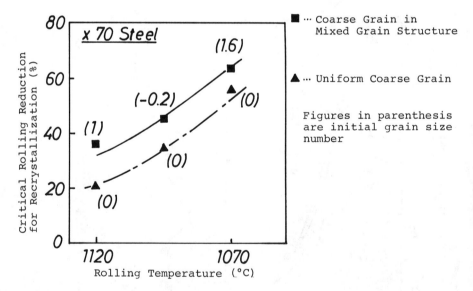

Fig.15   Critical rolling reduction for recrystallization of coarse grain in mixed austenite grain structure and uniform coarse grain

Formation of coarse bainite

The ferrite-pearlite structure containing a slight amount of coarse
bainite has been sometimes observed in controlled rolled steels, as
demonstrated in Fig. 16. Since a slight amount of coarse bainite causes
the ductile-brittle transition temperature to rise markedly, as shown in
Fig 17, it is necessary to prevent the formation of coarse bainite in
order to obtain good toughness controlled rolled steels. The formation of
coarse bainite is attributed to coarsening of austenite grains prior to
transformation into coarse bainite due to an increase in hardenability.

Figure 18 shows the effect of recrystallized austenite grain size and
rolling reduction at non-recrystallization temperatures, on the formation
of coarse bainite. The volume fraction of coarse bainite decreased with a
refinement in recrystallized grain structure and with an increase in rolling
reduction. Coarse bainite formation was eliminated at the rollng reduction
of 60% in the case of grain size number 2 and at the rolling reduction of
40% in the case of grain size number 5. Coarse austenite with grain size
number below 1 was transformed into the structure containing more than 10%
bainite even at a rolling reduction of 65% in the non-recrystallization
temperature region. The results show that it is necessary to refine
austenite to grain size number above 2 in order to obtain coarse bainite-free
steels in industrial rolling. Therefore, the refinement in austenite grain
structure becomes one of the most important points in controlled rolling.

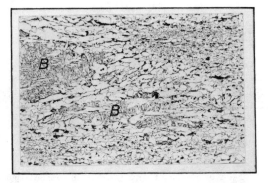

100μ

Fig.16 Bainite structure (B) formed from the coarse austenite
in the prior mixed austenite grain structure

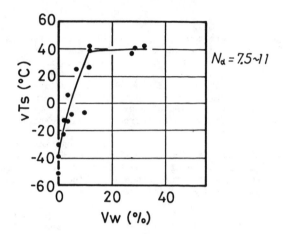

Fig.17   Effect of volume fraction Vw of coarse bainite
          on ductile-brittle transition temperature vTs

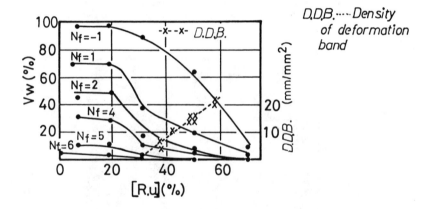

Fig.18   Effect of austenite grain size number Nf and rolling
          reduction Ru below the recrystallization temperature
          on volume fraction Vw of bainite

Conclusions

(1) The critical rolling reduction for recrystallization of austenite
decreased with a refinement in initial grain sizes and increased with a
lowering in rolling temperatures. As initial grain sizes became finer,
the critical rolling reduction became less dependent on the initial grain
size.

(2) Recrystallized austenite grain size numbers are linearly related to
initial grain size numbers for a fixed rolling reduction, but become
independent of initial grain size at rolling reductions above 70%.

(3)　At a fixed temperature the recrystallized grain size correlated with effective interfacial area per unit volume, defined as the sum of estimated interfacial areas of elongated grain boundaries and deformation bands.

(4)　In industrial rolling, where very high rolling reduction per pass is seldom possible, the refinement in austenite grain structure is attributed to static recrystallization.

(5)　Grain coarsening took place markedly at rolling reductions of 5 to 10% and at rolling temperatures above 1,050°C and 1,100°C in X 65 and X 70 steels, respectively.

(6)　The coarse grains in the mixed austenite grain structure are more difficult to recrystallize than the uniform coarse grains not to mention the fine grains in the structure. In industrial rolling, it appears to be difficult to eliminate the coarse grains once they are formed.

(7)　In controlled rolled steels with different ferrite grain sizes, the ductile–brittle transition temperature increases markedly with an initial increase in volume fraction of coarse bainite. Since coarse bainite forms from coarse austenite grains, it is necessary to prevent the formation of coarse austenite grains and to refine austenite grains in order to improve the toughness of controlled rolled steels.

References

1)　Y.E. Smith, A.P. Coldren, and R.L. Cryderman, "Manganese–Molybdenum –Niobium Acicular Ferrite Steels with High Strength and Toughness," pp.119–142 in Toward Improved Ductility and Toughness, Climax Molybdenum Development Company (JAPAN) LTD., Oct. 1971.

2)　H. Sekine and T. Maruyama, "Controlled Rolling for Obtaining a Fine and Uniform Structure in High Strength Steels," pp.85–88 in Proceedings of the Third International Conference on the Strength of Metals and Alloys, Vol 1, The Institute of Metals and The Iron and Steel Institute, Aug. 1973.

3)　T. Tanaka, T. Tabata　T. Hatomura　and C. Shiga, "Three Stage of the Controlled-Rolling Process," pp.107–119 in Microalloying 75, Union Carbide Corp., New York, N.Y , 1977.

4)　I. Kozasu, C. Ouchi, T. Sampei, and T. Okita, "Hot Rolling as a High-Temperature Thermo-Mechanical Process," pp 120–135 in Microalloying 75, Union Carbide Corp., New York, N.Y., 1977.

5)　M　Fukuda, T. Hashimoto, and K. Kunishige, "Effects of Controlled Rolling and Microalloying on Properties of Strips and Plates," pp.136–150 in Microalloying 75, Union Carbide Corp,, New York, N.Y., 1977.

6)　H. Kaji　M. Machida, and M　Katsumata, Tetsu-to-Haganè, 64(1978) pp.A219–A222 (In Japanese).

7)　H. Sekine and T. Maruyama, Seitetsu Kenkyu, 289(1976), pp 43–61 (In Japanese).

8)　J.P　Sah, G.J　Richardson, and C.M　Sellars, "Grain-Size Effects during Dynamic Recrystallization of Nickel", Metal Science, 8(1974) pp.325–331.

9)　J.J　Jonas, C.M. Sellars, and W.J McG. Tegart　"Strength and Structure under Hot-Working Conditions," Metallagical Review, 14(1969) pp.1–24.

10)　E.E　Underwood, "Surface Area and Length in Volume," pp.77–127 in Quantitative Microscopy, R.T. DeHoff and F.N. Rhines, ed.; McGraw-Hill Book Co., 1968.

11)　H　Kaji　M. Katsumata　M　Machida　and S. Kinoshita, Tetsu-To-Haganè, 60(1974) p.S295(In Japanese).

12)　T. Hatomura and T. Tanaka, Tetsu-to-Haganè, 62(1976) p.S648(In Japanese).

THE DEVELOPMENT OF MICROSTRUCTURE

DURING THE FIRST STAGE OF DEFORMATION OF X-70 STEEL

Klaus Hulka
Niobium Products Company
Dusseldorf, West Germany
(Formerly with Stahlwerke Peine-Salzgitter)

Constatin M. Vlad
Stahlwerke Peine-Salzgitter AG
West Germany

We would like to supplement some of the data shown at this conference on the behaviour of austenite during the thermo-mechanical processing by presenting the results of work carried out at Salzgitter some years ago.

The chemical composition of the steel used (X-70 grade Molytar type) was as follows:

0,06% C	0,23% Si	1,47% Mn	0,0032% P	0,007% S
0,040% Al	0,0050% N	0,07% Nb	0,33% Mo	

Within the framework of optimising the TM processing of this grade, hot rolling was carried out in a four high quatro laboratory mill. The influence of deformation upon the grain size and form and the precipitation behaviour as well as texture development was determined over a range of temperatures from 1300°C to 900°C. The degree of deformation selected for each pass was approximately 16% with a delay time between the passes of 10 sec. As seen from Fig. 1 the austenite grain grows during reheating to soaking temperature almost expon-entionally and reaches at 1300°C a value of 360μ in diameter. During the rolling process, (Fig. 2) the austenite grain size is refined through the subsequent multiple recrystallization steps accompanying high temperature deformation, whereas at lower temperatures only an elongation of the grains occurs. The development of the average grain size for the rolling con-ditions employed is shown in Fig. 3. It is evident, that, independent of the reheating temperatures, the average grain size reaches a saturated value of about 40μ if the total defor-mation is in excess of 70%.

Generally the form of the deformed and non-recrystallized austenite grain follows the relationship

$$\frac{length}{width} = \exp 2.n.\lambda$$

121

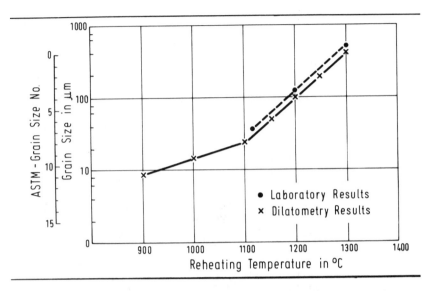

Figure 1. Austenite grain size of X-70 Molytar steel as a function of reheating temperature.

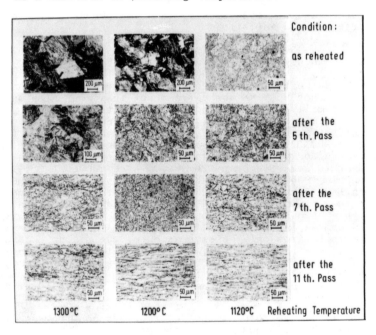

Figure 2. Austenite grain size and morphology as a function of processing.

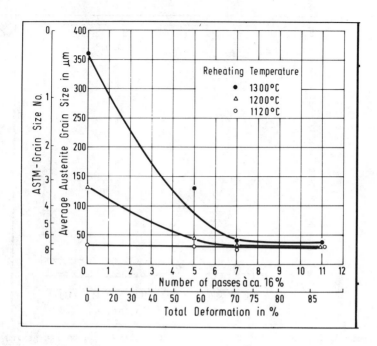

Figure 3. Influence of reheating temperature and hot deformation on the austenite grain size.

where n is the number of passes and λ is the logarithm value of the **strain per pass (0.174 for the deformation used).**
Its dependency on the deformation temperature is shown in
Fig. 4. It is evident that below a deformation temperature of about 1000°C no recrystallization of the deformed austenite grains occurs and the grains just elongated. The rolling conditions involved in this investigation consisted of 6 deformation passes in combination with a range of reheating temperatures of 1120°C to 1200°C and 4 deformation passes below the reheating temperature of 1300°C. These data agree very well with those of Kozasu[1.].

It should be pointed out that the elongated shape of the austenite grain remains for a long time at a temperature of 900°C (Fig. 5). New equiaxed grains begin to form only after a delay time of about 20 minutes. This static recrystallization is not completed even after a 2 hour hold.

The precipitation behaviour, as determined by chemical extraction, is depicted in Fig. 6. One can see that during the deformation process only a small amount of Nb is precipitated. There is a strong tendency at the end of the last passes for niobium to precipitate within a short delay time as carbonitrides. Depending on the reheating temperature, about 50 to 70% of the total Nb-content is precipitated at 900°C within 3 minutes.

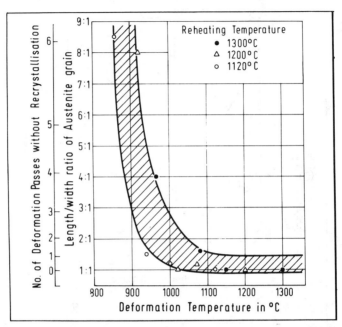

Figure 4. Dependency of the length/width ratio of austenite grains on the deformation temperature at a constant reduction of 16% per pass.

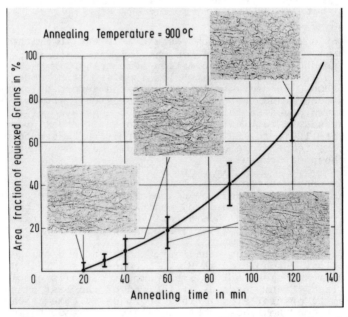

Figure 5. Increase in volume fraction equiaxed austenite grains with holding time at 900°C after deformation.

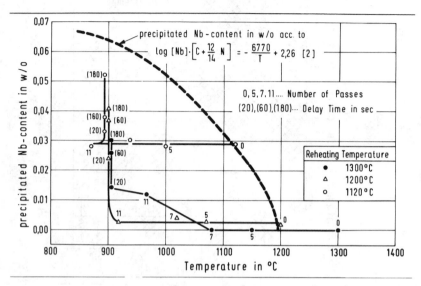

Figure 6. The amount of precipitated niobium as a function of reheating temperature, deformation and delay time.

The texture measurements made by way of modified Harries Inverse Pole Figures[3.] have shown that the original preferred orientation is maintained unchanged down to temperatures of 1100°C-Fig. 7. With a lowering of the deformation temperature, the volume fraction of the family planes of {111}, near {111} and {211} increases at the expense of the volume fraction of the planes corresponding to the zone <100>. If the deformation texture of the austenite lies in a transition from {110} <112> to {123} <412>[4.], the ferrite texture after transformation according to the Nishiyama/Wassermann relationship[5.] will comprise the major components {112} <110>, {111} <112>, and {332} <113> as well as the minor component {100} <011>. At deformation temperatures below 900°C only the above mentioned components have been found.

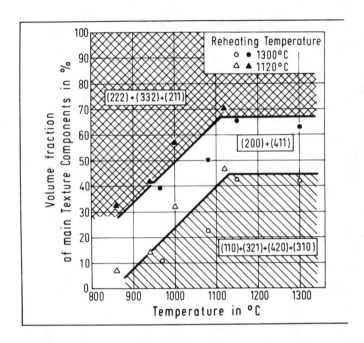

Figure 7. Volume fraction of main texture components as a function of deformation temperature.

In summary, the data obtained with the laboratory simulation of the first stage of deformation on X 70 strip with the Molytar steel grade type have given an insight into the development of texture during hot rolling, the precipitation behaviour and the microstructure of the austenite grain size and shape.

Literature

1. Kozasu, I., et.al., Microalloying 75, p.120, Union Carbide, New York, 1976.

2. Irvine, K.J., et.al., J. Iron Steel Institute, V205, 1967, 161.

3. Vlad, C.M., Salzgitter Report, TVW1/755/70.

4. Vlad, C.M. & Grzesik, Fourth International Conference on Textures, 311, The Metals Society, London, 1975.

5. Wassermann, G., Mitteilung, Kaiser Wilhelm Institute, V17, 1935, 149.

DISCUSSION

Comment: I'm interested in the early part of your paper, concerning measured austenite grain size and its change with rolling reduction, (Figure 3). Some time ago, we did similar work* on several steels, including aluminum-killed steel, and found that the austenite grain size did refine to a limiting grain size as you show. But it wasn't the same in all steels and it could vary from something like 6 ASTM down to about 10 ASTM. So there's quite a wide variation in the limiting fineness of grain size that you can get, according to the chemistry of your steel.

---

*R. Priestner, C. C. Earley and J. H. Rendall "Observations on the Behavior of Austenite during the Hot-Working of Some Low-Carbon Steels" J. Iron Steel Inst., V206 (1968) pp. 1252-1262.

THE EFFECT OF MICROALLOY CONCENTRATION ON THE
RECRYSTALLIZATION OF AUSTENITE DURING
HOT DEFORMATION

L. J. Cuddy
Research Laboratory
U. S. Steel Corporation
Monroeville, Pa.

In designing control-rolling schedules to produce the finest recrystallized grains while avoiding the duplex structures that develop in the
partial-recrystallization range, the temperature at which recrystallization
stops (the R-S temperature) must be known. Models of boundary pinning predict that the R-S temperature rises with increase in the initial solute
content of strong carbide- and nitride-forming elements, but few data are
available. To fill this need microalloyed steels containing ranges of Nb,
V, Ti, or Al were reheated to different temperatures to vary initial solute
content, and then deformed in a multipass simulation of plate rolling to
determine the R-S temperature for each steel in each initial condition.
Results confirm that the R-S temperature does increase with increase in
initial solute levels, but at markedly different rates for the several
solutes; the increase in the R-S temperature per gram atom of solute
increases in the order V→Al→Ti→Nb. Only the Nb effect is strong enough to
allow control rolling at high finishing temperatures. Two models predicting precipitate pinning of boundaries are in reasonable agreement with
the data, but direct observations of the temperature dependence of precipitation and the size and density of particles are needed for complete
confirmation.

Introduction

In a previous report (1) it was shown that, for a Nb steel, the size of the recrystallized austenite grains decreases in multipass deformation (as in strip or plate rolling) with decreasing deformation temperature, but levels off at 15 to 20 μm when realistic drafting schedules are used. This size austenite will produce 7- to 10-μm ferrite. Finer ferrite can be obtained if the austenite grains are flattened by rolling below the recrystallization temperature (control rolling), but care must be taken to avoid the duplex structures that are formed by rolling in the range of slightly higher temperatures in which partial recrystallization occurs.

It is known that the temperature that divides the ranges of complete recrystallization and partial recrystallization (the R-S temperature) depends on steel composition and the deformation schedule. For example an 0.07C-1.4Mn base composition, when deformed in a plate-rolling schedule, will recrystallize down to temperatures very near the Ar_3 (~790°C); by adding 0.03 to 0.05 percent Nb the R-S temperature is raised to 960 to 1000°C. Similar behavior is observed in other microalloyed steels, presumably because of a combination of boundary pinning by microalloy carbonitrides and solute effects.

Models that describe this phenomenon(2,3,4) predict that higher levels of microalloy solutes will raise the R-S temperature even more. This is an attractive idea where control rolling is needed for refinement of microstructure but where low-temperature finishing is impossible because of limitations in the mill capacity. Unfortunately the results of attempts to demonstrate this principle with steels containing high solute levels have been too erratic to permit the development of modified control-rolling schedules. This situation arises largely because there are few data that relate the initial level of microalloy solute (determined by steel composition and reheating temperature) to the R-S temperature during multipass deformation.(5) Consequently rolling is either stopped before pinning occurs, allowing excessive grain growth to occur, or rolling is carried into the partial-recrystallization range and coarsened duplex structures are produced.

To resolve these problems, the R-S temperatures during simulated multipass rolling were determined for a variety of microalloyed steels that had each been soaked at several temperatures to obtain a range of initial solute levels.

Procedure

The steels examined are listed in Table I. To obtain a range of solute levels available for precipitation during subsequent deformation, the steels were reheated to different temperatures. Table II lists the soak temperatures for each steel, its initial austenite grain size, and the level of the microalloy solute according to published solubility data.(6-9)

Table I Chemical Composition of Steels Investigated, Weight Percent

Steel Type	ID	C	Mn	P	S	Si	V	Nb	Ti	Al	N
CMn	J	0.09	1.45	0.03	0.01	0.30	-	-	-	<0.002	0.001
NbCN	A	0.071	1.40	0.028	0.008	0.245	-	0.011	-	<0.002	0.004
"	B	0.075	1.38	0.028	0.012	0.263	-	0.046	-	<0.002	0.003
"	D	0.064	1.43	0.025	0.009	0.289	-	0.048	-	<0.002	0.017
"	T	0.057	1.44	0.001	0.003	0.236	-	0.112	-	0.002	0.015
VN	Y	0.069	1.43	0.028	0.013	0.274	0.075	-	-	<0.002	0.017
"	Z	0.074	1.39	0.027	0.012	0.270	0.248	-	-	0.002	0.018
TiC	W	0.070	1.35	0.030	0.011	0.244	-	-	0.036	<0.002	0.003
"	X	0.070	1.35	0.029	0.011	0.257	-	-	0.108	0.003	0.004
AlN	A'	0.076	0.56	0.031	0.011	0.233	-	-	-	0.043	0.019

Table II Initial Conditions and Recrystallization-Stop Temperatures for Microalloyed Steels

Steel Type	ID	Soak Temp, °C	Initial Grain Diam, μm	Initial Microalloy Solute Content at %	wt %	Recryst-Stop Temp Range, °C
CMn	J	860	55	-	-	770-790
NbCN	A	1050	150	0.005	0.009	860-890
	B	1050	21	0.005	0.009	860-900
"		1200	100	0.025	0.043	980-1020
	D	1050	35	0.005	0.009	810-850
"		1150	120	0.016	0.026	920-970
	T	1050	30	0.007	0.011	800-840
"		1200	57	0.031	0.052	1000-1040
VN	Y	950	30	0.045	0.041	760-800
"		1100	165	0.082	0.075	810-850
	Z	1200	145	0.251	0.229	860-900
TiC	W	1050	45-75	0.042	0.036	860-900
	X	1050	40	0.061	0.052	860-890
"		1200	90	0.126	0.108	900-940
AlN	A'	1200	180	0.089	0.043	830-850

Because steels are repeatedly recrystallized during multipass deformation, we had to establish the temperature at which a previously recrystallized structure (not an as-soaked structure) was unable to recrystallize in a realistic interpass time (10 to 20 s) after a normal rolling reduction (10 to 15%). Such a determination cannot be made with a single, large-reduction pass. Therefore, the samples were control cooled at 0.5 to 1.0°C/s (similar to the cooling rate during slab-to-plate rolling) and subjected to a simulated multipass rolling comprising 5 sequential plane-strain compressions of 10 to 17 percent reduction per pass performed in a temperature interval of 40 to 50°C. The strain rate ranged from 5 to 10 s^{-1}; the interpass time was 10 to 16 s.

Each steel, for each reheat temperature, was deformed over a sufficient number of temperature intervals to reveal the transition between the completely recrystallized structure and the unrecrystallized structure. A typical schedule for the range 1100 to 1060°C is shown in Table III. After

Table III   Pass Schedule for a Sample Reheated to 1200°C and Deformed at 1100 to 1060°C

Pass	Time, s	Temp, °C	Reduction, %	
0	–	1200	–	
1	0	1100	10	
2	15	1090	11	
3	31	1080	13	
4	48	1070	14	
5	65	1060	17	
–	83	1050	–	Quench
		Total Reduction	65	

the last pass, the specimen was held for the normal interpass interval (10 to 18 s), then quenched and examined metallographically to determine the degree of recrystallization that occurred in the interpass time. To reveal the prior austenite structure, the specimens were tempered at 500°C for 15 to 20 hours, and etched by swabbing for ~1 minute in a saturated aqueous solution of picric acid heated to 80°C.

Results

Typical examples of recrystallized-refined, and unrecrystallized-elongated austenite grain structures are shown in Figure 1. The first appearance of a partially recrystallized structure, which is a mixture of these two, is used to define the R-S temperature. Table II lists the range of temperatures in which recrystallization was observed to stop for each steel in each initial condition. There was clearly an increase in the R-S

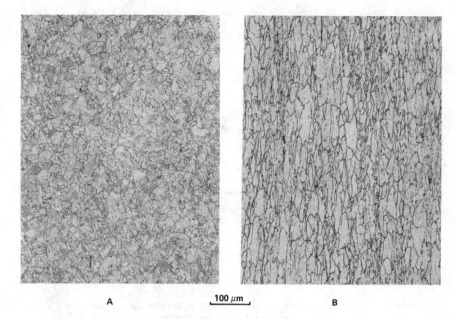

A   100 μm   B

Fig. 1 — Steel T reheated to 1200°C and reduced 55% in 5 passes to produce
A) Recrystallization at 1100-1070°C
B) Grain elongation at 1000-960°C

temperature with increase in initial solute level. However, as shown in
Figure 2, the rate of increase varied greatly with the type of solute. Nb,

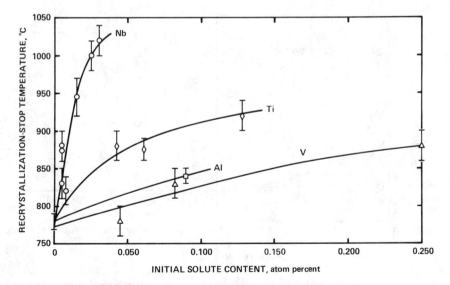

Fig. 2 — The increase in recrystallization-stop temperature with increase in the level of microalloy solutes
in 0.07C, 1.40Mn, 0.25Si steel

133

presumably through the formation of NbCN was the most effective in raising the R-S temperature; V, which forms VN in austenites of this range of compositions, was the least effective in impeding recrystallization of austenite.

The data in Figure 2 indicate that the increase in the R-S temperature (in degrees C) over that of the base steel $\Delta T_{RS} = (T_{RS} - 780)$ is related to the atomic percent solute X according to:

$$\Delta T_{RS} = \alpha X^{1/2}. \tag{1}$$

Here α is the measure of the strength of a solute's effect on R-S temperature. From the data in Figure 2, α is:

Solute	$\dfrac{\alpha}{^\circ C/(at. \%)^{1/2}}$
Nb	1350
Ti	410
Al	200
V	200

Because, for a given solute, the data for initially coarse- and fine-grained specimens fall on the same curves in Figure 2, it appears that initial austenite grain size had no effect on the R-S temperature. Previous studies of the change in austenite grain size in multipass deformation revealed that as a result of repeated recrystallization there was a rapid convergence of initially dissimilar grain sizes to nearly the same final grain size.(10) Examples are shown in Figure 3 for two different

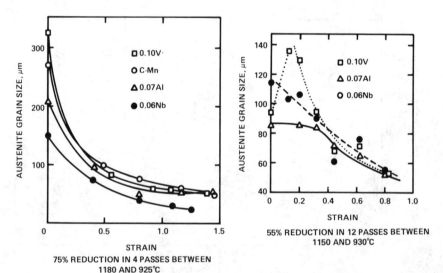

Fig. 3 — Refinement by repeated recrystallization of austenite grains in several steels during multipass deformation

134

pass schedules. Because the grain structures on which the R-S temperatures
are measured in this study have been repeatedly recrystallized, the effects
of initially dissimilar grain sizes have been eliminated before recrystallization ceases. Thus initial grain size is not a factor in setting the
R-S temperature in <u>multipass</u> deformation.

Discussion

From the data in Table II we can offer some suggestions concerning the
temperature ranges for processing some common alloyed steels with different
types and levels of alloy additions. These are summarized in Table IV.

Table IV   Control-Rolling Temperature Ranges for 0.07C, 1.4Mn Microalloyed Steels

Precipitate	Composition, wt % X	C	N	Solution Temp, °C	Lowest Allowable Roughing Temperature, °C	Finish* Release Temperature, °C
	–	0.07	0.005	1050	790	740**
NbCN	0.01Nb	0.07	0.005	1050	890	810
"	0.03Nb	0.07	0.005	1150	970	910
"	0.05Nb	0.07	0.015	1200	1040	990
VN	0.08V	0.07	0.015	1100	850	800**
"	0.20V	0.07	0.015	1200	900	840
TiC	0.05Ti	0.07	0.005	1050	900	850
"	0.10Ti	0.07	0.005	1150	940	880
AlN	0.04Al	0.07	0.020	1200	850	800**

* Highest temperature at which finish passes can avoid partial recrystallization.

** The observed Ar_3 of 0.07C-1.4Mn steels during deformation is $\sim$790°C. Therefore these
steels would have to be finished entirely intercritically.

First, there seems no good reason for reheating to a temperature higher
than that necessary to dissolve the microalloy additions; therefore, for
each steel we have set the reheat temperature equal to the solution temperature predicted by solubility studies previously reported.(6-10) To avoid
the duplex structure that results from partial recrystallization at temperatures just below the recrystallization range, we set the minimum roughing
temperature at the highest possible R-S temperature that could have
occurred in our studies (upper limit of error bars in Figure 2). To avoid
the partial-recrystallization range completely, finish passes were begun at
a temperature at least 50°C below the R-S temperature. This practice produced grain flattening only during finishing.

Clearly, with these restrictions control rolling is impractical in
those steels with low levels of the least effective solutes, V and Al
without considerable intercritial rolling, because the Ar_3 for these steels
is $\sim$790°C. Only Nb appears to be effective enough at reasonable solute
levels to allow control rolling (grain flattening) at higher temperatures. Data from Figure 2 predict that an 0.33 at %C-0.06 at % Nb steel
(0.07 wt % C-0.10 wt % Nb) would have an R-S temperature of $\sim$1050°C
provided that it was soaked at 1300°C to fully dissolve the microalloys.

Several models have been proposed to explain how the microalloy precipitates halt recrystallization of austenite.(2-4) These models are similar in that the driving force for recrystallization, F_R, arises from the difference in dislocation density across the moving boundary ($F_R = \mu b^2 \Delta\rho/2$), and the grain-boundary pinning force, F_p, arises from the area of boundary blanked off by precipitates ($F_p = \pi r \gamma N_s$). Pinning occurs when the net force on a unit area of boundary from these two sources is zero ($F_R = F_p$). Here μ is the elastic shear modulus of the matrix (4×10^4 MPa at elevated temperatures), b the Burgers vector (0.25 nm), ρ the dislocation density, r the average particle radius, γ the grain-boundary energy of austenite (0.8 J/m^2) (11), and N_s the average number of particles per unit area of grain boundary. The models differ in the manner in which N_s, and thus F_p, is computed.

In the simplest and least realistic model, a rigid boundary (not deflected by particles) of any shape interacts with only those particles of a random distribution with centers that lie within $\pm r$ of the boundary plane. In this case the number of particles per area of boundary is $N_{SR} = 3f_v/2\pi r^2$, where f_v is the volume fraction of pinning particles.

At the opposite extreme is the model with an infinitely flexible boundary that interacts with every particle in the 3-D array until it is fully pinned.(4) Here the interparticle spacing in the boundary is the same as that in three-dimensional space, which leads to $N_{SF} = 3f_v^{2/3}/(4\pi r^2)$. When, as is usual, $f_v \ll 1$, the particle density and pinning force are an order of magnitude larger in the flexible-boundary model than in the rigid-boundary model (for example, for $f_v = 0.0005$, $f_v^{2/3} = 0.006$).

A more realistic model by Hansen et al.(12) assumes that at the R-S temperature precipitates have formed on the substructure before recrystallization begins. This leads to $N_{SB} = 3 \ell f_v/8\pi r^3$ where ℓ is the average subgrain diameter. Because $\ell/r \gg 1$ this model also predicts a much larger pinning force than does the rigid-boundary model.

Our tests of these models are summarized in Table V. Equating F_R and F_p at the R-S temperature leads to an equation of the general form:

$$\pi r \gamma N_s = \frac{\mu b^2}{2} \Delta\rho \qquad (2)$$

where N_s is one of the three expressions given above. $\Delta\rho$ is estimated from Keh's relation(13) for increase in dislocation density during work hardening ($\Delta\sigma = 0.2 \mu b \sqrt{\Delta\rho}$). $\Delta\sigma$ is the observed increase in flow stress during that pass of a multipass schedule in which no recrystallization occurred, and f_v is estimated from the difference in solubility (obtained from published data) of the microalloy between the solution and recrystallization-stop temperatures. The value of ℓ is taken to be 0.8 μm on the basis of observations of substructure developed in similarly deformed stable austenites. Because we have insufficient direct observations of particle size r in these steels, we solved the equations for the maximum r that could cause pinning at the R-S temperature.

Results, in Table V show that the rigid boundary with its few pinning points frequently needs particles a fraction of an angstrom in radius to

Table V Maximum Particle Size That Can Pin Boundaries in Three Models

Steel		T_S °C	T_{R-S} °C	f_v x 10^4	$\Delta\sigma$ MPa	ρ x 10^{11} cm^{-2}	Maximum r for Pinning, nm Rigid*	Flex**	Subbdy***
CMn	J	860	780	-	207	9.9	-	-	-
NbCN	A	1050	875	0.9	144	4.9	0.017	0.314	1.86
	B	1050	880	0.9	158	5.8	0.015	0.265	1.72
	D	1050	830	0.9	89	1.9	0.044	0.809	3.00
	T	1050	820	1.3	179	7.4	0.016	0.266	1.82
	D	1150	945	2.7	130	4.0	0.063	0.800	3.58
	B	1200	1000	4.3	130	4.0	0.100	1.09	4.52
	T	1200	1020	5.0	138	4.4	0.106	1.09	4.64
VN	Y	950	780	3.5	193	8.6	0.038	0.442	2.78
	Y	1100	830	8.3	145	4.9	0.159	1.38	5.67
	Z	1200	900	8.7	172	6.9	0.118	1.01	4.89
TiC	W	1050	880	5.6	152	5.3	0.099	0.981	4.48
	X	1050	875	8.4	179	7.4	0.106	0.921	4.64
	X	1200	920	18.3	152	5.3	0.323	2.16	8.09
AlN	A'	1200	840	12.1	124	3.6	0.315	2.41	7.99

* Rigid boundary:

$$r_{max} = \frac{3 \, \gamma \, f_v}{\mu b^2 \, \Delta\rho}$$

** Flexible boundary[4]:

$$r_{max} = \frac{2.4 \, \gamma \, f_v^{2/3}}{\mu b^2 \, \Delta\rho}$$

*** Subboundary[12]:

$$r_{max} = \left(\frac{0.76 \, \ell \, \gamma \, f_v}{\mu b^2 \, \Delta\rho} \right)^{1/2}$$

halt the moving boundaries; this has no physical meaning. The flexible boundary that interacts with more points can be pinned by particles 3 to 20 Å in radius. The boundary that is coincident with subboundaries on which all the precipitates form can be pinned by larger particles 20 to 80 Å in radius. This last estimate seems most reasonable, and studies are planned to see whether its predictions can be confirmed.

The above models all assume that the equilibrium volume fraction of precipitates will form at each temperature during the deformation, and all require that a critical value of some function of (f_v/r) be attained to halt recrystallization. There is, however, another view, which is that no precipitation occurs until a critical supersaturation of microalloying elements provides sufficient driving force to cause an avalanche of precipitates to form and immediately pin the boundaries.(12) In this model, when the ratio of the solubility product at the soak temperature, K_{ST}, to that at the working temperature, reaches the critical value K_{RST} at the

recrystallization-stop temperature precipitation occurs and recrystallization is halted.

On the basis of limited data Hansen(12) suggested that at the R-S temperature the critial ratio K_{ST}/K_{RST} = 5 to 7. Using published solubility
data(6-9) we computed the value of this ratio for the combinations of soak
and R-S temperatures we observed. Results, summarized in Table VI, show
that K_{ST}/K_{RST} varies from 5 to 40. Considering the uncertainty in some of
the solubility-product determinations, particularly at the lower temperatures, it is not clear that this range of results invalidates this model.
Direct observations of the temperature dependence of precipitation <u>during
deformation</u> are needed. Such studies are under way.

Table VI   Supersaturation at the R-S Temperature

Steel		Soak T, °C	Soak K_{ST}* [wt %]^2	Recryst-Stop T, °C	Recryst-Stop K_{R-ST}* [wt %]^2	$\dfrac{K_{ST}}{K_{R-ST}}$
NbCN	A	1050	6.6E-4	875	6.1E-5	10.8
	B	1050	"	880	6.4E-5	10.3
	D	1050	"	830	2.8E-5	23.6
	T	1050	"	820	2.4E-5	27.5
	D	1150	2.0E-3	945	1.7E-4	11.8
	B	1200	3.3E-3	1000	3.6E-4	9.2
	T	1200	"	1020	4.6E-4	7.2
VN	Y	950	3.1E-4	780$^+$	3.6E-5	8.6
	Y	1100	1.3E-3**	830	7.2E-5	18.2
	Z	1200	3.0E-3	900	1.7E-4	17.6
TiC	W	1050	2.6E-3**	880	4.8E-4	5.4
	X	1050	2.9E-3	875	4.6E-4	6.3
	X	1200	8.0E-3**	920	7.6E-4	10.5
AlN	A'	1200	8.0E-4	840	2.0E-5	40.0

* K = Solubility product. For compound AB, K = [wt % A][wt % B].

** Microalloy completely in solution.

+ No increase over R-S temperature of CMn base steel.

Conclusions

The temperature at which complete recrystallization stops during
multipass deformation of microalloyed steels increases with an increase in
the level of solute present at the start of deformation. The strength of
the effect increases for the different solutes in the order V→Al→Ti→Nb.
Only the Nb effect is strong enough at reasonable levels of concentration
to allow control rolling at high finishing temperatures; and in this case a
1300°C soak would be required. Two models predict that the halt in recrystallization results from pinning by reasonably sized microalloy
carbonitrides (5 to 100 Å). Direct observations are needed to confirm
these hypotheses.

References

1. L. J. Cuddy, "Microstructures Developed During Thermomechanical Treatment of HSLA Steels," Met. Trans. A, Vol. 12A (1981), p. 1313.

2. T. Gladman, "On the Theory of the Effect of Precipitate Particles on Grain Growth in Metals," Proc. Roy. Soc. Series A, Vol. 294, (1966), p. 298.

3. M. F. Ashby, J. Harper, and J. Lewis, "The Interaction of Crystal Boundaries with Second-Phase Particles," Trans. TMS-AIME, Vol. 245 (1969), p. 413.

4. J. S. Lally, U. S. Steel Corporation, private communication.

5. A. T. Davenport and D. R. DiMicco, "The Effect of Columbium on the Austenite Structural Changes During the Hot Rolling of Low-Carbon Bainite and Ferrite-Pealite Steels," in Proceedings, International Conference on Steel Rolling, The Iron and STeel Inst. of Japan, 1980, p. 1237.

6. H. Nordberg and B. Aronsson, "Solubility of Niobium Carbide in Austenite," JISI, Vol. 206 (1968), p. 1263.

7. R. W. Fountain and J. Chipman, "Solubility and Precipitation of Vanadium Nitride in Alpha and Gamma Iron," Trans. TMS-AIME, Vol. 212 (1958), p. 737.

8. K. J. Irvin, F. B. Pickering , and T. Gladman, "Grain-Refined C-Mn Steels," JISI, Vol. 205 (1967), p. 161.

9. L. S. Darken, R. P. Smith, and E. W. Filer, "Solubility of Gaseous Nitrogen in Gamma Iron and the Effect of Alloying Constituents-- Aluminum Nitride Precipitation," Trans. AIME, Vol. 191 (1951), p. 1174.

10. L. J. Cuddy, J. J. Bauwin, and J. C. Raley, "Recrystallization of Austenite," Met. Trans. A., Vol. 11A, (1980), p. 381.

11. M. C. Inman and H. R. Tipler, "Interfacial Energy and Composition in Metals and Alloys," Met. Revs., Vol. 8, (1963), p. 105.

12. S. S. Hansen, J. B. VanderSande, and M. Cohen, "Niobium Carbonitride Precipitation and Austenite Recrystallization in Hot-Rolled Micro- alloyed Steels," Met. Trans. A., Vol. 11A, (1980), p. 387.

13. A. S. Keh, "Dislocation Arrangement in Alpha Iron During Deformation and Recovery" in Direct Observation of Imperfections in Crystals, J. B. Newkirk and J. H. Wernick, editors, Interscience, 1962, p. 213.

DISCUSSION

Q: In Table VI you show a big difference between the aluminum precipitates and the niobium precipitates in stopping recrystallization. You have a value of 40 for the critical ratio of the solubility products.

A large difference between the solute effect of Al in solution and niobium in solution has been reported, and I wonder whether, given that observation, there couldn't be an explanation for your data by adding a term to your pinning versus driving force equation (2).

A: Do you say a solute effect of Al effect? Either one is very small.

Q: Wasn't it the case that you have to go to a very low temperature to stop recrystallization in the aluminum steel as opposed to the other steels.

A: That's right, those results didn't include the natural fact that there might come a point where recrystallization stops and you have nothing in there. There's got to be some correction there.

Q: In relating the lowest recrystallization temperature to the content of the various solutes, I wasn't clear how you accounted for the effect of the different grain sizes, because that must also affect the recrystallization.

A: After 5 passes, deformation overcomes the difference in initial grain size has been wiped out.

Q: Doesn't the initial grain size affect the minimum temperature for completion of recrystallization?

A. No, by the time you get to the temperature where recrystallization is stopping there have been several recrystallization passes ahead of that. There is essentially no difference in the grain size, at that point, because you are at the very end of recrystallization.

GRAIN REFINEMENT
THROUGH
HOT ROLLING AND COOLING AFTER ROLLING

Hiroshi Sekine,
Tadakatsu Maruyama*
Hideaki Kageyama**
Yokimi Kawashima,

Fundamental Research Laboratories,
Nippon Steel Corporation
Ida, Nakahara-ku, Kawasaki, 211,
Japan

The influences of hot rolling and cooling conditions on the refinement of ferrite grain sizes in 0.1%C-Si-Mn-steels and a 0.1%C-Nb-containing-Si-Mn-steel were investigated using on experimental reversal rolling mill and a thermal cycling device.

The rolling conditions used to obtain recrystallized austenite grains finer than the initial size are discussed in connection with the change in the recrystallization behaviour from static to dynamic. In addition the critical rolling conditions were determined as a function of the grain size before the reduction and the strain rate of the reduction.

The ferrite grain refinement associated with air cooling and accelerated cooling through γ/α-transformation is discussed in terms of the size of the finally recrystallized austenite grains and the degree of their elongation below the no-recrystallization temperature.

A combination of low temperature reheating of slabs, increased draft for each pass and increased rolling speed in recrystallization temperatures, increased total reduction below no-recrystallization temperature, and accelerated cooling after rolling is known to save energy and alloying elements, reduce rolling time and carbon equivalent, and improve toughness and weldability.

Present Address
* Technical Department, Yawata Works, Nippon Steel Corporation, Edamitsu, Yawata-Higashi-ku, Kita-Kyushu, 805, Japan
** Technical Research Office, Yawata Works, Nippon Steel Corporation, Edamitsu, Yawata-Higashi-ku, Kita-Kyushu, 805, Japan

Introduction

Grain refinement is almost the only method available to obtain steels having high yield strength as well as good toughness. For this purpose a rolling technique called "controlled rolling" has been developed for ferrite pearlite steels (1), bainitic steels (2), and steels to be quenched and tempered (3). The main feature of controlled rolling is a shift in the temperature range of the final stage of rolling toward lower temperatures, whether alloying elements retarding the recrystallization of austenite are added (4) or not (1). This technique has developed from the experience with plain C steels in which thinner plates have finer grain structure, higher yield strength and better toughness. When hot rolling of these steels is started from the same slab thickness, thinner plates are finish-rolled with higher percentage reductions at lower temperatures and air cooled rapidly after rolling.

The authors have already described austenite grain refinement by the reductions in the recrystallization temperature range (RCR*) (5), if possible by utilizing dynamic recrystallization (6), to obtain fine and uniform structure even in Nb-containing steels. They found that only the total amount of reductions below the wholely no-recrystallization temperature (T_C) is practically important in the rolling schedule for the finishing reductions of Nb-containing steels (5 , 7), as did also Kozasu et al. (8). The main interest in the present paper is the refinement of austenite grains by reductions at recrystallization temperatures. Other important factors in controlled rolling are the reheating temperature which decides the austenite grain size before the first reduction and the whole temperature range of rolling, the no-recrystallization temperature (T_C), the total amount of reductions below this temperature (Σr) and austenite transformation characteristics during cooling. The influences of these factors and of chemistries on the structural changes of controlled rolled steels are also quantitatively investigated in this paper.

Experimental Procedure

Experiments were conducted mainly with hot-forged slabs of vacuum-melted 0.1%C-0.25%Si-1.4%Mn-0.005%N steels** (A) and small slabs cut from a commercial slab of a LD-melted and ingot-cast 0.1%C-0.25%Si-1.4%Mn-0.05%Nb-0.04%Al-0.004%N steel (B), shown in Table 1. The rolling experiments were made using an experimental 2 Hi-reversal plate mill with 350mm diameter rolls, and the rolling speed was 20 rpm. The measurement and control of the rolling temperatures were made by thermocouples inserted in the thickness center of each slab. Various transformed structures produced by various cooling processes were reproduced using a "formastor-F", fully automatic thermal cycle reproducing and transformation measuring device.

Refinement of Recrystallized Austenite Grains
by One Pass Hot Reduction in a Si – Mn – Steel

The change in the recrystallization behaviour of austenite grains after a single reduction was investigated on the 0.1%C-0.25%Si-1.4%Mn-steel (A) holding the thickness of specimens and the rotating speed of rolls constant

---

* Recrystallization Controlled Rolling or Rougher Controlled Rolling.

** Al was not added to these steels but as shown in Fig. 8, it was confirmed afterward that an Al addition did not alter the behaviour of austenite recrystallization.

Table 1   Chemical Compositions of Steels (in wt%)

Group	Steel	Melting	C	Si	Mn	P	S	Nb	V	Al	N	Other
A	0.10C-Si-Mn	VM	0.126	0.28	1.45	0.003	0.007	–	–	–	0.0045	–
			0.099	0.25	1.37	<0.003	0.006	–	–	–	0.0050	–
B	0.10C-Nb-V	LD	0.11	0.26	1.36	0.009	0.007	0.048	0.043	0.037	0.0037	–
	0.20C-Si-Mn		0.21	0.22	1.39	0.003	0.005	–	–	–	0.0049	–
	0.05C-Si-Mn		0.05	0.20	1.36	<0.003	0.005	–	–	–	0.0044	–
	0.20C-1.0Si		0.19	0.99	0.57	0.007	0.006	–	–	–	0.0061	–
	0.20C-0.6Mn	VM	0.17	0.24	0.56	0.004	0.006	–	–	–	0.0066	–
	0.20C-2.0Mn		0.19	0.25	1.96	<0.003	<0.005	–	–	–	0.-052	–
C	0.20C-1.5Ni		0.20	0.21	<0.02	<0.003	0.005	–	–	–	0.0063	Ni 1.45
	0.20C-1.8Cu		0.19	0.25	<0.02	<0.003	0.005	–	–	–	0.0066	Cu 1.80
	0.20C-1.3Cr		0.20	0.25	<0.02	0.003	0.006	–	–	–	0.0078	Cr 1.26
	0.10C-0.9Mo		0.11	0.26	0.80	<0.003	0.007	–	–	–	0.0057	Mo 0.91
	QT-60	LD	0.13	0.48	1.16	0.009	0.006	–	0.03	0.013	0.0074	Cr 0.22 Ti 0.01 B 0.0005
D	0.10C-Si-Mn-Al		0.092	0.25	1.37	0.003	<0.005	–	–	0.014	0.0058	–
	0.10C-0.1Mo		0.11	0.26	1.41	<0.003	0.005	–	–	0.021	0.0052	Mo 0.11
	0.10C-0.2Mo		0.10	0.25	1.39	0.003	0.006	–	–	0.019	0.0053	Mo 0.21
	0.10C-0.4Mo	VM	0.094	0.24	1.36	0.004	0.005	–	–	0.017	0.0056	Mo 0.38
	0.10C-0.6Mo		0.12	0.25	1.40	0.005	0.007	–	–	0.021	0.0062	Mo 0.61
	0.10C-0.02Nb		0.097	0.26	1.42	<0.003	0.006	0.026	–	0.020	0.0051	–
	0.10C-0.05Nb		0.10	0.26	1.38	0.003	<0.005	0.053	–	–	0.0060	–
	0.10C-Nb-Mo		0.11	0.24	1.39	<0.003	0.005	0.025	–	0.025	0.049	Mo 0.19

(H = 10mm and N = 20 rpm) and varying the amount and the temperature of the reduction (5,9). After reheating first to 1250°C the initial grain sizes ($N_{\gamma i}$) were varied by applying no, one or two reductions to 10mm at higher temperatures prior to the final rolling reduction. Within one second after reduction, the rolled specimens were iced-brine quenched. Austenite grain structures were revealed by electrolytic etching (10).

Fig. 1 summarizes the results of the coarse-grained austenite as-reheated to 1250°C ($N_{\gamma i}$ = 0.2)*. The circled figures are ASTM grain size numbers of recrystallized austenite ($N_{\gamma j}$)*. The whole specimen can be re-crystallized one second after the reduction, when the percentage reduction (r) is over the critical amount at a given temperature. The critical reduction for complete static recrystallization is a function of the thickness of the plate and the holding period, and is named r_{cs}'. The size of statically recrystallized grains is reduced by increasing the draft, but scarcely depends on the reduction temperature.

Larger drafts cause dynamic recrystallization. The size of dynamically recrystallized grains is mainly determined by the reduction temperature, and the lower the temperature the finer the recrystallized grain. The critical reduction for dynamic recrystallization is named r_{CD} and is intrinsically independent of the plate thickness and the holding time after reduction.

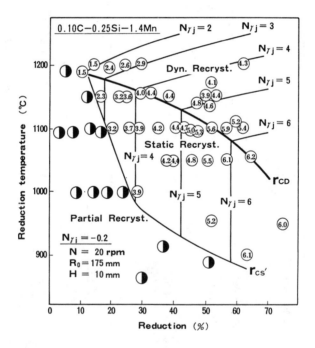

Fig. 1   Recrystallization behaviour of hot rolled austenite.
(Steel A)

* $N_{\gamma i}$:   Starting grain size in ASTM No.
  $N_{\gamma j}$:   Recrystallized grain size in ASTM No.

144

Equi-$N_{\gamma j}$-curves change their gradients at intersections with r_{CD}-curve, and become identical with equi-Zener-Hollomon-parameter-contours in the region of dynamic recrystallization (5,6).

Both r_{CS}' and r_{CD} are reduced by increasing the reduction temperature.

Fig. 2 shows the effect of the initial grain size ($N_{\gamma i}$) on the recrystallization behaviour. By reducing the starting grain size, r_{CD} and r_{CS}'are also reduced, that is, dynamic (5,6,9) and static recrystallization (5,8,9) more readily occur. The temperature below which recrystallization completely stops (T_c), is therefore determined not only by the steel chemistry and the amount of the reduction applied at the critical temperature range, but also by the complete draft schedule down to the reduction which decides the grain size prior to the reduction ($N_{\gamma i}$).

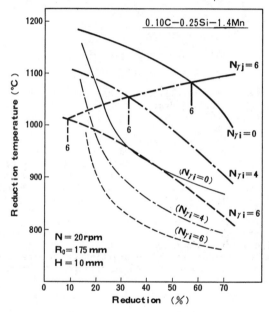

Fig. 2   Effect of initial grain size on the critical reductions for
static and dynamic recrystallization of austenite.
(Steel A)

The occurrences of dynamic and static recrystallization are favoured by decreasing the starting grain size and increasing draft. The size of dynamically recrystallized grains, however, is determined only by the temperature and strain rate of the rolling reduction, and does not depend on the initial grain size and the draft itself. On the other hand, the size of statically recrystallized grains does not depend on the reduction temperature and is refined by reducing the starting grain size and increasing the draft.

There is a linear relationship between the logarithm of the Z-value of the deformation conditions and the inverse of the diameter of the dynamically recrystallized grain (1/d) (5,6), as shown in Fig. 3. Fig. 3 shows the transition of the recrystallization behaviour from static to dynamic.

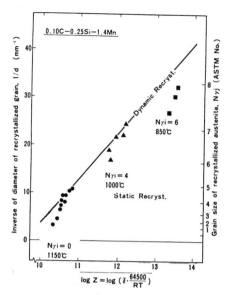

Fig. 3   Transition of recrystallization behaviour from static to
         dynamic by increasing drafts.
         (Steel A)

When the starting grain size and the reduction temperature are held constant,
the increase in the amount of the reduction brings about the successive refinement of recrystallized grains.   The inverse diameters of statically
recrystallized grains approach linearly to the straight line of dynamic
recrystallization from the lower side respectively corresponding to each
combination of the starting grain size and the reduction temperature.   When
the reduction amounts to the value of the r_{CD} for each experimental condition, the 1/d value reaches the straight line.   Further increase in the
amount of the reduction over r_{CD} changes the direction of the increase in
1/d and all the 1/d values increase along the single straight line of dynamic recrystallization independent of the starting grain size and the reduction temperature.   This is one of the characteristics of dynamic recrystallization.

Recrystallization Controlled Rolling
and Its Relation to Mill Capacity

    Fig. 1 shows that when the same reduction greater than r_{CS}' is applied
at various temperatures, the statically recrystallized grains are always
finer than or at least of equal size to the finest dynamically recrystallized grains.   Rapid static recrystallization can be used as the main
mechanism for the refinement of austenite grains during controlled rolling
("Recrystallization Controlled Rolling").

    The reduction conditions required to obtain recrystallized grains
finer than ASTM number 5.0 starting from the grain size of 4.0 are summarized in Fig. 4 as an example.   The favourable conditions are in the
region of higher drafts and lower temperatures than the equi-$N_{\gamma j}$-line of
5.0.

    The idea of Recrystallization Controlled Rolling can be simplified
and generallized as follows.   The values of the draft and the temperature

146

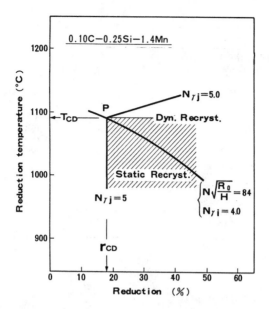

Fig. 4   Combination of lower critical reduction and upper critical
          reduction temperature to refine austenite grains by recrystal-
          lization.
          (Steel A)

at the point P of the intersection of r_{CD} curve for the starting grain size
with the equi-Z-contour corresponding to the aimed grain size are respectively named r_{CD} and T_{CD}, as shown in Fig. 4.   In order to obtain an
aimed recrystallized grain size starting from a given grain size, a reduction not smaller than r_{CD} has to be applied at a temperature not higher
than T_{CD}.   The range of these reduction conditions is hatched in Fig. 4.
The critical reduction for dynamic recrystallization, r_{CD}, depends on the
starting grain size, the reduction temperature and the strain rate of the
reduction.   The mean strain rate of a reduction in s^{-1}($\bar{\dot{\varepsilon}}$) is related to the
rotating speed of rolls in rpm (N), the radius of rolls (R_0), the slab

$$\bar{\dot{\varepsilon}} = \frac{\Pi}{30} N\sqrt{\frac{R_0}{H}} \cdot \frac{1}{\sqrt{r}} \ln \frac{1}{1-r} \tag{1}$$

thickness before the reduction (H) and the draft (r<1) and can be written
as a function of two terms as shown in Eqn. 1.   One term is referred to
as the mill capacity, $N\sqrt{R_0/H}$, and the other involves the reduction, r.
Experimental reports on the strain rate dependence of the critical deformation for dynamic recrystallization (ε_c), obtained by hot torsion tests
(11, 12) (Fig. 5), show that the gradient of linear relationships between
log ε_c and log $\dot{\varepsilon}$ is constant and does not depend on the deformation temperature and the chemical composition of the steel.   Using numerical results
on the critical reduction conditions as shown in Fig. 1 and 2 and the value
of the constant gradient in linear relationships in Fig. 5, the combinations
of the lower critical reduction (r_{CD}) and the upper critical reduction
temperature (T_{CD}) were evaluated as functions of $N\sqrt{R_0/H}$ for various combinations of starting and aimed grain sizes.   An example is shown in Fig. 6,
which gives the combinations of r_{CD} and T_{CD} required to obtain the recrystallized grains of ASTM number 5.0 starting from the grain size of 4.0
in the 0.1%C-Si-Mn steel (A).   The reduction by r_{CD} at a temperature higher

147

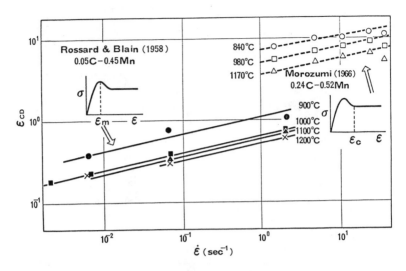

Fig. 5 Strain rate dependence of critical deformation for dynamic recrystallization.

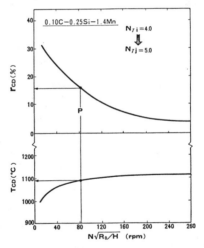

Fig. 6 Change in combination of lower critical reduction and upper critical reduction temperature with mill parameters. (Steel A)

than T_{CD} in Fig. 6 causes dynamic recrystallization and the resulting recrystallized grains becomes coarser than 5.0. Fig. 6 shows that higher rolling speeds can reduce the value of r_{CD} and raise the value of T_{CD}, or refine recrystallized grains (13).

It is proposed, that when the draft schedule is given and the initial grain size at reheating is known, the successive change in recrystallized grain sizes may be estimated using a series of figures such as Fig. 6. Such evaluations with existing conventional draft schedules show that

coarse austenite grains, reheated for example to 1250°C, can not be re
crystallized by the initial two or three reductions because of their too
low drafts, as Tanaka et al pointed out for Nb-steels (14). Austenite
grains may eventually be recrystallized statically or dynamically by the
accumulated strain of the several initial passes plus the increased drafts
of the successive passes due to the reduced slab thickness, H. The grain
size as first recrystallized could not be estimated.

Recommendations from this study to mill operations are the increases
in the rolling speed and draft of each pass, and decrease in the rolling
temperature of each pass. Another suggestion is to decrease the slab reheating temperature, which reduces the grain size before the first reduction,
facilitates recrystallization and refines the statically recrystallized
grains. It also reduces the temperatures of all successive reductions. This
processing may be called "Recrystallization Controlled Rolling" (RCR).

When the grain size before the first reduction is finer than ASTM
number 0, when the reduction temperature is lower than 1200°C and when the
rolling speed is increased to about one and half times the conventional
speed, then austenite grains can be recrystallized to finer new grains from
the first reduction. Furthermore by reducing the number of passes, con-
centrating drafts to a few final reductions, and reducing these reduction
temperatures by 30 to 50°C, the size of finally recrystallized grains in a
35mm thick plate became 6.4 in ASTM number. The finally recrystallized
grain size produce by conventional draft schedules were 3 to 5 in ASTM
number (15). The new draft schedule proposed here is very similar to those
of controlled rolling of plain C steels carried out in a few plate mills in
Europe (1). It is possible to roll further at temperature below T_c in plain
C steels, if necessary. This may be the controlled rolling of plain C steels
in Europe which demands only the total amount of reductions at lower temperatures (1).

The conventional roughing of Nb-containing steels terminates at temperature higher than T_c temperature. Applying the same principles as outlined
above, we have "Roughing Controlled Rolling" (RCR). The proposed rolling
procedure is found to give better toughness and to raise the rolling productivity. Especially in Nb-steels, the desired decrease in the slab reheating temperature sometimes difficult because of the relation between the
steel chemistry and the demanded strength of products, but lowering the
temperatures and increasing the drafts of the final several passes in roughing stage were effective when the steel chemistry did not permit the reduction of the rehearting temperature.

Figs. 1 and 6 show that the increase in the draft, the decrease in the
reduction temperature and the increase in the rolling speed make statically
recrystallized grains finer. A strong mill favours not only controlled
rolling in the no-recrystallization temperature range but also RCR.

Effects of Alloying Elements on Refinement
of Recrystallized Austenite Grains

Fig. 3 shows that dynamic recrystallization give the finest grains in
the austenite structures recrystallized by the reductions having the same
values of Z in a given steel. It is favourable to find the basic steel
compositions most suitable for RCR, to study the effects of alloying additions on the linear relationship in the log Z-1/d-diagram for dynamic recrystallization. When the same value of the activation energy of hot working is available for various steels, the Z-value of the abscissa is regarded

as the common parameter of the rolling action.

For this study, steels shown as Group C in Table 1 were used. Starting from a 0.2%C-0.25%Si-1.4%Mn-0.005%N-steel, effects of decrease in C to 0.05%, the increase in Si to 1.0%, the change in Mn from 0.6% to 2.0%, and the substitutions of 1.4%Mn by 1.5%Ni, 1.8%Cu or 1.3%Cr, on the dynamically recrystallized grain size, were studied by rolling experiments. In addition, a 0.1%C-0.25%Si-0.8%Mn-0.9%Mo-0.005%N-steel and a 0.13%C-0.48%Si-1.2%Mn-0.2%Cr-0.03%V-0.007%N-steel (a 60kg/mm² grade quenched and tempered steel) were studied in a similar way. Assuming that the activation energy of hot working is 64.5K cal/mol (6) for all steels, the results show no substantial deviations from the linear relationship between log Z and 1/d for the 0.1%C-0.25%Si-1.4%Mn-0.005%N-steel shown in Fig.3, except in the 0.05%C-Si-Mn- and the 0.1%C-0.9%Mo-steels (Fig. 7). An increase in Mn slightly refines

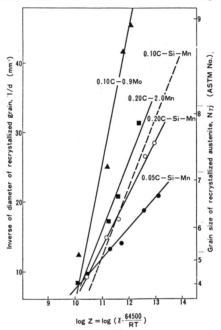

$$\log Z = \log \left(\dot{\varepsilon} \cdot \frac{64500}{RT} \right)$$

Fig. 7 Effects of alloying elements on refinement of austenite
by dynamic recrystallization.
(Steel A and Steels of Group C)

recrystallized grains. This all means that the chemistry of Si-Mn-steels is so favourable that the stable combination of strength and toughness can be expected in their as rolled state even with fluctuations in their chemistries, unless a serious change is brought about in their rolling conditions.

In the low-carbon 0.05%C-Si-Mn steel, the grain sizes given by the linear relationship are coarser than those of other steels at the same apparent Z-values (Fig. 7). This might mean that extra-low-C-steels are at a disadvantage for RCR. Amounts of Mn over 1.6% are generally added to extra-low-C-steels. Further experimental researches seem to be necessary on extra-low-C-high-Mn-steels.

r_{CS}' and r_{CD} are raised simultaneously by Mo addition, apparently because Mo in steels raises the activation energy of hot deformation and retards the recrystallization of austenite. Mo also apparently refines dynamically recrystallized grains (Fig. 7). Further studies were carried out to determine the detailed effects of Mo additions.

Nine 0.1%C-0.25%Si-1.4%Mn-0.005%N-steels containing various amounts of Al, Mo and Nb (Group D in Table 1), were reheated to 1250°C and reduced in five fixed successive passes and an additional sixth pass whose reduction temperature was varied to determine their full-recrystallization-stop temperature (T_c). Reduced specimens were quenched in iced-brine within one second after the fifth or sixth reductions. Observed austenite structures are summarized in Fig. 8. Figures in circles and squares are grain size

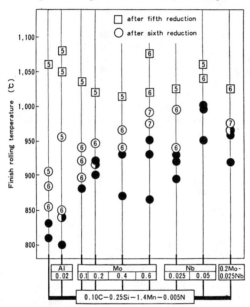

Fig. 8 Effects of Mo addition on recrystallization behaviour
of austenit.
(Steel A and Steels of Group D)

numbers of recrystallized austenite and half solid circles and solid circles mean partial recrystallization and perfect no recrystallization respectively. Addition of Mo retards the recrystallization of austenite. T_c temperatures of the 0.2%Mo-steel and the 0.6%Mo-steel are respectively comparable to those of the 0.025%Nb- and the 0.025%Nb-0.2%Mo-steels. Furthermore, Mo addition refines the grain size of statically recrystallized austenite. Such effects of Mo are considered to be the effects of Mo in solution, for Mo does not precipitate in austenite under the additions of interests (16). Fig. 8 also shows that partial recrystallization rarely appears when RCR has been effectively carried out and the considerable refinement of re-crystallized grains has been achieved.

Detailed study on the 0.2%Mo-steel shows that the Mo addition causes an increase in r_{CD} and T_{CD} when the starting grains are coarse but no considerable increase in r_{CD} when the starting grain size is finer than, for instance, $N_{\gamma i}$=3.5. It is not necessary to decrease the reduction temperature so much during RCR of Mo-containing steels. Mo can retard the austenite recrystallization by its effect in solution. Mo addition permits the reduction of Nb contents in Nb-containing steels and allows lowering of the slab reheating temperature for the dissolution of Nb(CN), which is also convenient for RCR. In addition, Mo changes the transformation behaviour of austenite not having been exhausted as precipitates in the manner of Nb, and produces the transformed structures having round stress-strain curves (2). Mo is an alloying element which is very favourable to controlled rolling of Nb-containing steels, although it is expensive.

Refinement of Ferrite Grains by
Transformation during Cooling

Various austenite grain size were obtained by changing the rolling conditions of 0.1%C-Si-Mn-steel plates (A) 7mm thick, and they were then allowed to air cool. Various austenite grain sizes were also obtained by changing the austenitizing temperatures of undeformed specimens, which were cooled by simulating a cooling path down to 300°C similar to that of the air cooled 7mm thick plates. The relation between austenite grain sizes and transformed ferrite grain sizes are shown in Fig. 9 along with similar results for a 0.05%C-0.25%Si-1.2%Mn-steel (10).

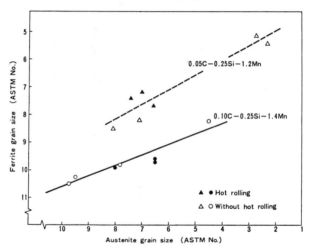

Fig. 9   Austenite grain size and ferrite grain size immediately
before and after transformation in 7mm thick plates.
(Steel A and Steel in ref. (10))

When the austenite grain size before the transformation and the plate thickness (or the cooling rate) are the same, there is no difference in ferrite grain sizes between the normalized and the as-rolled states (10). The finer grained austenite transforms to the finer grained ferrite-pearlite structure, for ferrite grains nucleate at austenite grain boundaries. Finer ferrite grains are obtained in the 0.1%C-1.4%Mn-steel than the 0.05%C-1.2%Mn-steel having the same finally recrystallized grain size, $N_{\gamma f}$, because of the difference in A_{r3} temperatures and C contents. A 0.05%C-2.0%Mn-steel having the A_{r3} temperature lower than that of the 0.1%C-1.4%Mn-steel transformed to coarser ferrite grains than the line of the line of the 0.1%C-1.4%Mn-steel which Fig. 9 would predict. It was experimentally confirmed that the nucleation of ferrite grains and the growth of ferrite grains into

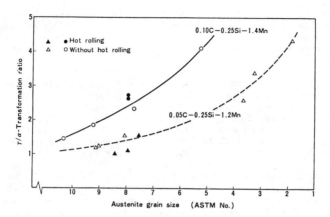

Fig. 10   Change in γ/α-transformation ratio with austenite grain size.
(Steel A and Steel in ref. (10))

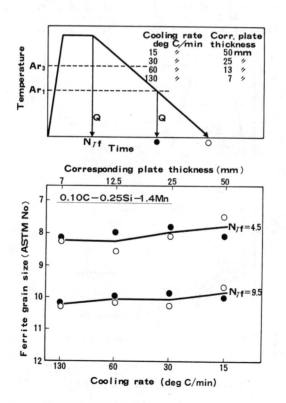

Fig. 11   Plate thickness dependence of grain refinement by
transformation   (Steel A)

153

the austenite matrix are still continued at the A_{r1} temperature even with
the cooling rate of the air-cooled 7mm thick plate. In addition, the
ferrite grain growth is disturbed by the presence of pearlite. Ferrite
grain sizes were more refined in Nb or V and N alloyed steels than in plain
C steels (10).

The ratio of the austenite grain diameter to the ferrite grain diameter
before and after the transformation is called the γ/α-transformation ratio
(17). The ratio is much larger than one when the austenite grain is com-
parably coarse as shown in Fig. 10. By reducing the austenite grain size,
the ratio is reduced and seems to approach one. There is a limit to the
ferrite grain refinement which is imposed by the available refinement of
recrystallized austenite grains in the case of air cooling (17).

The plate thickness dependence of the γ/α-transformation ratio was
studied with the 0.1%C-Si-Mn-steel (A). The results in Fig. 11 indicate
that there is only a slight cooling rate dependence of the γ/α-transforma-
tion ratio and no grain growth of transformed ferrite after transformation.
It became clear that the grain refinement and the improvement in yield
strength and toughness in thinner plates are not attributed to their faster
cooling rate but to the grain refinement accompanying the unintended RCR.

The effects of accelerated cooling and coiling after rolling were investigated on the 0.1%C-Si-Mn-steel (A) using similar simulation procedure
(Fig. 12). The effect of accelerated cooling on the ferrite grain

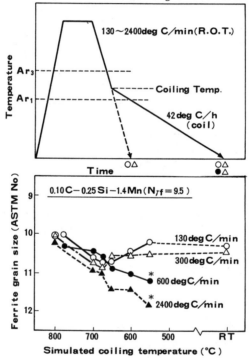

* ; Containing Upper Bainite

Fig. 12　Ferrite grain refinement by accelerated cooling and coiling
after hot rolling. (Steel A)

154

refinement becomes remarkable at cooling rates over 130 deg/min, corresponding to that of the air cooled plate 7mm thick. There may be an optimum
coiling temperature to obtain the finest ferrite grains at intermediate
cooling rates. This may be explained in terms of the suppressions of the
growth of ferrite grains into the austenite matrix by the rapid nucleation
and growth of pearlite just after the sudden decrease in the cooling rate
by coiling. The faster the cooling and the lower the coiling temperature,
the finer the ferrite grains at cooling rates in the range above the intermediate rate. A too drastic lowering of the coiling temperature brings
about bainitic structures and diminishes the toughness. The optimum coiling temperature is the A_{r_1} temperature for the adopted cooling rate. The
γ/α transformation ratio increases with the cooling rate and the change
when coiled at 600°C is shown by open triangles and the dotted line in
Fig. 13. Accelerated cooling removes the limit in the γ/α transformation
ratio of one.

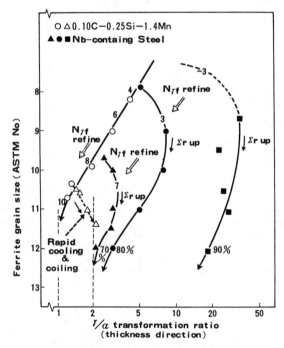

Fig. 13   Ferrite grain size and γ/α-transformation ratio in various
controlled rolling and controlled cooling.
(Steels A and B)

Transformation from Unrecrystallized Elongated Austenite

Using a LD-melted and slabed 0.1%C-0.25%Si-1.4%Mn-0.05%Nb-0.04%V-steel
(B), a study was made on the effects of the austenite grain refinement by
RCR and of the elongation of unrecrystallized austenite grains on the final
ferrite grain structure and the mechanical properties (5,18). Slabs of
various thickness were reheated to 1250°C, rolled in various two stage
rolling schedules having delay times between the end of roughing and 900°C
to plates 13 mm thick. The size of the finally recrystallized austenite

155

grains ($N_{\gamma f}$) was varied by controlling the rolling schedule down to 980°C and the total reduction at the temperatures of recrystallization inhibition below 900°C (Σr) was also varied by changing the intermediate slab thickness. Finally the plates were air cooled.

Results similar to those of Kozasu et al (8) were obtained as shown for example in Fig. 14. Figures in circles in Fig. 14 are the obtained ferrite

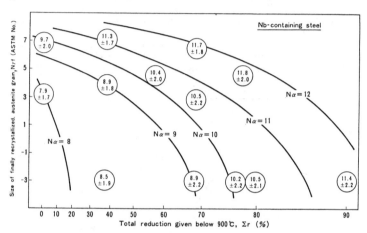

Fig. 14   Ferrite grain size in various controlled rolling (Steel B)

grain sizes in ASTM number (N_α) and their standard deviations.

The γ/α-transformation ratios were evaluated using the results in Fig. 14 and assuming the austenite grain size as the distance between grain boundaries in the through thickness direction. The results are also summarized in Fig. 13. By increasing Σr, the γ/α-transformation ratios in specimens having various $N_{\gamma f}$ are respectively increased, reached at their maximum values and finally seem to converge on two. The more refined the finally recrystallized austenite grains, the more rapid the convergence to the value of two. Cuddy's result that the γ/α-transformation ratio in a Nb-steel is two (13), is considered to be the consequence of the intensified RCR. The role of deformation bands as nucleation sites of ferrite grains (10) seems to be diminished by refining $N_{\gamma f}$ and increasing Σr. Only the elongated austenite grain boundaries may be serving as nucleation sites of ferrite grains (13). The difference in converged γ/α-transformation ratios between transformation from polygonal and elongated grains may be attributed to the difference in curvatures of their austenite boundaries. Thus the γ/α-transformation ratio limit of one obtained for air cooling of fine polygonal austenite grains does not apply if the final rolling reductions are made at temperatures below T_c.

Comparing standard deviations in ferrite grain sizes on every equi-N_α-contour in Fig. 14, the ferrite grain structures are always more uniform in specimens which started the reductions below T_c from the finer recrystallized austenite grains. It is worth noting that equi-transition temperature contours are very similar to Fig. 14 and the V-Charpy absorption energy at the ductile region is rapidly decrease when Σr is over 70% (5). Fig. 15 can be obtained using these equi-shelf energy-contours and Fig. 14.

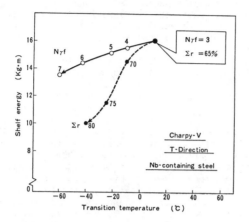

Fig. 15 Comparison of toughness between two types of controlled
rolling techniques. (Steel B)

Fig. 15 shows that the more refined $N_{\gamma f}$ gives the higher V-Charpy absorption
energy at the ductile region, comparing the specimens having the same
transition temperature. It is a consequence of the uniformity of their
ferrite grain sizes (5,6).

These results show that RCR and the reductions below T_c temperature
are equivalent and may be substituted for each other to obtain the same
transition temperature or used in combination to improve transition temperature. The improvement by the intensification of RCR, however, not only
gives higher V-Charpy absorption energies in the ductile regions, but trial
runs confirmed that RCR raises the rolling productivity when the same transition temperature is demanded. That is, it decreases the delay time between roughing and finishing and the rolling time in finishing, by reducing
the intermediate slab thickness and by diminishing the imbalance between
roughing and finishing rolling times.

Implications and Development of the Current Research Results

The traditional slab reheating temperature of 1250°C which has softened
slabs for easy rolling in mills driven by steam engines, can be lowered
with the introduction of rolling mills drived by motor generators and
strengthened for high speed production. This direction is coincident with
the energy saving and also with the direction in which controlled rolling
has been developing. The low temperature reheating gives the rolled
products a definite amount of improvement in their yield strength and toughness. Moreover, the energy saving may be easily combined with efforts to
save alloying elements and reduce heat treatments by the expanded application of controlled rolling and controlled cooling.

The coarsening temperature of austenite grains in Al-killed steels is
about 1050°C (19). Plain C steels not containing insoluble elements such as
Nb or Ti can be processed by reheating at temperatures below 1050°C to get
fine grained austenite before rolling (20). Fine grained austenite is
easily recrystallized and changed to finer grains by static recrystallization. This process can be the alternative of normalizing.

Reheating at extremely low temperatures is not appropriate for Nb-containing steels, because solution of Nb at reheating temperature for subsequent retardation of austenite recrystallization and strengthening of transformed ferrite is not guaranteed. The addition of a very small amount of Ti, about 0.01 to 0.02%, principally raises the coarsening temperatures of austenite grains (21) to 1200°C, especially in continuously cast slabs. Such a small addition of Ti to Nb-containing steels can suppress the austenite grain coarsening at the reheating temperature, ensuring that a sufficient amount of Nb remains in solution. Similar small additions of Ti are of course effective for plain C steels which are to be controlled rolled, when the reheating below 1050°C is impossible (22).

An addition of Mo to Nb-steels can reduce the amount of Nb necessary to retard the recrystallization of austenite and also reduce slab reheating temperatures, and further bring about our favourable changes in the stress-strain curves of Nb-steels having high yield ratios (23). Unfortunately, austenite grains are coarsened even at this reduced reheating temperature. Furthermore, Mo raises the lower critical reduction for the refinement of recrystallized austenite grains for these coarsened starting grains, but not so for the grains of intermediate size. To make sure that all the grains are of intermediate size, a small amount of Ti was added to extra-low-C-Mo-Nb-steels and new controlled rolled acicular ferrite steels (24) and bainitic steels (25) were developed.

Boron was added to the new bainitic steels (25) in an attempt to reduce or eliminate the Mo addition in the acicular ferrite steels (24). Another method for saving Mo is the introduction of accelerated cooling after controlled rolling (26,27). This technique was also applied to plain C steels and effectively reduced carbon equivalent (28,29). Accelerated cooling increases the γ/α-transformation ratio and amounts of pearlite and bainite (30). The latter two factors are not good for toughness. Controlled rolling preceding accelerated cooling refines ferrite grains, pearlite and bainite structures and decrease the amounts of the last two (5,8). These may be the factors which can increase strength without deterioration of toughness (30). The development of accelerated cooling combined with controlled rolling is inevitable for saving energy through eliminating heat treatment, saving alloying elements or carbon equivalent and improving weldability. In order to do that, it is also necessary to develop equipment and control systems for the cooling process, which will achieve the desired qualities of plate products.

From the viewpoint of saving energy, direct rolling of continuously cast slabs without reheating is most favourable. It is, however, least favourable for controlled rolling, because cast slabs retain very coarse grains down to hot reductions. Recrystallization of coarse austenite grains after successive hot reductions has to be studied quantitatively to determine which grades of steels can be produced as-rolled or as-rolled-and-cooled steels. This challenge of direct controlled rolling of cast slabs means a revival of the study of RCR. Otherwise the problems of RCR appear to have been solved by introduction of extra low temperature reheating and the addition of Ti.

References

(1) R.W. Vanderbeck, "Controlled Low-Temperature Hot Rolling as Practiced in Europe", Welding Journal, 37 (3) (1958) pp. 114s-116s.

(2) Y.E. Smith, A.P. Coldren, and R.L. Cryderman, "Manganese-Molybdenum-Niobium Acicular Ferrite Steels with High Strength and Toughness," Toward Improved Ductility and Toughness, pp. 119-142; Climax Molybdenum Development Company (Japan) Ltd., Tokyo, 1971.

(3) H. Gondoh, S. Yamamoto, M. Nakayama, H. Nakasugi, and H. Matsuda, "Super-Fine-Grain Microalloyed Steel", Micro Alloying 75, pp. 435-441; Union Carbide Corporation, New York, 1977.

(4) J. D. Jones, and A.B. Rothwell, "Controlled Rolling of Low-Carbon Niobium Treated Steels", Iron and Steel Institute Special Report, No.18, (1968) pp. 78-82.

(5) H. Sekine, and T. Maruyama, "Fundamental Research on Manufacturing of High-Tough, High-Tension Steels by Controlled Rolling", Seitetsu Kenkyu, No.209 (1976) pp. 43-61.

(6) H. Sekine, and T. Maruyama, "Controlled Rolling for obtaining a Fine and Uniform Structure in High Strength Steels", The Microstructure and Design of Alloys, Vol. 1, pp. 85-88; The Institute of Metals and the Iron and Steel Institute, London, 1973.

(7) H. Sekine, T. Maruyama, and Y. Kawashima, "Formation of Deformation Bands in Low Temperature Rolling of Nb-Containing Steels", Tetsu-to Hagane, 60 (11) (1974) p. S557, (Preprint for Annual Meeting of Iron and Steel Institute of Japan.

(8) I. Kozasu, C. Ouchi, T. Sampei, and T. Okita, "Hot Rolling as A High Temperature Thermomechanical Process", Micro Alloying 75 pp. 120-135; see ref. (3).

(9) H. Sekine, T. Maruyama, and Y. Kawashima, "Dynamic and Static Recrystallization of Austenite in Hot Rolling", Tetsu-to-Hagane, 59 (11) (1973) p. S636, (Preprint for Annual Meeting of Iron and Steel Institute of Japan.)

(10) H. Sekine, and T. Maruyama, "Retardation of Recrystallization of Austenite during Hot Rolling in Nb-Containing Steels", Transactions of Iron and Steel Institute of Japan, 16 (8) (1976) pp. 427-436.

(11) C. Rossard, and P. Blain, "Initial Results of Research on the Hot Deformation of Steels. Application of a Specially Designed Apparatus", Review de Metallurgie, 55 (6) (1958) pp. 573-594.

(12) F. Morozumi, "Study on the Evaluation of Hot Workability of Steels by Hot Torsion Tests", Tetsu-To-Hagane, 52 (13)(1966) pp.1859-1899.

(13) L. J. Cuddy, "Microstructures Developped during Thermomechanical Treatment of HSLA steels", Metallurgical Transactions, 12A (7) (1981) pp. 1313-1320.

(14) T. Tanaka, T. Funakoshi, M. Ueda, J. Tsuboi, T. Yasuda, and C. Utahashi, "Development of High-Strength Steel with Good toughness at Arctic Temperatures for Large Diameter Line Pipe", Micro Alloying 75, pp. 399 -409; see ref. (3).

(15) Private communication, K. Kawamura, M. Sagara, and T. Shinobe, Hirohata Works, Nippon Steel Corporation, Apr., 1974.

(16) J.J. Jonas, and M. Akben, "Retardation of Austenite Recrystallization
by Solutes: A Critical Apparaisal", Metals Forum, 4 (1 and 2) (1981)
pp. 92-101.

(17) S. Sekino, N. Mori, and S. Tamukai, "Behaviours of Recrystallization
and Transformation observed by Hot-Rolling of Nb-Containing Steels",
Tetsu-To-Hagane, 58 (8) (1972) pp. 1044-1053.

(18) H. Sekine, and T. Maruyama, "Effects of Refinement of Recrystallized
Austenite grains in Hot Rolling of Nb-Containing Steels", Tetsu-To
Hagane, 60 (11) (1974). p. S558, (Preprint for Annual Meeting of Iron
and Steel Institute of Japan.)

(19) T. Gladman, and F.B. Pickering, "Grain Coarsening of Austenite",
Journal of Iron and Steel Institute, 205 (6) (1967) pp. 653-664.

(20) H. Takeuchi, T. Suzuki, T. Hashimoto, and T. Yokoi, "New Technology
for High Toughness Plates (Sumitomo High Toughness Process - SHT
Process), Proceedings of International Conference on Steel Rolling,
pp. 957-969; Iron and Steel Institute of Japan, Tokyo, 1980.

(21) S. Kanazawa, A. Nakashima, K. Okamoto, and K. Kanaya, "Improvement of
Weld Fusion Zone Toughness by Fine TiN", Transactions of Iron Steel
Institute of Japan, 16 (9) (1976) pp. 486-495.

(22) M. Nagumo, H. Morikawa, N. Okumura, Y. Kawashima, Y. Sogo, and
O. Mantani, "Rolling Parameters and Their Significance on the Combi-
nation of Strength and Toughness in HSLA Steels Rolled at Austenite-
Ferrite Two Phase Regions", Proceeding of International Conference on
Steel Rolling, pp. 1075-1084; See ref. (20).

(23) G. Tither, and M. Lavite, "Beneficial Stress-Strain Behavior of Moly-
Columbium Steel Line Pipe", Journal of Metals, 27 (9) (1965)
pp. 15-23.

(24) H. Gondoh, H. Nakasugi, H. Matsuda, H. Tamehiro, and H. Chino,
"High Grade Line Pipe - New Acicular Ferrite Steel", Seitetsu Kenkyu,
No.297 (1979) pp. 49-58.

(25) H. Nakasugi, H. Matsuda, and H. Tamehiro, "Development of Controlled-
Rolled Bainitic Steel for Large-Diameter Line Pipe", Proceedings of
International Conference on Steel Rolling, pp. 1028-1039; See ref.(20)

(26) T. Okita, C. Ouchi, and I. Kozasu, "The Effects of Accelerated Cooling
after Controlled Rolling", Tetsu-To-Hagane, 63 (11) (1977) p. S798,
(Preprint for Annual Meeting of Iron and Steel Institute of Japan.)

(27) K. Matsumoto, T. Okita, and C. Ouchi, "The Effects of Accelerated
Cooling after Controlled Rolling", Tetsu-To-Hagane, 65 (9) (1979)
pp. A181-A184, (Preprint for Symposium in Annual Meeting of Iron and
Steel Institute of Japan.)

(28) K. Tsukada, Y. Yamazaki, K. Matsumoto, T. Nagamine, K. Hirabe, and
K. Arikata, "Application of On-Line Accelerated Cooling (OLAC) to
Steel Plate", Tetsu-To-Hagane, 67 (4) (1981) p.S340 (Preprint for
Annual Meeting of Iron and Steel Institute of Japan.)

(29) S. Tamukai, Y. Onoe, H. Nakajima, M. Umeno, T. Iwanaga, and S. Sasaji, "Production of Extremely Low Carbon Equivalent HT-50", _Tetsu-To-Hagane_, 67 (13) P.S 1334, (Preprint for Annual Meeting of Iron and Steel Institute of Japan.)

(30) T. Hasegawa, H. Morikawa, T. Fujii, H. Sekine, Y. Onoe, "Strengthening Mechanism of Utra Low C_{eq} HT-50", _Tetsu-to-Hagane_, 67 (13) P.S1335, (Preprint for Annual Meeting of Iron and Steel Institute of Japan.)

DISCUSSION

Q: Referring to Figure 15, I'm not sure I completely understood your finishing conditions in determining your shelf energy values. Did you change the amount of finishing reduction that went in for each corresponding austenite grain size? If so, did you affect the amount of splitting that may have been occuring in your Charpy samples at the same time? The splitting might explain the change in shelf energy, for it's well established that as you increase the number of splits in your Charpy specimens, the shelf energy will decrease. I wonder if the effect you are showing is one due to grain size or to splitting.

A: Refer to Figure 14. This is the amount of the total reduction given below 900°C and the grain size that you get just before these reductions. In this case, the ASTM No 8 austenite grain size contour is as shown in the bottom left. The transition temperature contour is very similar to this.

Q: Is that an effect of austenite grain size or the low temperature reduction in the two phase austenite-ferrite region?

A: All rolling was done in the gamma phase.

Q: Did you notice splitting increasing in your Charpy samples with increased reduction in the austenite range?

A: Yes, in this case, the shelf energy is sharply reduced when the total amount of reduction below 900°C is more than 70 percent, as shown in Figure 15. In the same Figure we see that if the controlled grain size of 4, 5, 6 and 7 just before finishing, the transition temperature is reduced, but the accompanying reduction in shelf energy is not so large. However, if this is followed by a total reduction greater than 65 percent at 900°C in the finishing train then the shelf energy will be reduced more rapidly.

THE INFLUENCE OF PROCESSING ROUTE AND NITROGEN CONTENT ON MICRO-

STRUCTURE DEVELOPMENT AND PRECIPITATION HARDENING IN VANADIUM-

MICROALLOYED HSLA-STEELS

Tadeusz Siwecki, Alf Sandberg, William Roberts and
Rune Lagneborg,

Swedish Institute for Metals Research,
Drottning Kristinas väg 48,
114 28 STOCKHOLM, Sweden

Four 0.09V steels were subjected to laboratory simulations of
normalizing, normal rolling, and controlled rolling, with various
cooling rates imposed during the $\gamma \rightarrow \alpha$ transformation. Three of the
steels were deoxidized with Al and had different nitrogen contents.
It is concluded that nitrogen refines the grain structure and increases precipitation strengthening ($\Delta\sigma_y(p)$). Furthermore, V-
microalloyed steels can be control-rolled to achieve an attractive
combination of properties, equivalent or superior to that of Nb-
steels. This can be done to a lesser degree rolling normally. An
increased cooling rate enhanced both grain refinement and $\Delta\sigma_y(p)$.
This double enhancement makes acceleration cooling especially
attractive for the processing of V-microalloyed steels because of
the need to compensate, via grain refinement, for the deleterious
effect that precipitation strengthening has on toughness. Small
amounts of Al had a positive influence on both grain refinement and
the yield strength of hot-rolled material.

PREAMBLE

In a recent paper (1), a thermodynamic analysis of the Fe-V-C-N
system is presented whereby the composition of VC_xN_y ($x+y=1$)
precipitates might be evaluated as a function of temperature, the
amounts of V,C and N remaining in solution, type of matrix (austenite,
ferrite) and precipitation mode (random, interphase). For microalloyed steels containing less than 0,20 wt.%C and normal nitrogen
contents, the computations predict that most of the vanadium consumed

during random decomposition of austenite or ferrite appears as particles having a composition which is quite near VN. For interphase precipitation, the situation cannot be predicted with the same certainty; however, the likelihood is that even in this case, the average composition of the particles is close to VN. These predictions are confirmed semiquantitatively by experiments in which the composition of chemically-isolated V (C,N) particles has been estimated via X-ray lattice-parameter determination.

The results obtained in the aforementioned investigation highlight the importance of the V-N interaction in V-microalloyed steels. In addition, other workers have demonstrated that the precipitation-strengthening contribution to the yield stress $(\Delta\sigma_y(p))$, at a fixed vanadium level, increases dramatically as the nitrogen content of the steel is raised (2). This suggests that even though VC is stable in ferrite, the greater driving force for precipitation of a nitrogen-rich carbonitride affects the kinetics of precipitation such that the rate of decomposition decreases markedly as the quantity of available nitrogen is reduced. In other words, a greater particle volume fraction is formed in high-N material even though equilibrium considerations forecast that vanadium not consumed as VN should combine with carbon. A corollary from this reasoning is that the N-level in a V-microalloyed steel, which derives its strength from precipitation reactions proceeding during continuous cooling, should be relatively high if the costly microaddition is to be utilized effectively.

The work described here was initiated with the aim of elucidating to what extent the level of nitrogen influences the development of microstructure and the magnitude of the yield strength contribution from precipitation strengthening in V-microalloyed steels. Special attention will be paid to:

(i)   the influence of nitrogen content and processing route (normalizing, normal rolling, controlled rolling) on the development of the austenite microstructure, and on certain features of the $\gamma \rightarrow \alpha$ transformation; and

(ii) the particle-hardening contribution to the yield strength and its relation to the V:N ratio of the steel, the conditions of processing and the dispersion parameters for V(C,N) precipitates.

EXPERIMENTAL DETAILS

Steels examined

The investigation has been limited to four vanadium-microalloyed steels originating from Svenskt Stål AB (Swedish Steel Corp.), Oxelösund (50 kg laboratory melts).The compositions are indicated below (wt.%):

Steel	C	Mn	Si	P	S	V	Al	N
A	0,12	1,35	0,35	0,002	0,008	0,094	0,023	0,006
B	0,12	1,34	0,30	0,003	0,009	0,091	0,018	0,013
C	0,12	1,28	0,28	0,003	0,008	0,090	0,021	0,019
D	0,12	1,27	0,38	0,003	0,009	0,087	-	0,012

The steels were received as 40 mm ingot-rolled slabs from which cylindrical compression specimens were machined ($H_o x \emptyset_o$ = 15x10 mm). These were used for simulations of normal and controlled rolling. Specimens for normalizing experiments had dimensions 10x10x15 mm. All materials were given a prior austenitization treatment comprising 30 min. 1250°C followed by water quenching.

Thermomechanical treatments

The hot-compression equipment is built up around a 100 kN MTS servo-hydraulic testing machine. A detailed description of the apparatus has been given in earlier publications, e.g. (3). Quenching of a deformed specimen can, if required, be effected by rapidly withdrawing the lower tool into a water spray located beneath the furnace.

In the rolling simulations, a specimen was first austenitized 3 min. 1150°C (which is sufficient to dissolve the fine particles of V(C,N) which might have precipitated during reheating after the prior austenitization treatment). For normal-rolling (NR) simulation, a simplified deformation scheme was adopted involving four individual steps at 1100, 1050, 1000 and 950°C; the corresponding reductions were 25%, and 3x22%. Cooling during the deformation phase of the thermomechanical schedule and under subsequent $\gamma \rightarrow \alpha$ transformation was such as to simulate air cooling of 40 mm, 12 mm plate and 10 mm $\emptyset$ bar, the average rate of temperature fall between 750 and 550°C being 0,24, 0,88 and 5°C.s^{-1}, respectively. The following schedule was used for simulation of controlled rolling (CR): 1100°C/25%, 900°C/22%, 850°C/22% and 800°C/22%. The cooling scheme adopted was the same as for NR. For cooling at the fastest rate, it was necessary to simplify the schedules for both NR and CR even further; this is due to the difficulty of performing a large number of deformation steps in a controlled way when cooling is fast. For NR, we used 1100°C/25%, 1025°C/32% and 950°C/32%; corresponding data for CR are 1100°C/25% and 850°C/52%. The overall reduction in all experiments is identical (65%; ε=1,05). The thermomechanical cycles for the various rolling simulations are illustrated in Fig.1. The strain rate during compression was maintained constant at 2s^{-1} which corresponds to the average value used in plate rolling.

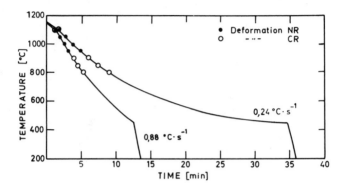

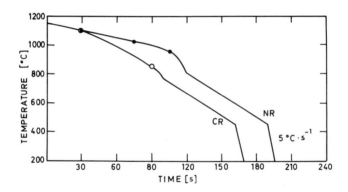

Fig.1: Thermomechanical treatments adopted for simulation of controlled rolling and normal rolling.

Normalizing (N) was carried out for 15 min. 950°C with subsequent cooling such that the average rate of fall of temperature between 750°C and 550°C was the same as that adopted in the rolling simulations (0,24, 0,88 and 5°C.s^{-1}). The as-normalized austenite microstructure was examined using specimens water quenched after heat treatment.

RESULTS FROM EXPERIMENTAL WORK

Austenite microstructure prior to transformation

The austenite linear intercept grain sizes observed in the various steels, after preliminary heating for 3 min. 1150°C prior to thermomechanical treatment, are given below:

Steel : W_N	$D_\gamma(\mu m)$
A : 0,006	144 ± 9
B : 0,013	115 ± 27
C : 0,019	116 ± 18
D : 0,012(no Al)	158 ± 21

The austenite microstructure, existing in material processed by hot-rolling simulation prior to γ→α, was examined via quenching fully-deformed specimens from 800°C. Austenite grain sizes in all four steels after NR and CR are presented as a function of cooling rate in Fig. 2. For processing by both NR and CR, D_γ(800°C) decreases with increasing cooling rate and to a lesser extent with increasing nitrogen level; the grain size in Al-free material is significantly coarser than in a corresponding steel containing Al. In general, D_γ is rather small, even (surprisingly) for NR. Faster cooling is obviously beneficial, the greatest effect accruing between 0,24 and ~1,5°C.s^{-1}. Optical microscopy revealed elongated γ-grains in specimens subjected to CR, thus proving that no recrystallization occurred after the final passes of CR. However, since in the latter case, the size of the equivalent equiaxed grains is less than with NR, one is forced to the conclusion that some further grain refinement, in addition to that from the predeformation, is effected via recrystallization after the initial (900°C) finishing step; recrystallization is then inhibited following deformations at lower temperatures.

Austenite grain sizes in normalized specimens quenched from 800°C, after cooling from the normalizing temperature at a rate corresponding to 12 mm plate in air, are tabulated below.

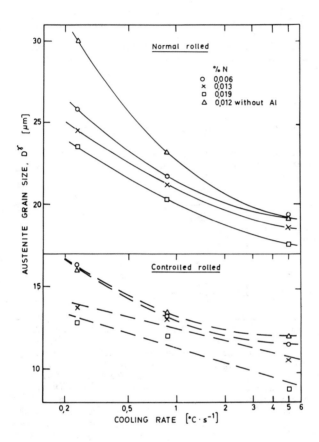

Fig.2: Mean austenite grain size as a function of cooling rate for all four steels after processing by NR or CR (specimens quenched from 800°C).

Steel : W_N	D_γ (μm)
A : 0,006	14,1
B : 0,013	13,9
C : 0,019	13,8
D : 0,012 (no Al)	16,1

Again, increasing nitrogen level has a small beneficial effect. As might be expected, AlN makes an important contribution to grain-growth inhibition during normalizing and the particle-limited grain size is coarser in an Al-free steel than in corresponding Al-bearing material.

Ferrite grain size

Complete information regarding the relationships between ferrite grain size, nitrogen level, processing route and cooling rate is summarized in Fig. 3. In general, the trends observed regarding the variation of austenite grain size with the aforementioned parameters are maintained, i.e. increasing the cooling rate imparts the greatest benefit and there is a small, but significant, effect from increased W_N. The ferrite grain sizes in steels processed by normalizing and controlled rolling are similar; the grains in normal-rolled material are somewhat coarser but still acceptably small, especially for 5°C.s^{-1} cooling.

The relationship between the grain size of austenite and that of the polygonal ferrite to which it transforms is given in Fig. 4. The data fall into three distinct bands for different cooling rates, which is to be expected. It is not possible to resolve any further trends within these groupings of points. It would appear that the ferrite grain size is determined primarily by austenite grain size and cooling rate. In other words, the various processing procedures and steel chemistries affect D_α only insofar as they influence D_γ; there is apparently no indirect influence on the nucleation frequency of ferrite. This conclusion is substantiated further by Fig. 5(a) in which D_α is plotted as a function of the austenite grain-boundary area per unit volume (S_V), for cooling at $0,88^\circ$C.s^{-1}. The data fall accurately onto a single curve which is very close to that reported by other workers (4,5) who have studied the transformation from recrystallized austenite in Nb- and V-microalloyed steels. These latter investigators also found that, for a given S_V, deformed grains of Nb-microalloyed austenite engender finer ferrite than do equiaxed ones. For V-steels, the present work indicates that elongated austenite grains (CR) transform to ferrite with exactly the same grain size as that produced from equiaxed grains (NR,N) of equivalent size (same S_V). This interesting deviation in behaviour between Nb- and V-variants will be discussed further below.

Another important point, in the context of ferrite grain refinement in V-N microalloyed steels, is illustrated by Fig. 5(b), which shows a D_α-S_V plot for cooling at 5°C.s^{-1}. It is clear that excellent ferrite grain refinement can be achieved, even from recrystallized austenite, provided that S_V is sufficiently high (>100mm^{-1}) and accelerated cooling is applied. The D_α which is attainable is quite comparable with that engendered in controlled-rolled Nb-steels following air cooling (12 mm plate).

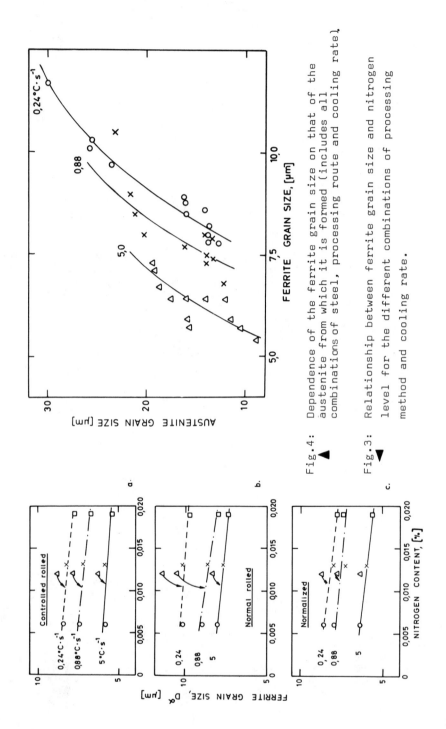

Fig.4: Dependence of the ferrite grain size on that of the austenite from which it is formed (includes all combinations of steel, processing route and cooling rate).

Fig.3: Relationship between ferrite grain size and nitrogen level for the different combinations of processing method and cooling rate.

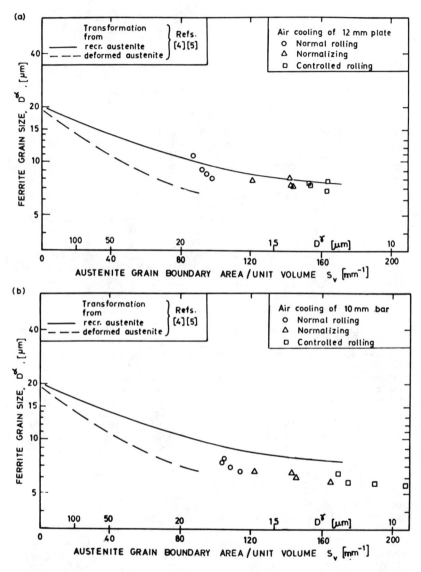

Fig.5: Dependence of ferrite grain size on the total area of austenite grain boundaries (plus, for deformed γ, deformation bands) per unit volume. Data points are from the present investigation; the curves, which refer to transformation from equiaxed, recrystallized γ or deformed, elongated γ, are taken from refs.(4) and (5) and are included for comparison.

Influence of chemical and process parameters on yield strength

In Fig.6 the measured values for the lower yield stress are plotted as a function of nitrogen content for the various thermal/thermomechanical treatments and cooling rates. The error bars indicate the spread over four separate determinations for each combination of steel, processing method and cooling rate. Points to note are:

(i) For a given W_N and cooling rate, the yield stress of normalized material is considerably less than that for specimens which have been hot deformed. In addition the dependence of σ_y on nitrogen level and cooling rate is different for, on the one hand, hot-rolled and, on the other, normalized steels.

(ii) All other things being equal, there is little difference between NR and CR as regards the level of yield stress. For these latter processing methods, the yield strength appears to be a linear function of W_N (fixed cooling conditions) or the logarithm of cooling rate (fixed W_N). Conversely, in normalized material, σ_y evidently saturates at high nitrogen levels; the increase in the former with cooling rate is also non-linear.

(iii) The Al-free steel exhibits consistently lower yield strength than the corresponding Al-treated variant, especially after processing by NR and CR.

We have already indicated that, for a given processing procedure, the ferrite grain size in the steels examined becomes smaller on increasing both nitrogen content and cooling rate. However, this alone cannot explain the rather sensitive dependence of σ_y on these parameters; neither, can grain size effects account for the much lower yield strength after normalizing. Clearly, the variations found in σ_y must originate principally from the influence of nitrogen level and process parameters on the precipitation-hardening component of the yield strength. We have evaluated $\Delta\sigma_y(p)$ from the yield strength values, making use of the regression formula for C-Mn steel ($W_N < 0,02$) recommended by Gladman et al.(6):

$$\Delta\sigma_y(p) = \sigma_y - \left[K + 37\,W_{Mn} + 83\,W_{Si} + 2920\,W_N(\text{free}) + 15,1.D_\alpha^{-1/2} \right]$$

where $K = 88\text{MPa}$ and D_α is in mm; in our evaluations, we assume that the amount of free nitrogen is negligible. The values for $\Delta\sigma_y(p)$ derived using the above regression equation are reviewed in Fig.7. Again, for NR and CR, the dependence on W_N and the logarithm of cooling rate is, all other things being equal, virtually linear over the ranges investigated; in normalized material on the other hand, the contribution from precipitation hardening, which is much lower than that after NR or CR, saturates at high nitrogen levels. These

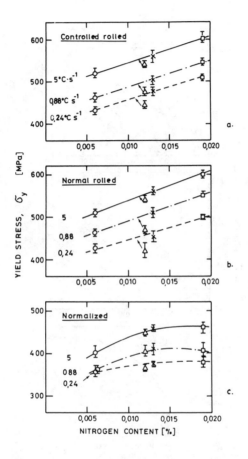

Fig.6:   Dependence of lower yield stress on nitrogen level for
         the various processing methods and cooling rates.

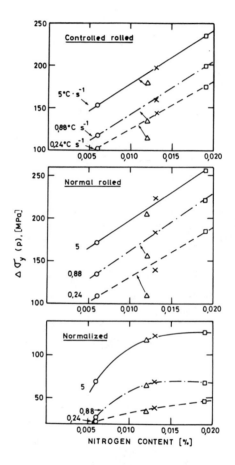

Fig.7: Precipitation strengthening contribution to lower yield
stress ($\Delta\sigma_y$(p)) plotted against nitrogen content for the
various processing methods and cooling rates.

differences, deriving from the various methods of thermal/thermomechanical treatment, are illustrated more cogently in Fig. 8. It is interesting to note that for a given W_N and cooling rate, NR engenders a $\Delta\sigma_y(p)$ which is about 20 MPa greater than CR (the yield stresses are about equivalent because of the finer grain size in controlled-rolled material). Observe also that after processing by both NR and CR, the Al-free steel is characterized by a significantly lower precipitation strengthening than corresponding Al-killed material. After normalizing, on the other hand, the precipitation strengthening is the same independent of whether Al is present or not.

Measurements of particle-size distributions

It is very difficult, on the basis of a qualitative electron microscope examination, to pin down any great differences between the various steels and thermal/thermomechanical procedures which have been studied. Various types of precipitate dispersions in specimens which have undergone hot-rolling simulations are shown in Fig. 9, interphase precipitates (Fig. 9 a,d), fibres (b,e) and random particles (c,f). It is clear that both hot-rolling routes engender, in a given steel, V(C,N) particles with essentially similar dispersion and size. The volume fraction of V(C,N) is, however, clearly diminished as the nitrogen level of the steel is reduced (Fig. 10).

In order to specify more accurately the influence of the various compositional and process parameters on the characteristics of the V(C,N) dispersion, we decided to establish particle-size distributions in a limited number of cases. Some quantitative deductions are possible from thin foils, e.g. inter-row spacings for interphase precipitates, but in general the coherence contrast hampers accurate size measurement. Size distributions were thus measured using extraction replica techniques.

The principal factor influencing the size distribution of precipitates is cooling rate. The greater proportion of smaller particles and the reduced average size as the cooling rate is increased, is clearly discernible in Fig. 11a. For a fixed cooling rate and processing scheme, increasing the nitrogen level in the steel engenders a small increase in average particle size while the form of the size distribution remains essentially unaltered (Fig 11b). The distributions in steels with and without Al-additions are also very similar. Various processing procedures are compared in Fig. 11c; for a given steel and cooling rate, CR and NR result in very similar distributions while for normalized specimens, the average particle size is significantly higher. The distributions of particle sizes was found to be log normal in all cases examined, i.e. a cumulative plot of the number of particles per unit area with size less than d against log d gives a straight line on normal probability paper (see Fig. 12). It will be noted that the particle-size distributions shown in Fig. 11

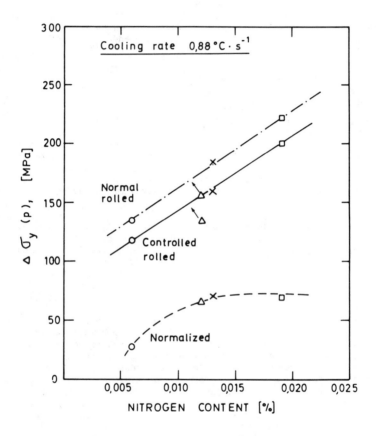

Fig.8: Illustrating the difference between processing by
normalizing and hot rolling as regards the magnitude of
$\Delta\sigma_y(p)$ and its dependence on nitrogen level (fixed
cooling rate).

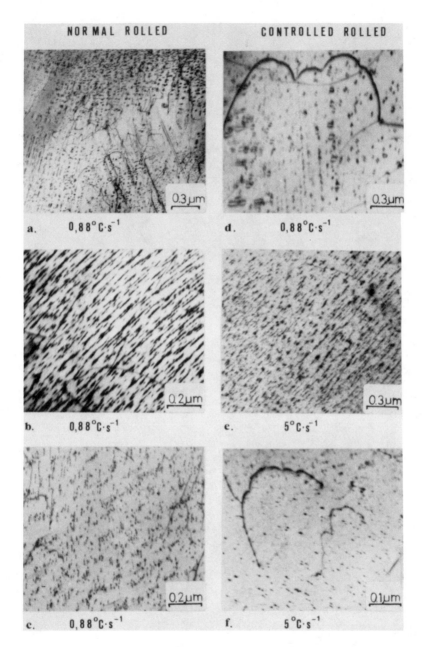

Fig. 9:

Precipitates in thin foils (TEM) of steels which have undergone
simulations of normal and controlled rolling. Note sharply-
curved dislocation segments in (d) and (f).

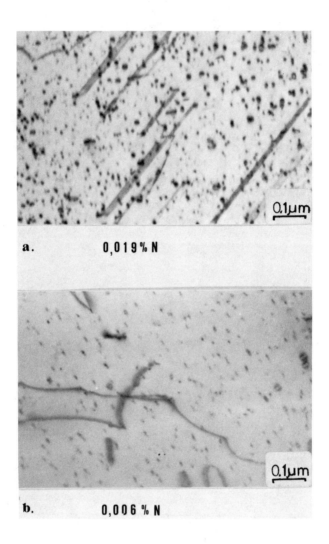

a.   0,019 % N

b.   0,006 % N

Fig.10:

Thin-foil electron micrographs illustrating the differences in
density and size of V(C,N) precipitates in identically-treated
(CR; $0,88^{\circ}$C.s^{-1}) specimens of high-and low-nitrogen steels.
The foil thickness is about the same in the areas chosen.

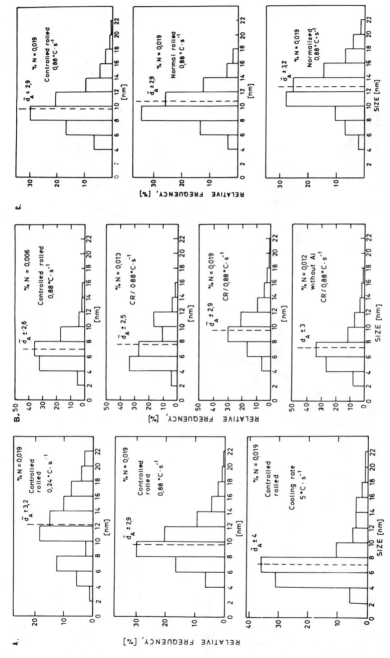

Fig.11: Measured V(C,N) particle-size distributions (d<22 nm) from extraction replicas for various combinations of steel, processing route and cooling rate. a) illustrates the effect of cooling rate, b) that of nitrogen, and c) of processing route.

179

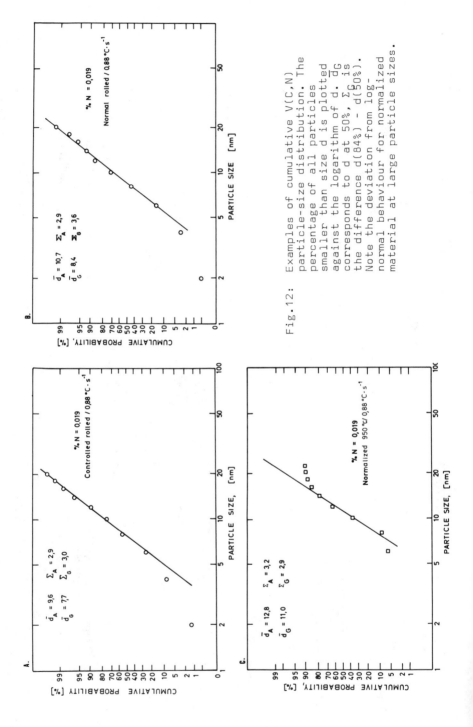

Fig. 12: Examples of cumulative V(C,N) particle-size distribution. The percentage of all particles smaller than size d is plotted against the logarithm of d. $\bar{d}_G$ corresponds to d at 50%, Σ is the difference d(84%) − d(50%). Note the deviation from log-normal behaviour for normalized material at large particle sizes.

refer to d < 20 nm. For NR- and CR-material, the range 0<d<20 nm includes over 99% of all particles, whereas in normalized steel 10% of the total number of particles have sizes >20 nm. This clear difference in distribution can readily be appreciated from the cumulative probability plots (Fig. 12); in the case of normalizing, the distribution is only log normal for d<20 nm. These cumulative probability distributions will be discussed further below in connection with the theoretical evaluation of the precipitation hardening contribution to the yield stress.

Fig.13 shows the decrease in the row spacing of interphase precipitates as the cooling rate is increased. Cooling rate exerts the strongest influence on this parameter, as it does with precipitate size. The effects from variations in W_N or processing route are secondary in comparison.

DISCUSSION

Evolution of grain structure during
processing of V-N microalloyed steels

After processing by CR and NR, the austenite grain size immediately prior to transformation (quenching from 800°C) decreases markedly with increasing cooling rate and, less spectacularly, with raised N-level (Fig.2). Both these effects are likely to stem principally from grain-growth inhibition, although the influence of nitrogen content on the recrystallized grain size, for a given temperature and strain, is an unknown factor. The presence of Al seems to be important in that corresponding Al-free materials are characterized by coarser D_γ after both NR and CR. Possible rationales for the inhibition of grain growth with increasing N-content are:

(i)   strain-induced precipitation of AlN, which can proceed rapidly even at 950°C according to Vodopivec (7);

(ii)   strain-induced precipitation of VN; or

(iii) a decrease in austenite grain-boundary mobility as the level of dissolved nitrogen increases.

Of these, only (i) is consistent with the coarser grains in the Al-free steel at all cooling rates. The positive influence of Al in reducing the austenite grain size is also evident from the D_γ-values after normalizing (see "Austenite microstructure prior to transformation"); in this case, however, increasing W_N has little influence on D_γ.

Clearly, Al plays an important role in V-N steels. It would appear that when W_N is high, strain-induced precipitation of AlN,

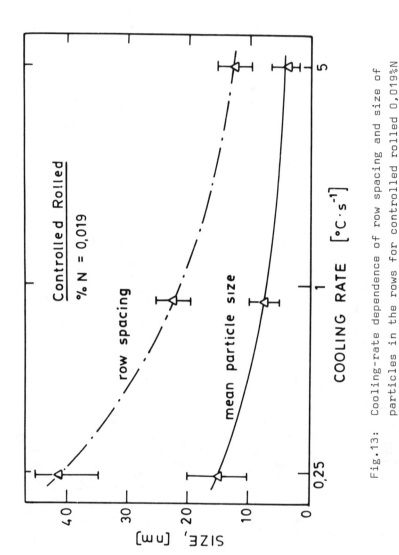

Fig.13: Cooling-rate dependence of row spacing and size of particles in the rows for controlled rolled 0,019%N steel.

in association with both NR and CR, effectively hinders austenite grain growth prior to transformation. Furthermore, if $W_{Al}<0,02$, the amount of nitrogen consumed as AlN, and which thereby cannot contribute to $\Delta\sigma_\gamma(p)$, is evidently rather small. The results of other workers (2) indicate that, for steels with 0,1V and 0,015N, the Al level needs to be in excess of 0,03 wt% before any serious reduction in the precipitation strengthening derived from V can be detected.

The reduced D_γ as cooling rate and W_N increase, results in concomitant grain refinement of ferrite. Faster cooling has, of course, the additional effect that the nucleation frequency of ferrite is increased for a specified D_γ. For NR and CR, the reduced D_γ with increased W_N, at fixed cooling rate, is wholly derived from the refinement of austenite grain size (see Fig.4).

It is interesting to note that despite the obvious grain elongation engendered as a consequence of controlled rolling, the relation between the specific area of austenite grain boundaries available for ferrite nucleation (S_V) and the ferrite grain size is the same independent of processing method (Fig.5). This is at variance with observations on Nb-steels which clearly show that for a given S_V, elongated austenite grains transform to a finer ferrite than do equiaxed ones (4,5). An explanation, which has been proposed for the results pertaining to the transformation of Nb-austenite, is that the grain surfaces of deformed austenite are rather irregular, as a result of local strain-induced grain-boundary migration, and that such irregularities (bulges) function as effective nucleation sites for ferrite (5). Depending on the local radius of curvature, grain-boundary bulges are at least as potent nucleation sites as grain edges, but rather less efficient than grain corners. Thus, for trans-formation from relatively large elongated grains, in Nb-steels for example (no recrystallization during finishing), the characteristic site density associated with bulges, which cover the entire grain surfaces, is greater than that of the sites of lower 'dimensionality', corners or edges. In V-steels, however, the grain size of controlled-rolled steels is rather small. This is because the recrystallization inhibition derived from V is weaker than from Nb. Hence, V-austenite recrystallizes during the first finishing passes giving fine grains, which are subsequently elongated in the final passes. The fine grain size of the deformed austenite in V-steels implies that grain edges and corners have high specific site densities/unit volume (proportional to $1/D^2$ and $1/D^3$ respectively) and can effectively compete with bulges on the grain surfaces as nucleation sites for ferrite; in other words, when the grain size is fine, the same nucleation sites tend to dominate for both equiaxed and elongated grains, and the relation between D_α and S_V is expected to be more or less the same, independent of grain shape. In practice, some minor differences will arise as a result of the geometrical limitations on the growth of ferrite imposed by the different grain shapes, but these are likely to be insignificant unless the deformed grains are very highly elongated. The full and dashed curves in Fig. 5 should therefore come together at high S_V's, unless the degree of deformation, characterizing the development of elongated grains during finishing, is very high. For $S_V \lesssim 100$ mm^{-1}, on the other hand, deformed austenite grains will always give finer ferrite than equiaxed ones with an equivalent specific ferrite nucleation area.

Influence of processing parameters and
nitrogen level on $\Delta\sigma_y(p)$

The principal findings on precipitation strengthening presented above are:

(i)   the precipitation hardening contribution to the lower yield stress of V-N microalloyed steels increases strongly with increasing nitrogen level and accelerated rate of cooling;

(ii)   for a given W_N and cooling rate, the $\Delta\sigma_y(p)$ induced by normalizing is much lower than that derived from NR or CR; and

(iii) for W_N=0.012-0.013, the absence of Al engenders a reduction in $\Delta\sigma_y(p)$ by between 0 and 30 MPa, depending on processing conditions.

Thermodynamic evaluations in a companion report (1) indicate that, for decomposition of microalloyed ferrite, the available V should form a high-nitrogen carbonitride unless the initial N-level is low ($W_N \lesssim 0,005$). However, when all nitrogen is consumed the remaining vanadium should theoretically combine with carbon. On this basis, we would expect $\Delta\sigma_y(p)$ to be independent of W_N, which is clearly not the case! One can conceive two possible explanations for this behaviour; either the lower thermodynamic stability of VC, relative to VN, is reflected in strongly decelerated precipitation kinetics for the former, or VC forms but coarsens more rapidly than VN (see (8), for example). The present observations pertaining to the effect of nitrogen level for a fixed processing procedure, on the size and distribution of V(C,N)(Figs. 11 and 12) indicate that the particle size tends, if anything, to be smaller as the nitrogen level is reduced. The absence of coarser, i.e. VC, particles in low-N steels lends support to the argument based on slower precipitation kinetics for VC compared with VN. Hence, the practical implication of this would appear to be that, unless sufficient time is available for VC formation, as during very slow cooling or during tempering, there is little point in having more vanadium than can combine with the available nitrogen. However, it should be pointed out that this argument will only be valid for steels with typical microalloy levels of V. If the V content is large enough there will be sufficient   left, after the completion of VN precipitation, to provide a large supersaturation with respect to VC, and thus make VC-precipitation possible. Hence, the present results and arguments do not contradict the observations that increasing V content beyond 0,1%, for a given N-level, augments the precipitation strengthening (9).

A beneficial effect from fast cooling via increased $\Delta\sigma_y(p)$ in micro-alloyed V-N steels is reported in several investigations (2,3,10); essentially similar behaviour has been found in this work (Fig. 7). The primary effect of increased cooling rate is a lowering of the temperature range over which nucleation of V(C,N) occurs. This is imposed in part directly by the increasing cooling rate and in part by the reduction in the Ar_3-temperature associated with faster cooling. This provides an increasing supersaturation with respect to V(C,N) and will therefore result in more prolific nucleation and smaller interparticle spacing. At the same time there is evidence that for the range of cooling rates investigated here, sufficient time is available for a complete precipitation of V(C,N), i.e. corresponding to the total N in the steel. This is supported by the calculations of $\Delta\sigma_y(p)$ reported below, and also by the fact that no maximum of $\Delta\sigma_y(p)$ was reached with increasing cooling rate (Fig. 7). The decreasing particle spacing will therefore necessarily be associated with a reduced particle size. In addition, the time available for particle coarsening will be limited progressively as the cooling rate is increased. Thus the observed reductions of the mean precipitate size with increasing cooling rate is due to the enhanced nucleation frequency and the virtual absence of particle coarsening. This results, in turn, in a gradual reduction of the precipitate spacing (and size) and explains the observed increase of $\Delta\sigma_y(p)$ with increasing cooling rate. Fig. 14 represents a striking illustration of the strong influence of the Ar_3-temperature on the spacing and size of V(C,N) particles discussed above. Interphase precipitation in steel B (W_N=0,013) from the present investigation and a 0,45%C V-micro-alloyed steel are compared; both materials were cooled at the same rate after processing ($\sim 1^{\circ}C.s^{-1}$). The high-carbon material with much lower Ar_3 (~ 680 as opposed to $\sim 780^{\circ}C$) is characterized by appreciably smaller row spacing and finer particle size.

Computation of $\Delta\sigma_y(p)$ characterizing V-N microalloyed steels

In view of the very high strength of vanadium carbonitrides and the circumstance that the slip planes of the ferrite and the precipitate are not likely to coincide, it seems reasonable to assume that the V(C,N)-particles are impenetrable for slip and that the precipitation strengthening therefore derives from the Orowan mechanism (dispersion hardening).

We have computed $\Delta\sigma_y(p)$ according to a recent treatment of the Orowan mechanism by Melander (11). The theory is based on a statistical model for the motion of a dislocation through a dispersion of obstacles and takes into account both the distribution of particle sizes and inter-particle spacings. The suggested relationship is:

$$\Delta\sigma_y(p) = 2\Delta\tau_y(p) =$$

$$\frac{3,56\ Gb}{4\pi l_s}\ \left(\frac{1+\nu-3\nu/2}{1-\nu}\right)\ \ln(l_s/b)\ \left\{\ln\left[\frac{2\bar{d}_G l_s}{(\bar{d}_G+l_s)b}\right]\ /\ln(l_s/b)\right\}^{3/2}$$

185

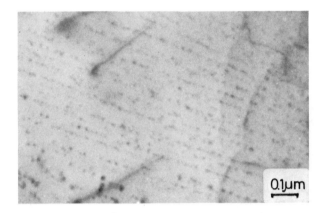

a.

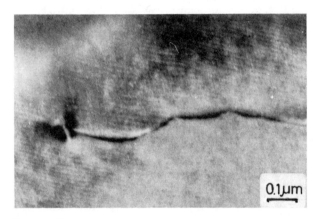

b.

Fig.14:

TEM micrographs illustrating the striking effect of
increased carbon level (lower Ar$_3$) on the size of the
particles in and the spacing of interphase rows. (a) Steel
from this work with 0,013%N. (b) Steel with 0,45%C, 0,10%V,
0,011%N.

where G is the shear modulus of ferrite ($8,03.10^4$ MPa), ν is Poisson's ratio (~0,3 for ferrite), b = 0,250 nm is the dislocation Burgers vector, l_s the average distance between obstacle centres in the glide plane and d_G is the <u>geometric</u> mean-particle diameter (spheres) as evaluated from an extraction replica. The above expression is derived on the assumption that the particle-size distribution is log normal, which seems to be the case for the V-N steels examined in this work (Fig.12). l_s is obtained via the following relationship for the mean nearest-neighbour distance between randomly-dispersed spherical particles (12)

$$l_s = 0.5 \left[\frac{\pi}{6 f_V} \{ (\bar{d}_A)^2 + (\Sigma_A)^2 \} \right]^{\frac{1}{2}}$$

where f_V is the volume fraction and $\bar{d}_A, \Sigma_A$ the <u>arithmetic</u> mean particle diameter and corresponding standard deviation.

In order to be able to compute l_s, and hence $\Delta\sigma_y(p)$, from the observed V(C,N) particle-size distributions, it is necessary to make some assumptions regarding the volume fraction of precipitates. We have taken the simple view that the volume fraction of V(C,N) is determined solely by the nitrogen content of the steel (W_N^S). Hence, f_V is given trivially by

$$f_V = \frac{\rho_s \cdot W_N^S \cdot M_{V_{VN}}}{100 A_N}$$

where ρ_s is the density of steel, $M_{V_{VN}}$ is the molar volume of VN and A_N the atomic weight of nitrogen. The above equation applies specifically to hot-rolled steel, i.e. all nitrogen is considered to remain in solution at the initiation of $\gamma \rightarrow \alpha$. In the case of normalization, the nitrogen contributing to $\Delta\sigma_y(p)$ is that which is dissolved at the normalizing temperature; this quantity can be evaluated using the methods described in ref.(1). For the three Al-steels investigated in this work with total N-contents of 0,006, 0,013 and 0,019, the amounts of dissolved N after normalizing at 950^0C are calculated as 0,0037, 0,0063 and 0,0072, respectively. f_V can in this instance be evaluated from the above expression using W_N(dissolved) instead of W_N^S.

The values of $\Delta\sigma_y(p)$ computed on the basis of the above premises are listed in Table 1. Data for the experimentally-determined geometric and arithmetic mean particle sizes plus the corresponding standard deviations, together with the estimated values for f_V and l_s are also included; the final column presents the observed $\Delta\sigma_y(p)$.

Table 1. Comparison of experimental and theoretical $\Delta\sigma_y(p)$'s

Steel & processing	$\bar{d}_A$ (nm)	Σ_A (nm)	$\bar{d}_G$ (nm)	Σ_G (nm)	$f_v \cdot 10^3$ (estimated)	l_s (nm)	Computed $\Delta\sigma_y(p)$ (MPa)	Measured $\Delta\sigma_y(p)$ (MPa)
$W_N^S = 0,019$ CR $0,88^\circ$C.s^{-1}	9,6	2,9	7,7	3,0	1,135	107,8	212	200
$W_N^S = 0,019$ NR $0,88^\circ$C.s^{-1}	10,7	2,9	8,4	3,6	1,135	119,0	197	222
$W_N^S = 0,019$ N $0,88^\circ$C.s^{-1}	12,8	3,2	11,0	2,9	0,430	230,2	107	70
$W_V^S = 0,019$ CR $0,24^\circ$C.s^{-1}	12,2	3,2	9,2	6,4	1,135	135,5	177	175
$W_N^S = 0,019$ CR 5°C.s^{-1}	7,1	4,0	5,5	4,4	1,135	87,5	234	236
$W_N^S = 0,006$ CR $0,88^\circ$C.s^{-1}	7,0	2,6	4,5	3,0	0,358	142,8	129	118
$W_N^S = 0,013$ CR $0,88^\circ$C.s^{-1}	7,6	2,5	4,5	3,3	0,777	103,8	181	160
$W_N^S = 0,012$ (without Al) CR $0,88^\circ$C.s^{-1}	7,3	3,0	4,9	4,1	0,717	106,6	182	135

When it is considered that the V(C,N) precipitates are neither spherical nor randomly distributed(e.g. interphase rows), the agreement between the experimental and computed $\Delta\sigma_y(p)$'s is really rather good. This indicates strongly that the evaluation of f_v directly from W_N^S constitutes an acceptable approximation, i.e. precipitates formed during or subsequent to $\gamma \to \alpha$, and which make the overriding contribution to $\Delta\sigma_y(p)$, do indeed consist predominantly of VN. Hence, the argument presented above that VC does not form during continuous cooling of V-microalloyed steels, even though it might be expected on thermodynamic grounds, is supported by the theoretical evaluations of $\Delta\sigma_y(p)$. It is clear from the calculations that the increases in dispersion strengthening engendered in hot-rolled material, on the one hand, via higher nitrogen and, on the other, via faster cooling, have quite separate origins. Increased cooling rate refines the particle size, while a raised nitrogen level evidently results in larger precipitates and the increase in $\Delta\sigma_y(p)$ is derived from the greater volume fraction of V(C,N). The linear dependence of $\Delta\sigma_y(p)$ on W_N which is found experimentally is probably only valid over the specific range of nitrogen contents examined $(0,006 < W_N < 0,019)$. What happens at lower nitrogen levels is not really known. It could be that the supersaturation of vanadium after the termination of VN precipitation is then high enough that VC forms at a significant rate during continuous cooling; if this is the case then one would anticipate a definite $\Delta\sigma_y(p)$ even when $W_N \to 0$.

The calculated $\Delta\sigma_y(p)$ for normalized steel ($W_N=0,019$; $0,88°C.s^{-1}$) is somewhat too high. This may be because the dissolution rate of V(C,N) at normalizing temperatures is rather slow and the equilibrium W_N (dissolved) is not attained in the short normalizing times (15 min.) used in the present experiments.

The reduced value of $\Delta\sigma_y(p)$ in Al-free steel can not be accounted for by the dispersion-hardening theory; the particle-size distributions in Al treated and Al-free variants are virtually identical and it is difficult to see why there should be any difference in f_v. Thus, small amounts of aluminium (~0,02%) seem to have an effect opposite to that one would expect in V-N steels, i.e. they actually enhance the precipitation strengthening derived from V(C,N). One possibility is that if only very limited amounts of AlN are formed in hot-rolled material (and we have not seen any except in normalized samples), then the Al-N interaction in solution should increase the activity of nitrogen and thereby encourage precipitation of nitrogen-rich V(C,N). An interesting corollary in this context is that since Al does not interact strongly with carbon, one can expect no corresponding increase in C-activity, i.e. the difference between the chemical potentials driving the precipitation of VN contra VC will be enhanced even further. This hypothesis is also in accord with the disappearance in normalized steels of the Al-effect on $\Delta\sigma_y(p)$ (Fig. 7 and 8); in this case, AlN can form in association with normalizing, Al is then no longer available in solution and the effect on the nitrogen activity is lost. Indeed, the formation of AlN during normalizing could well explain why the observed and calculated $\Delta\sigma_y(p)$'s do not concur especially well in this case (W_N(dissolved) and, hence, f_v, overestimated).

SUMMARY OF MAIN FINDINGS

Experiments on a series of V-N microalloyed constructional steels, processed via normalizing, normal rolling or controlled rolling and cooled at various rates, lead us to draw the following conclusions.

(i)   For processing by NR or CR, the austenite grain size prior to transformation decreases with increasing nitrogen level and accelerated cooling rate. In normalized material, D^γ is independent of nitrogen content. For all three processing methods, a steel containing Al (0,02%) is characterized by a finer grain size than corresponding Al-free material.

(ii) Finishing (900→800°C) of V-microalloyed austenite during CR is evidently characterized by recrystallization during the first finishing passes combined with grain elongation at the lower temperatures. The γ grain size of controlled-rolled V-steels is thus considerably smaller than corresponding Nb-variants, but the degree of grain elongation is less.

(iii) For a given cooling rate, the ferrite grain size is a unique function of S_v, the effective austenite grain-boundary area per unit volume. There is no dependence on nitrogen level or processing route, nor is there any difference between transformation from elongated or equiaxed austenite grains.

(iv) A corollary of the above conclusion is that the grain-refinement mechanism associated with CR of V-microalloyed steels is fundamentally different from that operative when Nb is present. V-steels owe their fine ferrite grain size principally to the fine D^γ engendered via recrystallization during the first finishing passes.

(v)   Controlled rolling of V-microalloyed steels engenders a degree of ferrite grain refinement equivalent to that attainable in Nb-variants.

(vi) In fully-transformed material, the contribution to the yield strength from V(C,N) precipitation, $\Delta\sigma_y(p)$, is considerably greater in hot-rolled than in normalized steels. This is related to the greater average particle size in the latter, which in turn derives from incomplete solution of V(C,N) at the normalizing temperature.

(vii) In hot-rolled material, $\Delta\sigma_y(p)$ increases markedly with nitrogen
content and cooling rate. With normalizing, on the other hand,
$\Delta\sigma_y(p)$ saturates at high nitrogen levels, while the relative
effect of cooling rate remains about the same as for hot-rolled
steel.

(viii) For NR and CR, $\Delta\sigma_y(p)$ is enhanced by up to 30 MPa if 0,02%Al
is present in the steel.

(ix) Measurements on the distribution of V(C,N) particle sizes
indicated that, all other things being equal, NR and CR
result in substantially identical size distributions, while the
average particle size in normalized material is considerably
greater. The mean size of V(C,N) precipitates decreases markedly
via faster cooling but <u>increases</u> somewhat as W_N is raised.
Characteristic differences exist between hot-rolled and
normalized steels as regards the shape of the cumulative size
distributions for V(C,N) particles.

(x) A calculation of $\Delta\sigma_y(p)$, based on recent refinements of the
Orowan theory for dispersion strengthening combined with the
experimentally-established V(C,N) particle-size distributions,
gives values in satisfactory agreement with the measured ones.
The volume fraction of precipitate, which is required in order
to perform the computations,was estimated via the assumption
that the particles are pure VN i.e. the volume fraction can
be deduced directly from the nitrogen content of the steel.

(xi) The experimental observations pertaining to the effect of
nitrogen level on $\Delta\sigma_y(p)$, the results from the theoretical
estimation of the latter and the predictions from the thermodynamic analysis of the Fe-V-C-N system reported previously,
collectively support the contention that precipitation strengthening in V-microalloyed steels ($W_V<0,1$) is governed by the interaction of vanadium and nitrogen. For material with higher
vanadium levels, the super-saturation of V, after all nitrogen
has been consumed, may well be sufficient for VC to contribute
effectively to $\Delta\sigma_y(p)$. Hence, for conventional processing of
steels with $W_V<0,1$ it seems reasonable to conclude that there
is little meaning in having more vanadium present than can
combine with the available nitrogen.

ACKNOWLEDGEMENTS

    Financial support for this investigation from Union Carbide is
gratefully acknowledged. It is a pleasure to thank Dr Michael Korchynsky
of that organization for many helpful discussions.

191

REFERENCES

1. W.Roberts, and A. Sandberg, "The composition of V(C,N) as precipitated in HSLA-steels microalloyed with vanadium". Swedish Institute for Metals Research Report No IM-1489 (1980).

2. J.D. Grozier, "Production of microalloyed strip and plate by controlled cooling". in Microalloying '75, Ed.M. Korchynsky (Union Carbide Corp.New York), vol. 1, pp.241-250 (1976).

3. W.Roberts, "Hot-deformation studies on a vanadium microalloyed steel". Swedish Institute for Metals Research Report No IM-1333 (1978).

4. I. Kozasu, C. Ouchi, T. Sampei and T. Okita, "Hot rolling as a high temperature thermomechanical process". As.ref. 2, pp. 100-114.

5. A.Sandberg, and W. Roberts, "Effect of thermomechanical treatment on the phase transformation of austenite in microalloyed steels". Swedish Institute for Metals Research Report No IM-1439 (1980), In Swedish. See also W. Roberts, H. Lidefelt and A. Sandberg, "Mechanism of enhanced ferrite nucleation from deformed austenite in microalloyed steels". in Proc. Sheffield Conf. on Hot Working and Forming Processes, Eds. C.M. Sellars & G.J. Davies (Metals Society, London), pp. 38-42 (1980).

6. T.Gladman, D. Dulieu, and I.D. McIvor, "Structure-property relationships in high-strength microalloyed steels". As.ref. 2, pp. 32-55.

7. F.Vodopivec, "Influence of precipitation and precipitates of AlN on torsional deformability of low-carbon steel". Metals Tech. 5 (1978), 118-121 .

8. N.K. Balliger, and R.W.K. Honeycombe, "Coarsening of vanadium carbide, carbonitride, and nitride in low-alloy steels". Met.Sci. 14 (1980), 121-133.

9. A.M.Sage, "Developments in controlled rolling of vanadium-bearing steels". in Proc. Sheffield Conf. on Hot Working and Forming Processes, Eds. C.M. Sellars and G.J. Davies (Metals Society, London), pp. 119-127 (1980).

10. T.Nilsson, private communication (Svenskt Stål AB, Oxelösund).

11. A.Melander, "The critical resolved shear stress of dispersion-strengthened alloys". Scan.J.Met. 7 (1978), 109-113.

12. R.Ebeling and M.F. Ashby, "On the determination of the number, size, spacing and volume fraction of spherical second-phase particles from extraction replicas". Trans.A.I.M.E. 236 (1966), 1396-1412.

DISCUSSION

Q: In Figure 13 you show that when the cooling rate is increased
from 0.25 to 5°C per second, the mean diameter of the vanadium carbonitride
particles decreases from 15 to 5 nm. What is the temperature range in
which the cooling was done, and can you explain the effect of cooling
rate on particle size?

A: The cooling size was measured over something like 750 to 550°C.
The main effect of the cooling rate is the decrease in the A_3 temperature
with increasing rate. Hence, the temperature at which precipitation
of the carbonitride can start is lowered. Then you will get more
prolific nucleation and if you precipitate all the nitrogen available,
the particles will be small and the particle spacing will also be small.

Q: Were any elongated austenite grains observed in the controlled-
rolled vanadium steel, and if so, did any of the elongated grains contain
deformation bands? In your plot of ferrite grain size versus austenite
grain boundary area S_v, Figure 5, there is no difference between the
normal control steel and the controlled-rolled steel. I wonder if there
is a great difference in the ferrite grain size of the controlled-rolled
steel deformed in a non-recrystallization region and that deformed in the
recrystallization region?

A: There was some grain elongation. We also point out that this
elongation, and hence inhibition of recrystallization, only took place
during the last one or two passes which we would react on as finishing
steps. Those passes took place somewhere between 850 and 800°C. Also,
as I recall, we did not find any deformation bands. This might be as
expected because some of the Japanese work reported here suggests a
critical reduction for deformation bands greater than the 25 percent we
worked with.

Q: I have a question about precipitation hardening. In the normal-
rolled material you showed higher strength than for the controlled
rolled (Figure 8). I can understand this. However, because of strain-
induced precipitation of vanadium nitride or carbonitride, the amount of
soluble vanadium in the austenite is decreased. Won't this decrease in
hardening potential also cause a smaller increase in the effective
strength of the controlled-rolled steel than in the normal steel?

A: I agree one hundred percent.

Q: The relationship between the austenite grain size and the
subsequent ferrite grain size is very important. What value of the
austenite to ferrite grain size ratio did you get and how does that
compare to some of the previous results?

A: The ferrite grain size depends on the factors, austenite grain
size and cooling rate. If we reduce the austenite grain size to 20 μm
and use the 0.88°C per second cooling rate we typically end up with a
ferrite grain size of 7.5 μm or a ratio of 3. If we use the lower cooling
rate, the ferrite will only be 9 μm and the ratio will be less. I don't
know how that compares to other work. Perhaps others can comment.
(Comment by Priestner) Over the past few years, I have measured austenite
grain sizes and the ferrite grain sizes produced therefrom. In a series
of fairly simple experiments I heated carbon-Mn steels and microalloyed
steels to various temperatures in the austenite range and cooled them at

193

various cooling rates. The austenite to ferrite grain size ratio does vary with austenite grain size and it does vary with cooling rate. The ratio gets very large for very fast cooling rates. In our work and that of others in England, the average austenite to ferrite grain size ratio is about eight to one but it may be as high as 12 or even more. The reason your ratio is so much smaller may be because your cooling rate is much slower.

Comment: For air-cooled thick plate there are two cases. In the first case, polygonal austenite transforms to ferrite and the grain size ratio is reduced by reducing the austenite grain size, converging to one. The convergence to one is faster in a steel having a higher transformation temperature, which is brought about by slower cooling and lower carbon content. The second case is transformation from elongated and unrecrystallized austenite, and the ratio between austenite grain height and transformed ferrite grain size is seen to converge to 2.

Comment: For materials in which the austenite was recrystallized, and at cooling rates corresponding to those for plate, we got austenite to ferrite ratios 8 and 4. With deformed austenite grains where the grain height was 30 μm or less, the ratio of that height to the ferrite grain size was 2, in agreement with the observation made in the previous comment.

FORMATION MECHANISM OF MIXED AUSTENITE GRAIN STRUCTURE

ACCOMPANYING CONTROLLED-ROLLING OF NIOBIUM-BEARING STEEL

T.Tanaka, T.Enami, M.Kimura, Y.Saito and T.Hatomura

Research Laboratories
Kawasaki Steel Corporation
Kurashiki, Japan

Controlled rolling produces finer ferrite grain size than not only conventional hot-rolling but normalizing. Unfortunately, controlled rolled steel very often exhibits a mixed structure consisting of fine and coarse ferrite grains and/or upper bainite, thereby deteriorating the low-temperature-toughness. Mixed grain structure of ferrite is attributed to that of austenite.

The purpose of the present experiment was to investigate the formation mechanism of mixed austenite grain structure, particularly in Nb-bearing steel. The general recrystallization behaviour was divided into three regions; recovery, partial recrystallization and recrystallization regions. The recrystallization region was subdivided into static and dynamic. The critical amount of deformation dividing each region increased rapidly with decreasing deformation temperature. It was also markedly influenced by the addition of microalloying elements and in particular Nb. Reduction in the recrystallization region produced fine and uniform, recrystallized grain structure. Reduction in partial recrystallization region produced mixed structure consisting of recrystallized grains and recovered grains. Reduction in recovery region produced locally very coarse grains due to strain-induced grain boundary migration, while most grains remained unchanged, releasing stored energy by recovery. Even if deformation was given repeatedly in the recovery region, very sluggish and incomplete recrystallization occured, while huge grains larger than the initial size were produced locally and persisted throughout later rolling passes, leading to the formation of mixed ferrite grain structure. However, if reduction was given in the partial recrystallization region, fraction recrystallized increased with increasing number of pass, reaching complete recrystallization. However, once huge grains were formed by slight reduction in recovery region, they persisted even after many passes in the partial recrystallization region. In summarizing the above description, if the initial rolling passes are given in recovery region, huge grains are locally formed due to the strain-induced grain boundary migration, which persist throughout the whole course of rolling. They transform to upper bainite and/or coarse ferrite grains, giving poor toughness. On the contrary, if reduction per pass more than that for partial recrystallization is given successively, uniform and refined grain structure is obtained. Once complete recrystallization takes place, successive recrystallization proceeds more easily.

Introduction

The major purpose of controlled rolling is to refine grain structure and thereby improve the properties of steel in the hot-rolled condition to a level equivalent to or better than those of highly alloyed or heat-treated steels. However, controlled rolled steel very often exhibits a mixed structure consisting mainly of very fine grains with some coarse ferrite grains and/or upper bainite. Though small in fraction, since coarse ferrite grains and upper bainite deteriorate toughness markedly, they must be suppressed during controlled rolling. As the mixed grain structure of ferrite results from that of austenite, the formation mechanism of the former must be sought to account for the latter.

Studies of the formation of mixed austenite grain structure which occurs during rolling at high temperatures, are few in number. Jones and Rothwell(1) reported that under the condition where partial recrystallization occurs in the transition temperature range from recrystallization to non-recrystallization, the mixed grain structure was produced by preferential recrystallization at austenite grain boundaries, grain interiors being left unrecrystallized.

When austenite grains are coarse or rapid grain growth takes place after recrystallization due to unsuitable deformation conditions, mixed austenite grain structure tends to be induced(2). Once a mixed austenite grain structure occurs, it can not be eliminated by later rolling passes(3). When rolled again, the recrystallized region in the vicinity of the prior grain boundaries tends to preferentially recrystallize since it has a smaller grain size than the unrecrystallized region of the grain interior. Based on the above understanding of the formation of mixed grain structure, suitable conditions of holding(2,4) or "continuous" schedules involving reduction in the whole temperature range(1), and rolling schedules based on the similar idea(5) have been proposed. To date, studies of the formation of austenite mixed grain structure have been confined mainly to the transition temperature range from recrystallization to non-recrystallization.

The present authors showed the occurrence of mixed grain structure when austenite was deformed slightly at the high-temperature recrystallization region(6). In the present investigation, the general restoration behaviour was studied by one-pass and multi-pass rolling in the temperature range of recrystallization and non-recrystallization. The relation between rolling schedule and formation of austenite mixed grain structure was made clear, and a rolling method which can give uniform, fine-grained structures of austenite and ferrite was established.

Experiment(1). Critical Reduction for Recrystallization

The materials used in the present experiment were melted in a basic converter and continuously cast. Chemical compositions are shown in Table 1. For the purpose of investigating the effects of the initial grain size and rolling temperature on the recrystallization behaviour, stepwise specimens shown in Fig.1 were heated for 60 min. at 1150°C, and immediately rolled in one-pass, which was intended to adjust austenite grain size via deformation-recrystallization. Following the first rolling pass, specimens were cooled to T°C, where the second rolling pass was given. The amounts of reduction were in the range of 13.8% to 50% for the first pass and 10% to 50% for the second one. At the period of ~5 sec after rolling the specimens were quenched in water, which enabled observation of the change in austenite grain structure accompanying hot-rolling. The austenite grain structure was revealed by etching specimens in a mixture consisting of picric-acid

Table 1 Chemical compositions of steels used (wt %)

Steel	C	Si	Mn	P	S	Ni	Nb	V	Al	N
Nb	0.11	0.24	1.35	0.014	0.016	0.30	0.035	0.038	0.034	0.0068
plain C	0.20	0.37	1.42	0.019	0.009	0.03	0.005	0.008	0.023	0.0047

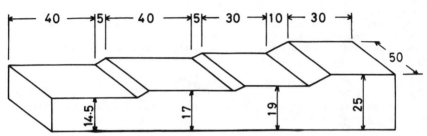

Fig. 1 Dimension of stepwise specimen employed for rolling experiment; mm

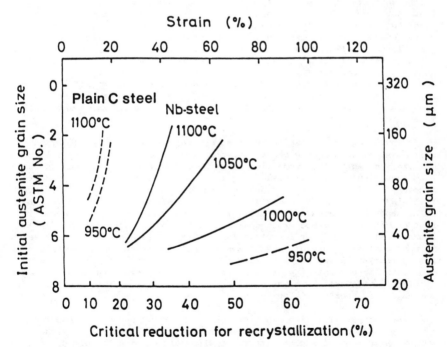

Fig. 2 Effects of deformation temperature and initial grain size on critical amount of reduction required for completion of recrystallization in plain carbon and niobium steels

saturated solution, cuprous chloride and surface-active reagent.

Figure 2 depicts the effects of the initial grain size and rolling temperature on the critical amount of reduction required for the completion of recrystallization in plain-carbon and niobium steels. In niobium steel, when the initial grain size is large, the critical reduction is extremely high and with decreasing grain size the critical reduction diminishes. The influence of rolling temperature is also strong, the critical reduction becoming very large with decreasing temperature. In plain-carbon steel, however, the effects of the initial grain size and rolling temperature on the critical reduction are slight. That is, there is a very large difference in recrystallization behaviour between niobium steel and plain-carbon steel. In the latter, recrystallization occurs almost under any rolling condition. While in niobium steel recrystallization does not always operate easily and under all conditions.

Experiment(2). Restoration Behaviour

General restoration behaviour was investigated under the condition where the reduction was less than that required for complete recrystallization.

Stepwise specimens shown in Fig.1 were heated for 60 min at $1150°C$, air-cooled to $T°C$ and subsequently rolled in one-pass, followed by water-quenching. As shown in Fig.3, restoration behaviour is divided into three regions, depending on the amount of reduction and rolling temperature; i) recovery region where recrystallization does not occur but recovery proceeds, ii) partial recrystallization region where recrystallization is incomplete, giving a mixed structure consisting of recrystallized grains and recovered ones and iii) recrystallization region where complete recrystallization occurs. The critical amount of deformation dividing each region increases rapidly with decreasing rolling temperature. In particular, the critical reduction for complete recrystallization becomes markedly large when rolling temperatures are low.

Figure 4 shows austenite grain structures which are obtained by one-pass rolling at $\sim1150°C$ in the recovery region (6% reduction), the partial recrystallization region (15% reduction) and the complete recrystallization region (30% reduction), respectively. Reduction in the recovery region produces recovered grains mixed with coarse grains much larger than the initial size (Fig.4(a)). The strain energy stored during reduction in the recovery region is not enough to cause recrystallization. The amount of strain energy is, however, enough to cause grain-coalescence due to the strain-induced grain boundary migration in preferably oriented grains, leading to the scattered formation of giant grains. Reduction in the partial recrystallization region causes recrystallization from place to place, thereby producing a mixed grain structure consisting of fine recrystallized grains and recovered grains (Fig.4(b)). Reduction in the complete recrystallization region gives a fine and uniformly recrystallized grain structure.

The recrystallization behaviour shown in Figs.3 and 4 was observed at the period of ~5 sec after one-pass rolling. In order to study the variation in austenite grain structure with holding time, specimens were heated to $1150°C$, rolled at $1100°C$, held at that temperature for a given time and subsequently quenched in water. Figure 5 depicts the influence of amount of reduction and holding time on the progress of austenite grain growth. The grain growth behaviour is divided into three regions: In region I (mixed-grain growth region), grain growth starts from a mixed

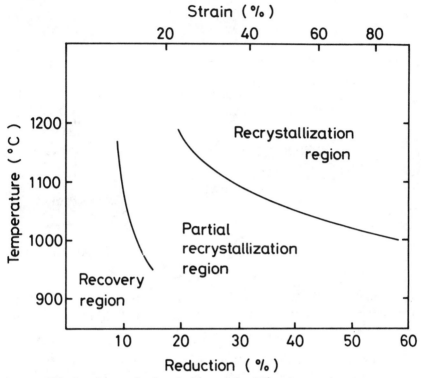

Fig. 3 Effects of amount of reduction and rolling temperature on restoration behaviour: niobium steel was heated to 1150°C which gave grain size ~180 μm, and one-pass rolled

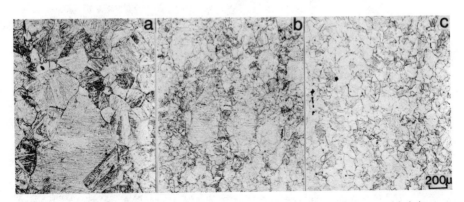

Fig. 4 Austenite grain structure produced by one-pass rolling of 6% (a), or 15% (b), or 30% (c) in niobium steel

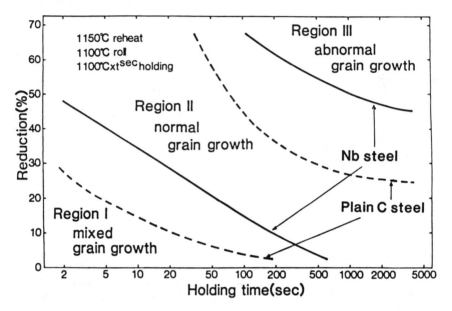

Fig. 5 Effects of holding time and amount of reduction on the progress of grain growth in plain-carbon and niobium steels

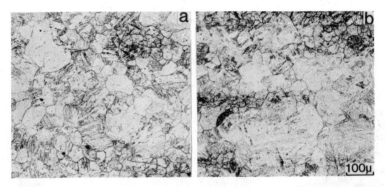

Fig. 6 Austenite grain structures produced by normal grain growth (a) and abnormal grain growth (b) in niobium steel which was rolled 30% or 65% and held for 600 sec

structure consisting of either recovered grains and giant grains or recrystallized grains and recovered ones, depending on the amount of reduction. Smaller grains have propensity to grow faster and thereby to arrive at uniform grain structure after long period of holding. In region II normal grain growth takes place according to Miller's equation $d = A t^n$, where d is grain size, A and n constants, and t time. n is in the range of 0.05 to 0.09, depending on the austenite grain size prior to isothermal holding.

In region III (abnormal grain growth region), huge grains suddenly develop in the midst of small grains, as shown in Fig.6. As is usually observed with such secondary recrystallization, the size of grain coalesced from small grain structure is much larger than that from large grain structure. It is worth noting that with the decrease in grain size prior to the commencement of abnormal grain growth, a critical holding time required for the start of abnormal grain growth decreases. Furthermore, the critical holding time required for abnormal grain growth is larger for niobium steel than for plain-carbon steel. Thus, even though the grain structure obtained by a large amount of deformation is refined and uniform, it has a propensity to cause abnormal grain growth, leading to the formation of mixed structure consisting of small grains and extremely large grains.

Experiment(3). Change in Grain Structure Accompanying Multi-pass Rolling

The austenite grain structures shown in Fig.4 were produced by one-pass rolling. The variation in grain structure produced by multi-pass rolling in each region of Fig.3 was also studied. Specimens were heated for 60 min at 1150°C and continuously rolled with 6% reduction per pass (recovery region), or 10% reduction per pass (partial recrystallization region), or 28% reduction per pass (recrystallization region). In production mills, for the purpose of broadside rolling and/or adjusting slab size, a rolling schedule of several initial light passes followed by heavy rolling in the later stage is often used. In order to more closely simulate the austenite grain structure accompanying a such rolling schedule, specimens were rolled with 6% reduction per pass (recovery region) and subsequently with 14% reduction per pass (partial recrystallization region).

Figure 7 shows grain structures produced by rolling with 28% passes (recrystallization region). One-pass rolling causes complete recrystallization, giving a uniform, refined grain structure. A rolling of two 28% passes causes further refinement in grain size. That is, when rolled continuously in recrystallization region, the austenite grain size decreases with increasing number of rolling pass.

Figure 8 exhibits austenite grain structures produced by rolling with 10% reduction per pass (partial recrystallization region). With one-pass rolling, recrystallization occurs partially and the recrystallized area increases with increasing number of pass, finally arriving at the complete recrystallization, shown in Fig.8(b). It is worth noting that once complete recrystallization is eventually attained in the partial recrystallization region, a uniform, fine-grained structure is obtained, instead of a mixed grain structure.

The austenite grain structure brought about by continuous rolling with light passes of 6% reduction per pass (recovery region) is shown in Fig.9. Rolling of five 6% passes (total reduction=27%) does not cause recrystallization, but exhibits mostly recovered grains mixed with coalesced large grains from place to place. This situation remains unchanged even after

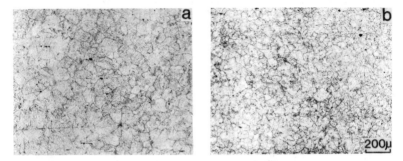

Fig. 7   Veriation in austenite grain structure with 28% rolling in niobium
steel: (a) one-pass, and (b) two passes

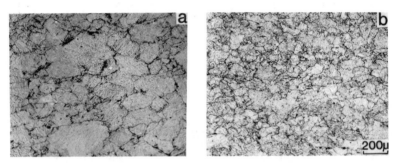

Fig. 8   Variation in austenite grain structure with 10% rolling in niobium
steel: (a) one pass, and (b) five passes (42% reduction in total)

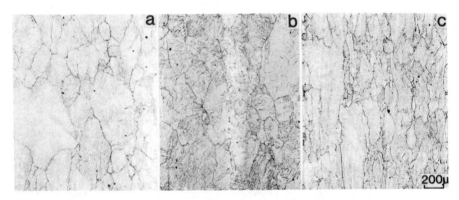

Fig. 9   Variation in austenite grain structure with continuous rolling of
6% pass in niobium steel: (a) five 6% passes (27% in total),
(b) seven 6% passes (36% in total), and (c) nine 6% passes
(43% in total)

rolling of seven passes (total reduction = 36%). When light nine passes
are given, recrystallization occurs only partially, while coalesced large
grains survive through rolling of multiple passes, though elongated in shape.
There is a remarkable difference in grain structure between Fig.8(b) and
Fig.9(c), even though the total reduction is the same between the two specimens. Let us examine the possibility of eliminating coalesced large grains
by giving large reduction at the later rolling, granting that large grains
are produced by slight reduction in the initial stage of rolling.   Figure
10 shows austenite grain structures which are produced by rolling of four
6% passes (recovery region) and subsequently multiple passes with 14% reduction per pass (partial recrystallization). Rolling of four 6% passes
gives mixed structure consisting of recovered grains and coalesced large
grains. When one 14% pass is added to the above rolling schedule, recrystallization occurs mostly in recovered grains, while preserving coalesced
large grain in elongated shape, exhibited in Fig.10(b). When three 14%
passes are added (total reduction = 50%), recrystallized grains cover the
whole area except coalesced large grains.

In summarizing the above experiments, when reductions greater than that
required for partial recrystallization are given successively, the recrystallized area increases with increasing number of pass and thereby produces
uniform, refined austenite grain structure. However, when reductions of less
than that for partial recrystallization are given successively, recovery and
coalescence of the initial grains occur, without attaining the refinement of
austenite grain structure. Once coarse grains are formed by small reductions
in the early stage, it is almost impossible to eliminate them in the later
rolling. Therefore, to attain a uniform, refined austenite grain structure,
it is necessary to hot-roll with the amount of reduction more than that
required for partial recrystallization.

Experiment(4). High Temperature Reduction

So far austenite grain refinement through recrystallization has been
shown to be attained only when multi-pass hot-rolling is done with reductions
of more than the critical amount for partial recrystallization. It is, however, not always easy to satisfy this condition during hot-rolling of
niobium-bearing steels. As shown in Fig.3, the critical reduction decreases
with increase in rolling temperature. Thus at higher rolling temperatures
partial recrystallization is more likely. On the other hand, high temperature rolling necessarily involves high-temperature heating of slab, thereby
causing grain coarsening. To know the optimum condition, the effect of
rolling temperature and initial grain size were studied.

Specimens were heated for 60 min at 1280°C to produce a uniform
austenite grain structure of ∼1200 μm. They were subsequently rolled 6% or
11%, with start and finish temperatures of 1280°C and 1100°C.   Figure 11
shows the change in austenite grain structure with increasing number of 6%
passes. The microstructure is a mixture consisting of recrystallized grains
and recovered ones. With increasing number of 6% pass, recrystallized area
increases. Seven passes, which total 36% reduction, cause almost complete
recrystallization, giving refined grain structure.   This means that even
though the initial grain size of ∼1200 μm is very coarse, 6% reduction per
pass is not within the recovery region but is in the partial recrystallization region. Figure 12 shows the change in austenite grain structure
with increasing number of 11% pass. One-pass rolling gives completely recrystallized grain structure. Three 11% passes and five 11% passes exhibit
more refined, uniform grain structure. That is, with the increase in rolling
temperature the critical amounts of reduction required for partial recrys-

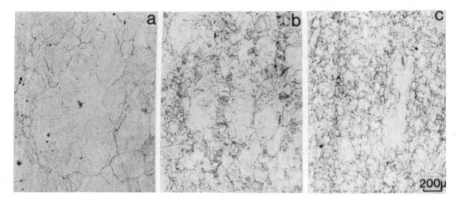

Fig. 10 Austenite grain structures produced by rolling of four 6% passes
followed by continuous rolling of 14% pass in niobium steel: (a)
four 6% passes, (b) (a) + one 14% pass and (c) (a) + three 14% passes

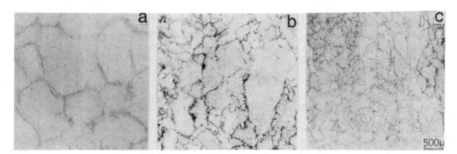

Fig. 11 Change in austenite grain structure with increasing number of 6%
pass in niobium steel which was heated to 1280°C and rolled at
temperatures between 1280°C and 1100°C: (a) one pass, (b) four
passes and (c) seven passes

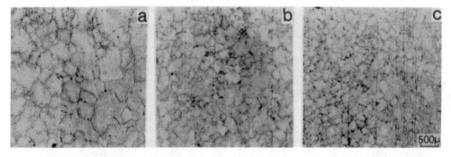

Fig. 12 Change in austenite grain structure with increasing number of
11% pass in niobium steel which was heated to 1280°C and rolled
at temperatures between 1280°C and 1100°C: (a) one pass, (b) three
passes and (c) five passes

tallization and complete recrystallization decrease, and thereby produces recrystallization more easily, even though the initial grain size may be very large.

As shown in Fig.2, the grain size dependence of critical amount of reduction for recrystallization decreases with increasing rolling temperature, implying that even if a mixed grain structure is formed at a high temperature, it is prone to undergo unselective recrystallization, leading to uniform grain structure.

Given that repeated deformation-recrystallization occurs easily, if the recrystallized grain size is not sufficiently small, then high temperature rolling can not fulfil the initial objective. Figure 13 depicts the relation between recrystallized austenite grain size and total reduction in specimens heated to 1150°C or 1280°C. In both sets of specimens, austenite grain structure becomes finer with increasing amount of reduction, reaching a certain limited value. However, there is much difference in grain size between the two sets. When compared at 50% reduction, grain size of $\sim$40 μm and $\sim$180 μm are obtained for 1150°C and 1280°C, respectively.

Recrystallization is more probable and rate of grain refinement is more marked with high temperature heating and rolling, whereas the lowering of heating and rolling temperatures suppresses recrystallization and decreases the rate of grain-refinement via recrystallization. However, it must be stressed that at any reduction the recrystallized grain size is finer for low temperature heating than for high temperature heating.

Experiment(5). Rolling in Non-recrystallization Region

So far it has been shown that, both single-pass rolling and multi-pass rolling with reductions of less than the critical amount required for partial recrystallization causes local grain-coalescence instead of grain-refinement due to strain-induced grain boundary migration. Higher heating and rolling temperatures can avoid the formation of coalesced grains, since the critical reduction for partial recrystallization becomes small. However, recrystallized grain size is much larger for high temperature than for low.

The next step is to investigate the variation in austenite grain structure accompanying rolling in the non-recrystallization region. As seen in Fig.2, with decrease in temperature, recrystallization becomes more difficult and it is virtually suppressed below $\sim$950°C. With increasing amount of deformation in this non-recrystallization region, austenite grains become elongated and deformation bands are generated within them, as shown in Fig.14.

For the purpose of investigating grain structure changes accompanying deformation in the non-recrystallization region, specimens were heated at 1150°C, air-cooled to 850°C and immediately rolled 65%. After hot-rolling, one specimen was water-quenched to observe austenite grain structure, another was air-cooled to obtain ferrite grain structure and the third specimen was air-cooled to 650°C and subsequently water-quenched for observation of ferrite-nucleation sites and homogeneity in ferrite formation, shown in Fig.15. Though 65% reduction causes a remarkably elongated austenite grain structure and at the same time a number of deformation bands in grain-interiors, there is much inhomogeneity in grain size and deformation band density (Fig.15(a)). As shown in Fig.15(b), air-cooling gives a mixed structure consisting mainly of fine ferrite-grains and large,

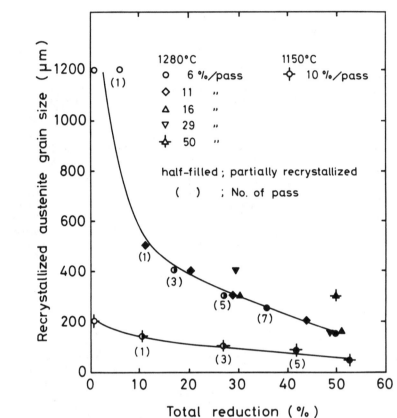

Fig. 13 Effect of total reduction on recrystallized austenite grain size in niobium steel which was reheated to 1280°C or 1150°C and subsequently multi-pass rolled

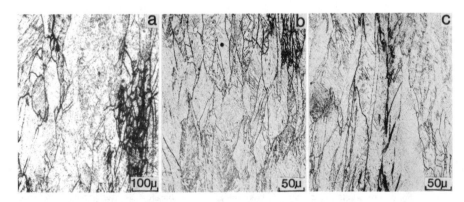

Fig. 14 Elongated austenite grains with deformation bands in grain-interiors in niobium steel deformed at non-recrystallization region alone: (a) rolled 30%, and (b) and (c) rolled 65%

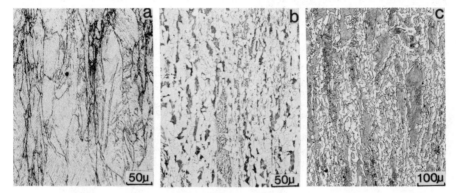

Fig. 15   Grain structures of austenite and ferrite and ferrite-nucleation
sites during ferrite transformation in niobium steel which was
heated to 1150°C, rolled 65% at 850°C and water-quenched (a),
air-cooled (b) or partially transformed (c)

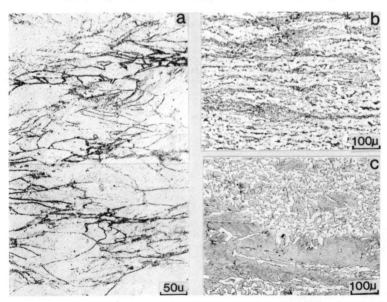

Fig. 16   Grain structures of austenite and ferrite and ferrite-nucleation
sites during ferrite transformation in niobium steel which was
heated to 1150°C, rolled 6% at 1100°C, rolled 65% at 850°C, and
water-quenched (a), air-cooled (b), or partially transformed (c)

elongated upper bainite, which resulted from the coarse austenite grains with none or few deformation bands. When air-cooled to a temperature between Ar_3 and Ar_1, ferrite grains are not formed uniformly. From place to place, there are elongated, untransformed austenite grains with no trace of ferrite nucleation or only a few ferrite grains within them, as exhibited in Fig.15(c).

Specimens were heated to $1150^\circ C$, rolled 6% at $1150^\circ C$ and 65% at $850^\circ C$, to simulate rolling conditions under which reheated slabs are slightly rolled in the initial several passes at high temperatures and subsequently heavily rolled in non-recrystallization region. As shown in Fig.16, heterogeneity in austenite grain structure and deformation band density is increased by rolling 6% at high temperature as seen from a comparison between Figs.15(a) and 16(a). When subjected to partial transformation, the coarse elongated austenite grains remain untransformed but have a small number of ferrite grains within them (Fig.13(c)). The air-cooled specimen gives a mixed structure consisting of mainly fine grains and elongated upper bainite (Fig.16(b)).

The heterogeneity of the grain structure of as-deformed austenite is enhanced in the transformed structure. This is considered to be due to the inhomogeneous distribution of deformation bands. Figure 17 shows the distribution of deformation bands in the same specimen as in Fig.16(a). In some grains a number of deformation bands are observed although they are infrequent in coarse austenite grains. As is well-known, ferrite nucleation occurs at deformation bands as well as prior austenite grain boundary, implying that a large austenite grain with none or few deformation bands necessarily leads to the formation of coarse ferrite and/or upper bainite.

Figure 18 shows ferrite grain structure produced by the following treatments; heated at $1150^\circ C$, rolled 60% at $1100^\circ C$, rolled 60% at $850^\circ C$ and subsequently either water-quenched or air-cooled. In this case a uniform, refined austenite grain structure is heavily deformed in the non-recrystallization region and the transformed ferrite grain structure is uniform and very fine.

Discussion

In the previous paper, one of the authors studied the influence of niobium on the ferrite-start temperature $Ar_3(7)$. He found that niobium in solution lowers the ferrite-start temperature, and the effect is more pronounced for coarse grain structure. Thus, in niobium-bearing steel, a large austenite grain structure suppresses ferrite formation and thereby causes upper bainite and/or coarse ferrite structure. Therefore it is necessary to attain fine austenite grain structure prior to the transformation-start, in order to suppress the formation of upper bainite and coarse ferrite, and to obtain a uniform, fine-grained ferrite structure.

Up to this time the mixed structure of small and coarse grains very often observed during hot-rolling has been considered to form mainly by partial recrystallization in the temperature range between the recrystallization and non-recrystallization regions, though the number of investigations is small(1-3,8). However, it is not yet clear how this type of mixed grain structure of austenite really affects the ferrite structure. Though large in size, unrecrystallized austenite grains contain several deformation bands within grain-interiors. Since deformation bands are equivalent to austenite grain boundaries with regard to ferrite-nucleation

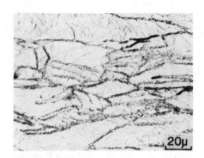

Fig. 17   Distribution of deformation bands in same steel
as shown in Fig. 16 (a)

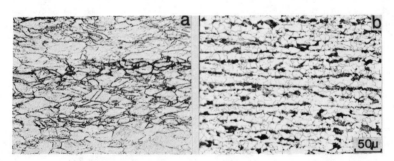

Fig. 18   Austenite and ferrite grain structures in niobium steel
which was heated to 1150°C, rolled 60% at 1100°C,
rolled 60% at 850°C and water-quenched (a) or air-cooled (b)

potentiality, a single unrecrystallized grain with deformation bands within it can be seen as an aggregate consisting of several grains which can possibly transform to a number of relatively fine grains.

The present study is perhaps the first to clarify the relation between the mixed grain structure formation and rolling conditions in the high-temperature, recrystallization region. When a reduction of less than the critical amount required for partial recrystallization is given, selective grain-coalescence occurs due to strain-induced grain boundary migration, creating a few coarse grains much larger than the initial size. This coalescence behaviour is quite similar with that observed in mild steel slightly rolled at room temperature and subsequently annealed. Though selective grain coalescence is considered to occur only in grains with preferable orientation, evidence confirming or otherwise has not been obtained in the present study.

After hot-rolling, the grain-coalescence due to strain-induced grain boundary migration proceeds very rapidly for a short time and further coalescence does not seem to take place even when the rolled specimen is held at higher temperatures. Instead, coalesced grains remain unchanged, while remaining recovered grains grow gradually with increasing holding time, finally reaching roughly the same size as the coalesced grains.

As shown in Fig.5, a uniform, fine-grained austenite grain structure produced by heavy rolling in recrystallization region exhibits normal grain growth for a certain period, and then suddenly exhibits abnormal grain growth. The critical time to the onset of abnormal grain growth decreases with decreasing grain size. For example, recrystallized austenite grains of 30 μm produced by rolling 65% suddenly grow to ~200 μm after a normal growth period of 600 sec. Also, coalesced grain-size increases with decreasing grain size prior to holding, as is generally observed in the secondary recrystallization. Therefore, it must be emphasized that even if a uniform, fine-grained structure is obtained by a suitable rolling schedule, a mixed structure consisting of refined grains and extremely large grains will form if the rolled material is held at high temperatures long enough for abnormal grain growth to occur. The commencement time for abnormal grain growth is pretty short in plain-carbon steels, but is longer in niobium steels. This delay is considered to be due to either niobium in solution or fine niobium carbonitrides formed by strain-induced precipi-tation.

If passes with reduction per pass greater than the critical amount for the partial recrystallization are given repeatedly, the fraction recrys-tallized increases with increasing number of pass (see Fig.8). In this case, austenite grains recrystallized in the previous rolling pass remain unchanged, while, unrecrystallized grains preferentially recrystallize. Thus, when complete recrystallization is attained, microstructure is refined and uniform. Since austenite grain boundaries are nucleation sites for recrystallization(9), the fact that nucleation-site density is much higher in the recrystallized region than in the unrecrystallized one suggess that recrystallization might occur preferentially during the follow-ing rolling pass. It has been proposed that when a mixed structure consisting of recrystallized and unrecrystallized grains is rolled again, deformation concentrates in the soft recrystallized region and subsequent recrystallization will occur there preferentially(1,8). The result of the present experiment denies this idea.

The fact that unrecrystallized grains undergo preferential recrystallization implies that strain energy will be accumulated gradually with increasing number of rolling passes at such high temperature range as 1150 to 1000°C, and finally reaches the critical amount required for recrystallization. The accumulation of strain energy was also observed during multi-pass rolling with reductions less than the critical amount for partial recrystallization. Though very sluggish, recrystallization proceeds gradually with each rolling pass, giving only after many passes mixed structure consisting mainly of recovered grains and small fraction of refined grains.

The critical amount of reduction required for partial recrystallization and that for complete recrystallization, exhibited in Fig.3, can be interpreted in terms of crystal orientation: Polycrystalline aggregates have a variety of grain orientation. The minimum reduction required for the recrystallization in the most favourably oriented grains is the critical reduction for partial recrystallization of Fig.3. With the increase in accumulated strain energy, the recrystallization process next operates in less favourably oriented grains. On the other hand, the minimum reduction required for recrystallization in the least favourably oriented grains is the critical reduction for complete recrystallization.

It is not clear whether the retardation of recrystallization results from the drag effect due to solute niobium or fine niobium carbonitrides produced by strain-induced precipitation. Jonas and Weiss(10) reported that there is a significant delay introduced by a solute effect. However, strain-induced precipitation of niobium carbonitrides was reported to occur even at such high temperatures at 1150°C to 1000°C(11,12), and generally the recrystallization-retardation effect is much stronger for fine precipitates than for solute drag effect.

It has been repeatedly stated that when rolled with reductions less than the critical value for partial recrystallization, grain coalescence instead of grain refinement occurs due to strain-induced grain boundary migration, producing much larger grains than the initial ones. Once large grains are formed, they persist through the whole course of rolling, even when the pass schedule of later rolling is favourable for recrystallization. As seen in Fig.2, the critical reduction required for recrystallization increases with the increase in the initial grain size at any temperature. For example, when a grain of 200 μm produced at 1150°C is grown to ~600 μm, which size is frequently observed in slightly deformed material, the critical reduction for the recrystallization of the ~600 μm grain is 60% at 1050°C. Because recovery is so rapid at 1050°C the 60% reduction can not be interrupted, and such large uninterrupted reductions are not common practice in multi-pass rolling.

The deformation bands produced by rolling in the non-recrystallized austenite region play a very important role in determining ferrite-grain refinement, since ferrite-nucleation occurs at deformation bands as well as austenite grain boundaries. Figure 19 depicts the effects of reheating temperature (austenite grain size prior to deformation) and the amount of subsequent reduction at 850°C on the occurrence and density distribution of deformation bands. In the reduction range of less than 15%, deformation bands are barely observed. When rolled more than 50%, deformation bands are observed in most grains. When rolled 30%, specimens previously heated to 1275°C (grain size = 500 μm) do not exhibit any deformation band, while in other specimens deformation bands are generated, though in not all grains. This indicates that deformation bands are difficult to generate in coarse grains. Sekine et al(13) reported that apparent grain boundary area (austenite grain boundary area plus deformation bands)

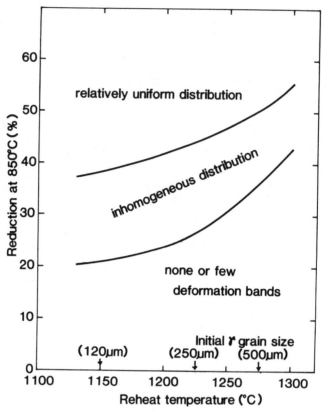

Fig. 19 Effect of amount of deformation on the ease of deformation band formation

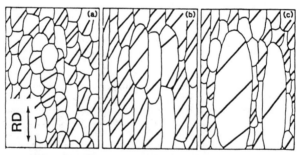

Fig. 20 Schematic illustration of deformation band distribution produced by deformation in non-recrystallization region:
(a) Finely recrystallized austenite is heavily deformed,
(b) as-reheated austenite is directly deformed, and
(c) recrystallized grains mixed with coalesced grains are deformed: RD is rolling direction

increases with increasing amount of reduction, reaching a steady level above 30% reduction. They also stated that even when the density of deformation bands reaches the steady level, about 10% of the austenite grains contain barely any deformation bands.

Once large grains are formed due to grain coalescence or abnormal grain growth, they persist through the whole course of rolling. In such grains deformation bands are more difficult to generate, and if deformation bands are never formed in some of them the result is coarse ferrite grains and/or upper bainite. This situation is schematically illustrated in Fig.20. When the microstructure consists of mixed grains, some refined grains and some coalesced, large grains are unfavourably oriented for the formation of deformation bands, and therefore contain no or very few deformation bands in their interiors. The average distance between effective boundaries, which include both austenite grain boundaries and deformation bands, is much smaller for fine grains than for coalesced, large grains. The latter provides a much smaller number of ferrite-nucleation sites and thereby favours the formation of coarse ferrite and/or upper bainite structure.

Inhomogeneity in grain structure is apparently the fate of as hot-rolled material. However, it can be minimized by reducing recrystallized austenite grain size. As shown in Fig.18, the best way to obtain a fine and uniform ferrite-grain structure is to attain recrystallized austenite grains as fine as possible, followed by a large amount of deformation in the non-recrystallization region(13).

Conclusions

The formation of mixed grain structure which is very often observed during rolling at high temperature region and in particular in niobium-bearing steel, has been studied by single-pass and multi-pass rolling and the following results obtained:

1) When one-pass rolled at high-temperatures, the restoration behaviour of austenite is divided into three regions, depending on rolling temperature and the amount of reduction; i) recovery region where the structure consists of mainly recovered grains and coalesced grains much larger than the initial size, ii) partial recrystallization region, and iii) complete recrystallization region.

2) The critical amount of deformation dividing each region increases rapidly with increasing grain size and decreasing rolling temperature.

3) When rolled with reductions greater than the critical amount for complete recrystallization, the recrystallized grain size decreases with increasing number of passes.

4) When rolled with reductions greater than the critical amount for partial recrystallization, the fraction recrystallized increases with increasing number of passes, eventually arriving at complete recrystallization with uniform grain structure.

5) When rolled with reductions smaller than the critical amount for partial recrystallization, grain coalescence instead of grain-refinement occurs locally, and thereby produces mixture of mainly recovered grains and extremely coarse grains much larger than the initial size.

6) When recrystallized grains are held at high temperature, normal grain growth occurs and after a certain incubation time abnormal grain growth intrudes giving a mixture of relatively small grains and very huge grains. The incubation time for abnormal grain growth decreases with reducing recrystallized grain size.

7) Once coarse grains are formed either by slight reduction or by abnormal grain growth, they can hardly be eliminated even by heavy rolling in the later stage. Therefore, draft schedules incorporating slight reductions in the initial stage followed by heavy rolling are not recommended.

8) Once coarse grains are formed, they are difficult to sub-divide by deformation bands.

References

1) J.D.Jones and A.B.Rothwell, "Controlled Rolling of Low-Carbon Niobium-Treated Steels", PP. 78-82 in Deformation under Hot Working Conditions, Iron Steel Inst. Publication 108, Iron Steel Inst., London, 1968.

2) A.le Bon and L.N. de Saint-Martin, "Using Laboratory Simulations to Improve Rolling Schedules and Equipment", PP. 107-119 in Microalloying 75 - Internl. Symposium on High-strength, Low-alloy Steel, Union Carbide Corp., New York, N. Y., 1977.

3) J.J.Irani, D.Burton, J.D.Jones, and A.B.Rothwell, "Beneficial Effects of Controlled Rolling in the Processing of Structural Steels", PP. 110-122 in Strong Tough Structural Steels, Iron Steel Inst. Publication 104, Iron Steel Inst., London, 1967.

4) J.H.Little, J.A.Champman, W.B.Morrison and B.Mintz,"The Production of High Strength Plate Steels by Controlled Rolling", PP. 85-88 in Microstructure and Design of alloys, vol. 1, The Metals Society, London, 1974.

5) M.L.Lafrance, F.A.Caron and G.R.Lamant, "Use of Microalloyed Steels in the Manufacture of Controlled-Rolled Plates for Pipe", PP. 367-371 in Microalloying 75 - Internl. Symposium on High-strength, Low-alloy Steel, Union Carbide Corp., New York, N. Y., 1977.

6) T.Tanaka, T.Funakoshi, M.Ueda, J.Tsuboi, T.Yasuda and C.Utahashi, "Development of High-Strength Steel with Good Toughness at Arctic Temperatures for Large-Diameter Line Pipe", PP. 399-409 in ibid.

7) T.Tanaka and N.Tabata, "Measurement of Austenite to Ferrite Trans-formation Temperatures in Niobium-Bearing Steels", Tetsu-to-Hagane, 64 (1978), PP. 1353-1362.

8) I.Kozasu, C.Ouchi, T.Sampei and T.Okita, "Hot Rolling as a High-Temperature Thermo-Mechanical Process", PP. 120-135 in Microalloying 75 - Internl. Symposium on High-strength, Low-alloy Steel, Union Carbide Corp., New York, N. Y., 1977.

9) I.Kozasu, T.Shimizu and H.Kubota, "Recrystallization of Austenite of Si-Mn Steels with Minor Alloying Elements after Hot Rolling", Trans. Iron Steel Inst. Jpn., 11 (1971), PP. 367-375.

10) J.J.Jonas and I.Weiss, "Effect of Precipitation on Recrystallization in Microalloyed Steels", Met. Sci., 13 (1979), PP. 238-245.

11) C.Ouchi, T.Sampei, T.Okita and I.Kozasu, "Microstructural Changes of Austenite during Hot Rolling and Their Effects on Transformation Kinetics", PP. 316-340 in The Hot Deformation of Austenite, John B. Ballance, ed.; AIME, New York N. Y., 1977.

12) T.Tanaka, T.Hatomura and N.Tabata, "Precipitation of Niobium-Carbonitride during Controlled Rolling", Tetsu-to-Hagane, 62 (1976), PP. S207

13) H.Sekine and T.Maruyama, "Fundamentals for the Manufacture of High-Strength, Good-Houghness Steel by Controlled Rolling", PP. 11920-11937 in Seitetsu Kenkyu, Nippon Steel Corp., Tokyo, 1976.

DISCUSSION

Q: How were the authors able to distinguish recrystallization twin boundaries from the deformation bands? According to my experience, it is very difficult to etch a twin boundary unless the austenite is deformed slightly.

A: You mean instead of deformation bands, twin boundaries will be formed?

Q: No. Recrystallization twins exist in the reheated austenite, but it is very difficult to etch them unless the austenite is deformed a little. Usually, a lot of nuclei then appear to be on the recrystallization twin boundaries.

A: Thank you.

Q: Did the authors obtain quantitative data on the influence of deformation bands on the ferrite grain refinement, compared to the influence of the austenite grain boundaries?

A: We don't have exact data at present, but the deformation bands are much more important then the elongated austenite grain boundaries, from the standpoint of obtaining very uniform ferrite grain structures. If the final reductions of the austenite are in the non-recrystallizing region, the area associated with deformation bands is much greater than the increase in area due to the elongated grain shape.

DYNAMIC RECRYSTALLIZATION BEHAVIOR OF AUSTENITE

IN SEVERAL LOW AND HIGH ALLOY STEELS

T. Maki, K. Akasaka[*] and I. Tamura

Department of Metal Science and Technology
Faculty of Engineering, Kyoto University
Sakyo-ku, Kyoto 606, Japan.

*Graduate School, Kyoto University
Sakyo-ku, Kyoto 606.
Present address,
Kobe Steel Ltd., Takasago, Japan.

The dynamic recrystallization behavior of austenite in several steels ranging from low alloy steels containing B, Nb or Cr to high alloy austenitic steels was studied mainly by the optical microstructural observation. The occurrence of dynamic recrystallization of austenite was determined as a function of temperature, strain rate and strain for each steel. Results indicate that the addition of micro-amounts of Nb and large amounts of Cr, Mn or Ni retard the dynamic recrystallization of austenite. The retarding effect of micro-amounts of B or small amounts of Cr on dynamic recrystallization was negligible.

The grain size ($\bar{D}$) of dynamically recrystallized austenite was uniquely related to the Zener-Hollomon parameter Z. The equation $\bar{D} = BZ^{-P}$ was valid for all the steels used. However, when $\bar{D}$ was compared under the same combination of T and $\dot{\epsilon}$, $\bar{D}$ of Nb steel and high alloy steels is smaller than that of other low alloy steels containing B or Cr.

Introduction

The deformation behavior and microstructural changes occurring during hot deformation of steels have become of interest in connection with the controlled rolling of HSLA steels, the hot workability of high alloy steels, the crack formation during continuous casting, etc. It is known that the dynamic recrystallization takes place in austenite during hot deformation under certain deformation conditions in mild or plain carbon steels (1-4), Nb steels (4,5) and 18-8 stainless steel (6-9). However, few attempts have been made to compare the dynamic recrystallization behavior of austenite among various steels.

The purpose of the present investigation was to examine the different dynamic recrystallization behavior of steels varying from low alloy steels containing micro-amounts of Nb or B to high alloy austenitic steels. The deformation conditions (i.e., temperature, strain rate and strain) for the occurrence of dynamic recrystallization of austenite and the grain size of dynamically recrystallized austenite were mainly studied by means of optical microscopy.

Experimental Procedure

Low alloy steels containing 0.001-0.002% B, 0.04% Nb or 1% Cr and high alloy steels containing Cr, Mn or Ni were used. Their chemical compositions are shown in Table I. All are commercial steels except the Fe-Ni-C alloy which was prepared by vacuum induction melting. The steels of group (a) in Table I are transformable during cooling from austenite, and the steels of group (b) retain an austenitic structure even at room temperature.

Table I.   Chemical Compositions of Steels

(Mass %)

		C	Si	Mn	P	S	
a	Fe-0.2C-0.002B	0.19	0.25	0.98	0.018	0.018	B:0.0016
	Fe-0.4C-0.001B	0.40	0.28	0.68	0.013	0.009	B:0.0012
	Fe-0.2C-1Cr	0.21	0.32	0.84	0.015	0.015	Cr:0.96
	Fe-0.4C-1Cr	0.42	0.18	0.81	0.013	0.014	Cr:1.12
	Fe-0.1C-0.04Nb	0.12	0.25	1.40	0.016	0.007	Nb:0.038, V:0.039
	18Ni maraging steel	0.003	0.006	0.02	0.002	0.001	Ni:18.51, Mo:5.08 Co:8.57, Ti:0.72
b	Fe-31Ni-0.3C	0.29	0.036	0.03	0.001	0.001	Ni:31.0
	Fe-14Mn-0.7C	0.68	0.78	13.94	0.033	0.005	Cr:2.35
	18-8 stainless steel	0.050	0.53	1.07	0.029	0.003	Cr:18.42, Ni:9.14

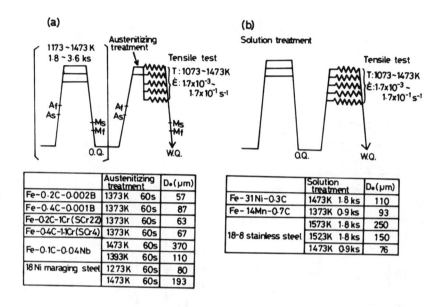

Fig. 1   Heat treatments, initial austenite grain size (D_0)
and deformation conditions.

Tensile specimens with a reduced central gage section of 10 mm in
length and 3.5 mm in diameter were machined from as-received bars. Conditions of heat treatment of specimens and tensile deformation are shown in
Fig. 1. In the case of group (a), specimens were at first austenitized in
a vacuum furnace in order to eliminate the thermal and mechanical histories
of specimen preparation. Next, they were set into tensile machine, austenitized at the temperature indicated in Fig. 1, and then immediately deformed
at the austenitizing temperature or at lower temperatures after atmosphere
gas cooling from the austenitizing temperature. The steels of group (b)
were solution treated in a vacuum furnace and were tensile deformed at
various temperatures lower than the temperature of solution treatment. The
austenitizing temperature or solution temperature for each steel was determined by preliminary experiments in order to obtain the almost same initial
grain size of austenite (D_0 = 60 - 90 μm). However, the Nb steel was
austenitized at higher temperatures compared with other low alloy steels to
dissolve the Nb(CN) precipitates completely or as much as possible. In some
steels, D_0 was changed by adjusting the temperature of austenitization or
solution treatment in order to study the effect of initial austenite grain
size.

Tensile deformation was performed with an Instron-type tensile machine
at various temperatures between 1073 and 1473K at strain rates of 1.7×10^{-3}
to 1.7×10^{-1}/s. Specimens were heated by induction heating in $90\%N_2$ +
$10\%H_2$ gas atmosphere. For the observation of microstructural change during
hot deformation, the specimens were rapidly quenched by water spray within
0.2 - 0.5 sec after the deformation to various strains. The microstructure
was observed by means of optical microscopy on the longitudinal section of
specimens. The austenite grain size was measured by the linear intercept
method.

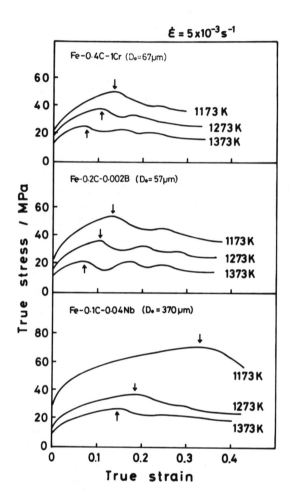

Fig. 2
Stress-strain curves in
0.4C-1Cr, 0.2C-0.002B
and 0.1C-0.04Nb steels
deformed at various tem-
peratures
($\dot{\varepsilon}$ = 5 x 10^{-3}/s).

Results and discussion

Stress-strain curves

Fig. 2 shows examples of the change in stress-strain curve with deformation temperature at strain rate of 5 x 10^{-3}/s in 1Cr-0.4C, 0.002B-0.2C and
0.04Nb-0.1C steels. All curves exhibit the peak stress followed by the
almost steady state flow stress. This type of curve is known to be obtained
when dynamic recrystallization takes place during deformation. It was
actually confirmed by the optical microstructural observation that dynamic
recrystallization occurred under the deformation conditions shown in Fig. 2
for each steel. Oscillations in the stress-strain curve are observed in the
case of deformation at higher temperature. The peak stress (σ_p) and the

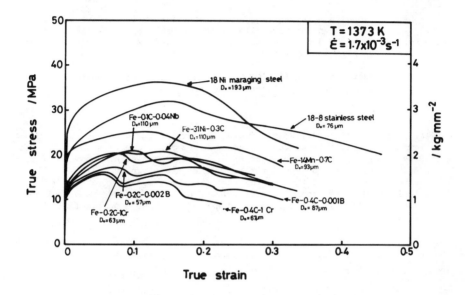

Fig. 3   Stress-strain curves of various steels deformed
at 1373K and at strain rate of 1.7 x 10^{-3}/s.

strain at peak stress (hereafter referred to as a peak strain (ε_p)) are
decreased with increase in deformation temperature. The differences in σ_p
and ε_p (indicated by the arrow in Fig. 2) between Cr steel and B steel are
hardly detected when compared at the same temperature. On the other hand,
in the case of Nb steel, ε_p is larger (especially at 1173K) and σ_p is
slightly larger in comparison with B and Cr steels. This indicates that the
dynamic recrystallization is retarded by the micro-alloying of Nb, and is
hardly influenced by the micro-addition of B or small amount of Cr, since
ε_p is an indication of appearance of dynamic recrystallization.

For comparison of the dynamic recrystallization behavior in all the
steels, Fig. 3 shows the stress-strain curves obtained at 1373K and 1.7 x
10^{-3}/s. It appears that the peak stress (σ_p) and the steady state stress of
high alloy steels are generally higher than those of low alloy steels. And,
the peak strain (ε_p) of high alloy steels and Nb steel is larger than that
of low alloy steels containing B or Cr. This indicates that the onset of
dynamic recrystallization is retarded by the addition of large amounts of
Cr, Mn or Ni, and a micro-amount of Nb. In the case of Nb steel, the ε_p is
intermediate between high alloy steels and other low alloy steels at 1373K
as shown in Fig. 3. However, as is evident from Fig. 2, it is characteristic
of Nb steel that the ε_p becomes fairly large at 1173K, unlike other low
alloy steels. Thus, the retarding effect of Nb on the dynamic recrystallization appears to be most effective at lower temperatures such as 1173K. This
may result from the increase in the fine precipitates of Nb(CN) formed
during deformation due to the decrease in the solubility of Nb.

Oscillations in stress-strain curves are frequently observed in the low alloy steels, although the high alloy steels usually exhibit a single peak. However, even in the high alloy steels, oscillations were observed when deformed at much higher temperatures.

It is generally known (3, 10) that the following equation relates $\dot{\varepsilon}$, T and σ_p when dynamic recrystallization occurs,

$$\dot{\varepsilon} = A\sigma_p^n \exp(-Q/RT) \quad \text{or} \quad Z = \dot{\varepsilon}\exp(Q/RT) = A\sigma_p^n \quad (1)$$

where n: stress exponent, Q: activation energy, R: gas constant and Z: Zener-Hollomon parameter. It was confirmed that the above equation was valid for all the steels used in the present study. Some examples are shown in Fig. 4. The values of n and Q obtained for each steel are given in Table II. The values of n are about 5 for all the steels. The values of Q are 60 - 70 kcal/mol for B or Cr steel, 78 kcal/mol for Nb steel and 80 - 100 kcal/mol for high alloy steels. The 18-8 stainless steel shows a particularly high value of Q.

Table II. Observed Values of n and Q in the equation of
$$\dot{\varepsilon} = A\sigma_p^n \exp(-Q/RT)$$

	n	Q kcal/mol (kJ/mol)
Fe-0.2C-0.002B	4.8	67 (281)
Fe-0.4C-0.001B	4.7	56 (230)
Fe-0.2C-1Cr	5.3	64 (268)
Fe-0.4C-1Cr	4.4	60 (251)
Fe-0.1C-0.04Nb	4.6	78 (327)
18Ni maraging steel	4.6	79 (331)
Fe-31Ni-0.3C	4.5	80 (335)
Fe-14Mn-0.7C	5.4	88 (369)
18-8 stainless steel	4.5	104 (436)

Deformation conditions for the occurrence of dynamic recrystallization

In order to confirm the deformation conditions for the occurrence of dynamic recrystallization of austenite, optical microstructural observations were carried out on specimens deformed to various strains at various combinations of T and $\dot{\varepsilon}$, and immediately quenched to room temperature. Fig. 5 shows optical micrographs of three deformation structures of austenite in a Nb steel, i.e., unrecrystallized (Fig. 5(b))(elongated along the tensile direction), partially dynamically recrystallized (Fig. 5(c)) and completely dynamically recrystallized (Fig. 5(d)). Dynamically recrystallized

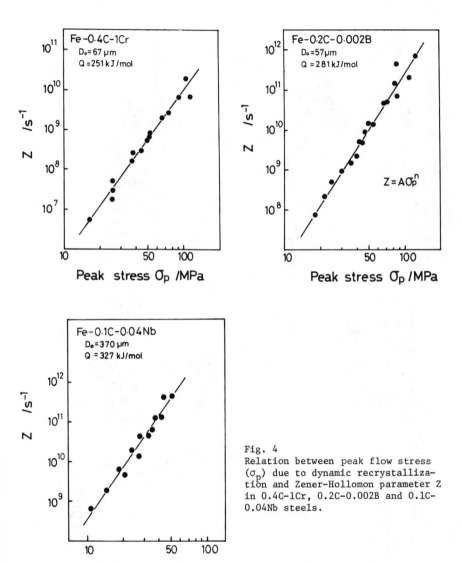

Fig. 4
Relation between peak flow stress
(σ_p) due to dynamic recrystallization and Zener-Hollomon parameter Z
in 0.4C-1Cr, 0.2C-0.002B and 0.1C-
0.04Nb steels.

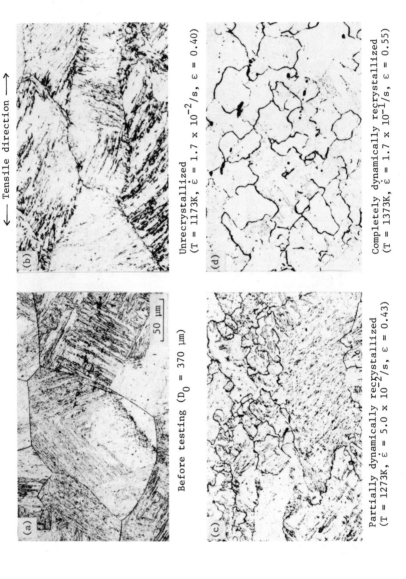

←— Tensile direction —→

(a)

50 μm

Before testing (D_0 = 370 μm)

(b)

Unrecrystallized
(T = 1173K, $\dot{\varepsilon}$ = 1.7 x 10^{-2}/s, ε = 0.40)

(c)

Partially dynamically recrystallized
(T = 1273K, $\dot{\varepsilon}$ = 5.0 x 10^{-2}/s, ε = 0.43)

(d)

Completely dynamically recrystallized
(T = 1373K, $\dot{\varepsilon}$ = 1.7 x 10^{-1}/s, ε = 0.55)

Fig. 5   Optical micrographs of 0.1C-0.04Nb steel deformed under various conditions, immediately followed by quenching. Etchant: 20ml saturated (in H_2O) picric acid + 5ml saturated (in H_2O) lauryl-benzenesulfonic acid sodium salt + 1.8ml 10%aqueous solution of $FeCl_3$ + 0.1ml HCl.

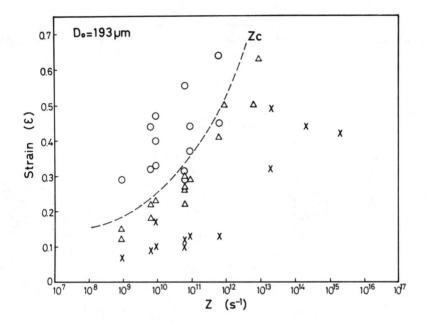

Fig. 6   Relation between high temperature deformation structure
          of austenite and deformation conditions (Z and ε) in
          18Ni maraging steel. 0, Δ and x indicate the completely
          dynamically recrystallized, partially dynamically recry-
          stallized and unrecrystallized, respectively.

austenite grains are characterized by the irregular shaped grain boundaries.
In this steel, although the structure was fully martensitic at room temperature, the austenite structure at high temperature deformation can be recognized by the observation of prior austenite grain boundaries.

    Based on the results of microstructural observations such as Fig. 5,
the high temperature deformation structure of austenite was determined as a
function of T, $\dot{\varepsilon}$ and ε for each steel. Fig. 6 is an example showing the
relation between deformation structure of austenite and deformation conditions of Z and ε in 18Ni maraging steel. At a given Z, with increase in
strain, the austenite structure changes from the unrecrystallized to the
partially recrystallized and finally to completely recrystallized. The
critical strain for obtaining the completely dynamically recrystallized
austenite is increased with increase in Z. Inversely, when specimens were
deformed to a given strain, the austenite structure changes from the completely recrystallized to the unrecrystallized with increase in Z, indicating
that there is an upper critical value of Z (Z_c) for obtaining completely
dynamically recrystallized austenite at a given strain. It is to be emphasized that Z_c increases with increasing amount of strain to be given.

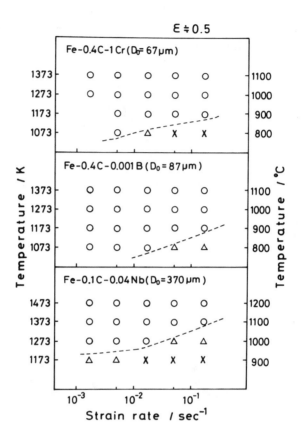

Fig. 7 Summary of deformation structure of austenite in 0.4C-1Cr, 0.4C-0.001B and 0.1C-0.04Nb steels deformed to strain of about 0.5 under various deformation conditions composed of different T and $\dot{\varepsilon}$. 0, $\triangle$ and x indicate the same meaning as in Fig. 6. Broken lines indicate the critical deformation condition for obtaining the completely dynamically recrystallized austenite.

As described in Fig. 6, the occurrence of dynamic recrystallization is influenced by the total amount of strain (ε) even at a given Z. Therefore, in this investigation, for the comparison of dynamic recrystallization between various steels, optical microstructures in specimens deformed to the fixed strain of about 0.5 under various combinations of T and $\dot{\varepsilon}$ were compared with each other. Fig. 7 shows the structural change in austenite under deformation conditions with different T and $\dot{\varepsilon}$ (ε was almost constant at about 0.5) in low alloy steels. It is evident that the dynamically recrystallized structure can be easily obtained by deformation at higher temperature or lower strain rate. Broken lines in Fig. 7 indicate the critical deformation condition for obtaining the completely dynamically recrystallized structure of austenite. In the case of Nb steel, the critical deformation condition for dynamic recrystallization is shifted to higher temperatures or lower strain rates in comparison with those of B or Cr steels.

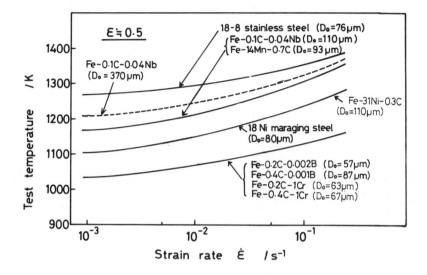

Fig. 8   Critical deformation conditions for completely
dynamically recrystallized austenite when deformed
to strain of about 0.5 in various steels.

In this investigation, such critical lines as shown in Fig. 7 were used
for the comparison of easiness of dynamic recrystallization of austenite in
various steels.   The results are shown in Fig. 8.   In the case of low alloy
steels containing B or Cr, the critical condition is at lower temperatures.
On the other hand, the critical condition in high alloy steels and the Nb
steel is at fairly high temperatures compared with other low alloy steels.
The deformation condition for dynamic recrystallization in the 18-8 stainless
steel is the highest among the steels used.   As can be seen in the Nb steel,
the critical condition slightly shifts to lower temperature with decreasing
the initial austenite grain size from 370 μm to 110 μm.   This indicates that
the initial austenite grain size is also one of factors influencing the
occurrence of dynamic recrystallization (7,8,11).   However, even in the case
of D_0 = 110 μm, the critical condition for Nb steel is at higher temperature
than for other low alloy steels (the temperature difference is about 150 –
200K).   It can be concluded from Fig. 8 that the addition of large amounts
of Cr, Mn or Ni and micro-amounts of Nb retard the dynamic recrystallization,
but the addition of micro-amounts of B and small amounts of Cr and the difference in carbon content (between 0.2 and 0.4%) have hardly any retarding
effect on the dynamic recrystallization of austenite.

Grain size of dynamically recrystallized austenite

It has been previously reported (3,8,12) that the dynamically recrystallized grain size ($\bar{D}$) is uniquely determined by Z, and is independent of
the initial grain size (D_0) and the amount of strain during steady state
deformation.   The same results were obtained in the present study for all the
steels used.   As an example, the variation of the structure of the dynamically
recrystallized austenite with Z in B steel is shown in Fig. 9.   It appears

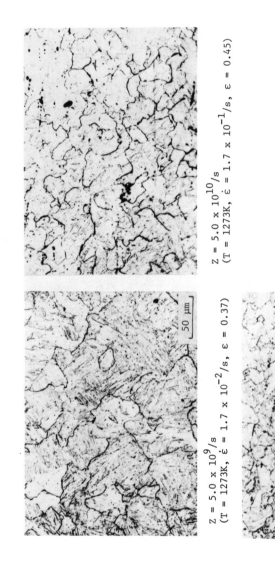

50 μm

Z = 5.0 × 10^{10}/s
(T = 1273K, $\dot{\varepsilon}$ = 1.7 × 10^{-1}/s, ε = 0.45)

Z = 5.0 × 10^9/s
(T = 1273K, $\dot{\varepsilon}$ = 1.7 × 10^{-2}/s, ε = 0.37)

Z = 6.9 × 10^{11}/s
(T = 1073K, $\dot{\varepsilon}$ = 1.7 × 10^{-2}/s, ε = 0.59)

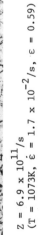

Fig. 9
Optical micrographs of dynamically recrystallized austenite formed under different
deformation conditions in 0.2C-0.002B steel.

Etchant: 20ml saturated (in H$_2$O) picric acid
+ 5ml saturated (in H$_2$O) laurylbenzenesulfonic
acid sodium salt + 1.8ml 10%aqueous solution
of FeCl$_3$ + 0.1ml HCl.

that the dynamically recrystallized austenite grains become finer with an increase in Z. Fig. 10 shows the relation between $\bar{D}$ and Z in some steels. $\bar{D}$ is independent of the initial grain size as can be seen in the results of Nb steel and 18-8 stainless steel. Fig. 10 indicates that the relation of $\bar{D} = BZ^{-p}$ is held between $\bar{D}$ and Z as has been previously reported (3). It was confirmed that the equation $\bar{D} = BZ^{-p}$ was also valid for all other steels used. The observed values of B and p are summarized in Table III. The values of p are about 0.3 -0.4 for all the steels. There seems to be a tendency for the high alloy steels and the Nb steel to exhibit a slightly higher p in comparison with low alloy steels containing B or Cr.

Table III.   Observed Values of B and p in Various Steels

$$\bar{D} = B \, Z^{-p}$$

	B ($\mu m \cdot s^{-p}$)	p
Fe-0.2C-0.002B	4.0×10^4	0.29
Fe-0.4C-0.001B	4.4×10^4	0.36
Fe-0.2C-1Cr	5.3×10^4	0.32
Fe-0.4C-1Cr	3.8×10^4	0.32
Fe-0.1C-0.04Nb	1.9×10^6	0.41
18Ni maraging steel	9.1×10^5	0.40
Fe-31Ni-0.3C	1.6×10^6	0.38
Fe-14Mn-0.7C	5.3×10^5	0.33
18-8 stainless steel	3.7×10^7	0.40

As described above, $\bar{D}$ is uniquely determined only by Z and is decreased with increase in Z in each steel. However, it was found that $\bar{D}$ obtained at the same combination of T and $\dot{\varepsilon}$ was markedly different from steel to steel. As an example, Fig. 11 shows the optical micrographs of dynamically recrystallized austenite obtained by deformation at 1373K and 1.7×10^{-2}/s for B steel, Nb steel and 18-8 stainless steel. With this deformation condition, $\bar{D}$ was 100 μm for 0.002B-0.2C steel, 75 μm for 0.04Nb-0.1C steel and 48 μm for 18-8 stainless steel, respectively. Fig. 12 shows the summary of the changes of $\bar{D}$ with strain rate at 1273K for various steels. It is evident that $\bar{D}$ of Nb steel and high alloy steels are smaller than that of other low alloy steels if compared at the same deformation condition. This indicates that the addition of large amounts of Cr, Mn or Ni and micro-amounts of Nb have a remarkable refining effect on dynamically recrystallized grains.

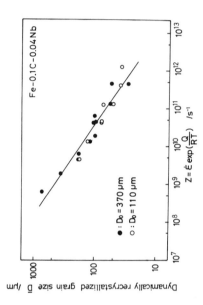

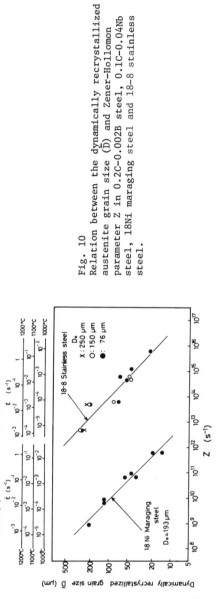

Fig. 10
Relation between the dynamically recrystallized austenite grain size ($\bar{D}$) and Zener–Hollomon parameter Z in 0.2C-0.002B steel, 0.1C-0.04Nb steel, 18Ni maraging steel and 18-8 stainless steel.

230

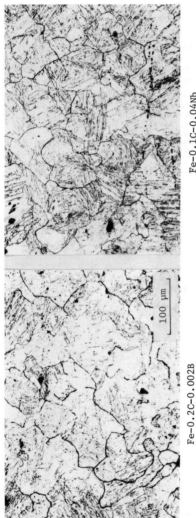

Fe-0.2C-0.002B

Fe-0.1C-0.04Nb

100 μm

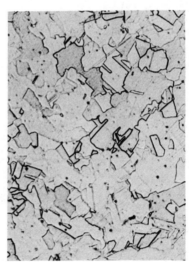

18-8 stainless steel

Fig. 11
Optical micrographs of dynamically
recrystallized austenite in 0.2C-
0.002B steel, 0.1C-0.04Nb steel and
18-8 stainless steel deformed under
the same deformation condition of
T = 1373K and $\dot{\varepsilon}$ = 1.7 x 10^{-2}/s.

Etchant: 20ml saturated (in H$_2$O) picric
acid + 5ml saturated (in H$_2$O)^{2}laurylbenzene-
sulfonic acid sodium salt ∓ 1.8ml 10%aqueous
solution of FeCl$_3$ + 0.1ml HCl for B and Nb
steels and HF 2 : HNO$_3$ 1 : glycelin 2 for
18-8 stainless steel.

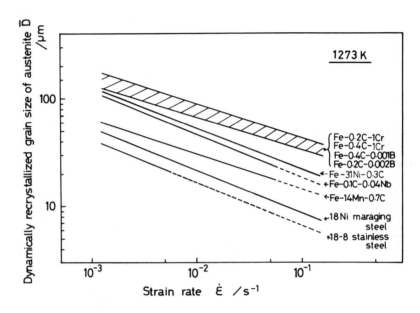

Fig. 12   Relation between $\bar{D}$ and $\dot{\epsilon}$ (at 1273K) in various steels.

As shown above, the Nb steel and high alloy steels exhibit fine dynamic-ally recrystallized grains compared with other low alloy steels containing B or Cr when deformed under the same deformation condition. However, it should be emphasized that, since in the case of low alloy steels containing B or Cr, the dynamic recrystallization can takes place at much lower temper-ature (or much higher strain rate) in comparison with the Nb steel or high alloy steels, the fine $\bar{D}$ can be obtained even in such low alloy steels when deformed under the higher Z conditions in which the Nb steel and high alloy steels do not usually exhibit the dynamic recrystallization. The minimum $\bar{D}$ obtained in the range of the present study was about 10 - 20 μm for all the steels used, as shown in Fig. 10.

Although $\bar{D}$ obtained under the same deformation condition is different between steels, it was found that the $\bar{D}$ is closely related to σ_p, irrespec-tive of the kind of steels and the deformation conditions. Fig. 13 shows the relation between $\bar{D}$ and σ_p for each steel, indicating that data of $\bar{D}$ in all the steels fall on the narrow straight band in log σ_p - log $\bar{D}$ plot.

Conclusions

The dynamic recrystallization of austenite was observed to occur in all the steels used, although the deformation condition for the occurrence of dynamic recrystallization is different between steels. Dynamic recrystal-lization in the Nb steel and high alloy steels was limited to higher temper-atures or lower strain rates in comparison with low alloy steels containing B or Cr. The addition of micro-amounts of Nb and large amounts of Cr, Mn or Ni retards the dynamic recrystallization of austenite. The retarding effect of Nb on dynamic recrystallization appears to be most effective at lower temperatures such as 1173K. On the other hand, the retardation of the dynamic recrystallization due to the addition of micro-amounts of B or small amounts of Cr was negligible.

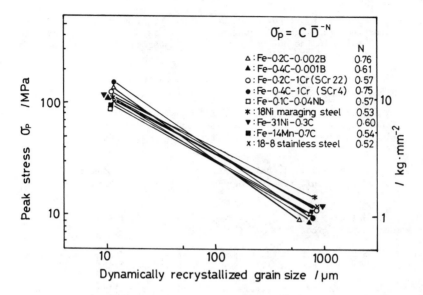

Fig. 13   Relation between flow peak stress (σ_p) and the dynamically recrystallized austenite grain size ($\bar{D}$) in various steels.

The dynamically recrystallized austenite grain size ($\bar{D}$) was uniquely related to the Zener-Hollomon parameter Z. The relation $\bar{D} = BZ^{-P}$ was valid for all the steels used. However, when $\bar{D}$ was compared under the same combination of T and $\dot{\varepsilon}$, $\bar{D}$ of the Nb steel and high alloy steels is smaller than that of low alloy steels containing B or Cr. Furthermore, it was found that $\bar{D}$ is closely related to the peak stress σ_p irrespective of the composition of the steel.

References

1. R. A. Petkovic, M. J. Luton, and J. J. Jonas, "Recovery and Recrystallization of Carbon Steel between Intervals of Hot Working," Can. Met. Quart., 14 (1975) 137.

2. T. Nakamura, and M. Ueki, "The High Temperature Torsional Deformation of a 0.06% C Mild Steel," Trans. ISIJ. 15 (1975) 185.

3. S. Sakui, T. Sakai, and K. Takeishi, "Hot Deformation of Austenite in a Plain Carbon Steel," Trans. ISIJ. 17 (1977) 718.

4. M. J. Luton, R. Dorvel, and R. A. Petkovic, "Interaction Between Deformation, Recrystallization and Precipitation in Niobium Steels," Met. Trans., 11A (1980) 411.

5. I. Weiss, and J. J. Jonas, "Dynamic Precipitation and Coarsening of Niobium Carbonitrides during the Hot Compression of HSLA Steels," Met. Trans., 11A (1980) 403.

6. H. J. McQueen, R. Petkovic, H. Weiss, and L. G. Hinton, "Flow Stress and Microstructural Change in Austenitic Stainless Steel during Hot Deformation," p.113 in The Hot Deformation of Austenite, J. B. Ballance, ed., AIME, New York, NY, 1977.

7. W. Roberts, H. Boden, and B. Ahlblom, "Dynamic Recrystallization Kinetics," Metal Sci., 13 (1979) 195.

8. T. Maki, K. Akasaka, K. Okuno, and I. Tamura, "Dynamic Recrystallization of Austenite in 18-8 Stainless Steel and 18Ni Maraging Steel and Its Related Phenomena," J. Iron and Steel Inst. Japan, 66 (1980) 1659. (in Japanese)

9. E. L. Brown, and A. J. DeArdo, "On the Origin of Equiaxed Austenite Grains that Result from the Hot Rolling of Steel," Met. Trans., 12A (1981) 39.

10. C. M. Sellars, and W. J. McG. Tegart, "Hot Workability," Int. Metallurgical Rev., 17 (1972) 1.

11. J. P. Sah, G. J. Richardson, and C. M. Sellars, "Grain-Size Effects during Dynamic Recrystallization of Nickel," Metal Sci., 8 (1974) 325.

12. H. J. McQueen, and S. Bergerson, "Dynamic Recrystallization of Copper during Hot Deformation," Met. Sci. J., 6 (1972) 25.

DISCUSSION

Q: You have shown a lot of recrystallization twins in an austenitic stainless steel which you claim has recrystallized dynamically. Normally, dynamic recrystallization does not produce recrystallization twins. Would you comment on that?

A: First, some reports suggest that 18-8 stainless steels cannot dynamically recrystallize, but my experiments show that austenitic stainless steels do recrystallized dynamically. You are right about the recrystallization twins. We could not find any recrystallization twins in the dynamically recrystallized grains. Initial austenite has a lot of annealing twins, but recrystallized grains do not have any twins.

Q: Did you include the twin boundaries as grain boundaries when you measured the grain size in the 18/8 steel?

A: For the initial grain size measurement, we used lineal analysis and included the annealing twin boundaries. After dynamic recrystallization we could not find any twin boundaries. Annealing twins exist in solution treated austenite grains in high alloy austenitic steels, especially in 18-8 stainless steel. However, annealing twin boundaries were not counted for the measurement of initial austenite grain size in our study.

Comment: I think it is accepted that the dynamically recrystallized austenite grains contain very few annealing twins compared with the statically recrystallized austenite. However, it is not true to say that the dynamically recrystallized austenite does not always contain the annealing twin at all. Annealing twins are sometimes observed in dynamically recrystallized austenite as shown in Fig. 11(c) in our paper. Judging from our recent results on the morphology of dynamically recrystallized austenite, there seems to be a tendency that the density of annealing twins in dynamically recrystallized austenite is increased with decrease in Z value (i.e., at higher temperatures or lower strain rates). However, even in the dynamically recrystallized grains formed at lower Z, the annealing twin density is usually lower than that of annealed austenite (i.e., static recrystallized austenite).

COMMENT: As to the measurement of initial austenite grain size, annealing twins exist in solution treated austenite grains in high alloy austenitic steels, especially in 18-8 stainless steel. However, annealing twin boundaries were not counted for the measurement of initial austenite grain size in our study.

COMMENT: As to the annealing twins in the dynamically recrystallized austenite grains of 18-8 stainless steel, I think it is accepted that the dynamically recrystallized austenite grains contain very few annealing twins compared with the statically recrystallized austenite. However, it is not true to say that the dynamically recrystallized austenite does not always contain the annealing twin at all. Annealing twins are sometimes observed in dynamically recrystallized austenite as shown in Fig. 11(c) in our paper. Judging from our recent results on the morphology of dynamically recrystallized austenite, there seems to be a tendency that the density of annealing twins in dynamically recrystallized austenite is increased with decrease in Z value (i.e., at higher temperatures or lower strain rates). However, even in the dynamically recrystallized grains formed at lower Z, the annealing twin density is usually lower than that of annealed austenite (i.e., static recrystallized austenite).

THE ROLE OF DYNAMIC RECRYSTALLIZATION
IN PRODUCING GRAIN REFINEMENT AND
GRAIN COARSENING IN MICROALLOYED STEELS

T. Sakai,* M.G. Akben and J.J. Jonas

Dept. of Metallurgical Engineering
McGill University
3450 University Street
Montreal, Canada, H3A 2A7

The critical strain criterion $\varepsilon_p = \varepsilon_x$ for the transition from cyclic to 'continuous' dynamic recrystallization is demonstrated to be invalid for the high temperature deformation of austenite in tension and in compression. The role of the strain and strain rate gradients present in solid torsion bars in raising the apparent torsion peak strain ε_p above the ε_p values obtained from homogeneous tension or compression testing is clarified. A similar, and larger, effect is shown to cause discrepancies in the values of the recrystallization strain ε_x derived from torsion data. An alternative criterion for the transition is deduced, based on the grain size measurements performed by Sakai and co-workers on a 0.16% C steel deformed in the γ range of temperatures. Their observations indicate that *cyclic* flow curves are associated with grain *coarsening* and that *single peak* flow curves are associated with grain *refinement*. The critical condition is $D_0 = 2D_s$, where D_0 and D_s are the initial and stable grain sizes respectively. The relative grain size model is confirmed by critical tests performed on a 0.115% V, a 0.035% Nb and a 0.040% Nb – 0.30% Mo microalloyed steel. The results indicate that single peak behavior is caused by the 'necklace' or 'cascade' recrystallization of coarse-grained materials which, because of its inhomogeneous and localized character, produces a large spread in the nucleation strain ε_c. By contrast, recrystallization is nearly completely synchronized in fine-grained materials, because the high density of recrystallization nuclei leads to a small spread in the nucleation strain.

*On leave from the University of Electro-communications, Tokyo, Japan.

237

Introduction

The conditions that favour the occurrence of dynamic recrystallization during controlled rolling are relatively well-known (1,2). They include the following: relatively high temperatures and low strain rates,
large accumulated strains (i.e. retained work hardening without static
recrystallization), and a low total concentration of alloying elements
(3). Although dynamic recrystallization does not always take place under
industrial conditions, when it *does* occur, it produces both textures (4)
and grain size distributions (5) that differ from those associated either
with static recrystallization between passes, or with the *absence* of static
recrystallization, i.e. under dynamic recovery or 'pancaking' conditions.

It has been reported by numerous authors (6-11) that when dynamic
recrystallization is initiated, it can lead to flow curves that are
either periodic (multiple peak) or 'continuous' (single peak) in nature.
Less well known is the close association between *cyclic* flow curves and
grain coarsening on the one hand, and between the flow softening that
follows a *single* stress maximum and *grain refinement* on the other. The
cause-and-effect relationship between the *sense* of grain size change and
the *form* of the flow curve was first pointed out by Sakai and co-workers
in Japan (12-15). It will be the first aim of the present paper to review
briefly the work of these researchers, with particular reference to its
application to the thermomechanical processing of austenite. After a
presentation of the general principles enunciated by these investigators,
some results will be described which were obtained recently on a series
of microalloyed steels. These will be shown to support the rationalization
proposed by Sakai et al. (12-15), and to confirm the key roles played by
'necklace' or 'cascade' recrystallization and size saturation during grain
refinement and by boundary impingement during grain coarsening (16).

Limitations of the Critical Strain
Model for Dynamic Recrystallization

The classical explanation for the transition from periodic to single
peak recrystallization was proposed by Luton and Sellars in 1969 (10).
On the basis of torsion testing results obtained on nickel and nickel-iron
alloys, they showed that *cyclic* recrystallization occurred when the peak
strain ε_p was *greater* than the recrystallization strain ε_x. (Here ε_x =
$\varepsilon_s - \varepsilon_p$, where ε_s is the strain at which the flow stress first goes through
its mean value at the start of the second flow stress cycle.) Conversely,
single peak recrystallization took place when ε_p was *less* than ε_x, where in
this case $\varepsilon_s = \varepsilon_p + \varepsilon_x$ is the strain at which the steady state of flow is
first attained. The critical strain criterion of Luton and Sellars appears
to work quite well for the *torsion* testing of steel, as illustrated in Fig.
1a. Presented here are values of ε_p and ε_x abstracted from the data of
Rossard and Blain (7), which were determined at 1100°C on a 0.25% plain C
steel. The intersection of the two curves at a stress value of about 46
MPa should be particularly noted.

It is of interest that the ε_p and ε_x curves do not intersect for
testing in tension or in compression, as clearly demonstrated by Sakai
and co-workers (12-15). Their *tension* results are displayed in Fig. 1b for
a 0.16% C steel deformed at various strain rates and at temperatures from
890 to 990°C. The absence of an intersection is evident in this plot, as
well as in the curves of Fig. 1c, which represent the critical strains
obtained by Petkovic et al. (11) on a 0.68% C steel which was strained in
compression.

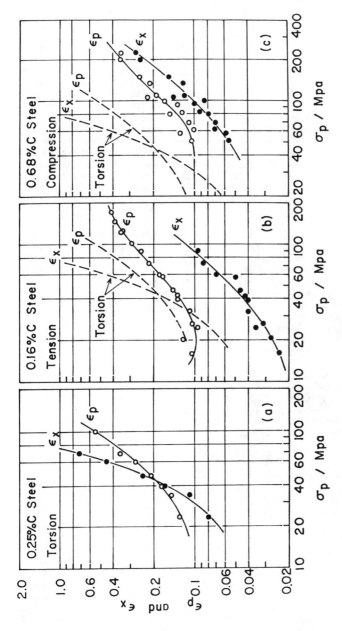

Fig. 1 Dependence of ϵ_p and ϵ_x on the peak stress σ_p. (a) Determined from the *torsion* flow curves of Rossard and Blain (7) (note intersection). (b) Determined from the *tension* data of Sakai et al. (13,14). Note that ϵ_p and ϵ_x do not intersect in this case. (c) Determined from the axisymmetric *compression* data of Petkovic et al. (11). Again the ϵ_p and ϵ_x curves do not intersect.

Primary Differences Between Torsion Testing and Deformation in Axisymmetric Tension or Compression

It is apparent from Fig. 1 that the peak strains determined in tension or compression are generally *lower* than those obtained in torsion testing. This can be attributed to the strain and strain rate gradients present in *solid* torsion samples. As a result of this gradient, the peak strain is attained first in the outermost layer of a cylindrical sample, the condition being satisfied later and later (in time) in shells of material of successively smaller radius. Because of this effect, and of the way in which the individual shells contribute to the macroscopic torque, the maximum in the couple is observed *later in time* than the moment associated with the peak stress in the outermost shell. For this reason, the stress/strain curves deduced from torque/twist data generally display maxima at strains *greater than* the peak strains determined in tension or compression at the same temperature and effective strain rate (13,14). Although this disparity is not likely to affect the results obtained from the torsion testing of *tubes*, it does apply to the differences between torsion testing and other homogeneous modes of deformation such as rolling or plane strain tension or compression (16).

There is a further component of difference between torsion-based flow curves and those determined in axisymmetric tension or compression which applies to the results of *both* tube and solid bar experiments. This is associated with the differences in the *textures* developed in these three modes of testing, as well as in the rates of *work hardening*. These two effects combine to further increase the differences between the peak strains deduced from torque/twist data and from axisymmetric testing (17, 18). By contrast, because torsion and rolling are both modes of plane strain deformation, the work hardening and texture differences between these two methods of straining are considerably smaller than between say torsion and either extrusion or continuous forging, both of which are axisymmetric in nature (17,18).

A similar but more complex consideration affects the apparent recrystallization strain ε_x determined from the torsion testing of solid bars (16). As is evident from Fig. 1, the differences between the torsion-based and axisymmetric values of ε_x are *even greater* than between the torsion-based and axisymmetric values of ε_p. This is because ε_x is obtained experimentally from the relation $\varepsilon_x = \varepsilon_s - \varepsilon_p$, and thus includes components of 'time averaging' from *both* ε_s and ε_p. This effect is discussed in greater detail in Reference 16.

The Evolution of Grain Size with Strain During the Dynamic Recrystallization of Austenite

An alternative to the critical strain condition as an explanation for the transition in recrystallization behaviour was proposed by Sakai and co-workers in 1976 (12). Their interpretation was based on results they obtained in tension on sheet specimens of a 0.16% C plain carbon steel. Their samples were deformed in vacuo in specially constructed tensile machines which were fitted with a H_2 gas quenching attachment. This facility permitted cooling rates of $2500°C/s$ to be attained and thus enabled the austenite grain structures present at each selected strain to be preserved for study at room temperature. Their results are summarized in Fig. 2, from which it is apparent that there were two distinct classes of behaviour.

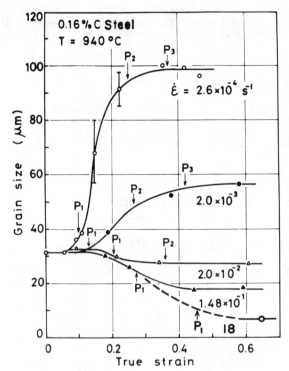

Fig. 2.
Effect of strain and
strain rate on the mean
grain size of a 0.16% C
steel deformed in tension
at 940°C. P_i identifies
the strain at the ith
flow stress peak. Note
that *multiple peaks*
are observed during grain
coarsening (and when grain
refinement produces less
than a 2:1 reduction in
grain size). Conversely,
a *single peak* is observed
when the grain *refinement*
ratio is more than 2:1.
Grain refinement is detectable prior to P_1 because
dynamic recrystallization
is nucleated well before
the peak strain (12,14).

When the testing strain rate was *low* (e.g. 2.6 x 10⁻⁴ or 2.0 x 10⁻³
s⁻¹), *grain coarsening* was observed. The coarsening was associated with
the occurrence of *multiple stress peaks*; the peak strains are identified
as P_1, P_2 and P_3 in Fig. 2. When the testing strain rate was relatively
high (e.g. 1.48 x 10⁻¹ or 18 s⁻¹), *grain refinement* took place. Under
these conditions, which are closer to industrial rates of straining, only
a *single stress peak* was apparent. The transition from multiple to single
peak behaviour did *not* occur, as perhaps might be expected, at $D_o = D_s$,
where D_o and D_s are the initial and stable grain sizes respectively. Thus
the transition in the nature of the flow curve did not coincide with the
transition from grain coarsening to grain refinement. Instead, multiple
peaks appeared to persist as long as the grain refinement ratio did not
exceed $D_o/D_s = 2$.

Thus the critical factor appears to be whether or not 'cascade' or
'necklace' recrystallization can take place (16). Under the latter conditions, a different critical strain ε_c is associated with each strand of
the necklace, leading to a large spread in the nucleation strains pertaining
to the different layers of material involved in the first full cycle of
recrystallization. By contrast, when $D_o/D_s < 2$, cascade growth is not
possible, and the initial structure is completely replaced in a single
concurrent cycle of recrystallization. In this case, the spread in the
nucleation strain is of course much smaller (16). The observations of
Sakai and co-workers suggest that there is a link between the spread in

241

the nucleation strain ε_c and the grain size ratio $D_O/2D_s$. This is a point to which we will return after a consideration of further experimental evidence regarding the characteristics of grain coarsening and grain refinement during deformation.

A Grain-Size-Based Critical Condition for
Cyclic Recrystallization

The two kinds of behaviour described above can be distinguished in terms of the microstructural mechanism map of Fig. 3. Here the testing conditions are represented by the vertical coordinate of temperature-corrected strain rate Z or by its equivalent, the peak stress σ_p, or the

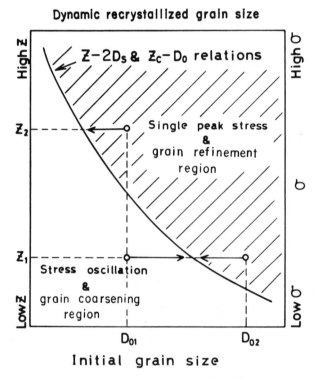

Initial grain size

Fig. 3. The curve describing the $Z-2D_s$ and Z_c-D_O relations distinguishes the single peak from the multiple peak region. When testing is carried out at a fixed temperature and strain rate (fixed Z), grain coarsening and repeated stress oscillations occur as long as each successive $D_O < 2D_s$ (e.g. at D_{O1} and Z_1). Conversely, when $D_O > 2D_s$, there is grain refinement and a single stress peak (e.g. at D_{O2} and Z_1). When a fixed initial grain size is used (e.g. D_{O1}), stress oscillations and grain coarsening are associated with $Z < Z_c$ (e.g. $Z_1 < Z_c$), and a single stress peak and grain refinement with $Z > Z_c$ (e.g. $Z_2 > Z_c$).

242

steady state stress σ_s. Two alternative horizontal axes are shown, the initial grain size D_0 and the equilibrium grain size D_S, both of which enter into an interpretation of the diagram. The full line dividing the cross-hatched from the plain region describes the *dependence* of $2D_S$ on Z (and therefore the *correlation* between $2D_S$ and σ_p) that is observed. The same curve also specifies the *dependence* of Z_c on D_0, where Z_c is the critical value of Z at which, for a fixed value of D_0, the nature of the flow curve changes from cyclic to single peak, or vice versa.

It is of interest that two distinct types of experiment are represented in this diagram. A 'vertical' series of tests corresponds to the classical case where a *fixed* initial grain size is taken, e.g. D_{01}, and where Z is varied (by changing the temperature or the strain rate) from say Z_1 in the grain coarsening region to Z_2 in the grain refinement region. As Z is increased, the oscillations in the flow curves gradually disappear, as has been demonstrated by numerous researchers (5-16). The second type of experiment corresponds to a 'horizontal' series of tests, in which the temperature-corrected strain rate is held constant at say Z_1 and the *initial* grain size is varied over the range D_{o1} to D_{o2}. To test the criterion represented in Fig. 3, a series of 'horizontal' tests was carried out, the results of which will now be described.

A Critical Test of the Grain-Size-Based
Criterion Performed on a Vanadium Microalloyed Steel

The microalloyed steel selected for the critical experiment contained 0.06% C, 1.2% Mn, 0.115% V and 0.006% N. In order to produce a range of initial grain sizes, individual samples were austenitized at temperatures of 975, 1000, 1030, 1100 and 1260°C. This led to grain sizes in the range 65 to 300 μm, as indicated in Fig. 4. The samples were hot compressed in a modified Instron testing machine (5) at a constant true strain rate of 1.4 x 10^{-3} s^{-1} and at 975°C. For this Z condition, D_S was determined metallographically to be ≈50 μm, so that $2D_S$ = 100 μm. It is apparent from Fig. 4 that when D_0 was less than $2D_S$, i.e. for the 65 and 80 μm material, multiple stress peaks were induced, which were not damped out in the 65 μm material by the time the experimental strain limit of 1.0 was reached. It should be noted, however, that the amplitude of the oscillations decreased with strain. In terms of Fig. 3, this means that the amplitude decreases as each successive wave of grain coarsening brings the mean grain size closer to D_S.

For the coarse-grained materials, i.e. when D_0 was greater than $2D_S$, a single marked flow stress peak was observed, followed by a small, but distinct, second peak. According to the results obtained on plain carbon steels (12-16), this is accompanied by the occurrence of grain refinement, which comes to an end when the stable mean grain size D_S is attained. This type of behaviour corresponds to industrial rolling after either continuous casting, or after reheating to a fairly high temperature.

Effect of Microalloying on Grain
Refinement and Grain Coarsening

The usual effect of the presence of microalloy elements on the shape of the flow curve is depicted in Fig. 5. Here it can be seen that, as the overall 'strength' of the alloying additions is increased, the oscillations evident in the case of the plain carbon material gradually disappear. The tests in Fig. 5a, which were conducted at 1075°C, were all completed in about 20 s; thus no precipitation of nitrides or carbonitrides occurred

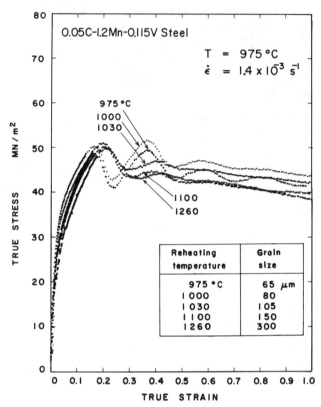

Fig. 4 Critical experiment designed to test the model of Fig. 3. A 0.05% C -
1.2% Mn - 0.115% V steel was austenitized at a series of temperatures ranging from 975 to 1260°C selected to produce initial austenite grain sizes of 65 to 300 μm, respectively (see inset).
Testing was carried out in axisymmetric compression at 975°C and
1.4 x 10^{-3} s^{-1}, for which $2D_S \cong 100$ μm. Cyclic σ/ε behaviour was
observed when $D_O < 2D_S$; by contrast, single peak behaviour was
obtained when $D_O > 2D_S$.

during the experiments (19,20). The differences between the curves are
therefore attributable entirely to the influence of the alloying elements
in solution.

     In this case, however, the rationalization based on the present model
differs somewhat from the one proposed above for Fig. 4. In that experiment, although the *final* grain size was the same for all five tests, the
initial ones were not, leading to grain size ratios $D_O/2D_S$ both larger
and smaller than one. By contrast, in the present instance, it was the
initial grain size which was held constant. The change in the $D_O/2D_S$
criterion was produced instead by varying the *stable* grain size D_S (at each
fixed temperature and strain rate) through the influence of the microalloying elements in solution on the recrystallization process. For example,

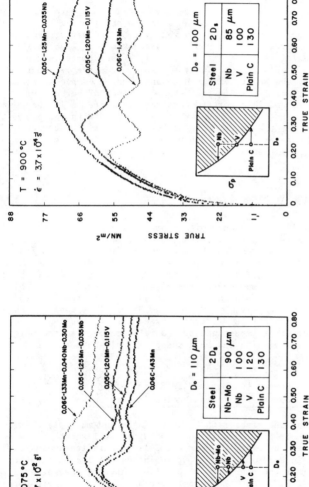

Fig. 5(a). Effect of microalloy addition on the shape of the flow curve. The plain carbon and 0.115% V steels undergo cyclic recrystallization because $D_0/2D_s < 1$. The Nb and Nb-Mo grades display single peak behaviour because $D_0/2D_s > 1$. The transition in the character of the curves can be ascribed to the effect of the solute elements on the stable grain size produced at a given temperature and strain rate.

Fig. 5(b). Effect of dynamic precipitation on the shape of the flow curve. The plain carbon steel exhibits cyclic recrystallization because $D_0/2D_s = 0.77$. Conversely, the Nb steel, with $D_0/2D_s = 1.18$, displays single peak behaviour. The V steel, with $D_0/2D \cong 1$, is at the transition between the two types of flow. The greater difference between the three steels apparent above than in Fig. 5(a) is attributable to the occurrence of dynamic precipitation during the compression of the two HSLA materials.

D_S for the C and V steels is about 65 and 60 μm, respectively, so that the relative grain size $D_0/2D_S$ is 0.85 and 0.91, which is consistent with the occurrence of grain coarsening and stress oscillations. For the Nb and Nb-Mo steels, on the other hand, D_S is about 50 and 45 μm, respectively, so that $D_0/2D_S$ is 1.10 and 1.22. As these values are greater than one, grain refinement and single peak behaviour are indicated.

A similar explanation applies to the curves of Fig. 5b. Here the initial grain size is again constant and the addition of alloying elements changes the character of the flow curves by reducing D_S. However, because these experiments were carried out at 900°C, and each test took about 2000 s to complete, the precipitation of Nb(CN) and of VN took place in the 0.035% Nb and 0.115% V steels, respectively (19,20). Thus the retardation of recrystallization by the solute effect was supplemented by a component of precipitate retardation. It is in fact the occurrence of carbonitride precipitation that is responsible for the considerable increase in the peak stress and strain in the HSLA steels over those associated with the plain carbon grade. This is why the difference in the behavior of the carbon steel on the one hand and the vanadium and niobium steels on the other is greater at 900°C (Fig. 5b) than at 1075°C (Fig. 5a), where no precipitation took place.

Cyclic Recrystallization in Microalloyed Austenite

Although the addition of elements with appreciable solute and precipitate effects generally promotes the transition from cyclic to single peak flow (cf. Fig. 5), this trend can be successfully reversed by making use of the principles described above. In Fig. 6, for example, *it is the Nb-Mo steel which exhibits oscillations and the plain carbon material which displays a single peak*. This somewhat paradoxical result was produced by ensuring that $D_0/2D_S$ for the plain carbon sample was *greater* than one (375/270 = 1.39), and that, conversely, $D_0/2D_S$ for the Nb-Mo alloy was *less* than one (150/270 = 0.55). Such a role reversal between the two types of steel helps to confirm the critical importance of the relation between the original and equilibrium grain sizes which has been the principal subject of the present paper.

The Link Between the Grain Size Ratio $D_0/2D_S$
and the Spread $\Delta \varepsilon_c$ in the Nucleation Strain

The conclusion that the transition between single and multiple peak behaviour occurs at the critical condition $D_0 = 2D_S$ is well supported by the observations reported by other workers (21-24). A detailed analysis of such results is given in Reference 16. We turn instead to a consideration of how it is possible for the *character* of the flow curve to change as sharply as it does in Fig. 4 when the temperature and strain rate are both held constant, and *only the initial grain size is changed*. Similar questions can be posed regarding the nature of the flow curves in Fig. 5. These were also determined at constant temperature and strain rate, but the alloy concentrations associated with the various curves are different at the *same* initial grain size.

Some light can be shed on these matters by considering the relative densities of dynamic recrystallization nuclei for the above experimental conditions. For simplicity, in what follows, the possibility of nucleating new grains at deformation bands or inclusions is neglected and all new

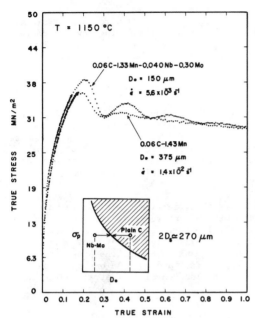

Fig. 6.
A further critical experiment designed to test the relative grain size model. Here the Nb-Mo steel exhibits *cyclic* behaviour and the plain carbon material undergoes a *single cycle* of flow softening. Although the stable grain sizes are similar for the two materials, the use of two different reheating temperatures (1260 and 1150°C for the plain carbon and Nb-Mo grades, respectively) led to relative grain sizes larger and smaller than one, and therefore to the observed inversion of the normal relationship between the two types of behaviour (Fig. 5a).

grains are assumed to form solely at the existing grain boundaries. Under these conditions, we can define $N_S = 2P/D_S$ as the equilibrium density of nuclei present during steady state flow, where P is the probability per unit surface area of a grain boundary site being activated (25). Thus the critical density associated with the criterion $D_O = 2D_S$ is $N_O^* = P/D_S$. When N_O (= $2P/D_O$) > N_O^* (i.e. in fine-grained materials), the flow is periodic; conversely, when $N_O < N_O^*$ (i.e. in coarse-grained materials), only one cycle of flow softening is observed.

A possible explanation of this behaviour can be based on the link between the nucleation strain ε_c (and particularly the *spread* in the nucleation strain $\Delta\varepsilon_c$) and the relative nucleus density N_O/N_O^*. When N_O/N_O^* is high, there are *many* potential nucleation sites within every future stable grain. Thus nucleation can be expected to have occurred in *all* regions within a short interval after the critical strain ε_c is attained in the most favourably oriented region. The *spread* in the critical strain $\Delta\varepsilon_c$ between the most and least favoured regions is likely to be small in this case. This is equivalent to the condition (16):

$$\Delta\varepsilon_c < \varepsilon_c \tag{1}$$

It signifies that the entire material is essentially at the same stage of the recrystallization process at a given moment. It is implicit in this view that several recrystallization cycles are needed to attain D_S because the initial nucleus density N_O is too high, so that too fine an intermediate grain size D_S' is produced in the first cycle. According to this model, the number of recrystallization cycles depends on the disparity between N_O and N_O^*.

By contrast, in the coarse-grained material, N_o/N_o^* is low, and there is a *scarcity* of potential nucleation sites associated with each stable grain. Indeed, there are NO potential nucleation sites in the regions corresponding to the inner strands of 'necklace' recrystallization until *after* the immediately preceding strand has been produced. Thus $\Delta\varepsilon_c$ generally exceeds ε_c under these conditions; i.e.

$$\Delta\varepsilon_c > \varepsilon_c \tag{2}$$

signifying that the different volume fractions of the material are at entirely different stages of the recrystallization process. According to this model, flow softening continues until the grain centers of coarse-grained materials have participated in the recrystallization process. Because of the relative scarcity of nucleation sites, if D_s is not attained in the first cycle of recrystallization, a faint second cycle is observed.

It is evident from the above that, in fine-grained materials, i.e. when $N_o > N_o^*$ and $D_o < 2D_s$, the progress of recrystallization is highly *synchronized* and the macroscopic flow curve readily reflects the *local* cycles of work hardening and recrystallization. Conversely, in coarse-grained materials, i.e. when $N_o < N_o^*$ and $D_o > 2D_s$, the progress of recrystallization is highly *unsynchronized* and only an *average* flow stress can be measured, which obscures the local cycles of strain hardening and restoration.

Conclusions

1. The critical strain model for explaining the transition from periodic to single peak flow does not apply to tension or compression, or to other homogeneous modes of deformation such as rolling and upset or continuous forging, because of the strain and strain rate gradients present during the torsion testing of solid bars. As a result of these gradients, the peak strain ε_p is attained progressively later as the radius of an elemental shell in the solid bar is decreased. In this way, the peak torque (and therefore the peak stress determined experimentally) is attained *later* (and at a larger effective strain) than in the outermost layer of the bar, or than in a tensile test performed at the effective strain rate associated with the outermost layer. This effect is compounded for the recrystallization strain ε_x, so that the differences between the values of ε_x determined in torsion and compression are even greater than the differences in the torsion and compression values of ε_p.

2. In a 0.16% C plain carbon steel tested in the γ range, periodic flow curves are associated with grain coarsening. By contrast, single peak flow curves are observed when the grain refinement ratio is at least 2:1. The criterion that distinguishes between the two types of behaviour is $D_o = 2D_s$, where D_o and D_s are the initial and equilibrium grain sizes, respectively. Under industrial conditions of austenite processing, because of the coarse grain size produced on reheating and the lower subsequent processing temperatures, single peak behaviour is generally expected, accompanied by grain refinement.

3. The critical grain size model was verified in a 0.115% V, a 0.035% Nb and a 0.040% Nb - 0.30% Mo microalloyed steel, under both 'solute retardation' conditions, as well as when there is an additional component

of 'precipitate retardation'. It is evident from these results that 'necklace' or 'cascade' recrystallization (in which grain growth is halted by concurrent deformation) is associated with single peak behaviour. Conversely, when grain growth is terminated by boundary impingement instead, as under grain coarsening conditions, the flow curve is cyclic.

4. An analysis of these two types of behaviour indicates that the necklace nucleation that takes place in coarse-grained materials leads to a large spread in the nucleation strain and therefore to a form of dynamic recrystallization which is highly unsynchronized. In fine-grained materials, on the other hand, there is a high density of nuclei present, as a result of which the spread in nucleation strain is small and all regions of the material are at approximately the same stage of the recrystallization process.

Acknowledgements

The authors are indebted to Dr. C.M. Sellars of the University of Sheffield for a stimulating discussion regarding the critical grain size criterion. They also acknowledge with gratitude the financial support received from the following sources: the Ministry of Education of Japan and the Hot Deformation Committee of the Iron and Steel Institute of Japan (TS); the Natural Sciences and Engineering Research Council of Canada, the Department of Energy, Mines and Resources, Canada, and the Quebec Ministry of Education (FCAC program) (MGA and JJJ). One of the authors (TS) expresses his thanks to the University of Electro-communications, Tokyo, for granting a period of sabbatical leave and to the Ministry of Education of Japan for providing a travelling fellowship.

References

1. H. Sekine and T. Maruyama, 'Controlled Rolling for Obtaining a Fine and Uniform Structure,' pp. 85-88 in 3rd Int. Conf. on the Strength of Metals and Alloys; Metals Soc., London, 1973.

2. I. Kozasu, C. Ouchi, T. Sampei, and T. Okita, 'Hot Rolling as a High-Temperature Thermo-Mechanical Process,' pp. 120-135 in Microalloying 75; Union Carbide Corp., New York, 1977.

3. A.H. Ucisik, I. Weiss, H.J. McQueen and J.J. Jonas, 'Multistage Hot Deformation with Decreasing Temperature of Two Plain Carbon and Two HSLA Steels,' Can. Metall. Quart., 19 (1980) pp. 351-358.

4. E.L. Brown and A.J. DeArdo, 'On the Origin of Equiaxed Austenite Grains that Result from the Hot Rolling of Steel,' Met. Trans., 12A (1981) pp. 39-47.

5. R.A. Petkovic, M.J. Luton, and J.J. Jonas, 'Recovery and Recrystallization of Polycrystalline Copper After Hot Working,' Acta Met., 27 (1979) pp. 1633-1648.

6. S. Steinemann, 'Experimental Investigations of the Plasticity of Ice (Experimentelle Untersuchungen zur Plastizität Von Eis),' Beiträge Geol. Schweiz (Hydrologie), 10 (1958) pp. 1-72.

7. C. Rossard and P. Blain, 'Evolution of Steel Microstructure during Plastic Deformation at High Temperature (Evolution de la structure de l'acier sous l'effect de la déformation plastique à chaud),' Mém. Sci. Rev. Mét., 56 (1959) pp. 285-299.

8. D. Hardwick and W.J. McG. Tegart, 'Structural Changes During the Deformation of Copper, Aluminum and Nickel at High Temperatures and High Strain Rates,' J. Inst. Metals, 90 (1961-62) pp. 17-21.

9. J.J. Jonas, C.M. Sellars, and W.J. McG. Tegart, 'Strength and Structure Under Hot-Working Conditions,' Metallurgical Rev., 14 (1969) pp. 1-24.

10. M.J. Luton and C.M. Sellars, 'Dynamic Recrystallization in Nickel and Nickel-Iron Alloys During High Temperature Deformation,' Acta Met., 17 (1969) pp. 1033-1043.

11. R.A. Petkovic, M.J. Luton, and J.J. Jonas, 'Recovery and Recrystallization of Carbon Steel between Intervals of Hot Working,' Can. Met. Quart., 14 (1975) pp. 137-145.

12. S. Sakui, T. Sakai, and K. Takeishi, 'Effects of Strain, Strain Rate, and Temperature on the Hot Worked Structure of a 0.16% Carbon Steel,' Tetsu-to-Hagane, 62 (1976) pp. 856-865.

13. S. Sakui and T. Sakai, 'Deformation Behaviour of a 0.16% Carbon Steel in the Austenite Range,' Tetsu-to-Hagane, 63 (1977) pp. 285-293.

14. S. Sakui, T. Sakai, and K. Takeishi, 'Hot Deformation of Austenite in a Plain Carbon Steel,' Trans. Iron Steel Inst. Japan, 17 (1977) pp. 718-725.

15. T. Sakai, 'Some Problems Involving Dynamic Recrystallization,' pp. 34-52 in Hot Deformation and Fracture, I. Tamura, ed.; Joint Society of Iron and Steel Basic Research, Tokyo, 1981.

16. J.J. Jonas and T. Sakai, 'The Transition from Multiple to Single Peak Recrystallization During High Temperature Deformation,' in Les Traitements Thermomécaniques: Aspects Théoriques et Applications, 24éme Colloque de Métallurgie, INSTN, Saclay, France, June 1981 (in press).

17. G. Canova, S. Shrivastava, J.J. Jonas, and C. G'Sell, 'Use of Torsion Testing to Assess Material Formability,' pp. 189-210 in Formability of Metallic Materials - 2000 A.D., J.R. Newby and B.A. Niemeier, eds.; ASTM, Phila., PA, 1982.

18. S.C. Shrivastava, J.J. Jonas, and G. Canova, 'Equivalent Strain in Large Deformation Torsion Testing: Theoretical and Practical Considerations,' J. Mech. Phys. Solids, in press.

19. I. Weiss and J.J. Jonas, 'Dynamic Precipitation and Coarsening of Niobium Carbonitrides During the Hot Compression of HSLA Steels,' Met. Trans. A, 11A (1980) pp. 403-410.

20. M.G. Akben, I. Weiss and J.J. Jonas, 'Dynamic Precipitation and Solute Hardening in a V Microalloyed Steel and Two Nb Steels Containing High Levels of Mn,' Acta Met., 29 (1981) pp. 111-121.

21. J.P. Sah, G.J. Richardson, and C.M. Sellars, 'Grain-Size Effects During Dynamic Recrystallization of Nickel,' Met. Sci., 8 (1974) pp. 325-331.

22. C. Ouchi and T. Okita, 'Dynamic Recrystallization Behaviour of Austenite in Nb-Bearing High Strength Low Alloy Steels and a Stainless Steel,' Tetsu-to-Hagane, 62 (1976) p. S208 (research abstract).

23. P.J. Wray, 'Rate and Recurrence of Recrystallization during the Tensile Deformation of Austenitic Iron,' pp. 86-112 in The Hot Deformation of Austenite, J.B. Ballance, ed.; AIME, New York, 1977.

24. T. Maki, K. Akasaka, and I. Tamura, 'Dynamic Recrystallization Behavior of Austenite in Several Steels,' Thermomechanical Processing of Microalloyed Austenite, AIME, Pittsburgh, PA, Aug. 1981, in press (this volume).

25. E.E. Underwood, Quantitative Stereology, p. 34; Addison-Wesley, Reading, Mass. 1970.

DISCUSSION

Q: Because you have chosen to introduce the question of dynamic recrystallization, I wonder if you would care to comment on the relative importance of dynamic and static recrystallization in thermomechanical processing.

A: This is an important point and the answer is not clear. In the lower alloyed steels or in plain carbon steels, it is easy to reach the peak strength at the onset of dynamic recrystallization. I understand from the work Priestner has done on plain carbon steels that dynamic recrystallization occurs in the roll bite. On the other hand, with the more highly alloyed steels, you can see that at the same temperature and strain rate the peak strength increases with the amount of alloying. In the microalloyed steels, I believe dynamic recrystallization is less important when the alloy content is higher.

Comments: I have two comments to make. First, I think involving grain size in this transition phenomenon is very important. Secondly, the normal grain sizes, the transition from multiple peaks to a single peak would occur at a rolling speed of about 5 ft. per day. Therefore, under normal circumstances, we are in the range where dynamic recrystallization is a very efficient way of refining grain size, if you can get a sufficiently large pass to exceed the peak strain. This may or may not be possible, depending on the alloy content and the rolling schedule, particularly the reduction in the passes.

As a comment to Dr. Sakai's paper, I would like to show one of our results on the relation between oscillation in flow curve and dynamically recrystallized grain size. Figure 1 shows the appearance of oscillation in true stress-true strain curve of several steels as a function of dynamically recrystallized grain size ($\bar{D}$) obtained at various conditions of Z values. In this figure, the mark of ⊚ indicates that the oscillated curve was observed and the mark of O shows the smooth s-s curve with single peak, and the dotted vertical line shows the initial austenite grain size (D_0) for each steel. This result shows that the shape of flow curve is closely related with some kind of relationship between $\bar{D}$ and D_0. This is essentially coincident with Dr. Sakai's proposal. Our results seem to indicate, roughly speaking, that the single peak flow curve is associated with grain refinement ($\bar{D} < D_0$) and the oscillated curve is associated with grain coarsening ($\bar{D} > D_0$), although Dr. Sakai proposed that the transition criterion is $D_0 = 2\bar{D}$. However, since our investigation was not focussed on this criterion from oscillated curve to single peak curve and it is generally difficult to determine the accurate transition from obtained s-s curves, I can not show the exact value of transition criterion from our observation at the present.

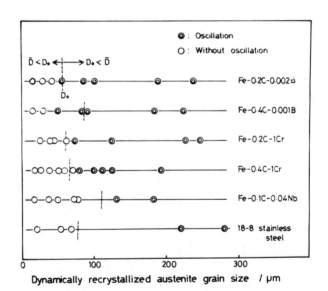

Figure 1

MODIFICATION OF THE AUSTENITE GRAIN STRUCTURE IN A Mn-Mo-Nb
STEEL BY Ti-N ADDITIONS

by

M.J. Godden, G.E. Ruddle, A.F. Crawley and J.D. Boyd
Physical Metallurgy Research Laboratories
CANMET
Ottawa, Canada

The origin of mixed ferrite grain size in a microalloyed steel, and
specific effects of Ti-N additions have been investigated by quenching specimens at various stages of the rolling schedule to follow the changes in
austenite grain structure. It is shown that the mixed ferrite grain size in
the as-rolled plate of the base steel results from a relatively large, mixed
austenite grain size which exists prior to transformation. Ti-N addition to
the base steel controls the austenite grain size throughout the rolling schedule to produce a finer and more uniform distribution of pancaked grains at
the end of rolling resulting in a fine, uniform transformed ferrite structure. The important implication for commercial plate rolling is that desirable microstructure and properties can be obtained in as-rolled plate of Ti-N
steel using a relatively simple processing schedule.

INTRODUCTION

Titanium-nitrogen additions are made to carbon-manganese steels (1,2) and microalloyed steels (3-5), primarily to refine the grain structure in the heat-affected zone (HAZ) of welds. However, it has been observed that the microstructure of as-rolled plate of Ti-N steels is finer and more uniform than in the Ti-free steels, and there is a corresponding improvement in strength and toughness (2-5).

An understanding of the origin of non-uniform, or mixed ferrite grain-size distributions in controlled rolled microalloyed steels is important in the design of optimum rolling schedules. A model has been described whereby such undesirable microstructures in as-rolled plate occur more frequently with increasing width of the austenite grains at the start of the ferrite transformation (6). It has also been shown that a mixed structure of fine and coarse austenite grains can develop when there is significant rolling deformation in the temperature range where the austenite grains only partially recrystallize between rolling passes (7).

The present work seeks to clarify the origin of mixed ferrite grain size in a microalloyed steel, and to identify the specific effects of Ti-N additions. The steels studied had a base composition of 0.06 C - 1.9 Mn - 0.3 Mo - 0.05 Nb - 0.010 N, and Ti contents of 0 and 0.013 weight percent. The microstructural changes during the complete rolling process were monitored for both steels by quenching specimens at various stages of rolling on an instrumented 457 mm reversing mill. The origin of the final microstructure was deduced from observations of the austenite grain structure during rolling, and measurements of the rolling mill parameters (temperature, reduction, interpass time and roll-separating force). In addition, a thorough study of the precipitate phases present at various stages of processing of these steels was carried out by conventional electron microscopy of extraction replicas, and scanning transmission electron microscopy. The electron microscopy results are presented in the subsequent paper by Houghton, Embury and Weatherly (8).

EXPERIMENTAL

The two experimental steels (Table I) were prepared as air-induction melts and cast as 68 kg rectangular ingots with approximate dimensions of 127 x 203 x 250 mm. A slice 127 x 35 x 250 mm was cut from one side of each ingot for the interrupted rolling experiments. The remainder, (referred to as the standard billet) was soaked at 1220°C for 2 h and rolled to 13 mm plate using the rolling schedule given in Fig. 1. The rolling mill was instrumented to record mill-drive power, roll-separating force, and roll-gap setting continuously during rolling, and temperature at the mid-thickness of the plate was monitored by an imbedded thermocouple. Most of the material from the standard billets was used for measuring mechanical properties of as-rolled plate and weld HAZs, and those results are reported elsewhere (5). For the present study, the optical microstructures of longitudinal sections of the as-rolled plate were recorded and characterized quantitatively by automatic image analysis.

The austenite structure after soaking was determined by quenching a small cylinder (13 mm diam. x 26 mm) which had been inserted in a dummy billet and given the same soaking treatment as the standard billets.

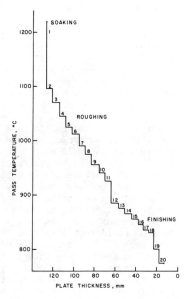

Fig. 1. Rolling schedule for standard billets.

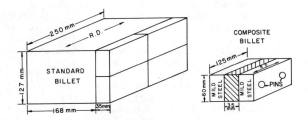

Fig. 2. Sketch of standard billet and composite billet used for
interrupted rolling experiments.

Table I - Chemical Analyses of Experimental Steels, wt %

Steel	C	Mn	Mo	Nb	Ti	N	Si	Al
Base	0.067	1.95	0.29	0.048	-	0.010	0.26	0.044
Ti-N	0.056	1.80	0.30	0.049	0.013	0.011	0.25	0.092

The austenite structure at various stages of rolling was determined on four samples approximately 35 x 60 x 125 mm cut from the 35 mm-thick ingot slice. Each sample was sandwiched between two mild steel plates of the same dimensions and fastened with press-fit steel pins to make a composite billet as shown in Fig. 2. The composite billets were rolled to duplicate the deformation schedule in Fig. 1, but the roll gap was reduced by one-half for each pass because the starting height of the composite billets was one-half that of the standard billets. The standard rolling schedule had to be modified further for the final few passes because the interpass times for the composite plate became too short owing to rapid heat loss from the thin sections. Hence, for the composite specimens, passes 12-18 were replaced with 4 passes of equal reduction and at equally-spaced temperature intervals. At selected points in the rolling schedule (after passes 2, 6, 11 and 18), the composite specimens were quenched into iced brine. Note that each specimen was quenched after a delay appropriate for producing the structure at the beginning of the next pass. Longitudinal metallographic sections were prepared from quenched samples at locations adjacent to the imbedded thermocouple. The prior austenite grain boundaries were revealed by etching in a mixture of saturated aqueous picric acid, wetting agent and HCl for times up to 45 min (9).

RESULTS

The microstructures of the as-rolled plate are shown in Fig. 3. The base steel shows pronounced banding and a mixed ferrite grain size, whereas the Ti-N steel has a finer and more uniform microstructure. The results of the quantitative metallography given in Table II support this observation. Both steels consist of bands of polygonal ferrite separated by acicular ferrite and carbon-rich constituents. There is a higher percentage of polygonal ferrite in the Ti-N steel compared with the base steel, and the mean polygonal-ferrite grain size is much smaller in the Ti-N steel. Measurements of the polygonal ferrite grain-size distributions show that there are more large grains in the base steel (max diam = 35 μm) than in the Ti-N steel (max diam = 23 μm). Athough there is considerable variation in the spacing of the polygonal ferrite bands through the thickness of the plate in both steels, the band spacing is larger at the mid-thickness than at the surface. However, the band spacing is less in the Ti-N steel than the base steel at all locations.

The rolling mill parameters for the standard billets are summarized in Table III, and the variation in roll-separating force during rolling is shown in Fig. 4.

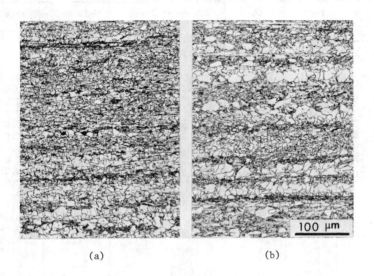

(a)   (b)

Fig. 3. Microstructures of plate rolled from standard billets, (nital etch).
a) Ti-N steel; b) base steel.

Table II - Quantitative Microstructural Measurements of Plate Rolled from
Standard Billets

	Base	Ti-N
Percent polygonal ferrite	65.0±8.2	80.6±3.2
Mean grain diameter (polygonal ferrite)	5.50±0.99µm	3.86±0.42µm
Aspect ratio (polygonal ferrite)	1.47±0.21	1.45±0.14
Polygonal ferrite band spacing: surface 1/4-thickness 1/2-thickness 3/4-thickness surface	47µm 51 70 55 55	28µm 34 50 35 25

Table III - Rolling Parameters for Standard Billets

Pass No.	Pass Temp., °C	Mill Setting, mm	Interpass Time, s Base	Interpass Time, s Ti-N	Unit Roll-Separating Force, MN/cm Base	Unit Roll-Separating Force, MN/cm Ti-N
1	1200	127				
2	1095	121				
3	1070	114			0.055	0.061
4	1045	108			0.062	0.068
5	1025	102			0.066	0.073
6	1010	95	16	17	0.070	0.076
7	990	89	22	24	0.075	0.079
8	975	83	17	19	0.077	0.082
9	955	76	23	24	0.080	0.085
10	940	70	18	20	0.082	0.086
11	925	64	26	21	0.085	0.088
12	885	57	54*	69*	0.087	0.091
13	875	51			0.090	0.093
14	865	44	14	16	0.094	0.098
15	855	38	15	16	0.099	0.101
16	845	33	11	16	0.106	0.097
17	835	28			0.113	0.101
18	830	23	47*	39*	0.104	0.102
19	800	18			0.113	0.113
20	775	13	11	26	0.136	0.133

*elapsed time

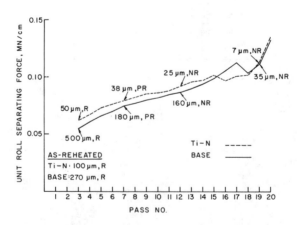

Fig. 4. Roll-separating force for standard billets. Mean austenite grain sizes from quenched specimens are shown; R - completely recrystallized, PR - partially recrystallized, NR - nonrecrystallized.

The austenite grain structures at various stages of rolling, as determined from the quenched specimens, are shown in Figs. 5 and 6. The corresponding mean austenite grain sizes, as measured in the through-thickness direction by the linear intercept method, are given in Table IV. The mean austenite grain size of the Ti-N steel is much smaller than for the base steel at all stages of processing, and there are other notable differences in the austenite grain structure of the two steels. The temperature for rapid grain coarsening was determined to be 1150°C for the base steel and 1250°C for the Ti-N steel (5), but the austenite grain size after soaking varied with the time to reach the reheat temperature. Consistent results were obtained for heat-up of 1 h or more. In the as-reheated condition, the mean grain size of the Ti-N steel is less than for the base steel by approximately a factor of 3, but the Ti-N steel has a duplex grain structure with some very large grains (Fig. 5a). After the second rolling pass (Fig. 5b), the Ti-N steel is completely recrystallized, and the grain size is further refined and has acquired a uniform distribution which persists throughout the remainder of the rolling schedule. The corresponding specimen of the base steel also exhibits a completely recrystallized, uniform grain structure (Fig. 6b), but the grain size is almost doubled from the as-reheated condition (Fig. 6a). At 990°C, both steels show partial recrystallization between passes, although the percentage of recrystallized grains is higher in the Ti-N steel, and the base steel develops a mixed grain size during rolling in the partial recrystallization regime (Figs. 5c and 6c). For 890°C and below, neither steel recrystallizes between passes, and pancaked austenite grains are produced in both (Figs. 5d and 6d). The 825°C-quenched structures approximate the austenite condition prior to ferrite transformation. In the base steel there is a wide range of austenite grain widths and evidence of deformation bands, whereas in the Ti-N steel, the mean grain width is smaller by a factor of 5, and much more uniform (Figs. 5e and 6e). These observations are summarized in Fig. 4 where the variation in mean austenite grain width and morphology are shown on the roll-separating force curves.

Table IV - Mean Austenite Grain Diameters from Quenched Specimens, μm*

Quench Temperature, °C	Base	Ti-N
1230**	270	100
1070	500	50
990	180	38
890	160	25
825	35	7

*linear intercept method, through-thickness direction
**from cylinders in dummy reheat billets

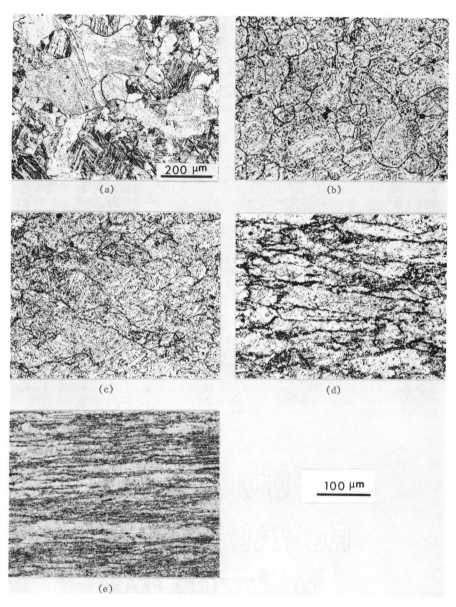

Fig. 5. Microstructures of quenched specimens of Ti-N steel (saturated
aqueous picral etch).
a) 1220°C; b) 1070°C; c) 990°C; d) 890°C; e) 825°C.

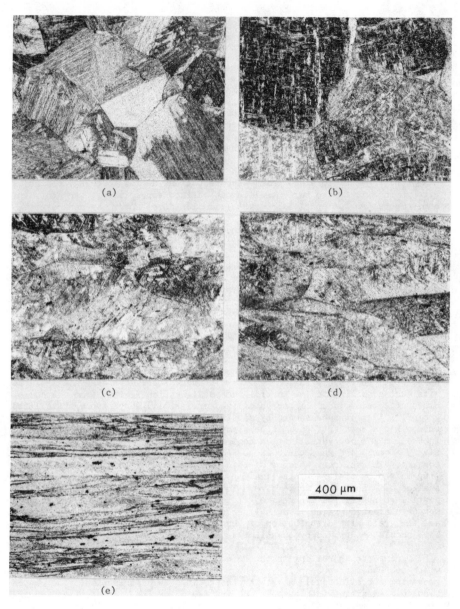

Fig. 6. Microstructures of quenched specimens of base steel (saturated aqueous picral etch).
a) 1220°C; b) 1070°C; c) 990°C; d) 890°C; e) 825°C.

DISCUSSION

The most striking difference in the austenite grain structures of the two steels is that the mean austenite grain size at the beginning of rolling is much smaller in the Ti-N steel than the base steel, and this difference persists throughout the whole rolling schedule. This observation conflicts with recent work of Cuddy (7), who found that for a given rolling schedule, the recrystallized grain size after several passes was independent of the starting grain size. However, the total rolling reduction and the reduction per pass used by Cuddy were much higher than in the present investigation.

With our rolling schedule, which simulates a commercial schedule with a high reheat temperature and small pass reductions during roughing, the austenite grain size in the base steel is not refined appreciably before the minimum temperature for complete recrystallization is reached. By contrast, in the Ti-N steel, the stable TiN precipitates which form during reheating (8) limit grain growth during soaking, and grain refinement occurs by repeated cycles of deformation and recrystallization from the start of rolling. The mean austenite grain size is reduced by a factor of approximately 2 during rolling in the complete recrystallization regime.

The interrupted rolling experiments indicate that both steels have similar temperature ranges for complete recrystallization, partial recrystallization and no recrystallization. However, the higher percentage of recrystallized grains in the Ti-N steel (compare Figs. 5c and 6c) indicates that the temperature range for partial recrystallization is lower for the Ti-N steel than the base steel. This is to be expected because the minimum temperature for complete recrystallization varies as the starting grain size (7). Nevertheless, it appears certain that both steels had considerable deformation in the partial recrystallization regime. The base steel developed a mixed grain size during rolling in the partial recrystallization regime, from the combined effects of i) small recrystallized grains, ii) large nonrecrystallized grains, iii) grains coarsened by strain-induced boundary migration (7). Once this mixed structure forms, it persists because with continued deformation in the partial recrystallization regime the grain-size distribution progressively widens, and when there is no recrystallization, all grains are reduced equally. The apparent absence of this effect in the Ti-N steel despite rolling in the partial recrystallization regime is attributed primarily to the smaller mean austenite grain size. That is, at any given temperature in the partial recrystallization regime, it is expected that the number and size of unrecrystallized grains is much lower for the Ti-N steel than the base steel, and the grain size distribution of the Ti-N steel is always much narrower. This should be checked by quantitative metallography.

During finish rolling in the no recrystallization regime, the grain width decreases in proportion to the rolling reduction, but there is no mechanism for effectively narrowing the grain-size distribution. The reduction in mean grain width between 890 and 820°C for the Ti-N steel is comparable to the corresponding rolling reduction (~3:1), but the value for the base steel is higher (~5:1). This could be a consequence of the development of deformation bands in the base steel (Fig. 6e) which generate high-angle boundaries by subgrain growth and coalescence within the bands (6).

The roll-separating force measurements provide additional information about the austenite deformation which is valuable in designing commercial rolling schedules and can be compared with microstructural observations. For rolling temperatures down to 850°C, the roll separating force is consistently higher for the Ti-N steel than the base steel (Fig. 4). The difference in force can easily be accounted for by the difference in flow stresses associated with the different grain sizes of the two steels. The roll-separating force curves exhibit a decreasing slope at the beginning of rolling (passes 1-10) and an increasing slope at the end (passes 11-20). This reflects the transition from complete recrystallization to no recrystallization, but these data are not sufficiently precise to define the limiting temperatures for each regime. The irregular results for rolling below 850°C are not well understood. It may be indicative of ferrite transformation, but there was no positive metallographic evidence of ferrite formation in the specimens quenched from 825°C.

Finally, to correlate the ferrite microstructure with the austenite grain structure prior to transformation, we consider Figs. 3a, 3b, 5e and 6e. It is not possible to make a direct quantitative correlation, because of the differences in total deformation and interpass times for the composite specimens quenched from 825°C (Figs. 5 and 6) and the plate rolled from the standard billets (Fig. 3). However, it is clear that the mixed ferrite grain size in the as-rolled plate of the base steel is a result of the large mean austenite grain size and the mixed austenite grain size which exist prior to transformation. The higher volume percent of polygonal ferrite in the Ti-N steel (Table II) is also consistent with its smaller austenite grain size, which would raise the transformation temperature range compared with the base steel. In addition, the austenite-polygonal ferrite transformation must be different for the 2 steels. The mean ferrite grain size is approximately 1/2 the mean austenite grain size for the Ti-N steel, indicating that ferrite nucleates predominantly at austenite grain boundaries. For the base steel, the mean ferrite grain size is approximately 1/5 the mean austenite grain size, which suggests there is ferrite nucleation within austenite grains.

In summary, it has been shown that Ti-N additions to a microalloyed steel control the austenite grain size during rolling to produce a finer and more uniform distribution of widths of pancaked grains at the end of rolling. This results in a corresponding fine, uniform, transformed ferrite structure, and enhanced strength and toughness. These microstructures were obtained in the Ti-N steel using a high reheat temperature (1220°C), small reductions per pass during roughing (5-8%), and approximately equally spaced passes throughout the rolling schedule. The same result likely could be obtained in the base steel by employing a low reheat temperature (e.g., 1150°C), higher drafts (10-20% per pass) during roughing and no deformation in the partial recrystallization regime (3,7, 10). However, this would require special reheat equipment, high mill loads and extremely precise process control. The use of Ti-N steel may be a commercially attractive alternative.

ACKNOWLEDGEMENT

The authors are grateful to M.J. Shehata and B.R. Casault for the quantitative metallography measurements and to D. Linkletter and Y. Lavoie for technical assistance.

REFERENCES

1. T. Horigome, S. Kanazawa, E. Tsunetomi, H. Mimura, A. Nakashima, K. Shinmyo and K. Okamoto, "Newly Developed 50 kg/mm² Ship Steel and its Welding Materials", pp. 679-705 in Welding of HSLA (Microalloyed) Structural Steels, A.B. Rothwell and J.M. Gray, eds., ASM, Metals Park, 1978.

2. H. Nakasugi, H. Matsuda and H. Tamehiro, "Properties of High Strength, Titanium Bearing Steel for Large Diameter Pipeline", pp.51-67 in Pipeline and Energy Plant Piping: Design and Technology, Pergamon, Toronto, 1980.

3. C. Ouchi, T. Tanaka, I. Kozasu and K. Tsukada, "Control of Micro-structure by the Processing Parameters and Chemistry in the Arctic Line Pipe Steels", pp. 105-125 in MiCon 78: Optimization of Processing, Properties, and Service Performance Through Microstructural Control, H. Abrams, G.N. Maniar, D.A. Nail and H.D. Solomon, eds., ASTM, Philadelphia, 1979.

4. H. Gondoh, H. Nakasugi, H. Matsuda, H. Tamehiro and H. Chino, "Development of Acicular-Ferrite Steel for Arctic-Grade Line Pipe", Nippon Steel Tech. Report No. 14, Dec. 1979, pp. 55-65.

5. D.B. McCutcheon, J.T. McGrath, M.J. Godden, G.E. Ruddle, G. Weatherly and D.C. Houghton, "Improvement in Submerged-Arc Weld Heat Affected Zone Toughness by Titanium Additions to Line Pipe Steels", MiCon 82, to be published by ASTM.

6. W. Roberts, "The Evaluation of Microstructure During Controlled Rolling of Microalloyed Steels", Scand. J. of Met., 9 (1980) pp 13-20.

7. L.J. Cuddy, "Microstructures Developed During Thermomechanical Treat-ment of HSLA Steels", Met. Trans., in press.

8. D.C. Houghton, J.D. Embury and G.C. Weatherly, "Characterization of Carbonitrides in Ti-Bearing HSLA Steels", this conference.

9. A. Brownrigg, P. Curcio and R. Boelen, "Etching of Austenite Grain Boundaries in Martensite", Metallography, 8 (1975) pp. 529-533.

10. P.J. Heedman and A. Sjostrom, "Controlled Rolling Schedule for Increased Plate Production", Scand. J. of Met., 9 (1980) pp. 21-24.

DISCUSSION

Q: In your Ti-N steel, the Ti:N ratio was about 1:1. Also, it is quite well established by Japanese work that the effectiveness of Ti nitride precipitation on grain coarsening temperatures is a sensitive function of the particle size. In your 70 Kg heats, I presume the cooling rate was comparable to that obtained in continuous-cast slabs. But first, could you tell us something about the particle size distribution? Secondly, do you feel that this Ti:N ratio in your steel of 1:1 is the optimum or perhaps should be modified? The last question is, as far as mechanical properties are concerned, would you feel that addition of titanium and nitrogen to a base steel of Mo-Nb is really necessary or perhaps these steels could be further modified and perhaps made leaner.

A: The first part of your question deals with precipitation on dislocations and will be dealt within the next paper. With regard to the last part, we commented right at the end of the paper that, yes, certainly the base steels could be improved by modifying the processing and composition. As for optimizing the Ti:N ratio, we are sure that can be done, but it is not clear yet exactly what rationale should be followed. We originally used different Ti:N ratios. It turns out this particular Ti:N steel had the best results in terms of heat-affected zone toughness so we selected that as a "good" steel. We are now studying the effect of the Ti:N ratio in more detail.

Q: What is the reason for the difference in roll-separating force between the Ti-N and base steels (Figure 4)? I believe that Ti, in tieing up the N, will lower the pancaking temperature in that steel and lead to lower separating forces. The less likely alternative is that ferrite is being formed. It should be easy to check both possibilities.

Q: Looking at the curves in Figure 4, there is no obvious indication that when you leave the recrystallization range there is a change in the separating force. Have you compared these results with those of a plain carbon steel in which the non-recrystallization region is absent?

A: We found that the roll forces for the Ti-bearing steel, compared to the base steel, were 10-20 percent higher, at least during the roughing part of the rolling schedule, and that the Ti-bearing steel roll-separating forces compared to plain C-Mn were 50-70 percent higher. This would be for a C-Mn steel that has somewhat lower C and Mn levels than these steels shown here.

Q: What is the slope?

A: Generally, for the plain carbon steel the slope is the same and the irregular characteristics during finish rolling are the same, but the level of the whole curve is about 50 percent lower.

CHARACTERIZATION OF CARBONITRIDES

IN Ti BEARING HSLA STEELS

D.C. Houghton, G.C. Weatherly and J.D. Embury
University of Toronto-McMaster Joint STEM Facility
Dept. of Metallurgy & Materials Science
McMaster University
Hamilton, Ontario, Canada

A series of H.S.L.A. steels containing titanium additions in the range 0 to 0.039 w/o have been prepared using a base composition of 0.06 w/o C, 0.01 w/o N, 0.05 w/o Nb, 0.3 w/o Mo, 1.85 w/o Mn and 0.3 w/o Si. The laboratory heats were chill cast using cooling rates closely similar to those operative in continuous slab casting. Subsequently some samples were control rolled under instrumented conditions which simulated commercial practice. The distribution, morphology and structure of the carbonitride precipitates were characterized using optical, CTEM, and STEM techniques for various conditions in order to delineate the changes occurring between the initial cast structure and the reheated, as rolled and welded conditions. The detailed microchemistry of the precipitation sequences was investigated using carbon extraction replicas. These studies included consideration of factors, such as reheat procedure and thermal cycles analogous to those of submerged arc welding, on the precipitate composition and morphology.

A dedicated STEM was used to determine the composition of particles in the size range 2 nm to 5 μm using both energy dispersive analysis and diffraction methods. The variation of the precipitate size and distribution with thermomechanical processing can be used to discuss the role of titanium in the grain refinement of austenite using simple models to predict the volume fraction of available precipitates as a function of thermal history.

Introduction

In considering the role of microalloying elements in the control of austenite grain size it is essential to realize that the sequence of casting, reheating, control rolling and subsequent weld cycles may produce a complex series of precipitate reactions. To delineate these reactions in a quantitative manner it is important to attempt to define both the microchemistry of the precipitate phases produced in the various processes and to relate their size and distribution to the macroscopic distribution of elements after the initial solidification process. A principal objective of the present study was to use the unique capabilities of a dedicated STEM to characterize the detailed composition and structure of the precipitate phases containing microalloying elements. The detailed chemistry of the various precipitates was then used to develop a model of the precipitation sequence based on the effective concentrations of microalloy elements available in a given portion of the precipitation sequence.

The role of precipitate species having solution temperatures of the order of 1100°C has been established by the classical work of Gladman [1-3]. With additions of titanium a stable nitride is formed which extends the grain coarsening temperature up to approximately 1300°C as shown by the previous work of George et al. [4-6] and Matsuda [7]. Currently there is interest in the use of titanium additions to suppress grain coarsening in the heat affected zone and the improvement of the toughness of this zone [8,9]. In order to understand and control the structure of the heat affected zone, consideration must be given to both the coarsening and the dissolution of precipitates during weld cycles and hence a knowledge of the detailed microchemistry of the precipitates is of importance. Further, for high heat input welding operations, the efficiency of titanium as a grain refiner will depend on the amount and distribution of titanium precipitated in the solidification process. This is of particular importance in steels produced by a continuous casting operation in which considerable segregation can occur. Hence, the approach in the current work has been to clarify and analyze the type of precipitate formed at various stages of the fabrication sequence from the cast structure through to simulated weld cycles. The base composition used was 0.06 w/o C, 0.01 w/o N, 0.05 w/o Nb, 0.3 w/o Mo, 1.85 w/o Mn and 0.3 w/o Si and titanium additions in the range 0 to 0.39 w/o were used in order to provide a wide range in the Ti:Nb ratio operative in the steels.

Experimental Procedures

a) Casting Practice

Cast steels of dimensions 203 mm x 127 mm x 254 mm were prepared in the laboratory using electrolytic iron as a base material. The cooling practice for the slab was adjusted [9] to simulate that normally achieved in commercial continuously cast steels, varying between 9.4°C min⁻¹ and 16.2°C min⁻¹ through the slab thickness. The analyses of the materials are shown in Table I.

The cast slabs were reheated to ∿1200°C for 4 hours prior to control-rolling. The drafting schedule was designed [9] to simulate the commercial rolling practice of nominal 13 mm gauge plate. Rolling was started at 1100°C and finish "roughing" was between 925°C and 900°C followed by 8 final rolling passes between 885°C and 760°C.

Table I. Compositions of Heats Used in this Study

Heat No.	Aim Ti	N	C	Mn	P	Si	Final Ti	N	Ti/N (wt)
1	-	0.006	0.071	1.97	0.002	0.27	-	0.009	-
2	-	0.012	0.061	2.01	0.007	0.20	-	0.010	-
3	0.010	0.006	0.055	1.96	0.002	0.25	0.011	0.005	2.2
4	0.020	0.012	0.070	1.91	0.002	0.34	0.033	0.007	4.7
4(a)	0.020	0.012	0.065	1.81	0.007	0.27	0.011	0.009	1.22
5	0.020	0.006	0.066	1.87	0.002	0.38	0.015	0.005	3.00
6	0.040	0.012	0.076	1.98	0.002	0.33	0.039	0.010	3.9

In all steels the Nb level was between 0.050 and 0.054 w/o, the Mo level 0.29 to 0.33 w/o, and the S level 0.003 to 0.005 w/o.

b) Reheat Practice

In addition to the standard reheat practice outlined above, additional reheat experiments were performed on Heat 6, the alloy containing the highest titanium level.

Test pieces measuring ∿13 mm x 13 mm x 13 mm were machined from an as-cast slab and sealed off in previously evacuated and argon back-filled quartz capsules. A high purity argon atmosphere at a partial pressure sufficient to prevent collapse of the capsule at anneal temperatures between 110 and 1450°C provided a protective atmosphere sufficient to preserve a polished surface even after 4 hours at 1450°C. The quartz capsules were placed on zirconia supports on silicon carbide trays and annealed for periods up to 16 hours prior to quenching in iced brine.

Isothermal reheat experiments were also performed on as-rolled plate samples in the temperature range 1150°C - 1260°C.

c) Studies of Heat Affected Zones

One important purpose of titanium additions is to provide control of the austenite grain size during the thermal cycles associated with submerged arc welding (SAW). Thus, in the present study attempts were made to ascertain the changes in the nature and morphology of the titanium bearing precipitates due to weld cycles both by direct observation of submerged arc welded plates and Gleeble simulated samples.

Charpy sized test coupons (10 mm x 10 mm x 55 mm) were machined from control-rolled plates. A single thermal cycle by Gleeble resistive heating was designed to simulate the first welding pass of a SAW in a pipe mill seam weld, with a heat input of ∿25 kJ cm^{-1}. The Gleeble cycle gave a peak temperature of ∿1350°C which was attained in 6 s. The time above 1000°C was ∿8 s and cooling from 800°C to 500°C took ∿25 s. This thermal profile is typical of that experienced in the CGHAZ of commercial submerged arc welded plates [9].

In addition to the simulation studies, observations were made of the heat affected zone of SAW plates. Extraction replicas were made in strips parallel to the heat flow direction and close to the fusion line in the HAZ of submerged arc welded plates. In this manner the characterization of precipitates influenced by a rapidly changing temperature field could be performed using STEM microanalysis. Through masking unwanted regions with Lacomit and patterned scoring of each replica section a peak temperature-displacement identification of precipitates could be determined. Essentially a finger print of each weld fusion zone was made relating precipitate chemistry to distance away from the fusion line. Details of the analysis of the HAZ structures are discussed elsewhere [10].

d) Metallography and Microanalysis

(i) Optical Microscopy. Conventional specimen preparation was achieved using ¼ μm diamond polishing prior to etching with saturated aqueous picric acid containing ∿1 v/o Teepol to delineate the prior austenite grain boundaries. Austenite grain sizes were determined by the linear intercept technique. The grain size and morphology of the ferrite was revealed by etching in 2 v/o nital for ∿10 s.

(ii) CTEM SAD Microanalysis. Conventional carbon extraction replicas were prepared from the selected samples identified for analysis in Table II. Etching in 2 v/o nital prior to carbon evaporation and replica stripping by 5 v/o nital produced a precipitate distribution suitable for microanalysis. A layer of evaporated gold was used as a standard to determine precise lattice parameters by selected area diffraction in CTEM in the Philips 300. Lattice parameters obtained by this technique were determined to an accuracy of ±0.5% or ±0.02 Å.

(iii) STEM Microanalysis. Individual precipitates in the size range 2 nm to 5 μm were quantitatively analyzed by EDX X-ray analysis, and microdiffraction techniques using a Vacuum Generators HB 5 dedicated STEM instrument. Thin film standards of niobium and titanium prepared by evaporation were employed to enable a high degree of accuracy in determining precipitate compositions.

Closely similar preparation methods were employed for each sample, in addition to checking each microanalytical measurement against standard spectra to ensure minimal instrument influence. Samples from various stages through the thermal-mechanical processing sequence were replicated for STEM microanalysis to follow the evolution of the precipitate chemistries (see Table II).

The EDX microanalysis of microalloying elements such as Ti and Nb in the carbonitride precipitates was established to better than 5% by weight and a minimum detectable mass of $\sim 10^{-20}$ g was typically achieved, which corresponds to the presence of approximately 100 atoms of microalloy element. A point to point microchemical resolution of ±5 nm was frequently obtained and the minimum size of particle analyzed was 2 nm. The uncertainty in EDX X-ray analysis was in most cases found to be of the order of ±5%.

Results

A summary of all the microanalysis data for the various stages of thermomechanical processing is given in Table II.

270

Table II. Details of Precipitate Particles Analyzed after Various Thermal and Mechanical Cycles

Heat	As Cast	As Control Rolled	Gleeble thermal cycled	HAZ of SAW
Heat #1 0 0.009		Few large spheroidal precipitates >1000Å Fine dispersion of Nb(C,N) Mean diameter ∿200Å	No precipitates detected	
Heat #3 0.011 Ti% 0.005 N%	Fine dispersion of irregular precipitates 50-2000A (Nb,Ti)N Nb ≥ 90%	(Ti,Nb)N, Nb+31%Ti±13 cuboids 500-2000Å, d=1100Å Nb(C,N)spheroidal precipitates 50Å-5000Å, zero Ti	$Ti_xNb_{1-x}N$ x 60%±6 clusters of cuboids; size range 300÷1000Å d ∿ 500Å no spheroids	
Heat #4a 0.011 Ti% 0.009 N%	Dendrites and fine cuboids. Size range 30-2000Å Mean composition Nb+7.6%Ti±0.8	Cuboids (Ti,Nb)N 200-500Å 31% Ti±4 750-2000Å 69%Ti±4 Nb(C,N) fine dispersion of precipitates ≤200Å, d ∿ 100Å		
Heat #4 0.033 Ti% 0.007 N%		Cuboids (Ti,Nb)N 1000-2000Å 66%Ti±3 500-1000Å 48%Ti±6 (Nb,Ti)C,N) spheroids (Nb+20%(Ti+Mo))(C,N) Size range 30-300Å		$Ti_xNb_{1-x}N$ cuboids d ∿ 400Å (100Å-1000Å) x̄=63%±7 No spheroids
Heat #6 0.039 Ti% 0.010 N%	Large (μm) cubes of TiN Austenite grain boundary dendrites Size range 200Å → 5μm; mean composition Nb+17%Ti±1.5	(Ti,Nb)N cuboids 500-2000Å 61%Ti±7 (Nb,Ti)(C,N) spheroids 30-300Å Nb+12%Ti±1 Mo detected in spheroids ≤ 200Å, heterogeneities in ppts.; TiN cuboids >1μm	$Ti_xNb_{1-x}N$ x̄ = 56%±6 Random distribution of cuboid clusters; size range 200-1500Å d ∿ 520Å no spheroids	$Ti_xNb_{1-x}N$ x̄ = 70%±9 100Å skin, x=82±4 d = 1200Å (size range 100Å-3000Å) No spheroids

a) As-cast Microstructures

The as-cast microstructures were composed of a mixture of polygonal and acicular ferrite together with rows of dendritic niobium-rich particles which delineated the prior austenite grain boundaries as shown in Fig. 1b, 1c and 1d. In all titanium bearing heats, large titanium nitride cuboids, several microns in edge length were also observed, as illustrated by Fig. 1a. The size, shape and distribution of these types of precipitates varied between heats but within a given heat a consistent titanium level was detected in the precipitates which was independent of their size and geometry.

b) Isothermally Reheated Microstructures

Precipitates extracted from isothermally reheated specimens (Heat #6) were seen to adopt a more facetted cuboidal geometry with prolonged annealing times. Fig. 2a shows a row of titanium-rich cuboidal precipitates which appear to have evolved from an array similar to that of Fig. 1b. An isothermal anneal at 1400°C for 4 hours produces well defined cuboids enriched in titanium at a consistent composition of 87% ± 4 Ti (Fig. 2b). This indicates that the dendritic niobium-rich particles are metastable

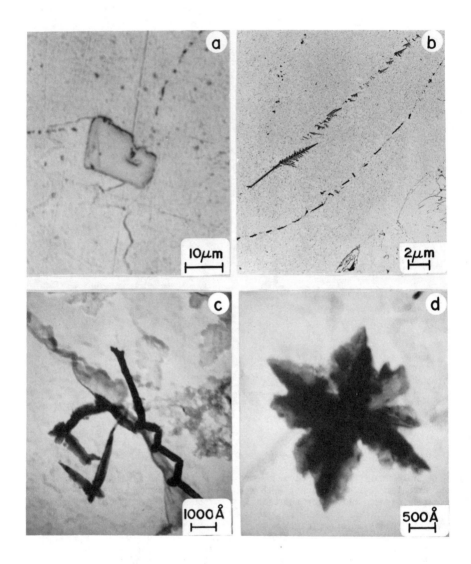

Fig. 1 As-cast microstructures Heat #6.
  (a) Optical micrograph of a large TiN cuboid at the intersection
       of as-cast grain boundaries.

  (b) Transmission electron micrograph of carbon extraction
       replica showing rows of dendrites along as-cast grain
       boundaries.

  (c) STEM micrograph showing fine dendrites nucleating on as-cast
       grain boundaries.

  (d) STEM micrograph showing typical dendrite morphology, in
       Heat #4(a).

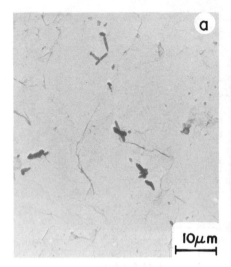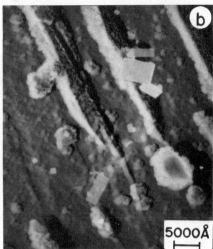

10μm

5000Å

Fig. 2 Isothermally reheated microstructure, Heat #6.
   (a) CTEM micrograph of extracted particles after 16 h at 1200°C.
        The dendrite morphology and row arrangement is still apparent.

   (b) STEM micrograph of cuboids after 4h at 1400°C.

with respect to the titanium-rich cuboids at high temperatures in austenite.
At austenitizing temperatures below 1100°C no change in particle morphology
was observed even after 16 h. At higher reheat temperatures the titanium
concentration in the cuboids increases as shown in Fig. 13 and Table II.
Accurate lattice parameter measurements of precipitates in the as-cast
and isothermally reheated conditions are presented in Fig. 6. These data
indicate that the stable compounds forming at temperatures close to the
melting point are of the form $Ti_xNb_{1-x}N$, having the B1 rock salt crystal
structure [16]. Electron energy loss spectroscopy confirmed that these
particles contained N, but a quantitative evaluation of their N content
could not be obtained because they were extracted using an amorphous C
film as a substrate. In the present work it was not possible to determine
the degree of stoichiometry in the precipitates; consequently ideal
stoichiometry of the various compounds had to be assumed in the modelling
of the precipitation process presented below.

c) Microstructural Features of Control-Rolled Materials

    The precipitate sizes and morphologies produced after thermomechanical
processing are shown in Fig. 3. The precipitate distribution in control
rolled microstructures is complicated by the coexistence of two major
precipitate species. Fig. 3c shows a set of precipitate size distributions
observed in as rolled plates. Fig. 3d shows a schematic histogram indicating that at soaking temperatures where most of the carbon and niobium
are taken into solution, an array of precipitates having the composition
$Ti_xNb_{(1-x)}N$ forms with a size range ∿50-200 nm. During control-rolling
these pre-existing cuboids are apparently sites for early nucleation of
$Ti_xNb_{1-x}-(C_yN_{1-y})$ and thus become coated with a niobium-rich layer, see

273

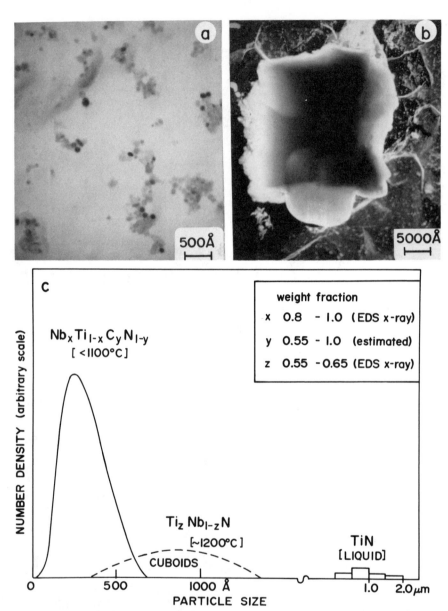

Fig. 3 Precipitate characterization in the control-rolled condition
   (a) STEM micrograph of fine niobium-rich carbonitrides in Heat #4.

   (b) STEM micrograph of large titanium rich cuboid with niobium-
        rich skin and fine dispersion of niobium carbonitrides Heat #6.

   (c) Schematic histogram showing the size, shape and chemical
        compositions of a typical precipitate sequence. The probable
        formation temperatures are noted.

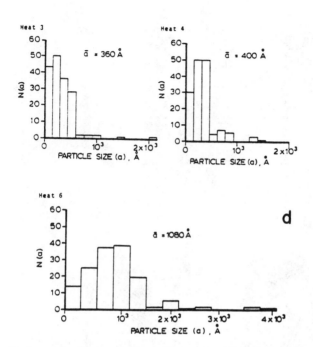

Fig. 3(d) Histograms of actual precipitate sizes found in control-rolled plates from Heats #3, 4 and 6.

Figs. 3b and 4. A fine dispersion of niobium-rich carbonitride precipitates nucleate and grow in the surrounding austenite matrix during cooling through the temperature range $\sim(1100^{\circ}C - 800^{\circ}C)$. The minimum diameter particle detected was 2 nm and a typical niobium-rich precipitate array is shown in Fig. 3a.

Large titanium nitride cuboids measuring several microns in edge length were found in the as-rolled microstructures. Their size and the thermo-dynamic stability of TiN precludes any further role in the precipitation sequence for these large particles but it is interesting to note that the volume of a one micron cube is equivalent to 10^6 precipitates 10 nm in diameter, making any quantitative estimate of the effective volume fraction of small particles available for grain size control very difficult.

The detection of molybdenum in the niobium rich carbonitride particles (< 20 nm) at approximately 20% of the detectable mass fraction was consist-ently encountered during STEM X-ray microanalysis in as-rolled Heats #4 and #6. In these steels the precipitates of diameter less than 10 nm were frequently found to have an apparent Mo concentration of the order of 25%.

The chemically heterogeneous nature of precipitates such as those shown in Fig. 4 is surprising since it clearly indicates that steep chemical concentration gradients are present within particles only a few hundred atomic diameters in size. The microdiffraction patterns show that no significant crystallographic change is found within a given precipitate,

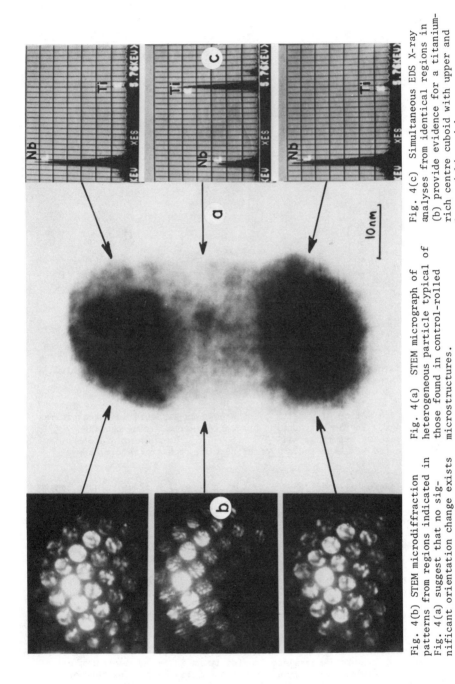

Fig. 4(b) STEM microdiffraction patterns from regions indicated in Fig. 4(a) suggest that no significant orientation change exists along the particle.

Fig. 4(a) STEM micrograph of heterogeneous particle typical of those found in control-rolled microstructures.

Fig. 4(c) Simultaneous EDS X-ray analyses from identical regions in (b) provide evidence for a titanium-rich centre cuboid with upper and lower niobium-rich caps.

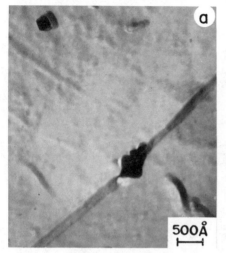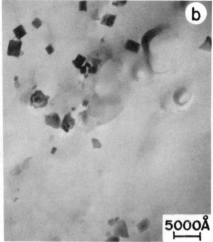

500Å

5000Å

Fig. 5 Precipitate morphologies resulting from Gleeble simulation of
weld thermal cycles.
(a) STEM micrograph of titanium-rich nitride particles on a prior
austenite grain boundary in Heat #3.

(b) STEM micrograph showing a cluster of titanium-rich nitride
precipitates in Heat #6.

suggesting that the growth of the niobium rich "caps" on the titanium rich
substrate is epitaxial in nature. This two stage precipitation sequence
indicates that the precipitate morphology evolves by first the decay of a
metastable solidification product to yield cuboids of the type $Ti_xNb_{1-x}N$
followed by the heterogeneous precipitation of niobium-rich carbonitrides
over a range of temperatures during thermomechanical processing. This
tendency to form layered or 'cored' precipitates is discussed in detail in
a later section.

d) Thermally Cycled Structures

The rapid thermal cycling induced by Gleeble simulation of weld HAZ
conditions gave rise to the precipitate morphologies shown in Figs. 5a and
5b. After thermal cycling the cuboids have lost their niobium rich envelopes,
regained a well-defined facetted geometry, and now exhibit a titanium concentration closely similar to that found in particles extracted from Heat
#6 after soaking at 1200°C for 16 hours. No significant difference in
titanium concentration or size distribution was observed in precipitates
extracted from Heats #3 and #6.

The austenite grain sizes resulting from the peak temperature zone of
Gleeble heat-treatments are presented in Table III. The growth of austenite
grains in as-cast samples of Heat #6 isothermally reheated above ~1300°C is
shown in Fig. 7. The temperature at which this abnormally high grain growth
begins to occur is decreased at longer annealing times. Austenite grain

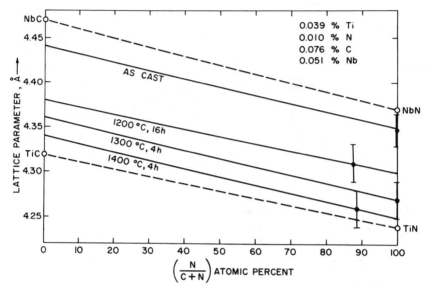

Fig. 6 Plot of CTEM electron diffraction measurements of lattice parameters and STEM EDS X-ray analysis for as-cast dendritic precipitates and cuboidal precipitates which appeared after isothermal reheating at various times and temperatures in Heat #6.

growth to sizes larger than ∿450 μm is not observed even at 1450°C.

Table III. Average Austenite Grain Sizes $\bar{d}$

Steel Heat Number	$\bar{d}$ μm	Ti,w/o	N,w/o
1	100	0	0.009
2	94	0	0.010
3	64	0.011	0.005
4	69	0.033	0.007
4(a)	48	0.011	0.009
5	75	0.015	0.005
6	66	0.039	0.010

The grain coarsening behaviour of the various heats subjected to thermal cycling after control rolling are summarized in Fig. 8. Titanium additions clearly prevent excessive grain growth.

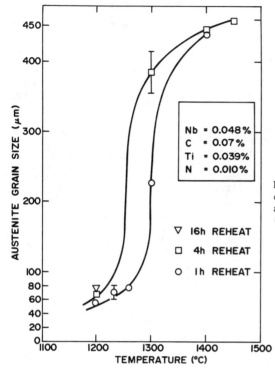

Fig. 7 Grain growth in as-cast and reheated Heat #6, as a function of isothermal anneal time and temperature.

Modelling of the Sequence of Precipitation

From the microanalysis results outlined in the previous sections, it is apparent that in addition to the two major types of precipitates, many precipitate particles contain concentration gradients and that one precipitate may serve as a nucleation site for further precipitation of other microalloy constituents. In order to model the sequence of precipitation and estimate the volume fraction of precipitated phases we can consider two cases. In the first the co-precipitation of two phases containing microalloy elements is considered but no allowance is made for homogenization or for mixing of the phases. In the second model complete mixing of the micro constituents, i.e. Ti and Nb, within the precipitating phase is allowed. Only the basic assumptions in the two models are outlined below, [11-13]; a more complete discussion of the models [10] is presented elsewhere.

Two types of simplified models were considered in order to rationalize the observed precipitation sequence. These are: -

Model 1   Co-precipitation of Separate Microalloy Phases

This model assumes that zero mixing of the precipitating phases occurs. Each compound is assumed to be stoichiometric and it is assumed that equilibrium is maintained in austenite throughout the precipitation process.

279

Model 2   Complete mixing of pairs of microalloy compounds

In this model the precipitating species are assumed to have complete
mutual solubility [16]. The enthalpy of mixing is taken to be negligible
[12,13] and the entropy of mixing is assumed to be ideal. In order to
simplify the calculations the compounds are assumed to maintain stoichiometry and the ternary solubility products are used to define solute concentrations in the austenite once precipitation begins. The free energy
of formation of the mixed compounds are assumed to vary linearly with mole
fraction.

Throughout this study the solubility product data due to Narita [14]
has been used as a data base although alternative sources of solubility
product data exist in the literature [11,13]. The aluminum levels
used in this study provide adequate protection of the titanium in regard
to deoxidation but have been assumed not to enter into other reactions
such as the formation of aluminum nitride.

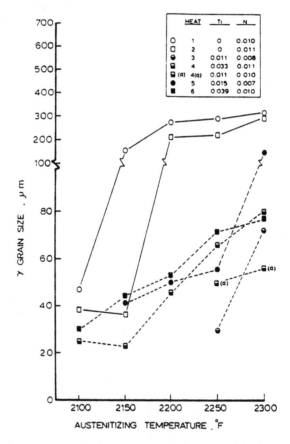

Fig. 8   Austenite grain size of control-rolled plates after austenitizing
for 1 h at various temperatures.

The co-precipitation model 1 effectively provides a lower bound for precipitate yield since it excludes any mixing of precipitate phases. The model can be used to provide an approximate guide in defining the temperature ranges where each microalloying element is of importance. The model in essence scans the possible precipitation reactions as the temperature is reduced from a temperature high enough to contain all elements in solution in austenite. As the temperature is reduced the compounds able to form are selected from reference to the solubility product isotherms as shown in Fig. 9. The program for model 1 continues calculating the cumulative amounts of each phase precipitated from the depleted austenite, although the assumption of co-precipitation is clearly inappropriate at high temperatures where diffusion is sufficiently rapid to allow homogenization of precipitates. However, in the controlled-rolling temperature range the sluggish reaction kinetics [17,18] prevent complete homogenization of the precipitates.

Thus model 2 which assumes complete mixing of pairs of precipitates is complementary to model 1, and model 2 can be used to estimate precipitate yields, solution temperatures and ratios of the elemental species involved in precipitates in any range.

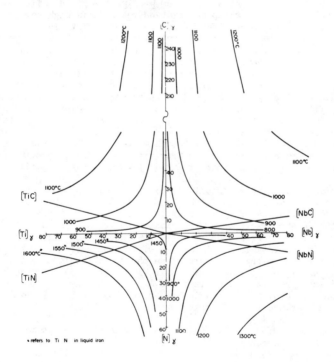

Fig. 9   Precipitate map indicating as diagonals the stoichiometric positions of possible precipitates. Solubilities in weight percent using solubility product data of Narita [14].

Prior to describing the application of the precipitation models outlined above it is important to state two important shortcomings of the models. For simplicity the effect of 0.3 w/o Mo, 0.3 w/o Si, 2.0 w/o Mn in these steels have not been considered in the analysis. These elements will however influence the activity of interstitial solutes and the processes of subsequent decomposition of the austenite on cooling. A second important feature which is not treated in the model is the influence of the pattern of initial solidification. It is not established whether the initial dendrites are of austenite or δ ferrite and also the amount of titanium or niobium precipitated from the liquid. Clearly the solidification process must define the 'available level' of microalloy elements available for precipitation in the austenite but at present it is difficult to treat this in a truly quantitative manner.

Application of the Precipitation Models

The approach in Model 1 can be considered with reference to Fig. 9. Here both the liquid and solid solubility products [14] for the four possible binary compounds expected in Ti, Nb microalloyed austenite are plotted as isotherms, where the weight percentages of solutes in austenite are varying. The stoichiometric lines shown for each compound define the weight ratio of solute to give a maximum precipitate yield. Titanium nitride clearly dominates the precipitation reaction in high temperature austenite and may also precipitate from the liquid. Titanium concentrations in excess of the stoichiometric ratio of nitrogen will allow the possibility of titanium carbide formation when titanium nitride precipitation becomes negligible, whereas titanium concentrations lower than the stoichiometric nitrogen ratio will allow excess nitrogen to precipitate at lower temperatures in the form of niobium nitride. Clearly the solute concentrations must be computed throughout the precipitation sequence to determine which compounds will form and the mole fractions of each precipitate during cooling. The precipitation sequences in Figs. 10 and 11 are computed using the assumptions of Model 1 for a high and low titanium level respectively. The temperature regimes within which each micro-constituent plays an important role are evident, and the precipitation sequence and the extent of overall precipitate yield differ markedly with a change in titanium level.

The effect of allowing mutual solubility of niobium and titanium nitrides to form the compound $Ti_xNb_{(1-x)}N$ as in Model 2 may be seen in Fig. 12. The influence of homogenization of a pair of compounds is to increase the weight fraction of precipitate and the precipitation of niobium as $Ti_xNb_{1-x}N$ occurs over a wider temperature range. The variation of the weight fraction of titanium in $Ti_xNb_{1-x}N$ with temperature is shown in Fig. 13. At titanium levels greater than the stoichiometric nitrogen equivalent, negligible solution of NbN in TiN is anticipated from Fig. 12. The experimental results for Heats 4 and 6 show that the cuboids contain both Nb and Ti, pointing out the important role played by the "available" Ti content as defined in the discussion.

The influence of excess titanium on the precipitation of niobium carbide is shown for both Models I and II in Fig. 14. In this case the curve for niobium carbide in the absence of other microalloying elements is shown for comparison. The variation of niobium concentration in the compound $Nb_xTi_{(1-x)}C$ is shown as a function of "available" titanium concentration in Fig. 15, based on Model 2.

The precipitation of niobium carbonitride over the range of temperatures experienced during hot working is presented in Fig. 16 using both

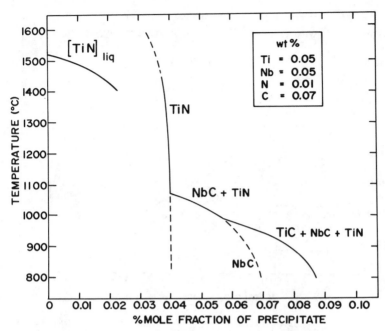

Fig. 10 Precipitation sequence in a 0.05 w/o Ti steel using the assumptions
of Model 1. Precipitation begins in the melt with the formation of
TiN which goes to completion in high temperature austenite.
Niobium carbide and subsequently titanium carbide are anticipated
at control-rolling temperatures. The niobium carbide yield is
reduced by the presence of titanium.

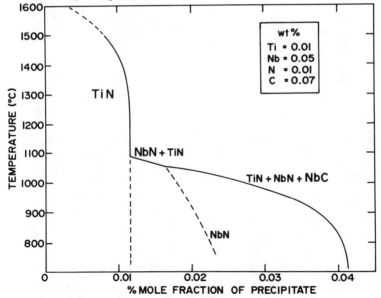

Fig. 11 Precipitation sequence in a 0.01 w/o Ti steel using the assumptions
of Model 1.

283

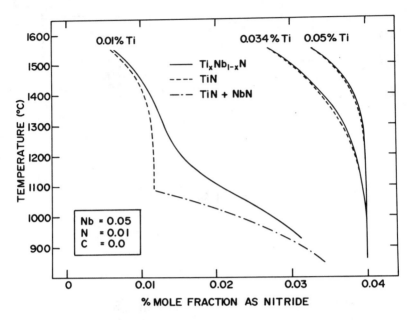

Fig. 12 Solubility curves predicted by both coprecipitation (dotted lines) and homogeneous solid solution (solid lines) models for high temperature austenite.

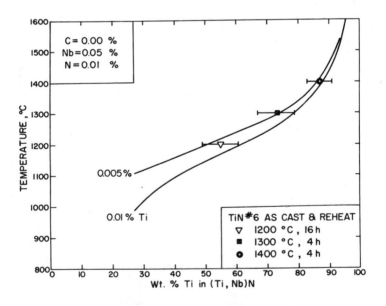

Fig. 13 Variation of titanium content in $Ti_xNb_{1-x}N$ as a function of temperature and the 'available' titanium concentration in austenite.

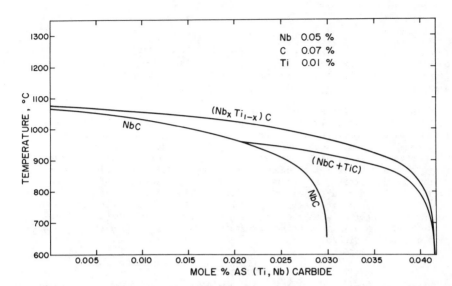

Fig. 14 Comparison of coprecipitation of NbC and TiC with the formation of the compound $Nb_xTi_{1-x}C$ for an 'available' titanium concentration of 0.01 w/o. A comparison of the NbC curve and the $Nb_xTi_{(1-x)}C$ curves indicate the increase in precipitate yield due to TiC.

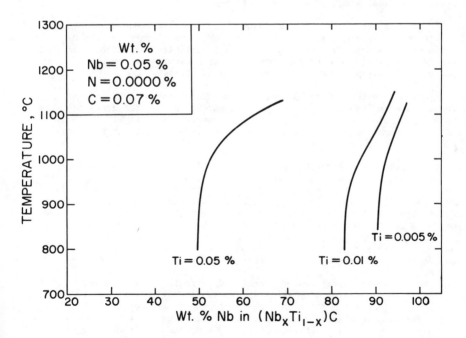

Fig. 15. Calculated values for the niobium content in the compound $Nb_xTi_{1-x}C$ for a range of 'available' titanium concentrations in austenite.

models and inputing two nitrogen levels. The two approaches to precipitation modelling clearly provide an upper and lower bound for the precipitate yield and converge at lower temperatures.

The formation of an homogeneous solution of nitrides in the form
$Ti_xNb_{(1-x)}N$ dramatically alters the solubility of nitride precipitates as
a function of temperature. As can be seen from Fig. 17, reducing the
available titanium concentration not only limits the mole fraction of
nitride available at high temperatures but greatly expands the temperature
range over which precipitation or dissolution of a mixed nitride would
occur. To illustrate the influence that the solid solution of niobium
nitride might have on the dissolution of titanium nitride precipitates,
the variation in austenite grain growth after 1 hour is plotted in Fig. 18
for Heat #6 together with the variation of the mole fraction of the compound $Ti_xNb1-xN$ with increasing temperature for an available titanium concentration of 0.01 w/o.

Discussion

Measurements of the distribution and microchemistry of the precipitates containing microalloy elements are important both from a fundamental
and practical viewpoint. The data is of value in elucidating the stability
and nucleation characteristics of the precipitates and from the practical
viewpoint of relating the efficiency of titanium in the weld cycle to the
previous thermal and mechanical processing.

The work described in the present paper indicates that a variety of
precipitate morphologies may occur in the presence of combinations of
titanium and niobium and that even in small particles marked compositional
gradients may occur due to the heterogeneous nucleation of one phase on a
phase previously precipitated at higher temperatures.

Let us consider first the precipitates formed at high homologous
temperatures and which are therefore present in the as-cast condition. The
combination of precise CTEM SAD lattice parameter and STEM EDX compositional
data, as summarized in Table II, clearly indicates that both the dendrites
and the large cuboidal precipitates are nitrides of the form $Ti_xNb_{1-x}N$.

The origin and distribution of the dendritic nitrides shown in Figs.
lb, c, and d, influences the subsequent distribution of nitride precipitates
which are desired for grain control during thermomechanical processing.
The appearance of a niobium-rich phase in interdendritic regions has been
well documented [20-22]. It has been suggested that the dendritic nitrides
may form at high temperatures [22] and are then pushed by the primary
solidifying phase [21] to form highly segregated interdendritic colonies.
However, dendritic nitrides can also be formed by solid state reactions [23]
and dendritic particles [24] have been seen in other precipitation systems.

The dendritic particles were observed to contain titanium levels which
decreased in proportion with the titanium level available at the freezing
temperature (see Table II). The degree of undercooling operative during
solidification influences the formation of the large primary TiN cuboids
shown in Fig. la. Large undercoolings are expected to yield a larger
amount of primary TiN directly from the melt as shown in Fig. 10.

On prolonged exposure at temperatures above 1100°C the dendritic

286

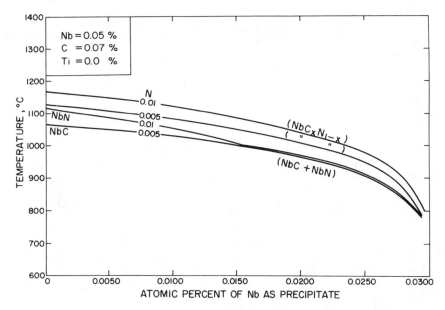

Fig. 16 Computed values of precipitate yield for the coprecipitation of $(NbC + NbN)$ and the compound NbC_xN_{1-x} at two nitrogen levels.

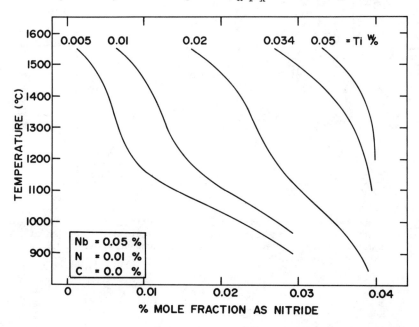

Fig. 17 Variation of the mole fraction of the compound $Ti_x-Nb_{1-x}N$ with temperature as a function of the 'available' titanium concentration in austenite.

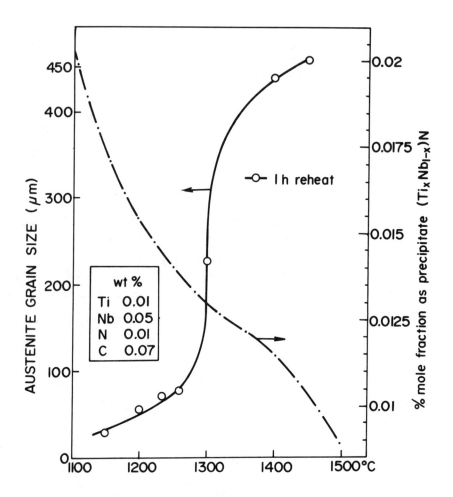

Fig. 18 Superposition of austenite grain growth characteristics, and mole
fraction of grain refining nitride precipitate as a function of
temperature for an "available" titanium concentration of 0.01 w/o.
Data for Heat #6.

288

morphology transforms to an array of titanium rich cuboids, see Figs. 2a and b. At 1200°C, after 16 hours only partial transformation of the cast morphology is seen (Fig. 2a) whereas at 1400°C after only 4 hours well defined facetted precipitates have formed (Fig. 2b). The titanium enrichment of these particles with increasing austenitizing temperature is in accord with the predictions of Fig. 13, assuming an "available" titanium concentration of 0.01%. However, the calculation of microalloying element concentrations "available" for further precipitation at any given temperature requires a knowledge of the total volume fraction already precipitated. However, the determination of the volume fraction of small particles is difficult in practice making it difficult to compare theoretical models of precipitate yield with experimental observations.

An important feature of the observed carbonitrides is their chemical inhomogeneity. The detection of steep concentration gradients in these particles has been reported previously [25]. The current STEM study has given high precision analysis with a resolution of the order 5 nm inside the precipitates as shown in Fig. 4. The layered nature of individual particles illustrated in Figs. 3 and 4 is in accord with the precipitation sequence described in Figs. 10 and 11.

The compositional variation within the very fine precipitates indicates that the degree of homogenization occurring during the control-rolling operation is very limited which is in accord with the co-precipitation Model 1.

The detection of molybdenum in the niobium-rich spheroids of the order of ≤ 20 nm in diameter observed in Heats #4 and #6 was very reproducible at 20-25% of total microalloy concentration. The precipitation of molybdenum carbide during the decomposition of austenite is well documented, (see for example Honeycombe) [25]. The production of a skin of molybdenum carbide only two atomic layers in thickness will increase the apparent molybdenum concentration of a 10 nm (Nb, Mo)C particle by 20%. A possible mechanism for the apparent coprecipitation of molybdenum and niobium carbides may be the production of a molybdenum-rich carbide layer onto pre-existing niobium carbonitride precipitates, because no pure molybdenum carbonitride particles were detected.

Titanium levels of about 20% were observed in niobium-rich spheroidal precipitates, notably in the high titanium Heats #4 and #6 (see Table II). This suggests that the effective titanium concentration is approximately 0.01 w/o if the compound $Nb_xTi_{1-x}C$ is considered as in Fig. 15. This again emphasizes that it is the 'available' levels of all solutes in each temperature range which must be used to predict the composition of precipitating compounds.

The influence of dissolved niobium on the formation of $Ti_xNb_{1-x}N$ is shown in Figs. 12, 13 and 17. As the 'available' titanium level is reduced the pseudo-solvus line is flattened, allowing the dissolution of mixed nitride precipitates to dissolve at temperatures well below that at which pure titanium nitride would. This progressive dissolution of the grain refining phase occurs over the same temperature range where austenite grain growth becomes more rapid (Fig. 18). The dissolution of nitride precipitates is likely to be linked with rapid austenite grain coarsening both at high anneal temperatures and in the CG HAZ of welded plates.

In contrast, if the niobium is present as an envelope of niobium rich carbonitride it may delay the dissolution of the titanium rich cuboids and hence help to retain a sufficient volume fraction of grain growth controlling particles.

Matsuda et al. [7] have advanced the argument that the grain coarsening of the austenite is affected by Ostwald ripening of the TiN particles. However, inserting values of annealing data such as 100 hours at 1200°C into the Lifshitz-Wagner equation [27,28] for particle growth indicates that the particles should approximately double in size from that observed in the as-rolled condition (500 Å). Thus Ostwald ripening results in a slow change in particle diameter. If the role of niobium is to decrease the stability of TiN then the change in volume fraction due to the change in solubility may lead to much faster austenite grain growth than that predicted by a particle coarsening process.

The similar chemistries of the compounds found in the CG HAZ of welded plates and in the Gleeble thermally cycled specimens both indicate the stability of titanium rich nitrides at temperatures close to the melting point of austenite. The measured titanium concentrations are not significantly different in the cuboidal precipitates found on thermal cycling of Heats #3, 4 and 6. The consistent level of niobium $\sim$40 w/o in the cuboids reflects the 'available' titanium level at the original soaking temperature (1200°C), Fig. 13, since little time is available for equilibration during a weld pulse as indicated by the 'skin' effect seen in the cuboids in the HAZ of Heat #6 (see Table II). The absence of niobium rich spheroidal precipitates close to the fusion line is indicative of the rapid cooling rate to ambient temperatures from what is essentially a solution treatment for niobium carbonitrides.

It should be emphasized that local segregation due to the original casting practice may lead to marked differences in the microchemistry of the precipitates formed in microalloyed steels. These differences in chemistry together with the variation in local particle volume fraction can produce severe inhomogeneities in grain size both in the as-rolled plate and in the samples subjected to weld thermal cycles.

The results described above indicate the value of the STEM method in characterizing the complex precipitates present in micro-alloyed steels and represent an initial attempt to use quantitative microchemical analysis to deduce the specific role of different microalloying elements at various stages in the overall thermal mechanical processing.

Conclusions

The main conclusions can be summarized as follows.
a) Titanium additions improve the resistance to austenite grain coarsening both during isothermal soaking and thermal cycling.
b) The stable precipitate responsible for grain boundary pinning in high temperature austenite is $Ti_xNb_{1-x}N$, where x is of the order of 0.6 and increases with increasing solution temperature.
c) The precipitates present in control-rolled microstructures are of two major species; high temperature nitride compounds $Ti_xNb_{1-x}N$, formed during soaking and carbonitrides of the general form $Nb_xTi_{1-x}(C_yN_{1-y})$ which form over a range of temperatures during thermo-mechanical treatment. Precipitates containing marked compositional gradients are frequently detected.

d) The 'available' titanium levels in heats containing different bulk titanium concentrations are very similar. This indicates that titanium additions in excess of $\sim$0.01 w/o do not substantially increase the grain pinning precipitate yield, but serve only to increase the amount of large TiN cuboids.

e) Dendritic particles formed during solidification persist in various forms throughout the annealing treatments and can determine the ultimate size and distribution of grain refining agents.

f) Solution temperatures in the region of $\sim$1200°C are necessary to ensure optimum precipitation of niobium carbonitride during cooling. Reheating to below $\sim$1100°C niobium remains 'locked' in the large dendritic clusters formed during solidification and can play no role in further precipitation.

g) A precipitate map and two simple thermodynamic models adequately describe the evolution of the complex precipitates observed at each stage of processing.

h) The detection of steep concentration gradients within precipitates, $\sim$100 A in size is reported for the first time. Evidently homogenization does not readily take place when precipitation occurs over a wide temperature range.

References

1. T. Gladman and F.B. Pickering, J.I.S.I. 205, (1967) 653.
2. T. Gladman, Proc. Royal Soc. A, 294 (1966) 298.
3. K.J. Irvine, F.B. Pickering and T. Gladman, J.I.S.I., 205 (1967) 161.
4. T. George and I. Irani, J. Aust. IWST Metals, 13 (1968) 94.
5. T.J. George, G. Bashford and J.K. MacDonald, J. of Aust. Inst. of Metals 16 (1971) 36.
6. T.J. George and N.F. Kennon, Journal of Australian Institute of Metals, 17 (1972) 73.
7. S. Matsuda and N. Okumura, Trans. I.S.I.J. 18 (1978) 198.
8. S. Kanasawa, Nakashima, K. Okamoto and K. Kanaya, Trans. I.S.I.J., 16 (1976) 486.
9. D.S. McCutcheon, J.T. McGrath, G. Ruddle, G.C. Weatherly, M. Godden, J.D. Embury, and D.C. Houghton, MICON (1982).
10. D.C. Houghton. Unpublished research.
11. V.K. Lakshmanan, Ph.D. Thesis, McMaster University (1977).
12. R.C. Hudd, A. Jones and M.N. Kale, J.I.S.I. 209 (1971) 121.
13. H. Nordberg and B. Aronsson, J.I.S.I. 206 (1968) 1263.
14. K. Narita, Transactions of I.S.I.J., 15 (1975)147.
15. B. Aronsson, Climax Molybdenum Symposium (1969) 77.
16. H.J. Goldschmidt, "Interstitial Alloys", Plenum Press, N.Y. (1967)
17. H. Watanabe, Y.E. Smith, and R.D. Pehle, "The Hot Deformation of Austenite", A.I.M.E. 1 (1977) 140.
18. M.G. Akben, I. Weiss and J.J. Jonas, Acta Met., 29 (1981) 111.
19. D.V. Doane and J.S. Kirkaldy, "Hardenability Concepts with Applications to Steel", Met. Soc. AIME (1978) 82.
20. E.E. Fletcher, A.R. Elsea and E.C. Bain, Trans. ASM. Quart., 54 (1961) 1.
21. V. Heikkinen and R. Packwood, Scan. J. Metal, 6 (1977) 170.
22. C. Parsons, Ph.D. Thesis University of Cambridge, quoted by R. Honeycombe, Scand. J. of Metal 8 (1978) 21.
23. J.M. Silcock, Metal Science J. (Dec. 1978) 561.
24. L.A. Nesbit and D.E. Laughlin, J. of Crystal Growth 51 (1981) 273.
25. T. Shiraina, N. Fujino and J. Murayama, J. Trans. I.S.I.J. 10 (1970) 406.

26. R.W.K. Honeycombe, Metallurgical Trans. A, 7A (1976) 915.
27. I.M. Lifshitz and V.V. Slyozov, J. Phys. Chem. Solids 19 (1961) 35.
28. C. Wagner, Z. Electrochem. 65 (1961) 581.

DISCUSSION

Q: After 20 years, it is very refreshing to see that we are finally asking the basic question, what is the nature of the precipitates in microalloyed steels? We can appreciate that microalloying may have an effect in solution, but the main function of microalloying is by precipitation. And I think this paper is extremely valuable to us. I would like your comments on three points. First, in processing steels by rolling it's important to precipitate in the early stages compounds which are particularly stable. Here the role of nitrides, as opposed to carbides, seems to be very pronounced. Nitrides have lower solubility, higher stability, and for that reason are perhaps the preferred compounds. Also, if the effectiveness of any kind of precipitation is related to the stoichiometry, we should keep in mind that under practical conditions it is much easier to approach the stoichiometric ratio for nitrides than it is for carbides. For instance, very few steels contain as little as 0.02 C. The second point concerns your observation of the coring effect and the mixed nature of precipitates. I wonder how really important this is. What seems to be the controlling factor is the nature of the nuclei, and in this case, it is the stable nature of the nitride nuclei that is important. That other elements subsequently join the nuclei is probably far less important. The third point is that to improve the toughness of microalloyed steels the carbon content has been reduced to the point where we are discussing 0.02-0.03 carbon levels. At the same time we have begun to appreciate the need for relatively high nitrogen levels. Our task in the future will be to learn to weld these low carbon-high nitrogen steels.

A: The second point is the easiest one to respond to. To some extent you are right in that once a particle is completely enclosed in a carbide cap it performs for the rest of the precipitation sequence as if it was all of that composition. However, our estimate of the thermodyanmics from that point on is going to be in error because we're assuming that the point is homogeneous. Also, the thermodynamics, and our understanding of the precipitate sequence, is influenced by what's left in solution.

EFFECT OF MOLYBDENUM ON DYNAMIC PRECIPITATION AND RECRYSTALLIZATION IN NIOBIUM AND VANADIUM BEARING STEELS

B. Bacroix, M.G. Akben and J.J. Jonas

Dept. of Metallurgical Engineering
McGill University
3450 University Street
Montreal, Canada H3A 2A7

A series of four molybdenum microalloyed steels was compressed at constant true strain rates in the temperature range 875 to 1150°C. The 0.06% C - 1.43% Mn-based series contained additions of: (i) 0.30% Mo; (ii) 0.30% Mo + 0.115% V; (iii) 0.30% Mo + 0.040% Nb; and (iv) 0.30% Mo + 0.115% V + 0.035% Nb. Measurement of the recrystallization peak strains over a range of strain rates permitted the determination of the dynamic precipitation-time-temperature (PTT) curves for the ternary and two binary steels. The results show that the presence of molybdenum in solution *retards* the precipitation of both VN and Nb(CN); this is attributed to the decrease in the activity coefficients of carbon and nitrogen caused by the addition of molybdenum. From the peak strains, the dynamic recrystallization-time-temperature (RTT) curves were also determined for the four materials, both *above* and *below* the solution temperature of the respective carbonitrides. *Above* the solution temperatures, the niobium had a considerably greater retarding effect than molybdenum, and molybdenum in turn exerted more influence than vanadium (on an equal atom fraction basis). *Below* the solution temperatures, although the occurrence of precipitation led to breaks in the RTT curves, and therefore to increases in the recrystallization start times, the relative 'no recrystallization' temperatures followed the rank order of the solute effects.

The influence of niobium, molybdenum, and vanadium in solution on the yield strengths of the steels was measured over the experimental temperature and strain rate range, and *prior to* the occurrence of strain-induced precipitation. On an equal atom fraction basis, the order of effectiveness followed that established for the retardation of recrystallization. In the ternary alloy, the combined presence of all three microalloying elements raised the $\gamma \to \alpha$ transformation temperature pertaining to concurrent deformation, and the ferrite transformation was initiated within about 40 s of the initiation of straining at 875°C.

Introduction

Molybdenum is generally added to microalloyed steels containing niobium and/or vanadium to enable higher strength levels to be attained. Despite its widespread use, the detailed effect of molybdenum addition on the microstructural processes occurring during control rolling, i.e. on the recrystallization of austenite and on the strain-induced precipitation of carbonitrides, is not well known. The primary objective of the present study was therefore to determine the effect of molybdenum in solution on the initiation of dynamic recrystallization during the high temperature deformation of microalloyed austenite. A subsidiary aim was to investigate the influence of molybdenum addition on the kinetics of carbonitride precipitation in niobium and vanadium bearing steels. Although the investigation was concerned with the initiation of *dynamic* recrystallization and *dynamic* precipitation, as shown in some earlier experiments conducted in the current manner (1,2), the relative influence of the various alloying additions on the *static* processes of greater interest under industrial rolling conditions can be established in this way.

The experimental design involved the preparation of seven microalloyed steels and a reference plain carbon steel. Each of the first three microalloyed steels contained a *single* addition of 0.30% Mo, 0.035% Nb or 0.115% V; three further steels were based on *binary* additions at the above nominal levels; the final composition was a *ternary* steel in which all three additions were present concurrently. The results obtained with the reference steel and with the Nb, V, and Nb-V additions have been described elsewhere (3-5) but will be utilized here for an assessment of the individual and combined effects.

In addition to the light shed on the recrystallization and precipitation behaviour, the experiments permitted the relative influence of the various alloying combinations to be determined when the elements are *fully in solution*. This was done: (i) by testing above the solution temperatures of the respective carbonitrides; and (ii) by measuring the influence of the additions on the yield strength (and therefore on the ease of dynamic recovery) prior to the occurrence of strain-induced precipitation. Finally, an incidental result of the current investigation was the observation that, for a fixed experimental time, the *ternary* composition had a higher transformation temperature during concurrent straining than the other combinations of alloying elements. This was indicated by its transformation to ferrite during deformation at the lowest experimental temperature, as will be described in greater detail below.

Experimental Materials and Procedure

The materials investigated were prepared in the Physical Metallurgy Research Laboratories of the Department of Energy, Mines and Resources, Ottawa. The chemical compositions of the alloys are given in Table I. The results for the steels identified with an asterisk (*) were reported earlier (3-5) and are presented here for reference purposes only. The samples were machined from 13 mm thick plates with the compression axis aligned along the rolling direction. The ends of the samples were grooved to retain the glass lubricants used to minimize friction between the end faces of the sample and the SiN compression platens. The sample diameter and height were 7.6 and 11.4 mm respectively. To eliminate the texture present in the rolled plates, all the steels were vacuum annealed at 1000°C for two hours and then air cooled or water quenched.

Table I. Compositions (wt%), Austenitization Temperatures, Austenite Grain Sizes, and Ac$_3$ Temperatures for the Steels Tested

Steel	C	Mn	Al	Si	Nb	V	Mo	Austenit. Temp.'1' (°C)	Grain Size'1' (μm)	Austenit. Temp.'2' (°C)	Grain Size'2' (μm)	Ac$_3$ (13)
* Pl C	.06	1.43	.025	.24	–	–	–	1030	110	1140	200	845
* V	.05	1.20	.03	.25	–	.115	–	1045	100	1160	200	870
* Nb	.05	1.25	.03	.27	.035	–	–	1100	130	1200	210	860
Mo	.05	1.34	.065	.20	–	–	.29	1060	110	1140	190	868
* Mo-Nb	.06	1.33	.025	.21	.040	–	.30	1100	130	1200	200	855
Mo-V	.045	1.21	.05	.15	–	.115	.28	1100	120	1150	200	870
Mo-Nb-V	.04	1.16	.025	.20	.034	.115	.31	1100	115	1200	200	880
All	P ≅ .006	N ≅ .006	S ≅ .012	Cr ≅ .04								

* A more detailed description of these steels is presented in references 3 - 5.

Immediately prior to testing, each sample was austenitized for half an hour in an argon atmosphere in the test chamber of a modified Instron compression machine. The austenitization temperatures (see Table I) were chosen to promote the complete dissolution of the carbonitrides and to lead to approximately the same initial austenite grain size for all the steels. For the first series of tests, i.e. for all the results presented below, with the exception of the RTT curves, the austenitization temperatures are identified by the number '1'. These led to mean grain sizes of approximately 100 μm. Following this heat treatment, the samples were cooled to the test temperature at an average rate of 1°C/second and then tested at 875, 900 or 925°C. For the RTT series of tests, the austenitization temperatures are identified by the number '2' and were chosen to give an initial grain size of approximately 200 μm. In this case, the test temperatures were in the range 875 to 1150°C.

The experiments were carried out in compression at constant true strain rates ranging from 5.6 x 10^{-5} to 1 s^{-1}. The manner in which the Instron testing machine was modified for this purpose is described in a previous publication (6). The method used to determine the precipitation kinetics (1-4) involves the measurement of the strain to the peak stress, which is indicative of the initiation of recrystallization. Although dynamic recrystallization is in fact nucleated somewhat before the peak strain (7), the latter is a useful approximate measure of the onset of recrystallization. The peak strain is sensitive not only to strain rate, temperature and solute concentration, but also to the occurrence of precipitation during straining. By determining the dependence of the peak strain on strain rate and temperature and searching for departures from the behaviour of systems in which no precipitation takes place, the precipitation start time can be detected quite sensitively (1-4,8). In the present case, this method worked quite well for the compositions containing niobium and/or vanadium, and also indicated that the dynamic precipitation of microalloy carbonitrides did not occur at all in the grade containing the single addition of molybdenum.*

Results

A typical set of flow curves for the molybdenum steel compressed at 925°C is shown in Figure 1. A very well defined peak strain which increases with strain rate can be seen on most of the curves. As will be demonstrated later, this strain is also a function of the temperature, the concentration of microalloying elements, and whether or not dynamic precipitation is initiated prior to the peak.

A selection of flow curves for all the steels tested is presented in Figure 2, also for 925°C. It is apparent that when molybdenum is added to the plain carbon steel (Fig. 2b), there is an increase in both the peak stress and peak strain. In the absence of precipitation effects, there is a further delay in the initiation of dynamic recrystallization when vanadium is added, and an even greater retardation when niobium is added to the molybdenum steel (Fig. 2b). The greatest retardation occurs when the ternary addition is made. It is also evident that dynamic recrystallization

---

*At 875°C, there was a suggestion that AℓN was precipitating out at a strain rate of about 10^{-3} s^{-1} (see Fig. 3c below). This can be attributed (8) to the relatively high aluminum concentration present in this material (0.065% - see Table I). This possibility is being investigated by means of transmission electron microscopy.

296

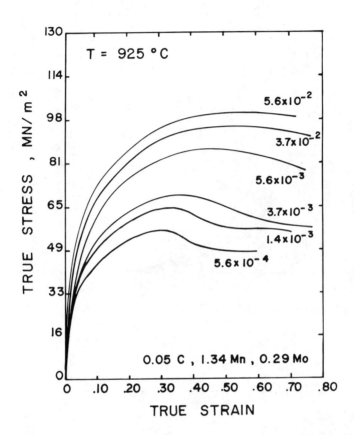

Fig. 1 Effect of strain rate on the flow curve of the 0.29% Mo microalloyed steel compressed at 925°C.

is the most strongly retarded by the single addition of niobium, followed
by that of molybdenum and only slightly retarded by the single addition of
vanadium (Figs. 2a and 2b). The order of these elements is unchanged, but
the effects are magnified if the amount of the retardation is normalized,
first for an equal *weight* of addition, and then for an equal *atom fraction*.
If the testing strain rate is reduced (Figs. 2c and 2d), permitting the
initiation of precipitation during straining, there is a contribution from
both solute and precipitate effects to the increase in the peak strain in
the vanadium and four niobium-containing steels. By contrast, solute effects
alone are responsible for the retardations observed in the molybdenum and
Mo-V steels.

The strain rate dependence of the peak strain ϵ_p is illustrated in
Figs. 3a to 3c for the three temperatures investigated. From these curves,
it is clear that the peak strain increases smoothly with strain rate

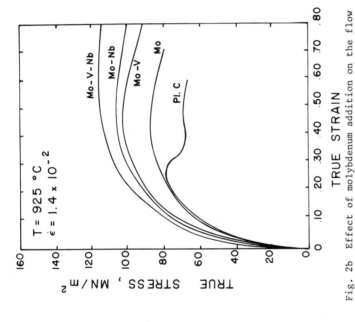

Fig. 2a Influence of vanadium and niobium addition on the flow curves of a series of micro-alloyed steels tested at a strain rate of 1.4×10^{-2} s⁻¹ and 925°C (3,4). The peak strains in all these experiments were attained in less than 42 s and so depend only on solute, as opposed to precipitate, effects. (The Nb-V steel contains 0.05% C, 1.18% Mn, 0.035% Nb and 0.115% V (3-5))

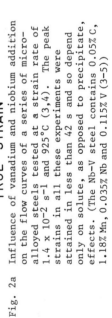

Fig. 2b Effect of molybdenum addition on the flow curves of the steels of Fig. 2a. At the testing strain rate of 1.4×10^{-2} s⁻¹, the peak strains were all attained prior to the commencement of carbonitride precipitation. The differences between the curves can there-fore be attributed to the influence of niobium, molybdenum and vanadium *in solution*.

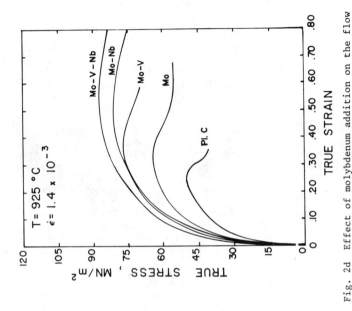

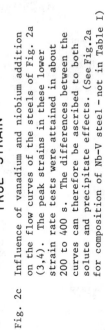

Fig. 2d Effect of molybdenum addition on the flow curves of the steels of Fig. 2c. At the testing strain rate, the peak strains were attained in about 200 to 400 s. As in Fig. 2c, the differences between the curves are associated with both solute and precipitate effects.

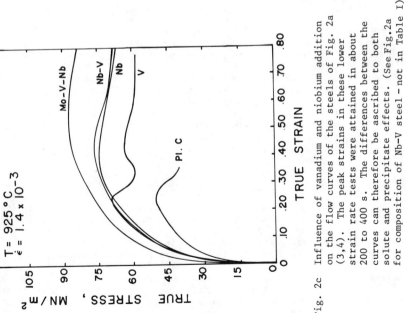

Fig. 2c Influence of vanadium and niobium addition on the flow curves of the steels of Fig. 2a (3,4). The peak strains in these lower strain rate tests were attained in about 200 to 400 s. The differences between the curves can therefore be ascribed to both solute and precipitate effects. (See Fig.2a for composition of Nb-V steel – not in Table I)

for the plain carbon steel and that the dependences are qualitatively similar at the three temperatures. In the microalloyed steels, there is an overall delay in the initiation of dynamic recrystallization associated with the shift of each curve to a higher level of peak strain than displayed by the reference material. This delay, due to a solute effect (1-4), is appreciable for the molybdenum steel, somewhat larger for the Mo-V alloy, still larger for the Mo-Nb grade, and the largest for the Mo-V-Nb material. Moreover, in the curves corresponding to the binary and ternary steels, 'humps' or 'bulges' are evident which are indicative of an interaction between dynamic recrystallization and dynamic precipitation. For these materials, the delay in the initiation of dynamic recrystallization cannot be attributed to a solute interaction alone (3,4,8-10).

From the data displayed in Fig. 3, it is evident that molybdenum, when added alone, has a strong solute effect, but that no detectable precipitation of carbonitrides occurs,(except for the reservation expressed above). This is in agreement with the literature, in that no data have yet been reported concerning the precipitation of molybdenum carbonitrides in high temperature austenite. Although some authors have found that molybdenum *is* present in precipitates in microalloyed austenite, this seems to occur only in steels containing niobium as well (11). The latter observation may not be completely general, however, as other researchers, e.g. Watanabe et al. (12), have not been able to detect *any* molybdenum in the niobium carbonitrides they studied by extraction replication in similar materials.

The influence of molybdenum as a single addition agent bears a strong contrast to that of vanadium or niobium. Although molybdenum can retard recrystallization *solely* through a solute effect, both vanadium and niobium exert their influence on delaying recrystallization both in solute as well as in precipitate form. For vanadium, the solute effect is small, whereas for niobium, it is appreciable (3,4). When niobium and molybdenum are added jointly, the retardation produced is *less* than the sum of the individual effects of niobium and molybdenum when they are present alone. A similar observation applies to the case of the binary addition of molybdenum and vanadium, as well as to the ternary addition of niobium, molybdenum and vanadium. These results suggest that the retardation of *recrystallization* by solutes saturates more quickly (i.e. at a lower total atomic concentration) than the retardation of *recovery* by solutes. As will be demonstrated below, when the microalloying elements are present *in combination*, the increase in the yield strength (which is controlled by dynamic recovery processes, see below) is *greater* than the sum of the effects on the yield strength of the individual alloying elements.

Effect of Molybdenum Addition on the Austenite-to-Ferrite Transformation

It is of interest that the ε_p versus $\dot{\varepsilon}$ results for the ternary steel at 875°C (Fig. 3c) do not display the expected dependence. Normally such a curve will be parallel to the one for the Mo-Nb steel, but at a higher overall level. Here the usual behavior is observed at the *high* strain rates, at which the difference in peak strain between the two curves is appreciable. The difference is much less at the *low* strain rates, and this unexpected trend cannot be due to the precipitation of VN in addition to the precipitation of Nb(CN). Instead, it appears to be caused by the *premature transformation* from austenite to ferrite during the course of the slower (i.e. longer time) experiments. This can be explained by the

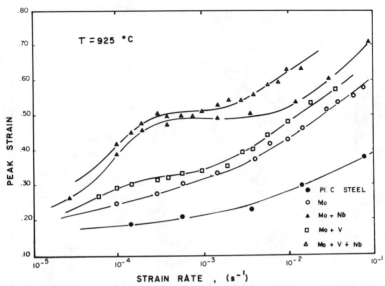

Fig. 3a Dependence of the peak strain on strain rate at 925°C for the four molybdenum-based materials. The differences in the general levels of the curves are attributable to a solute effect. The 'humps' located in particular strain rate ranges are associated with the occurrence of dynamic precipitation during testing.

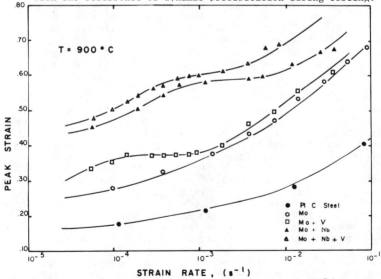

Fig. 3b Dependence of the peak strain on strain rate at 900°C for the four molybdenum-based materials.

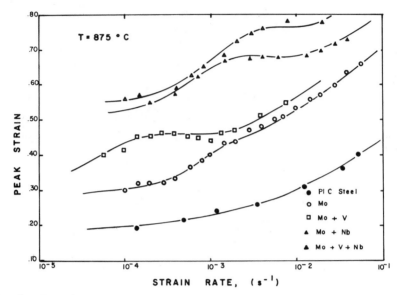

Fig. 3c Dependence of the peak strain on strain rate at 875° C for the four molybdenum-based materials.

observation that niobium, molybdenum and vanadium are austenite stabilizers when added in small quantities, but are ferrite stabilizers when present in greater amounts. Thus, although they *decreased* the Ac_3 temperature when added individually, when added together, they *raised* the Ac_3 temperature above the 875° test temperature. This interpretation is in agreement with the values of Ac_3 calculated from the literature (13,14) which are listed in Table I.

The hypothesis concerning the transformation to ferrite was confirmed metallographically, as indicated in Fig. 4, in which micrographs are presented corresponding to samples tested at three decreasing intermediate strain rates. These specimens were quenched within 4 s of the completion of each experiment. For the sample tested at the highest strain rate, the transformation has had sufficient time to begin during the 108 s required by the test and the structure is therefore partly ferritic and mostly martensitic. At the middle strain rate, the experiment extends over 216 seconds and there is accordingly more ferrite present. At the lowest strain rate, 2700 seconds are required for completion, and the structure is almost entirely ferritic. By contrast, microstructures for the other steels compressed at 875°C over the same range of strain rates and quenched in the same way show martensitic structures only.

From similar micrographs and test data, it is apparent that the ferrite start time is about 40 seconds. This is also the approximate time $t = \varepsilon^*/\dot{\varepsilon}^*$ at which the ε_p vs $\dot{\varepsilon}$ curve for the ternary dips *below* its normal course. Here ε^* and $\dot{\varepsilon}^*$ are the strain and strain rate at which ferrite begins to be detected in quenched samples. However, a ferrite start

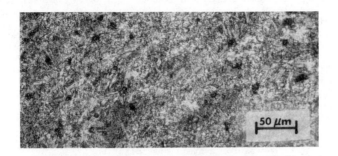

a) Strained for 108 s at a strain rate of 7.4×10^{-3} s^{-1} to a strain of 0.8.

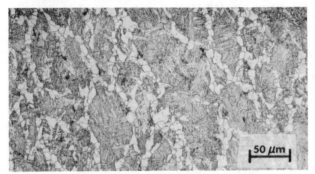

b) Strained for 216 s at a strain rate of 3.7×10^{-3} s^{-1} to a strain of 0.8.

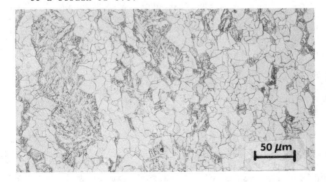

c) Strained for 2700 s at a strain rate of 3.7×10^{-4} s^{-1} to a strain of 1.0.

Fig. 4   Microstructure of the ternary Mo + Nb + V steel deformed at 875°C and quenched within 4 s of straining. Note that the volume fraction of ferrite increases with the time of deformation. Magnification: 300 X.

time of 40 s at a test temperature only about 5°C below the estimated
Ac_3 temperature (see Table I) seems unusually short, particularly when
compared with estimates based on conventional *static* TTT curves. Thus
it appears that the austenite-to-ferrite transformation is accelerated
considerably if the sample is being deformed during holding below the Ac_3
temperature. The observation that the transformation to ferrite during
testing affects the shape of the ε_p vs $\dot{\varepsilon}$ curve has the further implication
that it can be used to assess the effect of concurrent deformation on
the transformation diagram. By comparing the 'dynamic' and 'static'
diagrams, the amount by which the transformation is accelerated by
straining can be readily estimated [15].

PTT Curves for Dynamic Precipitation

The rate of *dynamic* precipitation appears to represent the upper
limit to the rate of *static* precipitation in heavily deformed materials
[1-3,8]. The start and finish times P_s and P_f for the precipitation of
VN and Nb(CN) during deformation were therefore deduced from the ε_p vs $\dot{\varepsilon}$
curves of Fig. 3 according to the method of References 1 to 4 (see Figs. 5
and 6). Here $P_s = \varepsilon_p^s / \dot{\varepsilon}_s$, where ε_p^s is the peak strain at which, as the
strain rate is reduced, the curve departs noticeably from the smooth
interpolated curve representing the retarding effect of solutes alone.
The strain rate at this point, $\dot{\varepsilon}_s$, is that at which dynamic precipitation
is initiated just prior to ε_p^s. In a similar manner, $P_f = \varepsilon_p^f / \dot{\varepsilon}_p$, where
ε_p^f is the peak strain at which the difference between the 'humped' and
interpolated 'solute' curves is at a maximum and $\dot{\varepsilon}_p$ is the strain rate at
this point. At $\dot{\varepsilon}_f$, dynamic precipitation is terminated at ε_p^f. As the
strain rate is decreased below $\dot{\varepsilon}_p$, dynamic precipitation is completed at
smaller and smaller strains and the amount of precipitate coarsening taking
place during deformation to the peak stress increases, thus reducing the
peak strain [9]. From Fig. 5 it can be seen that the addition of molybdenum leads to a considerable retardation in the rate of precipitation of
VN. The effect on P_f is greater than on P_s. The retardation can be attributed to the decrease in the activity coefficients of carbon and nitrogen caused by molybdenum addition, as reported by Wada et al. [16,17] and
by Nishizawa [18]. The change in the activity leads to an increase in the
solubility of VN, which in turn decreases the driving force for precipitation. The influence of molybdenum addition on the kinetics of VN precipitation is qualitatively similar to that of manganese addition on the rate
of Nb(CN) precipitation, as reported in more detail elsewhere [3]. It is
also possible that molybdenum addition decreases the rate of vanadium diffusion in austenite by analogy with the decrease in the niobium diffusion
rate attributable to manganese addition [19].

It can be seen from Fig. 5 that the shape of the curve for the
ternary steel differs considerably from that for the two binary materials.
This is directly associated with the transformation that occurs during
deformation and the observation that VN precipitation is faster in
ferrite than in austenite, as discussed for example by Gray and Yeo [20].
The PTT curves for the niobium-bearing steels presented in Fig. 6 lead to
similar conclusions. Here it can be seen that the presence of molybdenum
retards the precipitation of Nb(CN) and that, as before, its influence is
greater on the P_f than on the P_s times. The effects on carbonitride
solubility and on the diffusion rate of the precipitating substitutional
solute are more marked in the ternary steel, and thus parallel the trend
illustrated in Fig. 5 above for the vanadium series of steels.

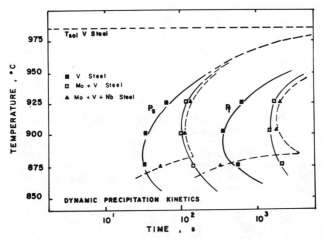

Fig. 5 Dynamic PTT curves for the *vanadium* series of steels. Precipitation
is retarded by about half an order of magnitude by the addition of
0.30% Mo to the simple 0.115% V steel. When 0.035% Nb is added to
the binary Mo-V composition, only a small amount of further retardation is produced, as Nb is a faster precipitating species (at these
temperatures) than V. The rapid increase in the rate of precipitation at 875°C in the ternary alloy is due to the transformation to
ferrite during deformation at this temperature (see Fig. 4) and to
the attendant higher supersaturation of the ferrite.

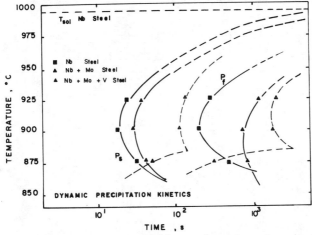

Fig. 6 Dynamic PTT curves for the *niobium* series of steels. Precipitation
is retarded by a factor of about two by the addition of 0.30% Mo
to the simple 0.035% Nb material. When 0.115% V is added to the
binary Mo-Nb composition, a significant amount of further retardation
is produced. As in Fig. 5, the rapid increase in the precipitation
rate at 875°C in the ternary alloy is due to the transformation to
ferrite during deformation at this temperature (see Fig. 4) and to
the attendant higher supersaturation of the ferrite phase.

Comparison of the Flow Stresses

Using an offset of 0.002, it is possible to determine the yield strengths of the individual steels from their respective flow curves. With the aid of the following formula:

$$\Delta S = \frac{\sigma_{ys} - \sigma_{ys}^{r}}{\sigma_{ys}^{r}} \times \frac{0.1}{at. \% \ S}$$

the strengthening increment attributable to each of the microalloying elements (per 0.1 at. %) can be assessed. Here σ_{ys} is the yield strength of the steel investigated, σ_{ys}^{r} is the yield strength of the reference steel and at. % S is the atom fraction of the element under consideration. For example, to calculate the strengthening produced per 0.1 at. % of molybdenum in a Mo-V steel, the yield strength of the vanadium steel is taken as the property of reference σ_{ys}^{r}.

The strengthening assessed in this way was calculated for each temperature and strain rate and the average values for each steel are presented in Table II. It is evident from this tabulation that molybdenum in solution has a slightly *larger* effect than solute vanadium, although the inequality is reversed if weight, rather than atomic, proportions are used. By contrast, molybdenum has a much *smaller* influence on the flow stress than niobium, no matter which way the normalization is carried out. It is also apparent that there is a kind of synergistic effect operating when the microalloying elements are added in combination. Molybdenum, for example, when added with niobium produces more strengthening than when it is present alone. That is, the strengthening produced by molybdenum in the Mo-Nb or the ternary steel is greater than the strengthening produced by molybdenum in a simple molybdenum steel.

Table II. High Temperature Strengthening Produced

By Mo, Nb and V Addition

Element	Per 0.1 at %	Per 0.1 wt %
Mo (in a Mo Steel)	9%	5%
Mo (in a Mo-Nb Steel)	10%	6%
Mo (in a Mo-V Steel)	14%	8%
Mo (in a Mo-V-Nb Steel)	14%	8%
Nb (in a Nb Steel)	70%	44%
Nb (in a Nb-Mo Steel)	100%	63%
Nb (in a Nb-V Steel)	80%	50%
Nb (in a Nb-V-Mo Steel)	90%	56%
V (in a V Steel)	7%	8%
V (in a V-Mo Steel)	11%	12%
V (in a V-Nb Steel)	8%	9%
V (in a V-Nb-Mo Steel)	9%	10%

Discussion

It was shown above that the addition of molybdenum leads to a decrease in the rate of precipitation of VN and Nb(CN) in microalloyed austenite. This effect is similar to that of manganese, which is known to *increase* the activities of niobium and vanadium, but to *decrease* those of carbon and nitrogen to a greater degree (3, 21). Somewhat similar results were obtained by Watanabe et al. (12), who reported that molybdenum addition reduces the nucleation rate of Nb(CN) by decreasing the nose temperature of the C-curve for precipitation. As the nucleation rate depends on the degree of supersaturation of niobium, carbon and nitrogen, as well as on the density of nucleation sites, the reduction in the rate of deposition of Nb(CN) is in direct consequence of the decrease in the activity coefficients of carbon and nitrogen.

For such steels, the reduction in the precipitation rate has a number of interesting effects. First of all, as there is generally less precipitation in the *austenite* under these conditions, a greater volume fraction of carbonitride is free to form in the *ferrite* range. Taking into account the lower temperatures at which the particles are formed in the ferrite, it is clear that they will be *smaller*, and therefore able to produce more particle strengthening than if deposited in the austenite range. It is worthy of note that when molybdenum is added to a microalloyed steel, the decrease in the amount of niobium precipitated in the austenite does not necessarily lead to a decrease in the 'pancaking' temperature. As more niobium remains in solution, its influence as a *solute* retarder of recrystallization is increased and cumulative to that of the molybdenum addition, which does not have a tendency to precipitate. These two additional solutes are then able to retard recrystallization, as well as to reduce the austenite grain size when recrystallization *does* occur (22).

Effect of Molybdenum, Niobium and Vanadium in Solution on Dynamic Recovery and Recrystallization

It is evident from the ε_p vs. $\dot{\varepsilon}$ curves presented in Fig. 3 that the addition of 0.30% molybdenum to a plain carbon steel increases the peak strain ε_p considerably more than does that of 0.115% vanadium. By contrast, the retarding effect on dynamic recrystallization of adding 0.035% niobium is still greater than that of 0.30% molybdenum. When these results are normalized for equal atom fractions (e.g. 0.1 atomic percent), the ranking of the three elements, in decreasing order of effectiveness, is still: (i) Nb; (ii) Mo; and (iii) V. This order is the same as that illustrated in Table II for the increase in yield strength. The yield stress at high temperatures depends, not only on composition, but also on temperature and strain rate. It thus differs from the yield strengths determined in the ambient temperature or athermal range, which are more directly related to the 'state' of the material. The temperature and strain rate dependence in turn indicates that the high temperature yield stress is controlled by a thermally activated process, for which the most likely candidate is dynamic recovery (36). In this way, the data of Table II indicate, not only the effects of the various alloying combinations on strength, but also their influence on the process of dynamic recovery.

These phenomena can be attributed to two distinct classes of mechanism. There is, on the one hand, the effect of niobium and vanadium (but not molybdenum, when present alone) as *precipitate formers*. Thus, when fine particles of VN or Nb(CN) are present, or are being deposited, the

307

rate of work hardening will rise above the normal level, and the nucleation of recrystallization will also be retarded. The effect of precipitation on the yield strength of *unstrained* austenite appears to be negligible (2), and only when precipitation occurs in *pre-strained* austenite are the particles fine enough to lead to an increase in the flow stress (23). Bearing in mind that the effect of precipitation on the yield stress of undeformed materials is small, and that the influence of molybdenum is greater than that of vanadium, it can be concluded that the rank order of retardation described above is *not* primarily due to the formation of fine particles.

Turning to the *solute* effects, these are usually ascribed to segregation, either to dislocations in the case of recovery, or to grain boundaries in the case of recrystallization (24). However, it is also possible that the addition of solutes leads to *bulk* changes, such as a decrease in the stacking fault energy, or in the mean atomic diameter or bond strength. At temperatures below 0.5 T_m, the atomic size and modulus effects are usually considered to be of over-riding importance (25). At elevated temperatures, on the other hand, it appears to be the differences in atomic size and electronic structure that are pre-eminent (3,8). The atomic size differences alone are not responsible for the rank order of the solute effects observed as can be seen from the following example. Aluminum and niobium have similar size differences with respect to γ-iron, and these are considerably greater than those pertaining to molybdenum or vanadium (see Table III). Nevertheless, aluminum has a much smaller solute effect in austenite than does niobium (8), and the solute retardation it produces is less even than that attributable to molybdenum on an equal atom fraction basis. The importance of taking both electronic and size differences into account is also supported by the extensive results of Yamamoto, Ouchi and co-workers (27,28).

Table III. Elements in Decreasing Order of Atomic
Size Difference with respect to Iron *

$r_x > r_{Fe}$			$r_x < r_{Fe}$		
Symbol	r_x (Å)	$(r_x-r_{Fe})/r_{Fe}$	Symbol	r_x(Å)	$(r_{Fe}-r_x)/r_{Fe}$
Zr	1.602	0.26	C	0.916	0.28
Nb	1.468	0.15	B	0.980	0.23
Al	1.432	0.12	Ni	1.246	0.02
W	1.402	0.10	Co	1.252	0.02
Mo	1.400	0.10	S	1.270	0.003
V	1.346	0.06			
Si	1.312	0.03	r_{Fe} = 1.274 Å		
Mn	1.304	0.03			
Cr	1.282	0.01	* Based on a coordination number		
P	1.280	0.005	of 12 and metallic bonding (26)		

Retardation of Recrystallization Above
the Carbonitride Solution Temperature

The amount of solute retardation attributable to the addition of molybdenum, as well as of niobium and vanadium, was described above (Fig. 3), together with the kinetics of precipitation of VN and Nb(CN) in these alloys (Figs. 5 and 6). A knowledge of these individual effects, together with information about the recrystallization kinetics (see Fig. 7), permits the interaction between the three phenomena to be seen more clearly. This is of interest because of the current debate concerning the relative roles of solutes and precipitates in retarding recrystallization. Some workers, for example, consider that the reduction in the recrystallization rate is due entirely, or almost entirely, to precipitate pinning (29, 30). According to this view it is the strain-induced deposition of VN or Nb(CN) on the dislocation substructure which is largely responsible for the observed effect. Others have contended that the microalloying elements in solution retard the mobility of the dislocations, sub-boundaries and grain boundaries via either a drag force (7, 24) or through bulk effects (31, 32). In this section, some new data will be introduced which may help to resolve this question.

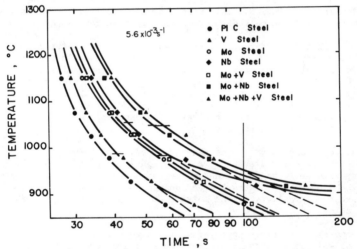

Fig. 7 Dynamic RTT curves displaying the recrystallization start times (R_s) for both sets of steels. The plain carbon and molybdenum grades do not contain precipitating species and display smooth, 'solute-controlled' behavior over the entire temperature range. The Mo–V grade also displays solute-controlled behavior (see Fig. 8). The curves for the other steels break away from the solute-controlled behavior (shown as broken lines below the break) at temperatures about 60 to 100°C below the solution temperature of the appropriate carbonitride (the solution temperatures are identified by the short horizontal bars). *Above* the solution temperature, the increase in the R_s time beyond that for the plain carbon material is a measure of the effectiveness of the particular single, binary or ternary microalloying addition in retarding recrystallization. Note that the relative 'no recrystallization' temperatures of these steels can be estimated from the R_s temperatures at a fixed deformation time of 100 s.

A measure of the possible importance of the precipitate effect is the degree of supersaturation of the elements forming the particle species (30, 33). Thus, in order to eliminate the likelihood of any contribution from the precipitation mechanism, some experiments were carried out *above the solution temperature of the respective carbonitrides* (see Table I). In these tests, the peak strains associated with the initiation of dynamic recrystallization were measured *both above and below the solution temperature*. The results are given in Fig. 7, where the R_s or recrystallization start times were calculated from the peak strains using the relation:

$$R_s = \varepsilon_p / \dot{\varepsilon}$$

where $\dot{\varepsilon} = 5.6 \times 10^{-3}$ s^{-1} was the strain rate employed for the tests. By comparing the R_s times for the microalloyed with those for the plain carbon steels, it is apparent from Fig. 7 that, even at temperatures well above the calculated solution temperatures, there is a *marked and significant delay* in the initiation of dynamic recrystallization. Because of the conditions of this experiment, the retardation of R_s with respect to the plain carbon material must be attributed to the presence of the microalloying elements in solution. This effect is the smallest for the vanadium, intermediate for the molybdenum, and the greatest for the niobium steel.

Note that the retardations produced by multiple additions, e.g. Mo + V, Mo + Nb, or Mo + V + Nb, are somewhat less than the estimates made on the basis of the simple additivity of the effects of the individual components. In both binaries as well as in the ternary alloy, the niobium addition has the greatest effect, followed by the molybdenum and then the vanadium. This order remains valid, except that the particular effectiveness of the niobium addition is underlined, when the results are normalized for a standard addition of 0.1 atomic percent.

At temperatures *below* the respective solution temperatures, the dynamic RTT curves break distinctly to the right for the three niobium-containing alloys, i.e. for the Nb, Nb-Mo and ternary compositions. For the vanadium grade, the break is less marked and only occurs at the lowest experimental temperature. Finally, for the molybdenum and Mo-V steels, there is no break at all in the R_s curve.

Some similar curves containing breaks were reported by Hansen et al. (33), as well as by Kreye and Hornbogen (34) and LeBon and Saint Martin (35). Although the latter workers investigated the interaction between *static* precipitation and recrystallization, their results are qualitatively similar to those described here. The interaction between *dynamic* precipitation and recrystallization was considered in some detail by Hansen and co-workers (33).

Here we present a somewhat similar, but more detailed, analysis, having both dynamic RTT *and* dynamic PTT data available for seven different steels. The results for the molybdenum steel, as well as for the three alloys containing vanadium are illustrated in Fig. 8; those for the three niobium grades are presented in Fig. 9. (The RTT and PTT curves for the seventh composition, the Nb-V binary, are omitted for the sake of clarity, but are similar to those for the niobium steel (4, 5).) Beginning with the straight molybdenum steel (Fig. 8), in which there is no deposition of microalloy carbonitrides, it is evident that there is *no* possible interaction between precipitation and recrystallization. Thus the RTT curve follows the

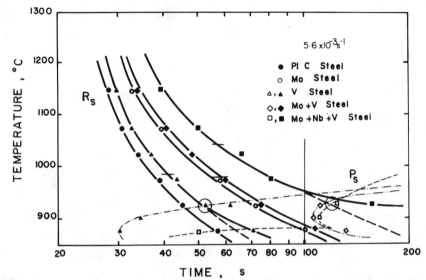

Fig. 8   Interaction between dynamic precipitation and recrystallization in
the *vanadium* series of steels. The breaks in the R_s lines for the
single vanadium and ternary Nb-Mo-V compositions occur near the
intersections of the RTT and PTT curves. There is *no* corresponding
intersection, and consequently no break, in the R_s curve for the
Mo-V grade. Note that the 'no recrystallization' temperature at
100 s increases systematically with the total concentration of the
microalloying elements in solution.

classical shape displayed in Fig. 7 by the plain carbon steel. For the
three remaining steels, it is possible to superimpose the PTT kinetics of
Fig. 5 on the RTT data of Fig. 7 leading to the intersections identified by
circles in Fig. 8. It should be noted that the points of intersection of
the RTT and PTT relations for the vanadium and ternary compositions are
in reasonable agreement with the positions of the 'breaks' in the R_s curves
for the two steels, as predicted by the models referred to above. In a
similar way, the *absence* of an intersection between the RTT and PTT relations
for the Mo-V composition is consistent with the lack of a break in the R_s
curve, although further data below 900°C are necessary before this conclusion can be drawn with full confidence.

In Fig. 9, the PTT kinetics for the three niobium alloys of Fig. 6
are superimposed on the appropriate RTT relations of Fig. 7. Once again
it is evident that the intersections of the PTT and RTT curves are in fair
agreement with the locations of the breaks in the R_s plots; nevertheless
there is a perceptible trend for the breaks to appear *before* the intersections. This discrepancy could result from the following two features
of the present test method:

(i) The PTT kinetics are established from tests carried out over
a *range* of strain rates (and therefore dislocation densities). At *constant*
dislocation density, the PTT curves are expected to be 'fuller' or rounder
than those of Fig. 9 and therefore to lead to *earlier* intersections.

311

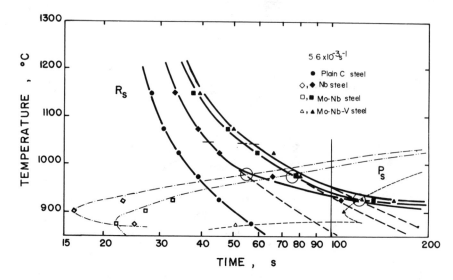

Fig. 9   Interaction between dynamic precipitation and recrystallization in
the *niobium* series of steels. The breaks in the R_s lines for the
three alloys occur slightly earlier than the intersections of the
PTT and RTT relations. As in Fig. 8, the 'no recrystallization'
temperature at 100 s increases systematically with the total
concentration of the microalloying elements in solution.

        (ii) The RTT curves are determined from tests carried out at a
fixed strain rate (and therefore at dislocation densities that *decrease* with
increasing temperature). At *constant* dislocation density, the R_s curves
would incline more steeply to the left; this change may also lead to earlier
intersections.

     A further type of interaction is also of interest. If the addition of
an alloying element were to lead *only* to a solute retardation of recrystallization (i.e. to horizontal motion of the RTT curve to longer times),
then, for *fixed* precipitation kinetics, the intersection between the relevant RTT and PTT lines (i.e. the 'break') would move to a higher
temperature. In fact, when an alloying element is added, the temperature
of intersection remains approximately *constant*, as seen in Fig. 9, because
the addition also generally *reduces* the activities of carbon and nitrogen,
and therefore *increases* the solubility of the precipitating species, and
reduces the driving force and rate of precipitation.

     Even though the break *temperatures* remain approximately fixed, the
break *times* generally increase with the total amount of alloying addition.
This permits a qualitative estimate to be made of the difficulty of preventing recrystallization during control rolling of the various steels discussed above. For example, the relative 'pancaking' temperatures can be
estimated from Figs. 7 to 9 by taking a fixed time (e.g. 100 s) and then

312

reading the R_S temperature associated with this time for each steel. For the materials studied, this approach leads to the following 'recrystallization stop' temperatures:

Steel	Recrystallization Stop Temp.(°C)
Mo + Nb + V	955
Mo + Nb	940
Nb	935
Mo + V	890
Mo	880
V	835

Although these temperatures will vary somewhat with the strain rate and amount of the accumulated prestrain, they remain consistent with the order of effectiveness of the solute elements discussed above.

Conclusions

1. The addition of 0.30% molybdenum to vanadium and/or niobium-bearing microalloyed steels *retards* the *precipitation* of the carbonitrides in austenite. This retardation, which in all cases amounts to less than an order of magnitude in time, is attributed to the *decrease* in the carbon and nitrogen *activity coefficients* associated with the presence of molybdenum. This change leads in turn to an *increase* in the *solubilities* of the particles, to a *decrease* in the *driving force* for austenite precipitation, and therefore to a corresponding *increase* in the propensity for *precipitation* in the *ferrite* phase.

2. In the temperature range 875 to 1150°C, and over the strain rate range 5.6×10^{-5} to 10^{-1} s^{-1}, the single addition of niobium, molybdenum, or vanadium leads to mean yield strength increases of 70, 9 and 7%, respectively, per 0.1 atomic percent of addition. (This is with respect to a plain carbon-manganese reference material and is measured prior to the occurrence of strain induced precipitation.) When these elements are added in combination, the overall effect is *larger than* the sum of the individual strengthening increments. The relative influence of these elements in solution can be interpreted in terms of changes in the ease of dynamic recovery, and is consistent with the relative magnitudes of their atomic size and electronic differences with respect to γ-iron.

3. When vanadium and/or niobium microalloyed steels are deformed *above* the solution temperatures of the respective carbonitrides, the addition of molybdenum leads to a distinct retardation in the initiation and rate of dynamic recrystallization. The solute retarding influence of molybdenum alone is intermediate between that of niobium, which has the greatest, and that of vanadium, which has the least effect on an equal atom fraction basis. When these elements are added in combination, the overall retardation is *less than* the sum of the individual effects.

4. When the same steels are tested *below* the solution temperatures of the respective carbonitrides, particle precipitation leads to a further component of retardation. The addition of molybdenum involves *two* opposing tendencies in this temperature range: one is the *increase* in the solute retardation due to its presence; the second is the *decrease* in the amount of precipitation retardation due to the decrease in the

carbon and nitrogen activities. The overall effect is one of retardation, and the relative ease or difficulty of recrystallization generally follows the rank order of solute effectiveness established *above* the solution temperatures. Accordingly, the highest 'no recrystallization' temperature was displayed by the ternary Mo-Nb-V steel, followed by the Mo-Nb binary, the single Nb grade, the Mo-V binary, the single Mo material, and finally the single V composition.

5. The combined presence of 0.30% Mo + 0.115% V + 0.035% Nb *raised* the dynamic transformation temperature of the ternary steel. At 875°C, because of the accelerating influence of concurrent deformation, the austenite-to-ferrite transformation was initiated within about 40 s of the commencement of straining.

Acknowledgements

The authors acknowledge with gratitude the financial support received from the following sources: the Physical Metallurgy Research Laboratories of CANMET, Department of Energy, Mines and Resources, Ottawa; the Natural Sciences and Engineering Research Council of Canada; and the Quebec Ministry of Education (FCAC program). Part of this investigation was performed under contract to PMRL, CANMET; the authors are indebted to Dr. G.E. Ruddle of CANMET for the encouragement and support received during the scientific administration of the contract. One of the authors (BB), expresses her thanks to the Quebec Ministry of Intergovernmental Affairs for the provision of a France-Quebec Research Fellowship.

References

1. I. Weiss and J.J. Jonas, " Interaction Between Recrystallization Precipitation During the High Temperature Deformation of HSLA Steels," Met. Trans., 10A (1979) pp. 831-840.

2. J.J. Jonas and I. Weiss, " Effect of Precipitation on Recrystallization in Microalloyed Steels," Met. Sci., 13 (1979) pp.238-245.

3. M.G. Akben, I. Weiss and J.J. Jonas, " Dynamic Precipitation and Solute Hardening in a V Microalloyed Steel and two Nb Steels Containing High Levels of Mn," Acta Met., 29 (1981) pp. 111-121.

4. M.G. Akben, B. Bacroix and J.J. Jonas, " The Effect of V and Mo Addition on the High Temperature Recovery, Recrystallization and Precipitation Behavior of Nb-based Microalloyed Steels," Submitted to Met. Sci.

5. M.G. Akben, " Precipitation, Recrystallization and Solute Strengthening in Microalloyed Steels," Ph.D. Thesis, McGill University, Montreal, 1980.

6. M.J. Luton, J.P. Immarigeon and J.J. Jonas, " Constant True Strain Rate Apparatus for Use with Instron Testing Machines," J.Phys.E.(Sci. Instrumen.) 7 (1974) pp.862-864.

7. C. Rossard, "Mechanical and Structural Behavior Under Hot Working Conditions," Proc. 3rd Int. Conf. Strength Metals and Alloys, Vol. II (1973) pp. 175-203.

8. J.P. Michel and J.J. Jonas, " Precipitation Kinetics and Solute Strengthening in High Temperature Austenites Containing Al and N," Acta Met., 29 (1981) pp. 513-526.

9. I. Weiss and J.J. Jonas," Dynamic Precipitation and Coarsening of Niobium Carbonitrides During the Hot Compression of HSLA Steels," Met. Trans., 11A (1980) pp. 403-410.

10. T. Chandra, I. Weiss and J.J. Jonas, " Influence of Dyamic Recovery and and Dynamic Recrystallization on the Coarsening of Nb(CN) in a Nb Microalloyed Steel, " Met. Sci., 15 (1981), in press.

11. S. Kanazawa, A. Nakashima, K. Okamoto, K. Tanabe and S. Nakazawa, " On the Behavior of Precipitates in the Nb-Mo Heat Treated High Strength Steel Having 80 kg/mm^2 Tensile Strength, " Trans. Iron and Steel Inst. of Japan, 8 (1967) pp. 113-120.

12. H. Watanabe, Y.E. Smith and R.D. Pehlke, " Precipitation Kinetics of Nb Carbonitrides in Austenite of High-Strength Low-Alloy Steels," pp.140-168 in The Hot Deformation of Austenite, J.B. Ballance, ed., AIME, New York, 1977.

13. K.W. Andrews, " Empirical Formulae for the Calculation of Some Transformation Temperatures," J. Iron and Steel Inst., 203 (1965) pp. 721-727.

14. G.T. Eldis, " A Critical Review of Data Sources for Isothermal and Continuous Cooling Transformation Diagrams," pp. 126-157 in Hardenability Concepts With Applications to Steel, D.V. Doane and J.S. Kirkaldy, eds. AIME, Warrendale, PA., 1978.

15. T. Chandra, J.P. Michel and J.J. Jonas, " Influence of Thermo-mechanical Treatment on the Austenite-to-Ferrite Transformation in Plain Carbon Steels Containing Aℓ and N," to be published.

16. T. Wada, H. Wada, J.F. Elliot and J. Chipman, " Activity of Carbon and Solubility of Carbides in FCC Fe-Mo-C, Fe-Cr-C, and Fe-V-C Alloys," Met. Trans., 3 (1972) pp. 2865-2872.

17. T. Wada, Y.E. Smith and W.E. Lauprecht, " Mn-Mo-V Steels for Pressure Vessels," pp.61-71 in Vanadium in High Strength Steel, Publication No.140, Vanitec, London, 1979.

18. T. Nishizawa, " Thermodynamic Study of the Fe-Mo-C System at 1000°C," Scand. Journal of Metallurgy, 1 (1972) pp. 41-48.

19. S. Kurokawa, J.E. Rozzante, A.M. Hey and F. Dyment, " Difusion de Nb en Fe y Algunas Aleaciones de Base Fe," (Diffusion of Nb in Fe and Some Fe Alloys) pp. 47-64 in XXXVI Congresso Annual de ABM, Vol. 1, Brasilian Association of Metals, Sao Paulo, 1981.

20. J.M. Gray and R.B.G. Yeo, " Cb Carbonitride Precipitation in Low-Alloy Steels with Particular Emphasis on Precipitate Row Formation," ASM Transactions Quarterly, 61, (1968) pp.255-269.

21. S. Koyama, T. Ishii and K. Narita, " Effects of Mn, Si, Cr and Ni on the Solution and Precipitation of Niobium Carbide in Iron Austenite," J. of the Japan Institute of Metals, 35 (1979) pp. 1089-1094.

22. I. Weiss, J.J. Jonas and G.E. Ruddle, " Hot Strength and Structure in Plain C and Microalloyed Steels During the Simulation of Plate Rolling by Torsion Testing, " pp. 97-125 in Proc. ASM Symposium on Process Modelling Tools, Cleveland, Oct. 1980, ASM, Metals Park, Ohio, 1981.

23. M.J. White and W. Owen, "Effects of Vanadium and Nitrogen on Recovery and Recrystallization During and After Hot-Working Some HSLA Steels," <u>Met. Trans.</u>, <u>11A</u> (1980) pp.597-604.

24. H.P. Stüwe, "Driving and Dragging Forces In Recrystallization," pp. 11-22 in <u>Recrystallization of Metallic Materials</u>, F. Haessner, ed., Dr. Rieder-Verlag GMBH, Stuttgart, 1978.

25. R.L. Fleischer, "Substitutional Solution Hardening," <u>Acta Met.</u>, <u>11</u> (1963) pp. 203-208.

26. W.B. Pearson, <u>The Crystal Chemistry and Physics of Metals and Alloys</u>, Wiley-Interscience, New York, N.Y. (1972) p. 151.

27. S. Yamamoto, C. Ouchi and T. Osuka, "The Effect of Microalloying Elements on the Recovery and Recrystallization of Deformed Austenite," this volume.

28. C. Ouchi, T. Okita, M. Okad and Y. Noma, "Controlled Rolling in Hot Strip Mill and Mill Designing of 96-Inch New Hot Strip Mill," pp. 1272-1985, <u>Proc. of the International Conference on Steel Rolling</u>, Iron and Steel Institute of Japan, Tokyo, 1980.

29. A.T. Davenport and D.R. DiMicco, "The Effect of Cb on the Austenite Structural Changes During the Hot-Rolling of Low-Carbon Bainitic and Ferrite-Pearlite Steels," pp. 1237-1248, in <u>Proc. of the International Conference on Steel Rolling</u>, Iron and Steel Inst. of Japan, Tokyo, 1980.

30. L. J. Cuddy, "The Effect of Microalloy Concentration on the Recrystallization of Austenite During Hot Deformation," this volume.

31. E.P. Abrahamson II and B.S. Blakeney, "The Effect of Dilute Transition Element Additions on the Recrystallization of Iron," <u>Trans. AIME</u>, <u>218</u> (1960), pp.1101-1104.

32. J.J. Jonas and M.G. Akben, "Retardation of Austenite Recrystallization by Solutes: A Critical Appraisal," in <u>Commemorative Issue for Dr. R.C. Gifkins</u>, Metals Forum, <u>4</u> (1981) pp.92-101.

33. S. S. Hansen, J.B. Van der Sande and M. Cohen, "Nb Carbonitride Precipitation and Austenite Recrystallization in Hot-Rolled Microalloyed Steels, "<u>Met. Trans.</u>, <u>11A</u> (1980), pp. 387-402.

34. H. Kreye and E. Hornbogen, "Recrystallization of Supersaturated Copper-Cobalt Solid Solutions," <u>Journal of Material Science</u>, <u>5</u> (1970) pp. 89-95.

35. A. Le Bon and L.N. de Saint Martin, "Using Laboratory Simulations to Improve Rolling Schedules and Equipment," pp. 90-100 in "<u>Microalloying 75</u>," Union Carbide Corp., New York, N.Y. (1977).

36. H. Mecking and G. Gottstein, "Recovery and Recrystallization During Deformation" pp. 195-222 in <u>Recrystallization of Metallic Materials</u>, F. Haessner, ed., Dr. Rieder-Verlag GMBH, Stuttgart, 1978.

DISCUSSION

Q: You found the strengthening effect of molybdenum by using the compression test, I found the same effect of molybdenum by using hot-rolling techniques. I also found that an addition of molybdenum raised the critical temperature below which the recrystallization of austenite stopped and raised the critical reduction for static recrystallization. When the same amount of reduction is used, the recrystallized grain size of the molybdenum steel is more refined compared with plain carbon steel. We can therefore replace part of the niobium in microalloyed steels with molybdenum, although I realize this may be expensive.

A: When you add the molybdenum it remains in solution. Secondly, molybdenum increases the solubility of the precipitates and therefore retards the precipitation. Molybdenum thus has a two-fold effect.

Q: We can get the same sort of retardation for vanadium that we get for niobium, but of course you have to go to a lower temperature to get the inhibition of recrystallization for vanadium because of the lower solubility. I think your results have to be put in that perspective.

A: We tried to normalize these effects for the same atomic percent of addition, and we also looked at possible effects above the solubility temperatures of the various precipitates. Above the solubility temperature, when things are in solution, and when you normalize per point one atomic percent, you get a sensible ordering of results. Of course, commercially you can use different levels of additions, but we were normalizing for a constant atomic fraction.

Q: I take your point that there is a difference when it's in solution but would you agree that when we have precipitation the temperature shift must be taken into account?

A: OK

Q: I would like to pursue the point about vanadium nitride precipitation. In this case 0.115V and 60 ppm N was in the steel. No more than a quarter of the vanadium could be used to precipitate the nitrides, and the rest is in solution. In a Nb-steel the corresponding situation would be 60 ppm C! Perhaps your conclusions regarding the effect of vanadium would be different if the nitrogen content was increased to say 200 ppm.

A: We kept the nitrogen content low because it is part of the base chemistry and we were interested in the effects of other elements. We still hold to the argument that when comparing the influence of elements in solution, vanadium was a smaller effect. We agree that at higher nitrogen levels one would see strong precipitate effects in the vanadium alloys, as shown in the paper of Siwecki and coworkers.

Q: Will you comment on the results at 875°C where you have ferrite forming at the low strain rates. Was recrystallization interacting with the ferrite formation?

A: Normally the peak in the flow curve happens because recrystallization starts. However, what we get here is a drop in the flow curve because the transformation to softer ferrite is occurring during deformation. In this case at 875°C, the flow curve peaks are not due to dynamic recrystallization. In fact, this technique can be used to determine

317

dynamic transformation curves, that is, to determine the times to the start of the gamma to alpha transformation. At 875°C it takes about 50 seconds to the start of transformation during deformation. That would mean, for example, in the reversing mill if you have 4 or 5 passes which take about 50 seconds in that temperature range, you could get ferrite formation because of the deformation enhanced transformation.

ALUMINUM NITRIDE PRECIPITATION IN C-Mn-Si AND MICROALLOYED STEELS

E. L. Brown*, and A. J. DeArdo**

*Colorado School of Mines, Golden, CO   80401
**University of Pittsburgh, Pittsburgh, PA   15216

Aluminum nitride precipitation was studied to determine its
morphology and crystallography in the thermomechanically treated austenite
of C-Mn-Si and microalloyed steels.   In addition, the effects of AlN
precipitation on austenite were studied as a function of steel composition and thermomechanical treatment.   Morphology and crystallography were
studied by means of transmission electron microscopy and selected area
electron diffraction (TEM-SAED).   The effects of AlN precipitation on
austenite microstructural behavior were monitored by means of optical
microscopy and correlated with microchemical analyses and TEM-SAED
observations.

Introduction

Aluminum nitride is a covalent compound with a Wurtzite crystal structure and actually belongs to a family of similar compounds in the Si-Al-O-N system (termed Sialons)[1-5]. AlN may also be viewed as a close-packed hexagonal structure with lattice parameters of a = 3.114Å and c = 4.986Å[5]. The covalent structure of AlN differs from the interstitial phase structures of other alloy carbides and nitrides found in steels[4]. In the latter, interstitial atoms order in various arrangements in the interstices of the close-packed metal structure.

Aluminum nitride can occur as a precipitating phase in Al-killed, nitrogen containing steels. In steels, the precipitate is a simple nitride with an equiatomic stoichiometry. The equilibrium solubility of AlN in austenite[6-8] and ferrite[9] has been determined. Solubility products of AlN in austenite and ferrite as a function of temperature are shown in Figure 1. The solubility of AlN is significantly lower in ferrite than in austenite.

AlN precipitation in ferrite has been extensively studied in connection with the development of alloys possessing favorable crystallographic textures for electrical properties[18] and formability [19,20]. The relationships which exist between precipitation (thermodynamics and kinetics), microstructure and processing parameters have been determined[19,20]. In these studies precipitation was generally inferred from structural observations and by monitoring the portion of the total nitrogen content associated with AlN precipitates by microchemical analysis[21]. The maximum in precipitation kinetics was found to occur in the vicinity of 700°C[26], consistent with earlier studies[15,19,20]. A significant amount of AlN forms during heating to a particular annealing temperature, with slower heating rates promoting precipitation[15,19]. Precipitation was less rapid during direct isothermal transformation at temperatures below the solution temperature[15]. Electron metallography has also been performed to determine the scale and morphology of AlN precipitation in ferrite under a variety of treatment conditions[18, 22-28]. Aluminum nitride precipitates were observed to nucleate on dislocation substructure in deformed ferrite[22, 23,27-29], and the kinetics were enhanced compared with undeformed ferrite [23-25].

Aluminum nitride precipitation was observed to retard recovery and recrystallization of a cold worked ferrite matrix when precipitation occurred just prior to or simultaneously with recrystallization[19,25,30]. The AlN particles that inhibit recovery and recrystallization of ferrite were found to be extremely fine[27,28,30]. A portion of the extremely fine AlN precipitates possessed a cubic NaCl-type (B1) crystal structure with a lattice parameter of approximately 4.00Å[22,24,25]. A dendritic appearing cubic AlN also had been observed earlier in nitrided steels[31]. Cubic AlN was found to precipitate at grain boundaries and deformation bands in cold worked specimens and to transform, in-situ, to the hexagonal form of AlN. It is unclear from the available literature whether cubic AlN precipitates exclusively in ferrite. The morphologies of coarsened AlN precipitates are thin plates and needles[18,26]. Taguchi and Sakakura[18] have determined the crystallographic orientation relationships of the two AlN morphologies in ferrite.

Considerably less investigation has been made into the precipitation of AlN in austenite. Aluminum-killed steels are generally termed fine grained because AlN precipitation refines the austenite grain structure and increases the resistance to austenite grain coarsening during austenitizing. The theory of grain growth inhibition by a dispersed second phase has been developed by Hillert[10] and Gladman[11] and has been applied to steels.[12]. Steels grain refined by a dispersed phase are prone to grain coarsening by abnormal grain growth or secondary recrystallization. The specific effects of AlN on grain growth have been considered by several investigators[12-17]. There appears to be some discrepancy as to whether the grain coarsening temperature of AlN grain refined steels is to be associated with dissolution[13-15] or coarsening[12,14,15] of AlN particles pinning grain boundaries. Both processes can result in the unpinning of boundaries in a given grain size distribution. The controlling process is actually determined by steel composition and prior thermal history[15]. The morphologies of AlN precipitation in austenite are also needles and plates[29]. AlN precipitates have been found along austenite grain boundaries[15,29,32,33]. The kinetics of AlN precipitation in austenite are generally slow and dependent upon thermal history[15,32,35]. Similar to observations on ferrite, precipitation is more rapid when a specimen is quenched to room temperature from a solution temperature and then reheated than when it is cooled directly to an austenite treatment temperature. The generally low AlN content of hot band has been rationalized in terms of the similarity in thermal cycle between hot rolling and direct isothermal transformation[15]. In explaining the effect of AlN on austenite grain refinement, precipitation during heating to the austenitizing temperature has been cited[35,41].

Hot deformation of austenite was found to significantly accelerate the kinetics of AlN precipitation[34,36,37]. The maximum in precipitation kinetics occurred between 800° and 900°C. Funnell and Davies[33] proposed that AlN particle dispersions, formed prior to deformation, retard recrystallization during hot tensile testing when the interparticle spacing is of the same order as the dislocation substructure. However, Vodopivec[32] has shown that the presence of AlN prior to hot rolling does not affect dynamic recrystallization. Using a mechanical testing technique, Michel and Jonas[37] determined the start and finish times for dynamic precipitation of AlN, during hot deformation. These investigators found that AlN precipitation was most rapid for Al-rich, off-stoichiometric compositions and rationalized this on the basis of a reaction controlled by aluminum diffusion. They also concluded that the interaction of dynamic AlN precipitation with dynamic recrystallization is dependent upon the magnitude of strain rate. At higher strain rates ($\dot{\varepsilon} > 10^{-2}$S-1) dynamic precipitation had no effect upon dynamic recrystallization, consistent with Vodopivec[32]. At the lowest strain rates, rapid coarsening of AlN was inferred.

The situation becomes somewhat more complex when the steel is microalloyed with other strong nitride-formers, such as V and Ti. An interaction between V and Al, in competition for nitrogen, has been proposed for Al-killed, V-microalloyed steels[38-41] with sufficient annealing time at elevated temperature. Variation in the carbon/nitrogen ratio of vanadium carbonitrides has been explained via this interaction[38].

The related studies, described herein, were conducted to enlarge upon the knowledge existent in the following areas:

i) The morphology and crystallography of AlN precipitation
in austenite.
ii) AlN precipitation kinetics and the evolution of austenite
microstructure during hot rolling.
iii) The effects of other alloying elements on AlN precipitation
in austenite during thermomechanical treatment.

Experimental Procedure

The compositions and designations of the materials
employed in these studies are described in Table I, below. All of the
materials were vacuum induction melted and cast into 25kg ingots, 10cm
x 10cm in cross-section. After a homogenization anneal at 1288°C, the
ingots were hot rolled to 1.6cm thick slabs. This provided the starting
material for all subsequent treatments. Specimens 5 cm x 5 cm were hot
rolled after an austenitization treatment of 60 minutes at 1288°C. Hot
rolling consisted of either single or double pass schedules with a 50%
reduction per pass. Specimens were held subsequent to deformation for
times of 0 to 1000 minutes at the various rolling temperatures. Optical
microscopy of the prior austenite microstructure was performed employing
a modified aqueous picric acid etch. Transmission electron microscopy
(TEM) and selected area electron diffraction (SAED) on carbon extraction
replicas and thin foils were employed to establish the morphology,
crystallography, and distribution of AlN in hot rolled steels. AlN
precipitation was monitored by Beeghly (ester-halogen) micro-chemical
analysis and V(C,N) precipitation in the V-steel via differential thermal
analysis-evolved gas analysis (DTA-EGA).

TABLE I						
	C	Mn	Si	Al	N	V
Fe-Mn-Al-N	0.01	1.43	0.03	0.330	0.037	-----
LoSi-C-Steel	0.15	1.56	Trace	0.041	0.018	-----
C-Steel	0.14	1.31	0.35	0.048	0.012	-----
V-Steel	0.14	1.32	0.35	0.035	0.020	0.130

Results

The results of these studies are concerned with the nature of
AlN precipitation in austenite and its effects upon austenite microstructural behavior.

A. Nature of AlN Precipitation in Austenite - The crystallography and morphology of AlN precipitation in austenite was characterized in the Fe-
Mn-Al-N alloy. As in earlier investigations, thin plates and needles or
rods were observed, Figure 2. The AlN in Figure 2(a) has precipitated
on substructure in the austenite. The rod axes were parallel to [00.1]
AlN (c-axis). Note the substructure present in the plate of Figure 2(b)
that resulted in a multi-zone electron diffraction pattern.

TEM-SAED studies performed on thin foils of the Fe-Mn-Al-N alloy
revealed the crystallography of AlN precipitation in austenite. Studies
of phenomena in the austenite of a transformable steel present difficulties
because observations can generally only be made on ferrite. Many AlN
precipitates observed possessed the following orientation relationship
with ferrite (Figure 3).

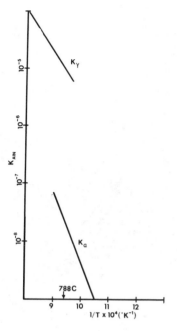

Figure 1 Solubilities of AlN austenite and ferrite.

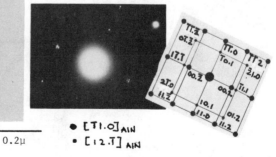

0.4μ

Figure 2(a)   AlN rods on austenite structure.   Carbon extraction replica electron micrograph. Hot rolled at 872°C, t = 360 min.

0.2μ

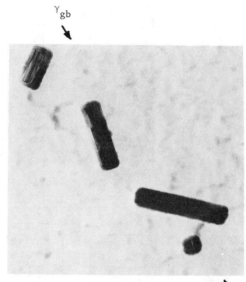

$\bullet$ $[\bar{1}1.0]_{AlN}$
$\bullet$ $[1\bar{2}.\bar{1}]_{AlN}$

Figure 2(b)   AlN plate with substructure and associated SAED pattern.   Extraction replica electron micrograph. Hot rolled at 872°C, t = 360 min.

0.4μ

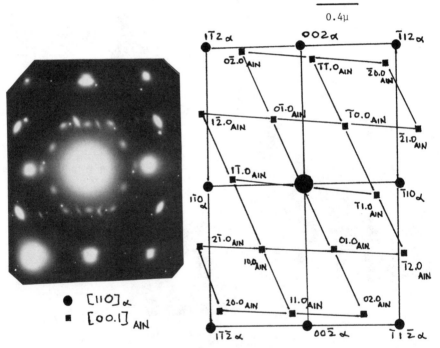

● [110]α
■ [00.1]AlN

Figure 3 AlN precipitation in Fe-Mn-Al-N alloy possessing orientation
relationship with ferrite. Thin foil electron micrograph.
Hot rolled at 872°C, t = 360 min.

(00.1) AlN // $(110)\alpha$
$[10.0]$ AlN // $[1\overline{1}\overline{1}]\alpha$

B. <u>Effects of AlN Precipitation on the Evolution of Austenite Micro-Structure Due to Hot Working</u> - Low silicon (LoSi-C-steel) and high silicon (C-steel) steels were employed for the bulk of the studies described in this section. Austenitizing at 1288°C for 1 hour produced coarse austenite microstructures for both steels. The grain structure of the LoSi-C-steel was coarser than that of the C-steel ($\sim$600μ vs. $\sim$400μ), and was sometimes characterized by a heterogeneous grain size distribution, indicative of abnormal grain growth. A thermocouple embedding roll pass of $\sim$20% at 1200°C resulted in a reduction in the disparity of austenite grain size existing in the austenitized condition. Rolling at 1038°C produced recrystallized austenite microstructures in both steels. The LoSi-C-steel possessed an equiaxed but duplex austenite microstructure, containing aggregates of coarse and fine grains, Figure 4(a). The C-steel possessed a more uniform equiaxed austenite microstructure directly after rolling. Figure 4(b). Numerous observations confirmed the difference in microstructural homogeneity noted. With holding time after rolling at 1038°C, the duplex austenite microstructure of the LoSi-C-Steel persisted. Some grain growth of both fine and coarse grains has occurred, Figure 5(a). The microstructure of the C-steel remained homogeneous with holding time and there was a small amount of grain growth, Figure 5(b).

When the rolling temperature was lowered to 872°C and below, the LoSi-C-steel exited the rolls with a mixed austenite microstructure, consisting of coarse elongated grains with very fine equiaxed grains surrounding them, Figure 6(a) The equiaxed austenite grains are indicated by the arrows in Figure 6(a) and Figure 6(b) is a higher magnification micrograph of a region of equiaxed grains. The extent of recrystallization diminished as the rolling temperature decreased and freshly nucleated grains were observed at boundaries and in deformation bands or twins, Figure 6(c). The C-steel, as rolled, possessed a predominantly equiaxed austenite microstructure with isolated regions of elongated grains, Figure 6(d). With holding time at rolling temperatures $\lesssim$872°C, the prior austenite microstructure of the LoSi-C-steel became more refined but very heterogeneous with respect to grain shape and size. There were fine equiaxed grains that appeared to separate colonies of coarser, irregularly shaped grains. Many of the irregularly shaped grains were elongated, Figure 7. As the rolling temperature was lowered, the evolution of the islands of irregularly shaped, coarse grains occurred less rapidly with holding time. The austenite microstructure of the C-steel exhibited similar behavior but generally greater stability with holding time.

The extent of AlN precipitation in these material conditions was monitored by chemical analysis of extracted residues (Beeghly ester-halogen) and the results are presented in Figure 8(a). The two steels exhibited kinetic curves of similar shape, generally characterized by an initial period of relatively rapid precipitation followed by a period of slower precipitation. Precipitation of AlN was slower in the LoSi-C-steel for all the treatment temperatures and the level of precipitation was low in both steels for a treatment temperature of 1038°C. The levels of AlN increased for both steels as the treatment temperature decreased. A maximum in kinetics occurred in the vicinity of 800°C.

100µ

Figure 4(a) LoSi-C-steel: 50% reduction; 1038°C, t = 0 minutes.
Light micrograph.

50µ

Figure 4(b) C-steel: 50% reduction; 1038°C, t = 0 minutes.
Light micrograph.

100μ

Figure 5(a) LoSi-C-steel: 50% reduction; 1038°C, t = 12 min.
Light micrograph

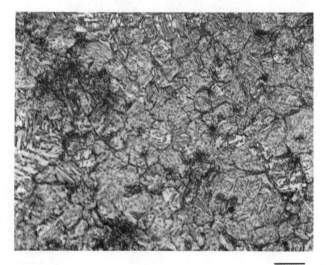

50μ

Figure 5(b) C-steel: 50% reduction; 1038°C, t = 12 min.
Light micrograph.

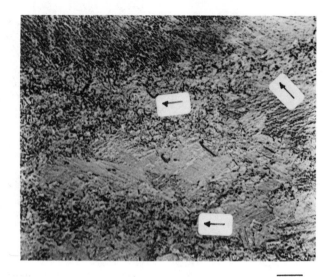

200μ

Figure 6(a) LoSi-C-steel: 50% reduction; 872°C, t = 0 min.
Light micrograph.

100μ

Figure 6(b) LoSi-C-steel: 50% reduction; 872°C, t = 0 min.
Light micrograph.

200μ

Figure 6(c) LoSi–C–steel: 50% reduction; 788°C, t = 0 min.
Light micrograph.

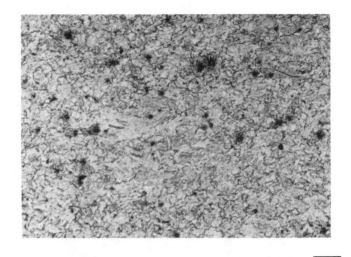

100μ

Figure 6(d) C-steel: 50% reduction; 872°C, t = 0 minutes. Light micrograph.

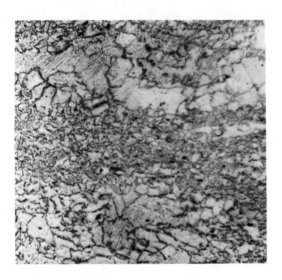

200μ

Figure 7 LoSi-C-steel: 50% reduction; 872°C, t = 30 minutes. Light micrograph.

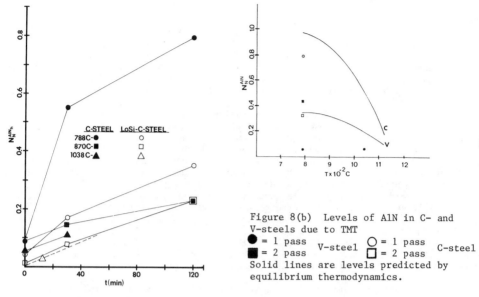

Figure 8(a) AlN precipitation kinetics.
N_N^{AlN} = fraction of nitrogen in AlN.

Figure 8(b) Levels of AlN in C- and
V-steels due to TMT
● = 1 pass V-steel ○ = 1 pass C-steel
■ = 2 pass □ = 2 pass
Solid lines are levels predicted by
equilibrium thermodynamics.

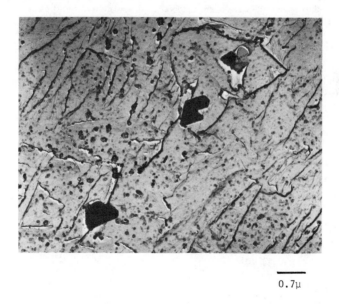

0.7μ

Figure 9(a) Large AlN particles on prior γ grain boundary. LoSi–C-steel:
50% reduction; 1038°C, t = 120 minutes. Carbon extraction replica electron
micrograph.

The co-precipitation of AlN and V(C,N) was studied in the V- and C-steels for single and double pass thermomechanical treatment (TMT), Figure 8(b). The extent of V(C,N) was monitored via TEM and DTA-EGA. A detailed account of V(C,N) precipitation in the V-steel is given elsewhere[42]. However, TEM and DTA-EGA analysis showed extensive precipitation in the V-steel after single pass or double pass TMT in which the final reduction occurred in the lower austenite temperature range (788-872°C). The extent of precipitation decreased as the treatment temperature was raised. The V(C,N) precipitates were nitrogen-rich ($VC_{0.3}N_{0.7}$) at low temperatures and became generally richer in nitrogen as the treatment temperature was raised. Precipitation at lower temperatures occurred in unrecrystallized austenite. The level of AlN precipitation in the V-steel remained low for all temperatures in material subjected to single pass TMT. This is to be contrasted with the previously described results on the C-steel, which displayed significant AlN precipitation after single pass TMT in the vicinity of 800°C. However, when the roll reduction of the V-steel at 788°C was part of a double pass TMT, in which the first reduction took place at 1038°C, the AlN content after two hours of holding increased dramatically. When the C-steel was subjected to the same double pass TMT the AlN content decreased compared to a single pass TMT at 788°C.

The dispersions of AlN observed with holding time after rolling were generally very fine and from which it was difficult to obtain diffraction contrast. This can be traced to the large extinction distances associated with AlN reflections. At high temperatures large AlN particles were observed at prior austenite grain boundaries in the C- and LoSi-C steels, Figure 9(a). The dark spots in the background of Figure 9(a) are etch pits as there was no electron diffraction evidence for precipitation. At longer aging times, a homogeneous dispersion of fine particles was observable in the C-steel, Figure 9(b). No corresponding dispersion in the LoSi-C-steel was observed.

At lower treatment temperatures, AlN precipitation was more extensive in both steels and there was an overall denser dispersion in the C-steel, Figure 10. The AlN particles were more easily observed in the regions of proeutectoid ferrite adjacent to prior austenite grain boundaries. At the longest holding times at low treatment temperatures AlN was found to coarsen via the agglomeration of several particles to form a dendritic appearing morphology, Figure 11. The AlN precipitates observed in specimens of the V-steel subjected to double pass TMT were generally very coarse, located in the prior austenite grain boundary region and surrounded by a fine dispersion of V(C,N) precipitates, Figure 12, location A.

Discussion

The observed orientation relationship between AlN and ferrite in specimens hot rolled in the austenite range differs from the orientation relationship between AlN and ferrite determined by Taguchi and Sakakura[18]:

(00.1) AlN // (001)α
[$\bar{1}$1.0] AlN // [$\bar{1}$00]α

Of course, their heat treatments resulted in the nucleation and growth of AlN in ferrite. The apparent orientation relationship between AlN and ferrite observed in the present study can be explained by assuming a Kurdjomov-Sachs orentation relationship to exist between the prior austenite and the ferrite to which it transforms:

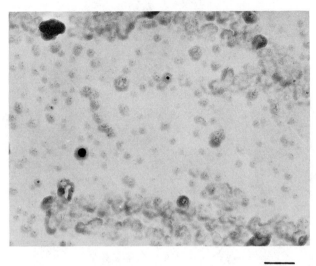

$\overline{0.2\mu}$

Figure 9(b)   Fine AlN precipitates in C-steel: 50% reduction; 1038°C,
t = 30 minutes.   Carbon extraction replica electron micrograph.

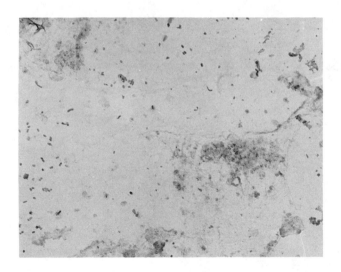

$\overline{0.4\mu}$

Figure 10(a)   AlN precipitates in LoSi-C-steel: 50% reduction; 788°C,
t = 120 minutes.   Carbon extraction replica electron micrograph.

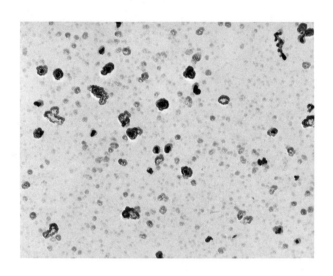

0.4μ

Figure 10(b) AlN precipitates in C-steel: 50% reduction; 788°C,
              t = 120 min. Carbon extraction replica electron
              micrograph.

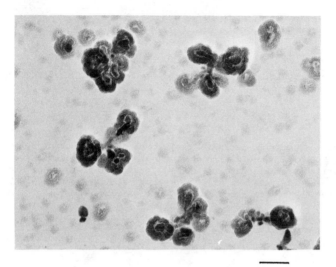

0.2μ

Figure 11   Coarsening of AlN in C-steel held at 788°C for 120 minutes after hot rolling.   Carbon extraction replica electron micrograph.

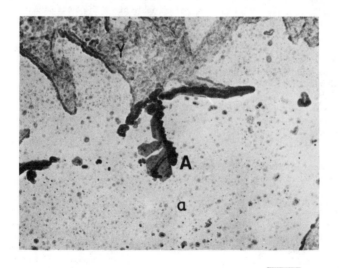

0.5μ

Figure 12   AlN and V(C,N) precipitation in V-steel after second 50% reduction at 788°C, t = 120 minutes.   Carbon extraction replica electron micrograph.

$$(111)\gamma \; // \; (1\bar{1}0)\alpha$$
$$[1\bar{1}0]\gamma \; // \; [\bar{1}\bar{1}1]\alpha$$

This implies a crystallographic relationship between AlN and the austenite in which it formed:

$$(00.1) \; AlN \; // \; (1\bar{1}1)\gamma$$
$$[10.0] \; AlN \; // \; [1\bar{1}0]\gamma$$

The inferred orientation relationship between AlN and austenite, in which densest packed planes and directions are parallel, is consistent with minimum lattice mismatch parallel and perpendicular to the c-axis of AlN, Figure 14. The analysis was made in a manner similar to that of Ryan, et. al.[43] for cubic carbides and nitrides in molybdenum.

The interatomic distances ($S\gamma$, S_{AlN}) along significant crystallographic directions in austenite and AlN were calculated. The fractional mismatch of the two lattices, ζ, was calculated along specific pairs of these directions:

$$|\zeta| = \frac{|S_\gamma - S_{AlN}|}{S_\gamma}$$

On Figure 13, the mismatch associated with a given pairing lies on the line drawn between the respective interatomic spacings, i.e. between the S_{AlN} and $S\gamma$ axes. For example, ζ between $<110>_\gamma$ and $<10.0>_{AlN}$ lies on the line zz' in Figure 13. This ζ value is associated with several symbols because the particular pair of crystallographic direction is associated with several possible orientation relationships. The pairings were grouped further by considering them in terms of the most probable orientation relationships between austenite and AlN. The symbols of Figure 13 are identified with the most probable orientation relationships. Three of the five smallest mismatches are consistent with the inferred orientation relationship. The assumption of this orientation relationship and associated minimum lattice mismatch can produce a minimum in volume strain energy. Nabarro[44] predicted minimum volume strain energy for thin discs and needles. Mayo and Tsakalakos[45] predicted discs when the strain energy function, associated with a hexagonal coherent phase, is minimized in a direction normal to the habit of the discs and needles when the function is minimized in a direction normal to the needle axis. The observations of needle axes parallel to the c-axis and plates parallel to the (00.1) AlN plane are then consistent with minima in strain energy along directions of minimum mismatch.

The microstructural behaviors of the C- and LoSi-C-steels were similar in some respects. Both steels recrystallized during high temperature hot rolling at 1038°C. The levels of AlN in both, as determined by TEM and chemical analysis, were low. However, there was a higher volume fraction of AlN precipitated in the C-steel at all times. The heterogeneous microstructure of the LoSi-C steel in this temperature range is partly due to a duplex grain structure inherited from austenitization and partly due to a lowered microstructural stability associated with lower levels of AlN. Alternatively, a lowered microstrucutral stability could result in lower precipitation kinetics. According to the results of Vodopivec[32] and Michel and Jonas[37], one would not expect interaction of AlN precipitation with

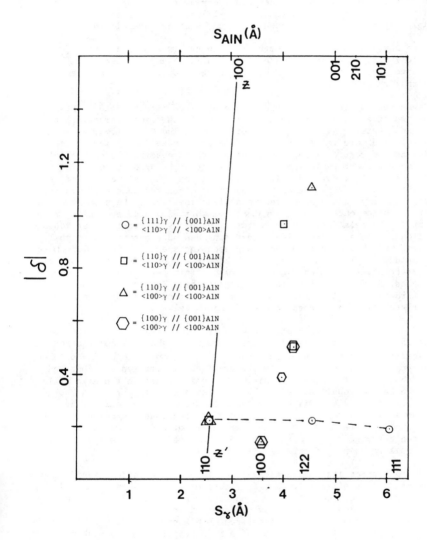

Figure 13 Mismatch (δ) along specific pairs of crystallographic directions in austenite and AlN.

$$|\delta| = \frac{|S_\gamma - S_{AlN}|}{S_\gamma}$$

dynamic recrystallization at these strain rates. However, some prior nucleation of AlN could have taken place during thermocouple embedding. Combinations of grain growth and recrystallization during and subsequent to high temperature TMT have been observed by other investigators[46].

The low levels of AlN observed in both steels are probably due, in part, to the recrystallized state of the matrix[34] and is consistent with diminished precipitation of AlN during TMT that approximates a direct isothermal treatment [15].

When the rolling temperature was lowered the microstructural behavior of the two steels was once again similar and the C-steel microstructure exhibited greater stability with holding time. The C-steel may have recrystallized more readily due to a fine grain size just prior to rolling. The islands of coarse, irregularly shaped grains observed in both steels appear to represent regions of the material that undergo preferential and directional grain growth. It is interesting that the islands are of approximately the same size and shape as the prior elongated austenite grains. These grains may have somehow recrystallized in-situ, perhaps by subgrain growth or coalescence. The irregular, somewhat elongated austenite grains present in the islands imply preferential boundary movement in particular directions. The growth of elongated grains has been observed previously[30] and it is well known that boundaries possessing special orientations move very rapidly[47]. It is also possible that the higher volume fractions of precipitate existent at this temperature participate by pinning all but the highest driving force boundary orientations. The relatively higher volume fractions of AlN precipitated in the C-steel are responsible for the greater stability in microstructure of this steel.

Above 800°C the more rapid and greater volume fraction of AlN precipitation in the C-steel are apparently due to its higher silicon content. Equilibrium thermodynamics would suggest that for the chemistries of these steels only a small difference in the fraction of nitrogen precipitated as AlN should exist. More extensive AlN precipitation in the high silicon steel can be reationalized on the basis of silicon's effect on the solubility product of AlN. Silicon is known to increase the activity of nitrogen in austenite[48] and has been found to decrease the solubility of BN[49]. Similar effects of Si on the solubility of NbC in austenite have also been observed[50]. The precipitation kinetics of these steels subsequent to hot rolling compare favorably with the results of Vodopivec[34].

The AlN precipitaion kinetics at 788°C cannot be rationalized merely on the basis of a silicon induced thermodynamic effect. As mentioned previously, some of the AlN nucleation and/or growth may be occurring in ferrite since 788°C is probably an intercritical temperature. There is a greater thermodynamic driving force for AlN precipitation in ferrite (Figure 1). Nucleation and growth of AlN in ferrite would be enhanced because of the greater diffusivities of Al and N in ferrite. Since silicon, and austenite grain refinement both would be expected to raise the Ar$_3$ temperature, more AlN precipitation may be expected for the C-steel, associated with higher levels of ferrite formed at temperature. It is for this reason that all results after rolling at 788°C must be considered very carefully, with precipitation of AlN in ferrite being a distinct possibilit

When vanadium is added to the C-steel (to form the V-steel) the precipitation of AlN is altered. Comparison can be made with the results predicted by equilibrium thermodynamics of AlN in austenite and between steels. After hot rolling in the upper austenite range, the AlN levels in the V- and C-steels were comparable, indicating no interaction between V and Al. When the rolling temperature is lowered, nitrogen-rich V(C,N) precipitation predominates over AlN. The AlN levels in the V-steel at low austenite rolling temperatures are much lower than in the C-steel after single pass TMT. The increase in AlN precipitation in the V-steel after double pass TMT is probably due to the higher defect density of the austenite in this material condition. Some of this AlN precipitation may also be occurring in isothermally formed ferrite since finer austenite microstructures and deformation enhance the kinetics of ferrite formation. The lower AlN level after double pass TMT of the C-steel, compared to single pass TMT of this steel, could be due to the greater extent of recrystallization during double pass TMT[42]. This is consistent with the previous observation that AlN does not precipitate extensively in recrystallized material[34]. Other investigators have determined that lowered Al and elevated N contents, similar to those of the V-steel, enhance V(C,N) precipitation[38]. It is interesting that after a single pass at 788°C the level of AlN in the V-steel is much lower than what the thermodynamics of AlN in austenite would predict for the V-steel composition. However, when the deformation at 788°C occurs as the second pass in a double pass TMT, the AlN level is approximately what thermodynamics would predict, Figure 8(b). The enhanced precipitation kinetics associated with double pass TMT, could therefore have allowed for a closer approach to equilibrium.

Conclusions

1. Aluminum nitride precipitates in austenite as needles and plates with the following orientation relationship:

$$(00.1)AlN \mathbin{//} (111)\gamma$$
$$<10.0>AlN \mathbin{//} [1\bar{1}0]\gamma$$

   Observed morphologies are consistent with this crystallography on the basis of minimum lattice mismatch.

2. The kinetics of AlN precipitation is enhanced in the presence of silicon and the higher levels of AlN attainable could be correlated with greater post-deformation microstructural stability. The effect of silicon may be due to an increase in the activity of nitrogen in austenite, thus decreasing the solubility of AlN and/or raising the Ar_3 temperature to produce more ferrite in which AlN can form easily.

3. With the vanadium, aluminum, and nitrogen levels of the V-steel, an interaction between vanadium and aluminum was observed at low treatment temperatures.

References

1. G. A. Jeffrey, G. S. Parry, and R. L. Mozzi: J. Chem. Phys., vol. 25, No. 5, (1956), p. 1024.
2. G. A. Jeffrey and V. Y. Wu: Acta Cryst., vol. 16, (1963), p. 559.
3. K. H. Jack: J. Met. Sci., vol. 11, (1976), p. 1135.
4. D. H. Jack and K. H. Jack: Mat. Sci. Eng., vol. 11, (1973), p. 1.

5. W. B. Pearson: "Handbook of Lattice Spacings and Structures of Metals and Alloys", Pergamon Press, (1951).
6. L. S. Darken, R. P. Smith, and E. W. Filar: J. Metals, (December 1951), p. 1174.
7. K. Narita: J. Chem. Soc. Japan Pure Chem. Sec., vol. 75, (1954), p. 1041.
8. K. J. Irvine, F. B. Pickering, and T. Gladman: J.I.S.I., vol. 205, (February 1967), p. 161.
9. H. A. Wriedt: Physics of Materials (Festschrift for H. Bass), (1979).
10. M. Hillert: Acta Met., vol. 13, (1965), p. 227.
11. T. Gladmann: Proc. Roy. Soc., vol. 294, (1966), p. 298.
12. T. Gladmann and F. B. Pickering: J.I.S.I., vol. 205, (1967), p. 653.
13. L. A. Erasmus: J.I.S.I., vol. 202, (1964), p. 32.
14. A. B. Chatterjea and B. R. Nijhawan: Iron and Steel, vol. 40, (1967), p. 208.
15. M. P. Sidey: Iron and Steel, vol. 40, (1967), p. 168.
16. D. Hall and G. H. J. Bennett: J.I.S.I., vol. 205, (1967), p. 168.
17. T. Gladmann and D. Dulieu: Met. Sci. J., vol. 8, (1974), p. 167.
18. S. Taguchi and A. Sakakura: Acta Met., vol. 14, (1966), p. 405.
19. R. L. Solter and C. W. Beattie: J. Metals, (September 1951), p. 721.
20. W. C. Leslie, R. L. Rickett, C. L. Dotson, and C. S. Walton: Trans. ASM, vol. 46, (1954), p. 1470
21. H. G. Beeghly: Analytical Chem., vol. 21 (1949), p. 1513.
22. S. Hanai, N. Takemoto, and Y. Mizuyama: Trans. I.S.I. Japan, vol. 11, (1971), p. 307.
23. T. Ototani, Y Katama, and T. Fukuda: Trans. I.S.I. Japan, vol. 12, (1972), p. 307.
24. Y. Yagi, T. Fukutsuka, and R. Ogawa: Trans. I.S.I. Japan, vol. 12, (1972), p. 233.
25. R. Ogawa, T. Fukutsuka, and Y. Yagi: Trans. I.S.I. Japan, vol. 12, (1972), p. 291.
26. T. Ichiyama: Trans. I.S.I. Japan, vol. 10, (1970), p. 429.
27. E. Furubayashi, H. Yoshida, and H. Endo: Metal Sci. J., vol. 7, (1973), p. 65.
28. E. Furubayashi, H. Endo, and Y. Yoshida: Mat. Sci. Eng., vol. 14, (1974), p. 123.
29. S. Hasebe: Tetsu-to-Hagane, vol. 3, (1963), p. 20.
30. J. T. Michalak and R. D. Schoone: Trans. AIME, vol. 242, (1968), p. 1149.
31. W. Koch, C. Ilschner-Gensch, and H. Rohde: Arch. Eisenhuttenwesen, vol. 11, (1956), p. 701.
32. F. Vodopivec: Metals Technol.,(April 1978), p. 118.
33. G. D. Funnell and R. J. Davis: Metals Technol.,(May 1978), p. 150.
34. F. Vodopivec: Metals Technol, (April 1978), p. 118.
35. Washburn: Trans. AIME, vol. 162, (1945), p. 658.
36. F. Vodopivec: J. Mat. Sci., vol. 10, (1975), p. 1082.
37. J. P. Michel and J. J. Jonas: Acta Met., vol. 29, (1981), p. 513.
38. L. Meyer, H. E. Buhler, and F. Heisterkamp: Thyssenforschung, vol. 3, (1971), p. 8.
39. K. J. Irvin in Symposium: Low Alloy High Strength Steels, sponsored by Metallurg Companies, Nuremberg BRD, (May 1970).
40. F. E. Lithuber, ibid.
41. R. R. Preston in Welding of HSLA (Microalloyed) Structural Steels, Rome, (1976).
42. E. L. Brown: Ph.D. Dissertation, University of Pittsburgh, (1979).

43. N. E. Ryan, W. A. Soffa, and R. C. Crawford: Metallography, vol. 1, (1968), p. 195.
44. F. R. N. Nabarro: Proc. Roy. Soc., vol. A175, (1940), p. 519.
45. W. E. Mayo and T. Tsakalakos: Met. Trans. A, vol. 11A (1980), p. 1637.
46. R. A. Petkovic, M. J. Luton, and J. J. Jonas: Acta Met., vol. 27, (1979), p. 1633.
47. P. Gordon and R. A. Vandermeer: "Grain Boundary Migration", Chapter 6 in Recrystallization, Grain Growth and Textures, ASM, (1965).
48. N. S. Corney and E. T. Turkdogan: J.I.S.I., vol. 180, (1955), p. 344.
49. H. C. Fiedler: Met. Trans. A, vol. 9A, (1978),p. 1489.
50. S. Koyama, T. Ishii, and K. Narita: Trans. Jap. Inst. Met., vol. 35, (1971), p. 1089.

DISCUSSION

Comment: There's an important clarification that needs to be made about some of the abnormally coarse grains that have been shown in some of the talks that have taken place to date. First of all let's take one of the slides that Elliot Brown showed, where he demonstrated the existence of abnormally coarse grains in structure. There we're dealing with the abnormal growth of austenite grains in a freshly nucleated, fully-recrystallized strain-free structure. This would be interpreted as classical secondary recrystallization. Some of the information that was presented yesterday concerning the production of abnormally coarse grains during rolling after very light reductions, indicated that we are dealing with the growth of grains in a strained matrix, essentially a nonrecrystallized or partially recrystallized matrix. Here we are dealing with a phenomenon that would be referred to as static recrystallization.

Q: I would like to ask two questions or make two comments about the difference in the Al levels in your two steels. I noticed that the base steel contains 0.04 Al and your other steel 0.09 Al. First I wonder whether that 0.09 Al might have not led to some Al nitride precipitation which also played a role in the grain modification results that you described. And, before you answer that, my second question is that, that amount of Al or extra Al might act as an extra ferrite stabilizer by some modest amount but enough to support your hypotheses about a difference in the ferrite transformation behavior during rolling.

A: I recognize the argument and you were sharp to pick that up. It's always difficult to reproduce steels identically, so my simple answer to your question is, really two answers. Firstly, on the Al austenite grain-refining effect, the results that I've given here are very simply pertaining to the observed austenite structures and we can rationalize that. What you see in the austenite structure corresponds to what you see in the ferrite, and so on. Any questions pertaining to precipitation that occurs during this I think I'll defer to the following papers. The second point is, what occurs here in the austenite-ferrite transformation, and that really hasn't been studied in detail.

PREDICTION OF PRECIPITATE PHASES IN MICROALLOY STEELS

CONTAINING NIOBIUM, CARBON, NITROGEN AND ALUMINUM

S. R. Keown
Department of Metallurgy
University of Sheffield
Sheffield, England

W. G. Wilson
Manager, High Strength Steel Development
Niobium Products Company Limited
Pittsburgh, Pennsylvania

Using available solubility product data, a system has been devised to predict the precipitation sequence of the compounds that may form in the complex system Nb, C, N and Al during the heating and rolling of steels containing these elements. The system is based on the application of the thermodynamic concept of "lines of equal solubility" for the various compounds. This concept has been used for predicting the precipitation sequence of various cerium compounds in steel melting and predicting the composition of the nucleating particles for graphite nodules in spheroidal graphite irons. A simple method for computing lines of equal solubility is included. For the purpose of illustration, the predictive system is used to explain microstructures and particle composition in Nb-containing microalloyed steels, and the limitations of the procedure are discussed.

INTRODUCTION

Over the last 20 years mild steels have been developed significantly, using the concepts of grain refinement and precipitation
strengthening, to meet the ever increasing requirements of higher
strength, toughness and ductility. The role of precipitate particles
is to refine austenite grains by pinning of the boundaries during
normalizing heat treatments or during thermomechanical working, and to
harden the ferrite grains on cooling after transformation. Using a mild
steel base composition, microadditions of about 0.1% of niobium, vanadium,
titanium and aluminum give the required precipitate particle distributions
of carbides, nitrides and carbonitrides. The solubility of these precipitate phases is such that the compounds are generally dissolved in
solid solution in the austenite during high temperature heat treatments
but the phases precipitate out partially during normalizing or hot rolling with further rejection of the same phase or different phases on
cooling to ambient temperatures.

Thus a sequence of precipitate phases can occur with a range
of precipitate particle shapes, sizes and distributions. Obviously
some knowledge of the solubility products of these phases is desirable
so that the mechanical properties can be scientifically controlled by
relating process temperatures and conditions to steel compositions
and microstructures. However, it is worth noting that most steels have
originated from empirical experiments rather than from scientific
prediciton of the optimum alloy additions and thermomechanical treatments. It is the purpose of this paper to present solubility data for
phases occurring in Nb-containing microalloy steels so that the information is readily available for predicting steel compositions, heat treatment, and rolling schedules.

Solubility Product Data

There are many examples of the application of thermodynamic
data to metallurgical processes. Specifically, solubility product
data has been determined for a large number of compounds in iron based
alloys and related to steelmaking reactions [1, 2] and mechanical property
data. [3, 4] The role of stoichiometry in optimizing mechanical properties is also of importance and both these concepts of solubility and
stoichiometry have been described in detail elsewhere. [5, 6]

Experimental difficulties in determining solubility product
data have resulted in widely varying values of the constants,
Figure 1. [7] Faced with this problem, the writers decided to use
the data of Narita [8] because he has determined the solubility products
of many of the compounds likely to be encountered in microalloy
steels, with the additional advantage that he used constant purity base
materials and standardized techniques of heat treatment and analysis.

In situations where more than one phase is likely to precipitate,
the concept of "lines of equal solubility" has been found to be of great
value. Previously this concept has been used to explain and predict the
deoxidation and desulphurization reactions of Ce in steels.

344

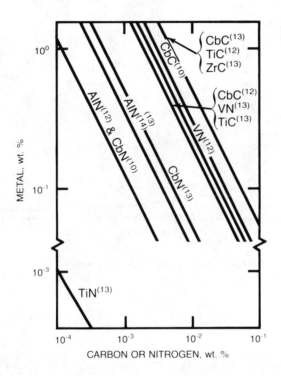

Figure 1.   Solubility products of selected
carbides and nitrides at 1100°C.

<u>Presentation of Solubility Data</u>

Figure 2a shows the presentation of the solubility data and stoichiometric ratio for NbC in austenite. The solubility product[8] is represented as:

$$\log (Nb)(C) = -7900/T + 3.42 \qquad (1)$$

where T is the absolute temperature.

The stoichiometric ratio, which is the ratio of the atomic weights of niobium and carbon is 92.91/12.1 = 7.74. The solubility data are shown as a normal plot of Nb, C, Figure 2a and as a log/log plot, Figure 2b for 1300°C, 1100°C and 900°C.

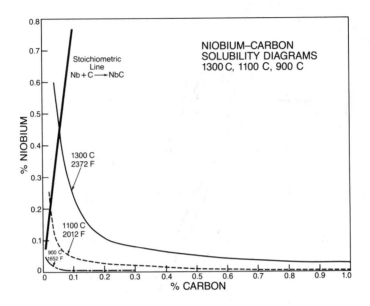

Figure 2a. Niobium carbon solubility diagrams.

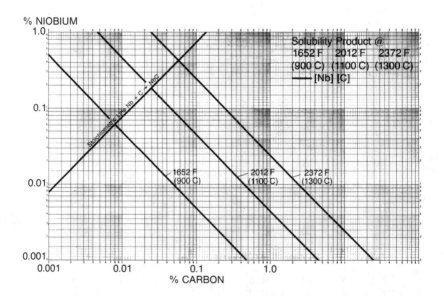

Figure 2b. Niobium carbon solubility diagrams, logarithmic scales.

Figure 3 shows the NbC data with that for NbN and AlN using the following solubility products at the same three temperatures in log/log form.

$$\log (Nb)(N) = -8500/T + 2.80 \qquad (2)$$

$$\log (Al)(N) = -7184/T + 1.79 \qquad (3)$$

The problem of identifying the exact species of precipitate occurring in niobium steels containing carbon and nitrogen has been discussed many times in the literature but so far experimental techniques have not allowed precise identification of NbC, NbN or Nb (CN) phases to be established. Narita's data[8] for NbC and NbN was produced in pure alloys where nitrogen and carbon respectively were eliminated. Whereas it is fairly clear that separate carbide and nitride phases occur in Ti steels it is generally accepted[9] that an isomorphous Nb(CN) phase occurs in commercial steels with a range of composition varying from carbon-rich to nitrogen-rich. The situation might then be represented by consideration of a range of solubility as shown in Figure 4. It is to be noted that the solubility of NbN is much less than that of NbC.

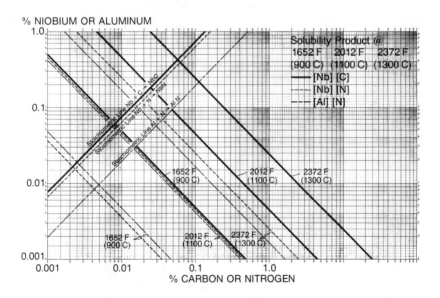

Figure 3. Niobium carbide, niobium nitride and aluminum nitride solubility diagrams.

Calculation and Presentation of "Equal Solubility Lines"

From solubility equations (1) and (2), the solubility constants at $1300°C$ can be calculated and give 2.4989×10^{-2} for NbC and 2.4906×10^{-3} for NbN.

Using these constants, it is possible to compute the amount of

347

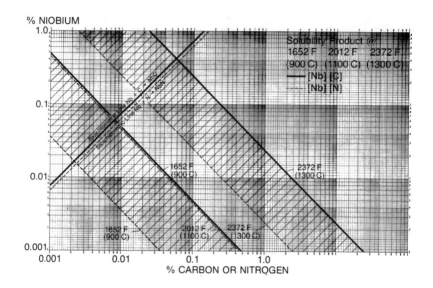

% NIOBIUM

% CARBON OR NITROGEN

Figure 4.   Niobium carbide, niobium nitride solubility diagrams.
Hatched regions represent solubility of the carbonitrides.

C and N in equilibrium with a given amount of Nb at 1300°C.   For
example, 2.5C and 0.25N are in equilibrium with 0.01Nb and 0.25C
and 0.925N with 0.10Nb.

   This data is then plotted as shown in Figure 5 to give a line
of equal solubility for niobium at 1300°C as a function of nitrogen
and carbon contents.   The Figure also shows the lines of equal solubility at 1100°C and 900°C calculated in the same way from the solubility product constants at these temperatures.

   This technique allows the presentation of the solubility data
for two compounds of the same metal, e.g., the carbide and nitride of
Nb in Figure 5 or for the nitrides or carbides of different metals,
e.g., Figure 6 shows lines of equal solubility of NbN and AlN generated
from equations (2) (3). The data calculated for plotting Figures 4 and 5
are given in Table I.

Use of Equal Solubility Data Diagrams

   To illustrate the application of these diagrams and taking
Figure 6, which shows the line of equal solubility (LES) for NbN and
AlN, any steel composition to the right of the LES will precipitate
AlN in preference to NbN (or a N-rich Nb carbonitride) and the Al
content will decrease towards the LES for that temperature at which
the reaction is occurring.   Meanwhile the Nb content remains constant
but the N content is also decreasing by an amount which can be calculated from the stoichiometric ratio of AlN and knowing the decrease
in the amount of Al indicated by the diagram.   At this point, when the
Al content reaches the LES there will be the possibility of both AlN
and NbN (NbCN) precipitating together as the temperature decreases

348

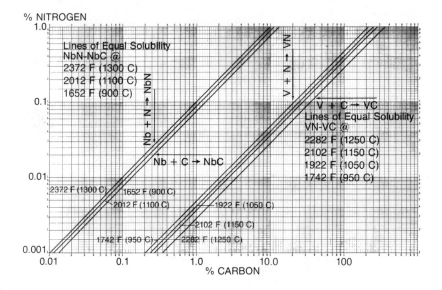

Figure 5. Lines of equal solubility (LES); niobium carbide and
nitride, and vanadium carbide and nitride.

TABLE I

		w/o Carbon	% Nitrogen
1300C	.01 Nb	2.50	.249
	.10 Nb	.25	.0249
1100C	.01 Nb	.463	.041
	.10 Nb	.0463	.0041
900C	.01 Nb	.0484	.0036
	.10 Nb	.0048	.00036
		% Aluminum	% Niobium
1300C	.001 N	1.67	2.49
	.010 N	.167	.249
	.100 N	.0167	.025
1100C	.001 N	.361	.407
	.010 N	.036	.047
900C	.001 N	.046	.036
	.010 N	.0046	.0036

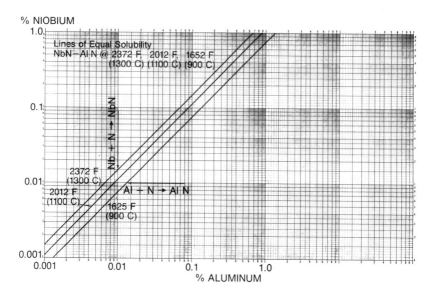

Figure 6. Lines of equal solubility (LES); niobium nitride, aluminum nitride.

until the solubility product curve, Figure 3, indicates no further rejection of nitride precipitates. It should be pointed out, however, that there is the strong possibility in Al-treated steels that all the nitrogen will be removed as AlN before the Al content reaches the LES and, therefore, the co-precipitation of AlN and NbN will never arise.

For a steel composition to the left of the LES of NbN and AlN in Figure 6 the situation is much more complicated because of the possibility of the Nb being reduced by NbC precipitation. Therefore, the equal solubility diagram for NbC and NbN, Figure 5 must also be considered. With a low C-high N steel, the composition is to the left of the LES for NbC-NbN in Figure 5, and NbN precipitation is favored. Taking next the Al content into account and using Figure 6, it is necessary to determine whether the steel composition is to the left or to the right of the LES for NbN-AlN precipitation. For a high C-low N steel which gives a composition to the right of the LES for NbC-NbN in Figure 5, NbC precipitation will be favored reducing the content as the reaction proceeds. This will in turn affect the nitride reaction in the equal solubility diagram for AlN-NbN in Figure 6 and also, of course, the Al content will have to be considered in deciding whether AlN or NbN is favored.

The whole of this complicated situation is further confused if we consider the kinetics of precipitation and the increase in speed of precipitation due to thermomechanical working. Additionally, segregation effects and the problem of not really knowing whether NbN, NbC or NbCN is the precipitation species, makes the interpretation of these diagrams uncertain. However, as will be shown, remarkable success in explaining the microstructures and properties of a variety of

microalloy steel compositions has been achieved by using these equal
solubility diagrams in association with solubility product diagrams
such as Figure 3. In some instances, the precipitation of one phase
such as NbC reduces the Nb content to a level where the possibility of
co-precipitation of all these phases NbC, NbN and AlN is predicted.
Such situations and interrelationships are extremely complicated and
computer simulation techniques are currently being attempted.

Examples of Precipitate Phase Predictions

The Role of Nitrogen in Nb Microalloy Steels. Current work at
the University of Sheffield is showing major effects of varying nitrogen between 0.002 and 0.02% in 0.08C, 1.2Mn, 0.06Nb steels. [10] With
the very low nitrogen content, acicular microstructures are produced
by normalizing at 1100°C and these give excellent strength due to
copious precipitation of NbC. The high nitrogen addition of 0.02N
gives ferrite plus pearlite microstructures with moderate dispersion
strengthening.

The difference in response to normalizing at 1100°C of these
steels with high and low nitrogen may be explained using the LES diagrams for NbC and NbN and solubility product calculations. First, the
position of the composition of these two steels on the LES diagram for
NbC-NbN is determined. The low nitrogen steel (0.002N) is to the right
and below the LES, indicating that the phase most likely to precipitate
is NbC or at least NbCN rich in carbon. The steel high in nitrogen
(0.020N) is to the left and above the LES, indicating that the phase most
likely to precipitate is NbN or high nitrogen NbCN. Computations can then
be made from the solubility product equations (1) and (2) to determine
the temperature at which the NbC and NbN all go into solution under
equilibrium conditions. These calculations indicate that all NbC should
be in solution at 1104°C, but the equilibrium temperature for the
solution of the NbN is 1212°C. At 1100°C only 0.035Nb is in solution,
with the balance of the Nb remaining as undissolved NbN particles.

When these two steels are air cooled from the 1100°C solution
temperature, most of the 0.06Nb in the steel with the low nitrogen
content is likely to remain in solution because NbC precipitation from
solution is very slow in the absence of deformation. Nb in solution
can shift the nose of a CCT curve to longer times and lower the γ to α
transformation temperature by as much as 100°C, thereby increasing the
hardenability of the steel significantly and favoring the acicular
structures. Solubility of the NbC in the ferrite is much less than
in the austenite and the cooling rate may be sufficiently slow for NbC
to precipitate. The combination of the acicular structure and precipitation in the ferrite accounts for the superior strength of the low
nitrogen steel.

In the presence of high nitrogen (0.02N), the computation based
on equation (2) indicates that only 0.035Nb went into solution at 1100°C.
Therefore, any increase in hardenability must be substantially less than
would be expected in the low nitrogen steel. Watanabe[11] has shown
the rate of precipitation of NbN (or NbCN high in nitrogen) in similar
steels is approximately a third faster than the precipitation of NbCN with
almost equal amounts of carbon and nitrogen. Furthermore, with almost
half of the Nb in the steel present as undissolved particles at the solution temperature, there is a possibility that NbN may be deposited
on the existing particles or may precipitate as particles of such a size
that they do not strengthen the steel. Little of the Nb taken into

solution is likely to remain in solution after the γ to α transformation. Therefore, the hardenability of the high nitrogen steel would not be high enough to favor the formation of acicular structures.

The Role of Aluminum in Microalloy Steels. Optimum ductility and maximum resistance to hydrogen induced cracking (HIC) are obtained in plates with a fine uniform grain size and a minimum of banding. To achieve such structures, some steel companies [12-14] have developed elaborate processing schedules involving heating and cooling through the transformation temperature and rolling in the two-phase region. (Figure 7)

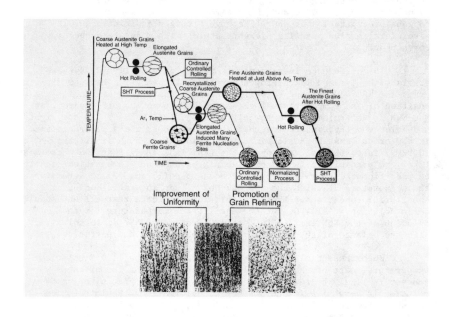

Figure 7. Schematic diagram of Sumitomo High Toughness (SHT) rolling process and resulting microstructures, Ref. 12

The type of banded structure commonly found in control rolled plates is shown in Figure 8a. Figure 8b shows a fine equi-axed structure free of banding produced by the Algoma Steel Corporation [13 14] for maximum toughness and HIC resistence. The LES diagram for NbN-AlN can be used to explain the rolling practice developed for such a structure.

The chemical composition of such a steel is 0.10C Max., 1.30/1.40Mn, 0.012S Max. (rare earth treated) 0.04Nb, 0.05V, 0.04Al. The patent of this steel [13] states that the rolling is continuous in the range 1900°F (1037°C) to 1650°F (900°C). Heating temperature is at least 2100°F (1149°C) and preferably 2250°F (1232°C). Productivity

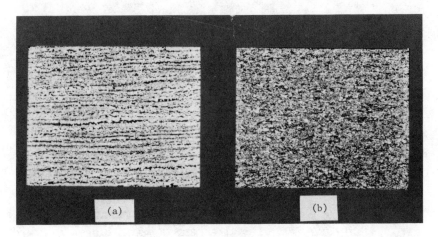

Figure 8. Microstructure developed at Algoma Steel; (a) conventional rolling practice, (b) patented procedure.

is only reduced to 85 to 90% of normal with these practices, which would imply that a delay in the rolling schedule would occur at a temperature higher than the top of the range at which continuous rolling must occur (1037°C).

The LES for NbN-AlN provides an important key to the explanation of the success of this rolling practice in achieving this desired microstructure at such a small sacrifice in mill productivity. When the Al-Nb analyses given by Acker and Bowie[14] are plotted on the LES for NbN-AlN they indicate that the nitride most likely to precipitate would be AlN. Jonas[15] and his co-workers at McGill University have stated that AlN particles coming out of solution at temperatures above 1038°C can prevent recrystallization, but the length of time AlN is available to prevent recrystallization is short compared to NbCN. Therefore, during the roughing schedule AlN particles could prevent recrystallization and the reductions in the roughing schedule would produce a fine austenite grain size. After recrystallization occurs during the delay between roughing and finishing rolling schedules, recrystallization of the austenite is likely to have homogenized the structure so that the pearlite bands commonly found in steels conventionally rolled no longer occur. Figure 9, from Ackert and Bowie's paper, shows hardness (relative rolling loads) during the finishing schedule and the Nb containing plates are only slightly stronger at the beginning (higher hardness multiplier) than low carbon steels indicating that recrystallization previously retarded by the AlN has now occurred. The resulting fine grained austenite is further refined in the finishing schedule with rolling loads increasing rapidly due to prevention of recrystallization by precipitation of NbC particles. Ackert and Bowie acknowledge that the patented process does not work without the aluminum addition.

Sumitomo have shown that optimum structures without banding are obtained when fine recrystallized austenite is achieved by cooling through the $\gamma \rightarrow \alpha$ transformation temperature and then reheating into the austenite region prior to final rolling reductions.

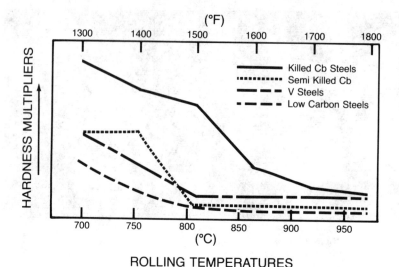

Figure 9.   Hardness multipliers versus rolling temperatures
of Algoma Steel, Ref. [16]

Confirming the benefits of such a schedule are the high CVN
values for such steels at temperatures as low as -100°C, Figure 10.

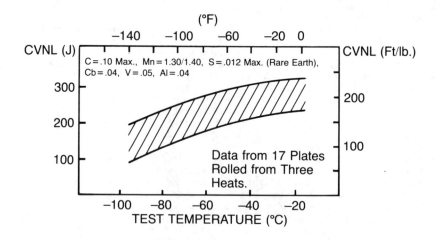

Figure 10.   Charpy transition curves of plates rolled with the
patented Algoma Steel practice.

CONCLUSIONS

1. Solubility product data are shown to be useful in predicting the phases that may precipitate during the processing of austenite.

2. Lines of equal solubility (LES) derived from such solubility data can be used to determine the particles most likely to precipitate in complex situations where there is a multiplicity of carbide and nitride formers.

3. Examples presented have shown the usefulness of LES diagrams in explaining the results obtained in laboratory and steel plant procedures.

BIBLIOGRAPHY

1. Langenberg, F.C. and Chipman, J., "Equilibrium Between Cerium and Sulfur in Liquid Iron", _Trans. AIME_ Vol. 212, p. 290.

2. Wilson, D.J., et. al., "Desulfurization, Deoxidation and Sulfide Shape Control with Nickel-Magnesium" _Proceedings Electric Furnace Conferences_, Vol. 33, 1975, pp. 196-209.

3. Wilson, W.G., et. al., "The Use of Thermodynamics and Phase Diagrams to Predict the Behavior of Rare Earths in Steel", _Journal of Metals_, May 1974, pp. 14-23.

4. Meyer, L., et. al., "Titanium as a Strengthening and Sulfide-Control Element in Low Carbon Steels", _Processing and Properties of Low Carbon Steels_, TMS-AIME, 1973, J.M. Gray, et.al.

5. Gladman, T., et. al., "Structure-Property Relationships in High Strength Microalloyed Steels", Micro Alloying 75, Union Carbide Corporation, New York, N.Y., 1977.

6. George T. and Irani, J.J., "Control of Austenite Grain Size by Additions of Titanium", _The Journal of the Australian Institute of Metals_, Vol. 13, No. 2, May 1968.

7. McLean, A. and Kay, D.A.R., "Control of Inclusions in High Strength, Low-Alloy Steels", Micro Alloying 75, Union Carbide Corporation, New York, N.Y., 1977.

8. Narita, K., "Physical Chemistry of the Groups IVa (Ti, Zr) Va (V, Nb, Ta) and the Rare Earth Elements in Steel", _Transactions ISIJ_, Vol. 15, 1975, pp. 145-152.

9. Storms, E.K. and Krikorian, N.H., "_Jr. Physical-Chemistry_", 1960, Vol. 64, p. 1471.

10. Private Communication, Shams, N., Sheffield University, Department of Metallurgy, Sheffield, England, December, 1980.

11. Watanabe, H., et. al., "Precipitation Kinetics of Niobium Carbo-Nitride in Austenite of High Strength Low-Alloy Steels", The Hot Deformation of Austenite, TMS-AIME, 1977, John Ballance, Editor.

12. Ikeshima, T., "Development of Sumitomo High Toughness (SHT) Process for Low Temperature Service Steels", Transactions ISIJ, Vol. 20, 1980.

13. Shaughnessy, R.N., et. al., "Method for the Production of High Strength Notch Tough Steel", Canadian Patent 966702, U.S. Patent 3,897,279.

14. Ackert, R.J. and Bowie, M., "Temperature Controlled Rolling of Plates as Developed and Practiced at the Algoma Steel 4220 mm (166 inch) Wide Plate Mill", International Conference on Steel Rolling, Tokyo, Japan, 1980.

15. Private Communication, Jonas, J.J., McGill University, Department of Metallurgy, Montreal, Quebec, Canada, June, 1979.

DISCUSSION:

Keown and Wilson have devised a system based on available solubility data to show which of several compounds will first precipitate during cooling of a steel of known composition. But the proposed system cannot account for the subsequent changes during cooling in the type and amount of precipitate that result from changes in composition of the solid solution caused by solute depletion. I propose a somewhat more complex program which is within the capacity of a desk-top computer, and which follows the effects of solute depletion on precipitation.

We begin with a system of L components, these being carbon, nitrogen, and (L-2) nitride and/or carbide-forming elements A, B, etc. The phase rule states that the number of variants or degrees of freedom V in a condensed system (pressure independent) is

$$V = L - P + 1$$

where P is the number of phases, which are the solid solution and all the precipitates, i.e., P = ppts + 1. Thus, in the case of interest, where all precipitates are forming from a single solution,

$$V = L - ppts.$$

From this relation we see that no more than (L-1) precipitates can be in equilibrium with the solution over a range of temperatures (V = 1). L precipitates can form only at an invariant point (V = 0).

As an example of the operation of the program consider the formation of three precipitates (one carbide and two nitrides) from four components (A, B, C for carbon, and N for nitrogen) over a range of temperatures:

$$1) \quad A + C \rightleftarrows AC$$
$$2) \quad A + N \rightleftarrows AN$$
$$3) \quad B + N \rightleftarrows BN$$

This leads to 10 simultaneous equations in 10 unknowns:

$$\left.\begin{array}{l} A_s\, C_s = K_1 \\[6pt] A_s\, N_s = K_2 \\[6pt] B_s\, N_s = K_3 \end{array}\right\}$$ Solubility products K_i of three compounds (known).

$$A_s + A_1 + A_2 = A$$
$$B_s + B_3 = B$$
$$C_s + C_1 = C$$
$$N_s + N_2 + N_3 = N$$

Material balances - soluble and precipitated weights on left (unknown), total weights on right (known).

$$C_1 = R_1 A_1$$
$$N_2 = R_2 A_2$$
$$N_3 = R_3 B_3$$

Ratios R_i of weights of elements in three compounds (known).

A, B, C, and N are the total weights of microalloys A and B and of carbon and nitrogen in the steel. A_s, B_s, C_s, and N_s are the weights of these elements in solution, and A_1, A_2,---, N_3 are the weights of these elements in the compounds formed in Equations 1, 2, and 3. R_1, R_2, R_3 are the stoichiometric ratios of the weights of the elements in the compounds 1, 2, and 3. K_1, K_2, K_3 are the solubility products of the elements in equilibrium with the compounds 1, 2, and 3.

These equations reduce to a single quadratic equation in one of the unknowns. The solution for soluble nitrogen is:

$$N_s = \frac{F + \sqrt{F^2 + 4GH}}{2G}$$

where $F = N - R_3 B - R_2 A + \dfrac{R_2}{R_1} C$

$$G = 1 + \frac{R_2}{R_1} \frac{K_1}{K_2}$$

$$H = R_2 K_2 + R_3 K_3$$

In solving the equation, the program tests before and after each computation to assure that there is (was) sufficient solute to avoid negative or imaginary quantities. If such are found, the program branches to avoid attempts to form compounds when adequate solute is not available. The solution gives the weights (real and positive) of the solutes and a maximum of three compounds in equilibrium in austenite at temperature T. These values are based on the solubility products K_i (T) of the compounds taken from published data. Effects of other elements on solubilities, or effects of mutual solution of the nitrides and carbides, can be incorporated by adjusting K_i (T).

By applying the program repeatedly to steels containing combinations Ti, Nb, and V along with 0.07 C and 0.007 N, we can demonstrate the order of precipitation of the various alloy carbides and nitrides on cooling, or their dissolution upon re-heating, Figure 1. On cooling, TiN forms first at ∿1650°C. If no Ti is present or if excess N remains in solution, NbCN forms next at 1200°C; but if N has been exhausted, TiC and NbC begin to form about 1070°C. Because there is a great excess of C available, each carbide forms almost as it would in the absence of the other carbide former. If N still remains, VN begins to form at 1000°C until N is depleted. The remaining V begins to form VC at about 700°C, in the ferrite.

Clearly these results are based on equilibrium solubility data which predict what "ought to form." No account has been taken of kinetic effects.

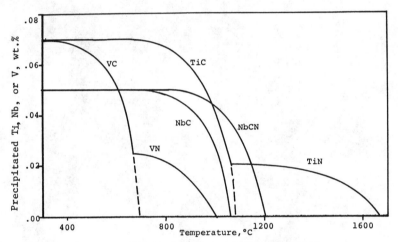

Fig. 1 - Amount of precipitate vs. temperature in an
0.07C - 0.007N steel containing 0.07 Ti and/or
0.05 Nb and/or 0.07 V.

CHANGES IN AUSTENITE DURING AND AFTER HOT DEFORMATION OF VANADIUM,
ALUMINUM, AND NITROGEN MICROALLOYED STEELS

Petre P. Ianc
and
Nicolae I. Dragan

ICEM, Bucharest, Rumania,
Laboratory for Corrosion and Physical Metallurgy

Kinetics of static recrystallization and nitride precipitation has been determined in three steels, microalloyed with vanadium, aluminum and nitrogen. Samples were tested isothermally in torsion over a temperature range from 760 to 1200°C. Results of metallographic examination and chemical analysis of combined nitrogen are presented as recrystallization (or precipitation) - temperature - time curves.

Using the same techniques, plate rolling schedules -- applicable to actual industrial conditions -- have been simulated.

1. Introduction

The desirable engineering properties of fine-grained structural steels, such as high strength, low-temperature toughness and weldability, are the result of microalloying with one or more elements: vanadium, niobium, titanium, and nitrogen, added -- generally -- in amounts not exceeding 0.2 percent (1,2).

In selecting the microalloying system, it is important to take into account the individual, or combined, effect of microalloying elements on microstructural changes taking place during the whole technological process. These changes include: change in shape and properties of non-metallic inclusions, solution and precipitation reactions of carbides, nitrides or carbo-nitrides, recrystallization after plastic deformation, and the gamma to alpha (or vice versa) phase transformation occuring on heating or cooling. All these structural changes have an effect on the final microstructure of the steel, and -- thus -- on mechanical property characteristics (3,4,5).

For a given chemical composition, grain refinement is a unique strength-ening mechanism, contributing both to increase of strength and improvement of toughness. Numerous equations have been derived to express this relat-ionship quantitatively (1,6,7,8,9,10).

The degree of grain refinement and other microstructural characteristics depend on the technological processing history and on the kinetics of structural changes (11, 12,13,14).

In microalloyed steels, the fine ferrite-pearlite structure is formed during the gamma to alpha transformation of fine-grained austenite. Factors influencing the degree of ferrite grain refinement are: (a) conditions of plastic deformation and degree of recrystallization in the austenite; (b) solution and precipitation reactions of microalloying elements, mainly vanadium and niobium, in combination with nitrogen and carbon.

Two solid state reactions: precipitation in the austenitic region, enhanced by plastic deformation, and the process of recrystallization con-tribute to grain refinement of both austenite and ferrite. The role of precipitates may be three fold:
  *act as nuclei for recrystallization of the deformed austenite;
  *inhibit growth of the recrystallized austenite grains;
  *provide nucleation sites for the ferrite during transformation
    on cooling.

The effectiveness of precipitates will depend not merely on volume fraction, but on their size and distribution (15,16).

2. Experimental Procedure

One of the objectives of this study was to determine the kinetics of the following reactions:
  *static recrystallization of the austenite after preceeding high-
    temperature deformation,
  *precipitation and solution reactions of vanadium and aluminum
    nitrides.

TABLE I	Chemical Analysis, Liquid Base, of the Steels Studied							
Steel No.	Elements, percent							
	C	Mn	Si	S	P	Ni	Nppm	Al	V
1	0.16	1.32	0.37	0.015	0.011	0.74	145	0.070	-
2	0.16	1.34	0.30	0.015	0.010	0.70	140	0.002	0.16
3	0.16	1.25	0.40	0.015	0.010	0.65	135	0.040	0.15

The high-temperature deformation of austenite was produced by torsion, in an automatic torsion machine, capable of being programmed up to maximum 12 deformation cycles.

As shown schematically in Figure 1, all samples were heated first to 1280-1300°C, to assure complete solution of microalloying elements (15,17). Plastic deformation was carried out in the temperature range: 760 to 1200°C at the maximum obtainable speed of 1500 RPM.

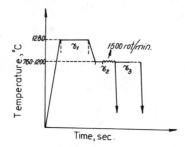

FIGURE 1

Schematic diagram of the heating and deformation cycle used to study the recrystallization and precipitation processes. Samples were water quenched either immediately after deformation, or after predetermined hold at the temperature of deformation.

The shape of the stress (torque) vs. deformation curve, shown in Figure 2, is related -- schematically -- to changes in the microstructure of austenite. It was assumed that the maximum hardening of austenite occurs at a strain, Cem, corresponding to 4/5 of the strain associated with the maximum torque (Cm). For each test temperature, the samples were deformed to the strain equal Cem, to attain a maximum of strain-hardening of the austenite prior to static recrystallization.

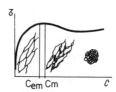

FIGURE 2

Schematic presentation of changes in the microstructure of
austenite during deformation and recrystallization. The
stress (torque) vs deformation curve is determined for
constant temperature and rotating speed. The condition
of austenite depends on strain C:

$0 \leq C \leq Cem$	strain hardened
$C = Cem$	maximally hardened
$Cm < C$	dynamically recrystallized

To reveal austenitic grain boundaries in samples quenched either immediately after plastic deformation, or after static recrystallization, the
samples were etched in saturated solution of picric acid containing 1%
$CnCl_2$, and 10-15% of an inhibitor based on sodium sulphowate alkyl.

Chemical analyses performed either on bulk sample or on extracted
residue, applied to determine the kinetics of precipitation of vanadium and
aluminum nitrides at various stages of deformation and recrystallization,
are shown schematically in Figure 3 (18,19,20,21).

Based on the results of isothermal torsion tests, a controlled rolling
schedule was simulated in the hot torsion apparatus. The simulation took
into account the actual conditions existing at the plate rolling mill of the
Iron and Steel Works at Galati. The objective of the simulation was to
develop an industrial rolling practice which would produce maximum grain
refinement in the austenite and the ferrite.

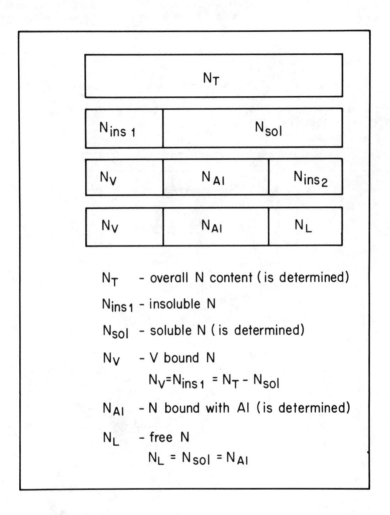

N_T - overall N content (is determined)

N_{ins1} - insoluble N

N_{sol} - soluble N (is determined)

N_V - V bound N

$$N_V = N_{ins1} = N_T - N_{sol}$$

N_{Al} - N bound with Al (is determined)

N_L - free N

$$N_L = N_{sol} = N_{Al}$$

FIGURE 3

Schematic presentation of analytical steps used to determine the
fraction of nitrogen present as vanadium nitride and aluminum
nitride. The free nitrogen is calculated as the difference
between the total and the combined nitrogen.

3.   Experimental Results and Discussion

3.1 Static Recrystallization

Samples deformed in the temperature range: 760 to 1200°C to a strain Cem (corresponding to the maximum strain hardening of austenite), were held at the deformation temperature for various times prior to water quenching (Figure 1).

The fraction of statically recrystallized austenite grains was determined on micrographs at 100x magnification.

The obtained results are presented in Figure 4, in the form of a Recrystallization-Temperature-Time (RTT) diagram (22). The data corresponding to steels 1,2, and 3, listed in Table I, are shown in Figures 4a, 4b, and 4c, respectively.

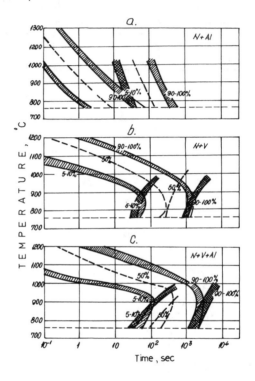

FIGURE 4

Diagrams depicting progress of static recrystallization and nitride precipitation as a function of time and temperature (RTT and PTT diagrams), for the three steels containing (a) aluminum, (b) vanadium, and (c) vanadium plus aluminum.

▨▨▨▨ Recrystallization

▩▩▩▩ Precipitation

Evaluation of diagrams in Figure 4 suggests the following:

The recrystallization diagram for steel 1, microalloyed with N + Al (Figure 4a) is typical for plain carbon steel. Rate of recrystallization is very rapid, decreasing gradually with the drop in temperature.

The vanadium steels, 2 and 3, recrystallize statically at much slower rate than steel 1. Microscopic examination revealed that the austenitic grain size of completely recrystallized vanadium steels is smaller than that observed in steel 1.

3.2 Isothermal Precipitation in Austenite after Deformation

The progress of nitride precipitation was determined analytically on samples deformed over the same range of temperatures (760 to 1200°C) to a strain Cem, held at the deformation temperature for various times, and water quenched. The methodology of analytical determination of the form of nitrogen: vanadium nitride, aluminum nitride, and free nitrogen is shown schematically in Figure 3. The results are plotted as Precipitation-Temperature-Time (PTT) diagrams, for steels 1, 2, and 3 in Figure 4a, 4b, and 4c, respectively.

It should be noted that in these diagrams, the combined nitrogen is shown as a percentage of the total nitrogen present in steel. For the two vanadium steels (2 and 3), nitrogen which combined during the plastic (strain induced precipitate) deformation, was excluded from calculations.

Analysis of samples of steel 1 quenched immediately after plastic deformation did not reveal any aluminum nitride. On the other hand, steels containing vanadium: steel 2: V+N and steel 3: V+N+Al, have shown, immediately after deformation at 1000°C, about 5% combined nitrogen, and up to 20% combined nitrogen after deformation in the 760 to 800°C region.

Examination of RTT and PTT diagrams in Figure 4 reveals the following:
*precipitation reaction in N-Al steel (Figure 4a) proceeds faster than in the two vanadium steels (Figure 4b and 4c).
*precipitation of aluminum nitride takes place after completion of the static recrystallization.
*in the vanadium steels, precipitation at low temperatures (760 to 900°C) commences either simultaneously, or before the static recrystallization.
*compared to V+N steel 2, precipitation in the V+Al+N steel 3 starts later, but then progresses more rapidly, as indicated by the position of the curve for 50% precipitation, located close to the starting time. In the presence of aluminum, however, completion of the precipitation reaction seems to be delayed.

4. Laboratory Simulation of Industrial Hot Rolling Schedule

Research summarized in Chapters 3.1 and 3.2 provided information on the kinetics of austenite recrystallization and precipitation of nitrides of vanadium and aluminum, after a plastic deformation at high temperatures.

In the course of an industrial hot-rolling, plastic deformation is applied in steps over a range of temperatures. This results in a sequence of interdependent structural changes.

To be of practical value, the results of laboratory investigations have to be correlated with the conditions existing in the rolling mill, aiming at development of optimum structural characteristics in the finished product.

Evaluation of a variety of rolling schedules in the mill is both difficult and expensive. To overcome these problems, a laboratory simulation
was adopted. It was expected that the obtained results will offer a useful
guidance for industrial plate rolling practice.

To reproduce industrial rolling process in the laboratory, it is important to use in the torsion test strain and strain rates applicable to the
rolling process (23).

$$n = 5.305 \ln \frac{h_1}{h_2} \qquad (1)$$

$$N' = 318.31 \; \frac{V}{R\alpha \sin} \; \ln \frac{h_1}{h_2} \qquad (2)$$

where:

n = the number of torsions (indicative of the degree of deformation)
N = torsion velocity
V = average speed of plate emerging from the rolls
$h_1 h_2$ = plate thickness before and after a reduction pass
α = angular velocity of rolls
R = working roll radius

Calculations of the deformation during rolling and of that developed
in a torsion test revealed that in an actual rolling process, the degree of
deformation was lower than strain corresponding to maximum strain hardening
(Cem), and the strain rate was higher than that obtainable with 1500 RPM.

The heavy plate mill under consideration has two reducing stands, and
slabs, 200 to 300 mm in thickness, are rolled to plates ranging in thickness
down to 6 mm.

In general, the slabs are reduced in the first stand to about 60 mm (the
actual intermediate thickness depends on the final plate thickness), and then
rolled down to the final dimension in the second stand.

The semi-finished product travels from the first to the second stand on
a roller table. Entry into the second stand can be delayed by holding on a
adjacent cooling bed.

The entry temperature at the second stand can be controlled by water
cooling between roughing presses, by varying the degree of deformation per
pass and the number of passes, and by adjusting the time between the two
successive deformations.

The rolling schedule used at the mill for processing 20 mm thick plates
involves a series of passes at relatively high temperatures, resulting in a
finishing temperature in excess of 900°C (Figure 5A).

Based on results of our studies, summarized in Figure 4, two modified
rolling schedules were designed for simulation (Figures 5B and C). These
modified rolling schedules would enhance precipitation of vanadium or/and
aluminum nitrides during austenite deformation and recrystallization, since
a significant part of the total deformation would be performed below 900°C.

After the completion of the three deformation schedules (A,B,C), the
samples were either water quenched -- to determine the austenitic grain size --
or air cooled -- to assess the ferrite-pearlite microstructures.

The results of the microexamination are presented as follows: Figure 6
for Steel 1 (N+Al), Figure 7 for Steel 2 (N+V), and Figure 8 for steel 3 (V+N+Al)

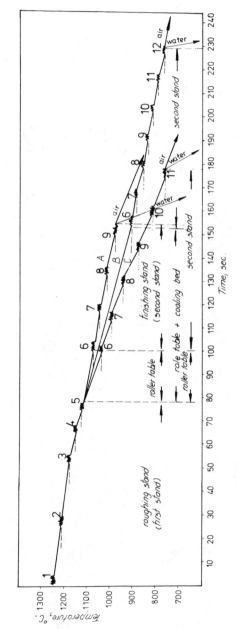

Fig.5   The hot torsion laboratory simulation scheme of the industrial heavy plate rolling.

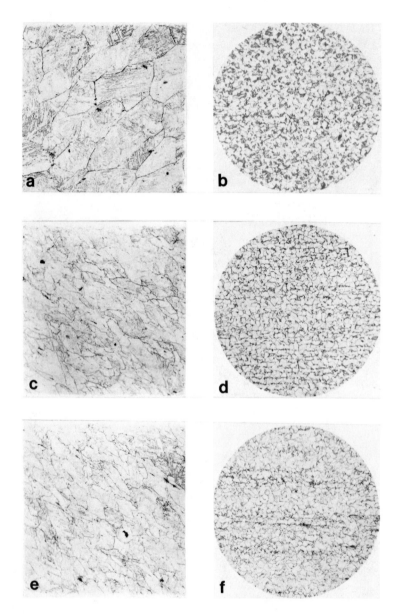

Figure 6 – Austenitic (left) and ferritic (right) grains for the N + Al microalloyed steel at the end of plastic deformation. (a,b) Plastic deformation according to diagram A, Fig. 5; (c,d) Plastic deformation according to diagram B, Fig. 5; (e,f) Plastic deformation according to diagram C, Fig. 5. Magnification 100X.

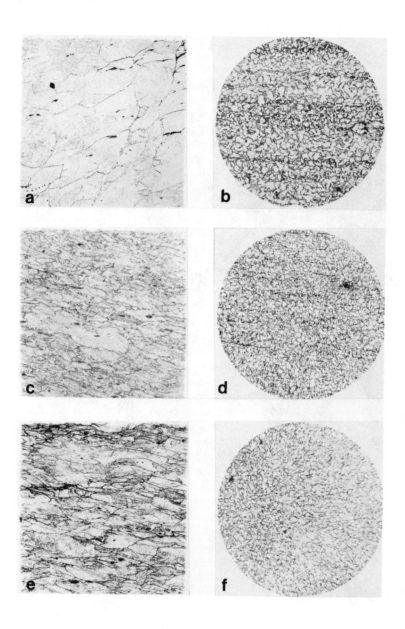

Figure 7 - Austenitic (left) and ferritic (right) grains for the N + V microalloyed steel at the end of plastic deformation. (a,b) Plastic deformation according to diagram A, Fig. 5; (c,d) Plastic deformation according to diagram B, Fig. 5; (e,f) Plastic deformation according to diagram C, Fig. 5. Magnification 100X.

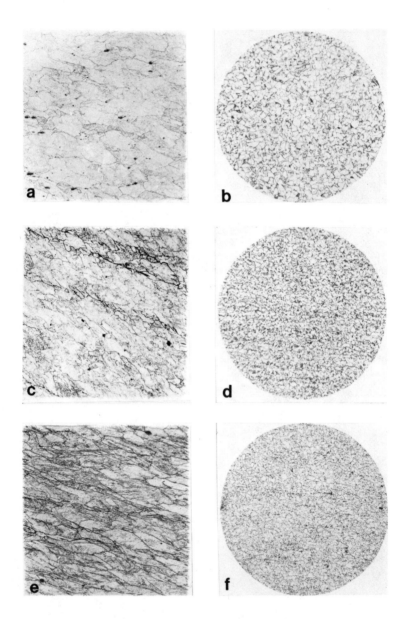

Figure 8 - Austenitic (left) and ferritic (right) grains for the N + V + Al microalloyed steel at the end of plastic deformation. (a,b) Plastic deformation according to diagram A, Fig. 5; (c,d) Plastic deformation according to diagram B, Fig. 5; (e,f) Plastic deformation according to diagram C, Fig. 5. Magnification 100X.

Samples deformed according to Schedulds A, B, and C (Figure 5) were water quenched for subsequent analytical determination of the form in which nitrogen was present. The results are shown in Table II.

Examination of austenitic and ferritic microstructures, shown in Figures 6 through 8, of torsion specimens processed according to three simulation schedules (A,B,C, in Figure 5), as well as data presented in Table II, lead to the following conclusions:

*control of the temperature range and the degree of deformation, pro-
vided in Schedules B and C, results in finer austenitic and ferrite-
pearlite structures, compared to microstructures resulting from
Schedule A, currently used by the industry.
*comparing the microstructures produced by Schedules B and C, the lat-
ter produces finer and more uniform grain structure, both of austenite
and ferrite; this is probably the result of a shorter hold time between
rolling in the two stands, which may contribute to a finer austenitic
grain structure at the start of rolling in the second stand.
*compared to Steel 1 (N+Al) (Figure 6), the two vanadium steels 2 and 3
have a much finer and more homogeneous austenitic and ferritic grain
structure, when processed according to Schedules B or C (Figures 7
and 8).
*some free nitrogen (determined as the difference between the total
nitrogen and nitrogen isolated as VN and AlN) seems to exist, especially
in N+V steel 2, being highest for processing according to Schedule A.

5. Conclusions

Kinetics of static recrystallization and nitride precipitation has been determined for steels microalloyed with vanadium, aluminum and nitrogen. The interaction between these two phenomena and mechanisms contributing to grain refinement have been clarified.

At high temperatures (above 900°C), the form of static recrystallization curves for vanadium-bearing and vanadium-free steels is similar, the recry-stallization process taking place prior to precipitation. Vanadium, however, contributes to some retardation of the recrystallization process.

The difference in the form of static recrystallization curves between vanadium-bearing (Figures 4b and c), and vanadium-free (Figure 4a) steels can be related to the progress of nitrides precipitation.

At lower temperatures (below 900°C), static recrystallization in vanadium steels takes place immediately after or simultaneously with the start of precipitation. It appears that as the vanadium is rejected from austenite to form the precipitate, the austenite -- depleted of the alloying element -- behaves during recrystallization in a similar way as vanadium-free austenite.

Formation of precipitates contributes to the acceleration of static recrystallization by providing surfaces supporting nucleation, and decreasing in this manner the critical dimensions of the nucleus and the activation energy for the initiation and progress of the recrystallization process.

These phenomena may explain the shape of recrystallization curves in vanadium steels, indicative of accelerated static recrystallization the temperature range of nitride precipitation.

Vanadium nitrides are formed both during plastic deformation, and before or simultaneously with the austenite recrystallization. They may contribute

Table II. Modality in which Nitrogen is Bound after Plastic Working According to Schemes A, B, and C

Crt. No.	Steel (Micro-Alloying System)	Hot Plastic Deformation Variant								
		A			B			C		
		Nitrogen Contents (per cent)			Nitrogen Contents (per cent)			Nitrogen Contents (per cent)		
		N_V^+	N_{Al}^{++}	N_L^{+++}	N_V	N_{Al}	N_L	N_V	N_{Al}	N_L
1	N + Al	–	0.009	0.0055	–	0.0115	0.0025	–	0.0120	0.0020
2	N + V	0.0030	–	0.0110	0.0120	–	0.0020	0.0125	–	0.0015
3	N + V + Al	0.0030	0.0050	0.0055	0.0075	0.0055	0.0005	0.0070	0.0055	0.0010

$+ N_V$ – Vanadium Bound Nitrogen

$++ N_{Al}$ – Aluminum Bound Nitrogen

$+++ N_L$ – Free Nitrogen

to austenite grain size refinement in the following manner:

   *through decrease in the size of the critical nucleus during static
    recrystallization;
   *by pinning the grain boundaries of newly formed austenitic grains,
    resulting from static recrystallization.

   Precipitation of aluminum nitrides takes place after completion of
static recrystallization, and their role is limited to inhibiting grain
growth of the recrystallized austenite.

   To obtain a maximum grain refining effect of vanadium, hot deformation
should be conducted in such manner as to assure maximum vanadium nitride
precipitation before and simultaneously with recrystallization.

References

1. K. J. Irvine, F. B. Pickering, T. Gladman, J.I.S.I., vol. 205, February 1967, pp. 161-182.
2. R. Phillip, W. E. Duckworth, F. E. L. Copley, J.I.S.I., July 1964, pp. 593-600.
3. L. Mayer, F. Heisterkamp, W. Mueschenborn, Micro-Alloying, October, 1976, Washington, U.S.A., pp. 153-167.
4. U. Lotter, L. Mayer, R. D. Knorr, Arch. Eisenhüttenwessen 47, 1976, pp. 289-294.
5. T. Greday, M. Lamberights, Micro-Alloying, October, 1975, Washington, U.S.A., pp. 172-187.
6. K. J. Irvine, F. B. Pickering, The Iron and Steel Institute's Special Report, No. 81, 1963, p. 10.
7. J. N. Petch, Fracture, New York, Wiley, 1959.
8. T. Prnka, Archiv fur das Eisenhüttenwessen, vol. 42, Nr. 12, 1971.
9. J. Little, J. A. Chapman, W. B. Marrison, B. Mintz, The Microstructure And Design of Alloys, Cambridge, England, August, 1973.
10. M. Korchynsky, H. Stuart, Symposium, Low-Alloy High-Strength Steels, Nüremberg, 21-23 May, 1970, pp. 17-27.
11. W. G. Marrison, B. Mintz, R. C. Coehrane, Structure Property Relation-ships in Controlled Processed Steel, British Steel Corporation, Control-led Processing HSLA Steels, Conference at University of York, September, 1976, England.
12. J. D. Baird, R. R. Preston, Processing and Properties of Low Carbon Steel, AIME, New York, 1973.
13. G. Melloy, J. D. Dennison, The Microstructure and Design of Alloys, Cambridge, England, August, 1973, p. 60.
14. M. Fukuda, T. Hashimoto, K. Kunishige, Micro-Alloying, October, 1975, Washington, U.S.A., pp. 172-187.
15. M. Hillert, Acta Met. Nr. 13, 1965, p. 227.
16. T. Gladman, Proceedings of the Royal Society, 1966 A, vol. 294, p. 298.
17. P. Ianc, Romanian-Japanese Symposium of Metallurgy, Bucarest, Romania, 1978.
18. H. J. Wiester, Stahl und Eisen, 1957, pp. 773-784.
19. M. D. Lintara, Zovodskaia Laboratoriia, Nr. 10, 1970, pp. 48-55.
20. H. F. Beeghly, Anal. Chem., 1949, pp. 1513-1519.
21. M. Făgărăsanu, N. Catuneanu, Metallurgical Researches, vol. 17, 1976, p. 139, Romania.
22. N. Drăgan, P. Ianc, Vanadium Steels "Technology and Applications of Structural Vanadium Steels", October, 1980, Krakow, Poland.
23. G. Radu, C. Teodosiu, E. Soos, T. Dumitrescu, "Metallurgy" Rev., 1979, 31, 580, Romania.

FERRITE FORMATION FROM THERMO-MECHANICALLY PROCESSED AUSTENITE

R.K. Amin

British Steel Corporation
Sheffield Laboratories
Swinden House
ROTHERHAM
England

F.B. Pickering

Department of Metallurgy
Sheffield City Polytechnic
SHEFFIELD
England

An investigation has been made of the complex interactions between solute content, austenite grain size, undissolved and precipitated carbides and nitrides, austenite grain morphology, and defect structure on the formation of ferrite in thermo-mechanically processed Nb and V steels. Nb has been shown to increase hardenability more than V, and N also increases the hardenability. Nb was more effective than V in retarding austenite recrystallisation at high rolling temperatures, but high N—V particularly was effective in retarding recrystallisation at lower rolling temperatures. Nb, and high N—V additions were effective in inhibiting recrystallisation of the ferrite formed during rolling in the critical range. There was a clear relationship between austenite and ferrite grain size, with a finer ferrite grain size being produced from unrecrystallised austenite. The various sites in thermo-mechanically worked austenite which act as ferrite nuclei have been identified, and the conditions which lead to mixed ferrite grain sizes have been investigated. Suggestions have been made for avoiding mixed ferrite grain sizes.

1. Introduction

Controlled rolling conditions the austenite for subsequent transformation to a fine, uniform ferrite grain size, thereby optimising strength and toughness. The ferrite nucleation rate is maximised by increasing the number of nucleation sites, and it is also necessary to restrict the growth of the ferrite grains during and after transformation. It is not possible to refine the austenite grain size to below 15-20µm diameter (1,2) by recrystallisation, which limits the degree of ferrite grain refinement achieved by nucleation at austenite grain boundaries. However, the presence of unrecrystallised austenite produced by deformation below the recrystallisation temperature increases nucleation at austenite grain boundaries and within the austenite grains (2-7). The relative importance of such intragranular ferrite nucleation is not well understood.

Nb and V additions are used in controlled rolled steels to refine the recrystallised austenite grain size (1, 8-10), and by retarding recrystallisation they enhance intra-granular nucleation of ferrite and thereby further refine the ferrite grain size. They also, when dissolved in austenite, increase the hardenability (11,12) and so lower the Ar_3 temperature and refine the ferrite grain size. Any undissolved carbo-nitrides however, can lower the hardenability by refining the austenite grain size and due to their nucleation effect for ferrite. Any thermo-mechanical treatment which leaves micro-alloy carbo-nitrides undissolved, or causes them to precipitate, would therefore decrease the hardenability, raise Ar_3 and produce a coarser ferrite grain size if the number of ferrite nuclei had not been greatly increased by the treatment (13,14). Such an effect is more prevalent at the stoichiometric micro-alloy : carbon or nitrogen ratio.

Mixed grain sizes often occur in controlled rolled steels, and are reported to impair toughness (5). Partial recrystallisation of austenite prior to transformation has been shown to be a major cause for these mixed grain sizes (15-18), but a detailed examination of such structures is required in order to identify the processing parameters which will minimise the effect. The present work describes the effects of thermo-mechanical processing variables and steel composition on the evolution of the ferrite structure from different austenite grain morphologies.

2. Experimental Procedures

Three series of steels were made with a base composition 0.08%C, 0.9%Mn, 0.3%Si, 0.02/0.05%Al, and various additions of either Nb or V. The Nb steels contained low N contents of 0.006/0.009%, whilst the V steels contained either this low N content or an enhanced N content of 0.015/0.020%, Table I.

Prior to the experimental thermo-mechanical processing, all the steels were solution treated at 1300°C to produce a uniform structure. A flow chart illustrating the hot rolling parameters used, is shown in Fig.1. Reheating temperatures of 1000°C, 1150°C and 1300°C were used with rolling temperatures of 700/750°C, 900/950°C and 1200/1250°C, and either 20% or 50% rolling reduction in one pass. Immediately after rolling, specimens were cut in two, one part being iced-brine quenched to reveal the prior austenite structure, whilst the other was held at 750°C, 950°C or 1250°C for times up to 1000s before again being cut into two. One part was ice-brine quenched to reveal the austenite structure whilst the other was cooled at about 400°C/minute, i.e. equivalent to air cooling a 12mm thick plate.

Table I. Analyses of Experimental Alloys

Series	Cast No	Analysis (Mass %)							
		C	Mn	Si	Al	Nb	V	N	M/C
Base Steel	1	0.083	0.76	0.24	0.045	-	-	0.008	-
Nb series low nitrogen	2	0.088	0.89	0.25	0.053	0.07	-	0.008	0.85
	3	0.093	0.93	0.27	0.066	0.11	-	0.009	1.16
	4	0.088	0.79	0.18	0.010	0.16	-	0.007	1.81
	5	0.088	0.82	0.23	0.033	0.46	-	0.006	5.22
V series low nitrogen	6	0.086	0.84	0.25	0.022	-	0.06	0.007	0.70
	7	0.087	0.87	0.28	0.020	-	0.11	0.007	1.21
	8	0.088	0.93	0.29	0.023	-	0.14	0.007	1.59
	9	0.083	0.86	0.26	0.030	-	0.22	0.008	2.65
	10	0.086	0.85	0.26	0.047	-	0.55	0.008	6.40
V series high nitrogen	11	0.075	0.98	0.29	0.05	-	0.05	0.016	0.66
	12	0.075	0.85	0.25	0.03	-	0.14	0.020	1.80
	13	0.074	0.91	0.24	0.03	-	0.24	0.015	3.31
	14	0.075	0.82	0.21	0.02	-	0.46	0.018	6.13

The isothermal transformation characteristics of the base steel and the 0.07%Nb and 0.14%V steels with low and high N, were studied after being thermo-mechanically treated to produce different recrystallised austenite grain sizes but similar dissolved solute contents. Deformation was carried out in a forging press after reheating at 1150°C. After 50% reduction at 1150°C, one specimen was iced-brine quenched to reveal the austenite grain size whilst other specimens were immediately quenched into a salt bath at 700°C and the progress of transformation followed for times up to 100s.

Estimates of the austenite grain size and the amounts of the various transformation products were made using standard quantitative metallographic methods, to give relative errors of <5% of the measured value.

The studies made required a knowledge of the amounts of Nb or V in solution, or as precipitated carbides/nitrides, at various temperatures. The solubility equations available in the literature (19,20) were used and the method for obtaining the Nb as coarse undissolved precipitates, as fine precipitates and in solution is illustrated schematically in Fig.2.

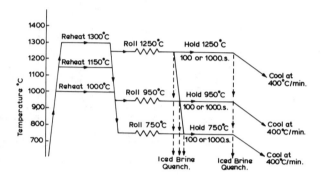

Figure 1   Flow sheet of experimental rolling programme.

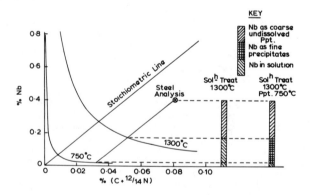

Figure 2   Method of obtaining various forms of
niobium in a 0.08%C, 0.4%Nb steel.

3. Experimental Results

3.1 Nb Steels

(a) <u>Reheated at 1300°C</u> Rolling at 1250°C caused complete recrystallisation of the austenite in the base steel after both 20% and 50% reduction,
the austenite grains being coarse and producing coarse ferrite and bainite
on cooling. Reheating at 1300°C dissolved all the Nb(CN) in steels containing up to 0.11%Nb, there being increasing amounts of undissolved Nb(CN) with
higher Nb contents, Tables II and III. Up to 0.11%Nb, the steels were partially recrystallised after 20% reduction but fully recrystallised after 50%
reduction. With the higher Nb contents, only partial austenite recrystallisation occurred after 50% reduction, and holding at 1250°C produced only a
slow increase in the extent of recrystallisation. Because the Nb in solution, together with the relatively coarse austenite grain size, increased
the hardenability, cooling at 400°C/min produced little ferrite, which also
decreased with increasing amounts of dissolved Nb. The main transformation
product was bainite the carbon content of which decreased with increasing Nb
due to the increasing amounts of undissolved Nb(CN). Holding at 950°C after rolling at 1250°C did not change the transformed structures, but the
effect of dissolved Nb on the hardenability was clearly shown by the refinement of the bainite, Figs. 3(a) and (b). Due to the large amount of austenite recrystallisation there was no precipitation of NbC (14). Further decreasing the holding temperature to 750°C, i.e. within the critical range,
allowed the recrystallised austenite to transform to ferrite. Above 0.16%Nb,
the increased hardenability gave less ferrite, giving only the nucleation of
ferrite at prior austenite grain boundaries, Fig.3(c).

Table II. Amount of Nb in solution and as Precipitate at Different
Temperatures for the Various Steels

% Nb (Mass) in Steel	% Nb (Mass) in solution or as precipitate											
	750°C		950°C		1000°C		1150°C		1250°C		1300°C	
	Sol.	Ppt.	Sol.	Ppt.	Sol.	Ppt.	Sol.	Ppt.	Sol.	Ppt.	Sol.	Ppt.
0.07	0.0	0.07	0.01	0.06	0.01	0.06	0.04	0.03	0.07	0.00	0.07	0.00
0.11	0.0	0.11	0.01	0.10	0.01	0.10	0.04	0.07	0.08	0.03	0.11	0.00
0.16	0.0	0.16	0.01	0.15	0.01	0.15	0.05	0.11	0.09	0.07	0.12	0.04
0.46	0.0	0.46	0.02	0.44	0.03	0.43	0.08	0.38	0.14	0.32	0.19	0.27

Decreasing the rolling temperature to 950°C produced only partially recrystallised austenite in the base steel after 20% reduction, but completely
unrecrystallised austenite in the Nb steels. 50% reduction at 950°C, however, caused the austenite to recrystallise partially with up to 0.16%Nb.
Transformation on cooling after holding at 950°C produced large amounts of
ferrite. This was due to the lower hardenability resulting from precipitation of Nb(CN) during rolling and holding at 950°C, together with the accelerating effect of unrecrystallised austenite on the rate of transformation.
With increasing Nb, the amount of pearlite decreased because of the undissolved and the freshly precipitated NbC. Decreasing the holding temperature
to 750°C, within the critical range, gave much slower precipitation of NbC
even in unrecrystallised austenite (13), so that the increased dissolved Nb

increased the hardenability and inhibited ferrite formation, which occurred at the unrecrystallised austenite grain boundaries, Fig.3(d).

Table III - Maximum Potential Amount of Nb Precipitated at Different Temperatures After Reheating

Nb % in Steel	% Nb (Mass) as precipitate						
Reheating Temp. $^{\circ}$C	1000		1150		1300		
Precipitated Temp. $^{\circ}$C	750	950	750	950	750	950	1250
0.07	0.01	0.01	0.04	0.04	0.07	0.07	0.00
0.11	0.01	0.01	0.04	0.04	0.11	0.11	0.03
0.16	0.01	0.01	0.05	0.05	0.12	0.12	0.03
0.46	0.03	0.01	0.08	0.06	0.17	0.17	0.04

Lowering the rolling temperature to 750°C, i.e. within the critical range, caused ferrite to form at the rolling temperature, and this ferrite was heavily deformed, Fig.4(a). Holding the base steel at 750°C allowed the ferrite to recrystallise to extremely fine grains, whilst the deformed but unrecrystallised austenite transformed to coarse ferrite and pearlite, Fig.4(b), the transformation being accelerated by the retained deformation. With increasing Nb, less ferrite was formed during rolling, Fig.4(c), and moreover, this ferrite was more reluctant to recrystallise during holding. During rolling and holding at 750°C, strain induced Nb(CN) precipitation is very slow (15) so that the hardenability was not impaired, resulting in the smaller amount of ferrite. An interesting observation was that during holding at 750°C and subsequent cooling at 400°C/min, multiple nucleation of very fine ferrite grains occurred at and near to the deformed austenite grain boundaries in the Nb steels, Fig.4(d). This will be discussed later, but should be clearly distinguished from the recrystallisation of deformed ferrite.

(b) Reheated at 1150°C There was much less Nb dissolved in the austenite at 1150°C than at 1300°C, Table II. Rolling to 20% or 50% reduction at 950°C, followed by holding at 950°C, produced partial austenite recrystallisation in all the Nb steels. Due to the decreased hardenability resulting from the lower Nb solubility and the finer but mixed austenite grain size produced by the lower reheating temperature, ferrite was produced in all the steels during subsequent cooling. Up to 0.16%Nb, however, there was a mixed ferrite grain even when rolled to 50% reduction, due to the variable austenite grain size produced by partial recrystallisation. Decreasing the rolling reduction to 20%, however, produced a uniform but somewhat coarser grain structure which was possibly because the recrystallised austenite grains in the partially recrystallised structure were somewhat coarser. Decreasing the holding temperature to 750°C gave fully ferrite-pearlite structures up to 0.16%Nb, but there were regions of coarser ferrite grains, Fig. 5(a). These resulted from the mixed austenite grain size prior to rolling which led to some coarser unrecrystallised austenite grains. At higher Nb contents, the large amount of undissolved Nb(CN) caused a more uniform austenite grain structure prior to rolling, and consequently a more uniform ferrite grain size, Fig.5(b). Somewhat coarser ferrite grains resulted from

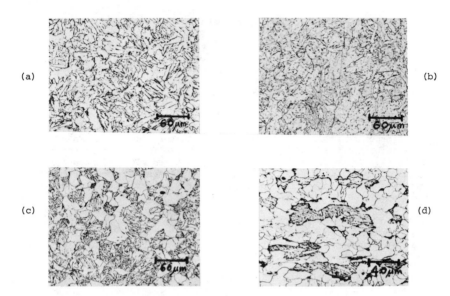

FIG.3   Microstructures of Nb Steels Reheated 1300°C

    (a)   0.11%Nb Rolled 1250°C - Held 950°C - 100s
    (b)   0.46%Nb Rolled 1250°C - Held 950°C - 100s
    (c)   0.16%Nb Rolled 1250°C - Held 750°C - 100s
    (d)   0.11%Nb Rolled 950°C - Held 750°C - 100s

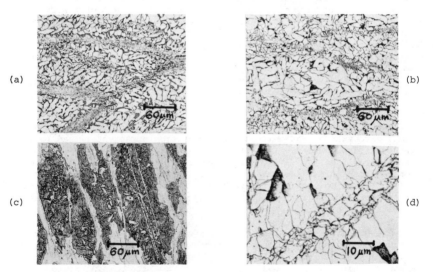

FIG.4   Microstructures of Steels Reheated at 1300°C and
       Rolled at 750°C

(a) 0%Nb Cooled after Rolling   (b)   0%Nb Held 750°C - 100s
(c) 0.16%Nb Cooled after Rolling   (d)   0.16%Nb Held 750°C - 100s

20% reduction during rolling, compared with 50% reduction.

Decreasing the rolling and holding temperature to 750°C, caused ferrite to be formed, and deformed, during rolling. The steels containing less than 0.16%Nb only partially transformed to ferrite during rolling, but the remaining austenite then formed ferrite and pearlite on cooling, Fig.5(c). The deformed ferrite did not recrystallise on holding at 750°C.

(c) Reheated at 1000°C The austenite grain size prior to rolling was very fine with little Nb in solution, Table II. Rolling and holding at 950°C produced a uniform ferrite grain structure in all the Nb steels, and decreasing the holding temperature to 750°C refined the ferrite grain size, Fig.5(d). These fine uniform ferrite structures resulted from the fine uniform initial austenite grain size, and the fact that transformation occurred immediately after rolling. Varying the rolling reduction had little effect.

Decreasing the rolling temperature to 750°C again caused ferrite to form from deformed austenite during rolling, but despite the fact that strain induced precipitation of NbC was slow at this temperature (13), the ferrite did not recrystallise even after holding 1000s at 750°C.

3.2 V-low N Steels

(a) Reheated at 1300°C At 1300°C, all the VC and VN were dissolved in the austenite. Rolling with 20 or 50% reduction at 1250°C produced complete recrystallisation of the austenite, which simply coarsened during holding at 1250°C. Ferrite formed only at the grain boundaries of this coarse austenite during transformation, bainite forming within the austenite grains. Lowering the holding temperature to 950°C allowed less austenite grain coarsening and therefore an increased amount of ferrite formed on cooling. The amount of ferrite increased with increasing rolling reduction at 1250°C due to the finer recrystallised austenite grain size decreasing the hardenability, and because increasing V content refined the austenite grain size. The amount of ferrite also increased with increasing V content. Further decreasing the holding temperature to 750°C allowed ferrite to form at the recrystallised austenite grain boundaries during holding, and again the amount of ferrite increased with increasing V content.

Rolling and holding at 950°C, depending on the reduction and holding time, gave varying austenite morphologies from unrecrystallised to fully recrystallised grains. Immediately after rolling, all the V steels were unrecrystallised, Fig.6(a), and the nucleation of ferrite occurred on cooling both at the austenite grain boundaries and within the grains, Fig. 6(b). It can be seen that ferrite nucleated on deformation bands and at undissolved precipitates and there was copious intragranular nucleation of ferrite, Fig.6(c).

At 950°C there was less ferrite after 20% than after 50% reduction, because the lower reduction did not cause so much elongation of the unrecrystallised austenite grains nor so much strain induced precipitation. Hence the hardenability was higher for the smaller rolling reduction. The higher rolling reduction entirely eliminated bainite formation after holding at 750°C, and all the V steels exhibited mixed ferrite grain structures, Fig.6(d).

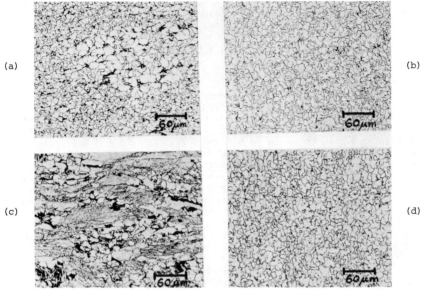

FIG.5 Microstructures of Nb Steels
      (a) 0.16%Nb Reheated 1150°C, Rolled 950°C, Held 750°C - 100s
      (b) 0.46%Nb Reheated 1150°C, Rolled 950°C, Held 750°C - 1000s
      (c) 0.16%Nb Reheated 1150°C, Rolled 750°C, Held 750°C - 100s
      (d) 0.46%Nb Reheated 1000°C, Rolled 950°C, Held 750°C - 100s

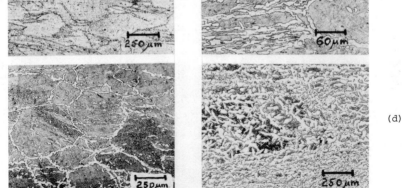

FIG.6 Microstructures of Low N – V Steels
      (a) 0.14%V Reheated 1300°C, Rolled 950°C, Quenched
      (b) 0.14%V Reheated 1300°C, Rolled 950°C, Held 750°C - 100s
      (c) 0.55%V Reheated 1300°C, Rolled 950°C, Quenched
      (d) 0.55%V Reheated 1300°C, Rolled 950°C, Held 750°C - 100s

Rolling at the lowest temperature, 750°C, led to the formation and deformation of ferrite during rolling, but this deformed ferrite recrystallised to very fine ferrite grains during rolling, Figs.7(a) and (b). The deformed austenite also transformed to coarser ferrite and pearlite. Hence mixed ferrite grain structures were observed in all the V steels.

(b) <u>Reheated at 1150°C</u>   At 1150°C all the VC and VN was dissolved. Rolling and holding at 950°C, for both 20% and 50% reduction, produced partially recrystallised austenite, but the extent of recrystallisation increased with increasing holding time and with increasing reduction. After 20% reduction, whilst the structures after transformation in steels containing up to 0.14%V were mainly ferrite-pearlite with some bainite, at 0.55%V there was no bainite. This was due to the greater supersaturation at the highest V content giving more strain induced precipitation which lowered the hardenability. Decreasing the holding temperature to 750°C, gave ferrite-pearlite structures and the 0.6%V steel had a very uniform ferrite grain structure. Higher V contents however, resulted in mixed ferrite grain sizes and increasing the holding time produced more mixed ferrite grain structures, particularly in the coarser unrecrystallised austenite. Holding at 750ºC, and allowing ferrite to form isothermally, so partitioned carbon to the untransformed austenite that the increased hardenability resulted in bainite/martensite structures.

Decreasing the rolling temperature to 750°C, i.e. within the critical range, gave similar structures to those already described for these conditions, and in contrast to the Nb steels, the deformed ferrite recrystallised during or immediately after rolling in the low N-V steels. Mixed ferrite grain sizes were therefore, observed.

(c) <u>Reheated at 1000°C</u>   At 1000°C, the VC was dissolved for all the V steels, but the VN was dissolved only up to ~0.13%V.   Above 0.13%V, increasing amounts of VN remained undissolved with increasing V content. Rolling and holding at 950°C produced similar structures after cooling to those observed for the higher reheating temperature of 1150°C. Small amounts of bainite were observed in the 0.06% and 0.11%V steels. In the steels containing >0.14%V, and rolled to 20% reduction, the undissolved and strain induced VN precipitation decreased the hardenability so that fully ferrite-pearlite structures were observed. However, in the higher V steels the ferrite grain structures were mixed due to the inhibition of recrystallisation. Increasing the rolling reduction to 50% gave more uniform ferrite grain structures because of the greater elongation of the unrecrystallised austenite grains and the finer recrystallised grain size Fig.7(c). Decreasing the holding temperature to 750°C increased the tendency for mixed ferrite grain structures at V contents up to 0.14%, Fig.7(d), due to the mixed morphology of the austenite grains and the different ferrite grain size formed by isothermal transformation at 750°C compared with that formed on cooling.

Rolling at 750°C within the critical range again resulted in the formation and deformation of ferrite during rolling. The austenite itself did not recrystallise at this low rolling temperature, and above 0.14%V the deformed austenite grains were much refined by the undissolved VN. The deformed ferrite quickly recrystallised.

3.3 V - high N Steels

(a) <u>Reheated at 1300°C</u>   In all the V steels the VN and VC were dissolved in the austenite.

Rolling to 50% reduction at 1250°C completely recrystallised the austenite which simply coarsened during holding at 1250°C. Due to high hardenability induced by the relatively coarse recrystallised austenite grains and the V in solution, a limited amount of ferrite formed at the austenite grain boundaries during cooling. Holding at 950°C allowed less austenite grain growth, and the structures formed on cooling were then mainly ferrite-pearlite due to the lower hardenability. This was despite the fact that the increased V should have increased the hardenability, and it is probable that the refinement of the recrystallised austenite grain size with increasing V more than offset the hardenability produced by the higher dissolved V contents. Holding at the still lower temperature of 750°C led to isothermal transformation of the recrystallised austenite, the ferrite forming mainly at austenite grain boundaries. Increasing holding time at 950°C resulted in more ferrite after cooling than did holding at 750°C. More V(CN) precipitated during holding at 950°C than at 750°C (14), which lowered the hardenability and allowed more ferrite to form.

Rolling at 950°C to 50% reduction produced completely unrecrystallised austenite in all the V-high N steels immediately after rolling, but holding at 950°C allowed recrystallisation to occur. Transformation on cooling occurred therefore, in partially recrystallised austenite, and gave rise to a mixed ferrite grain structure. Increasing the holding time at 950°C caused more complete recrystallisation which decreased the extent of the mixed ferrite grain structures. Holding within the critical range at 750°C led to some isothermal transformation to ferrite, but the remaining austenite transformed to ferrite on cooling so that mixed ferrite grain structures were again observed at all V contents.

Rolling within the critical range at 750°C again produced some deformed ferrite during rolling, whilst the remaining deformed austenite transformed to a fine ferrite grain size during subsequent cooling. In contrast to the situation in the V-low N steels, the deformed ferrite in these V-high N steels did not recrystallise during holding at 750°C. These steels were therefore, similar to the Nb steels.

(b) Reheated at 1150°C   At 1150°C, VC was dissolved in all the steels, but the VN was only dissolved for V contents up to 0.28%V, beyond which there was increasing amounts of undissolved VN. Rolling to 50% reduction at 950°C and holding at 950°C resulted in partial austenite recrystallisation in all the steels, the amount of recrystallisation decreasing with increasing V content. Due to the large amount of austenite recrystallisation in the low V steels, a uniform ferrite grain structure was produced after cooling but with increasing V content above 0.14%, the less recrystallisation resulted in more pronounced mixed ferrite grain structures. It was also observed that with increasing holding time at 950°C, the tendency for mixed ferrite grain structures became more pronounced. This appeared to be unusual as the amount of recrystallisation increased and therefore might have been expected to produce a more uniform ferrite structure. The probable explanation of the observation is that the increasing holding time caused more recovery in the unrecrystallised austenite grains, and this resulted in less intra-granular nucleation for ferrite. The result was that the unrecrystallised but recovered austenite, being relatively coarse grained, produced coarser ferrite grains on cooling. Holding at the lower temperature of 750°C increased the tendency for mixed ferrite grain structures in all the steels because of the different transformation temperature of the ferrite formed on holding from that formed from the remaining austenite on cooling. This was particularly the case where the structure prior to holding comprised coarse unrecrystallised austenite grains in a fine recrystallised austenite matrix. Allowing the recrystallised austenite grains to grow, however, produced more uniform, but somewhat coarser ferrite grain sizes.

Rolling at 750°C produced similar effects to those described previously, but again the deformed ferrite did not recrystallise.

(c) Reheated at 1000°C   The steels containing more than 0.05%V had undissolved VN at 1000°C, although the VC had dissolved in all the steels. Due to the undissolved VN particles, and the low reheating temperature, the austenite grain size prior to rolling was uniformly fine. Rolling at 950°C to 50% reduction and holding at that temperature gave a fine uniform ferrite structure in all the steels, but holding at the lower temperature of 750°C resulted in a mixed ferrite grain structure in the lower V steels due to differential transformation at 750°C and during cooling, Fig.8(a). In the higher V steels however, the ferrite grain size was uniform because of the increased intragranular nucleation rate for ferrite at undissolved VN particles, Figs.8(b) and (c).

As already described, rolling at 750°C produced deformed ferrite which did not recrystallise during holding at 750°C. The deformed austenite however, transformed to fine ferrite grains, but the amount of such ferrite decreased with increasing V content of the steel, which increased the hardenability.

3.4 The Isothermal Transformation of Austenite

The effects of Nb and V were investigated on the rate of isothermal transformation at 700°C, using two widely different austenite grain sizes. Variations in austenite grain size were produced by full recrystallisation after 50% rolling reduction at the reheating temperature of 1150°C. The relevant austenite grain sizes are given in Table IV for the four steels investigated. Due to the different potential of the four compositions for grain refinement by recrystallisation, varying grain sizes were observed.

Table IV - Austenite Grain Sizes of Various Steels After Reheating at 1150ºC, and After 50% Reduction at 1150ºC

Steel	Austenite Grain Size, μm	
	At 1150°C for ½h	After 50% reduction at 1150ºC
Base	90	25
0.07% Nb	63	12
0.14% V	97	18
0.14% V-N	85	15

For each steel, the deformation and recrystallisation was carried out at the same temperature as the reheating temperature, and because these temperatures were above those at which strain induced precipitation of the micro-alloy carbon-nitride occurred, the amount of dissolved Nb or V was constant for a particular steel and did not vary with the austenite grain size. Hence the effects of austenite grain size could be studied independently and the effect of alloying addition could be examined independently

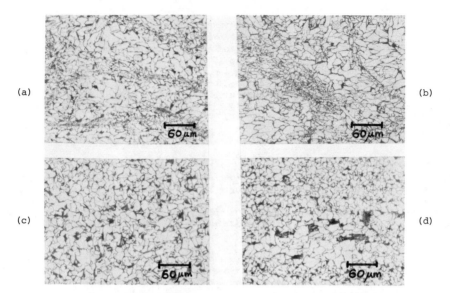

FIG.7   Microstructures of Low N —V Steels
         (a) 0.11%V Reheated 1300°C, Rolled 750°C Held 750°C - 100s
         (b) 0.22%V Reheated 1300°C, Rolled 750°C Held 750°C - 100s
         (c) 0.14%V Reheated 1000°C, Rolled 950°C Held 950°C - 1000s
         (d) 0.14%V Reheated 1000°C, Rolled 950°C Held 750°C - 1000s

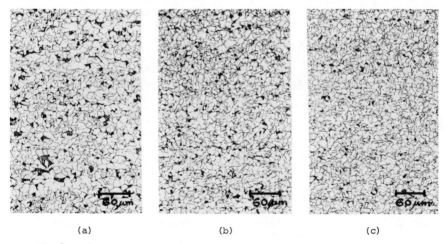

         (a)   (b)   (c)

FIG.8   Microstructures of High N —V Steels

         (a) 0.14%V Reheated 1000°C, Rolled 950°C, Held 750°C - 100s
         (b) 0.25%V Reheated 1000°C, Rolled 950°C, Held 750°C - 100s
         (c) 0.46%V Reheated 1000°C, Rolled 950°C, Held 750°C - 100s

389

of austenite grain size. Fig.9 shows that increasing the austenite grain size considerably retarded the transformation to ferrite, as expected. By interpolation, the results were corrected to a constant austenite grain size of 40μm, and the curves shown in Fig.10 depict the affect of the different microalloying additions. It can be seen that 0.14%V retards ferrite formation, and this effect is accentuated by the presence of an enhanced nitrogen content. This is in agreement with recent work on the effect of nitrogen on the hardenability of V steels (21). The addition of 0.07%Nb however, is more effective than that of 0.14%V in retarding ferrite formation and increasing the hardenability.

4. Discussion

The transformation of thermo-mechanically processed austenite to ferrite is complex, and is determined by:-

(i) the composition of the austenite, which controls the hardenability of the steel and thus the transformation temperature for a given cooling rate.

(ii) the micro-alloying addition, and the temperature dependence of the solubility of the micro-alloy carbide/nitride which will determine the solute content and the amount of undissolved carbides/ nitrides.

(iii) the reheating temperature, which will determine the dissolved solute concentration and the austenite grain size prior to thermomechanical processing.

(iv) the morphology of the austenite after thermo-mechanical processing which will control the ferrite nucleation rate through grain boundary surface area, serrations on grain boundaries, deformation bands, recovered sub-structures, etc.

(v) the presence of undissolved or strain induced micro-alloy carbide/ nitride precipitates which will decrease the solute concentration and also act as nuclei for ferrite formation. The occurrence, size, distribution and morphology of these particles will not only be controlled by the thermo-mechanical processing parameters, but will also greatly influence austenite recovery and recrystallisation during and after such processing, and the grain size of the austenite.

(vi) the rate of cooling after thermo-mechanical processing, which for a constant microstructure and composition of the austenite, will greatly affect the transformation temperature and thus the nucleation and growth of the ferrite, i.e. the ferrite grain size and morphology.

Many of the complex interactions between these various effects have been illustrated by the present results.

4.1 The Effect of Nb

Nb promotes bainite formation, and the effect of 0.07%Nb in decreasing the rate of transformation to ferrite, compared with 0.14%V, is seen in Fig.10. On an atomic basis, Nb has a considerably more powerful effect on hardenability than does V (12,22,23). In this respect Nb has often been compared with Mo, (22), and even with B (23,24). It is possible that Nb,

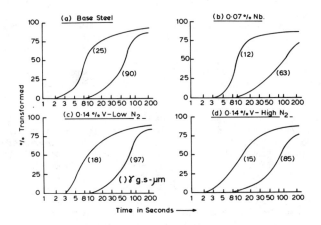

Figure 9 Effect of austenite grain size on transformation
kinetics of different microalloyed steels at 700°C.

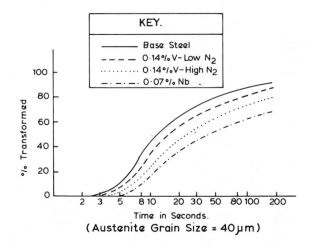

Figure 10 Effects of microalloying additions
on transformation kinetics at 700°C.

being the larger atom,segregates more positively to austenite grain boundaries than does V, and decreases the diffusion rate and activation energy of carbon diffusion in austenite more than does V because of its stronger affinity for carbon. These effects would all decrease the nucleation rate for ferrite as observed in the present work and that of others (25).

There is clear evidence that undissolved NbC nucleates for ferrite, Fig.4(c), but it is possible that there is a difference between undissolved NbC and strain induced NbC precipitates from the point of view of the very different particle sizes and the different Nb diffusion gradients around the precipitates. Because precipitation of NbC even from unrecrystallised austenite is relatively slow, (13,26), this effect has little influence on hardenability.

In order to avoid bainite, which has detrimental effects on the properties, particularly toughness,a low dissolved Nb content is required. This can be produced by a low reheating temperature which will also decrease the hardenability by virtue of the smaller austenite grain size. If strain induced NbC precipitation is encouraged by rolling in the range 900/950°C, this will also lower the hardenability slightly but at the expense of precipitation strengthening in the transformed ferrite (13). Heavy deformation below the recrystallisation temperature will likewise promote ferrite formation and increase the nucleation rate by virtue of intra-granular nucleation; this will refine the ferrite grain size. Some Nb dissolved in the austenite is clearly beneficial as this will refine the ferrite grain size by lowering the transformation temperature by virtue of its hardenability effect.   But bainite should not be formed, and for this reason the increased ferrite nucleation rate brought about by thermo-mechanical processing is important.   The optimum dissolved Nb content which can be tolerated increases with the decreasing austenite grain size and increasing ferrite nucleation rate brought about by thermo-mechanical processing. This dissolved Nb can also decrease the growth rate of ferrite by solute drag effects and by decreasing the diffusion rate of carbon.   It is suggested however, that the effect of dissolved Nb on the hardenability is less important in unrecrystallised austenite than in recrystallised austenite because in the former, the diffusivity of carbon is greater, the transformation temperature is higher and the intergranular nucleation of ferrite is more pronounced.

Nb was also effective in inhibiting the recrystallisation of the deformed ferrite produced by thermo-mechanical processing within the critical range.

4.2 The Effect of V

Dissolved V increased the hardenability of austenite, as found by other workers (27); the effect was not so pronounced as that of Nb. This is confirmed by Fig.10. Increasing the nitrogen content increased the hardenability by retarding ferrite formation, Fig.10, and this is in agreement with other work (21,30).   The effect does not however, increase the hardenability quite up to that produced by Nb. Increasing V, by refining the recrystallised austenite grain size, and thereby offsetting the effect of V on the hardenability, increased the amount of ferrite formed during transformation at the cooling rate employed. This may be due to V increasing the A$_3$ temperature by about 104°C per 1%V (31). V therefore, has the interesting effect of increasing the hardenability and thereby decreasing the transformation temperature for a given cooling rate, but at the same time raising the A$_3$ temperature. However, the presence of undissolved or precipitated VC or VN,particularly during low temperature rolling, can lower the hardenability by grain refining the austenite, by providing nuclei for ferrite, and by decreasing both the amounts of V and C dissolved in the austenite. These complex interactions produced variable effects on the ferrite grain size, but for advantage to be taken of the hardenability effect of V in producing finer ferrite grain sizes, a fairly high reheating temperature is required to dissolve both VC and VN. Because V is a weaker carbide and nitride former than Nb, it is less sensitive in relation to the hardenability effect observed after thermo-mechanical treatments. In the present work, although nitrogen increased the hardenability in the V steels, its effect in refining the austenite grain size was such that it had little overall effect on the transformation of austenite after thermo-mechanical treatment.

The effect of V was much less pronounced than that of Nb in inhibiting the recrystallisation of ferrite formed during deformation within the critical range, this ferrite being readily recrystallised. However, increasing the nitrogen content in the V steels completely prevented the recrystallisation of the ferrite, and the V-high N steels behaved very similarly to the Nb steels. It is often observed (13,14) that in terms of precipitation and recrystallisation behaviour, V-high N steels are not dissimilar from Nb steels.

4.3 The Relationship Between Austenite and Ferrite Grain Sizes

The relationship between the austenite and ferrite grain size for fully recrystallised austenite is shown in Fig.11. There was a general trend for decreasing austenite grain size to be associated with a refined ferrite grain size. This effect has been observed by Kozasu (5), whose data falls within the scatter band in Fig.11. There was also a trend for the V-high N steels to have finer ferrite grain sizes than either the V-low N steels or the Nb steels. A similar grain size relationship is shown for incompletely recrystallised austenite, Fig.12, and again this has been observed by Kozasu (5) whose data lies towards the top of the scatter band in Fig.12. In this case there was no systematic trend of ferrite grain size with the type of micro-alloying addition, possibly because of the effect of intragranular nucleation of ferrite. A comparison of the mean curves for recrystallised and unrecrystallised austenite is given in Fig.13, the effective austenite grain size taking account of grain elongation but not of grain boundary serrations or deformation bands. It can be seen that the incompletely recrystallised austenite produced a consistently finer ferrite grain size, and this is interpreted as being the result of an additional ferrite nucleation on:-

    (a) austenite grain boundary serrations
    (b) deformation bands
    (c) particles of micro-alloy carbide/nitride
and (d) recovered dislocation sub-grains possibly aided by
        precipitates occurring on the sub-grain boundaries.

The austenite grain size and morphology do not uniquely determine the ferrite grain size after transformation because it will critically depend on the transformation temperature, which will be controlled not only by the type of micro-alloy addition, and the dissolved solute concentration, but also by the Mn content of the steel and particularly the cooling rate used.

4.4 The Mechanism of Ferrite Formation

The observed sites for ferrite nucleation were in order of decreasing significance:-

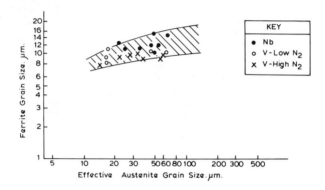

Figure 11   Relationship between austenite and
ferrite grain sizes for fully
recrystallised austenite.

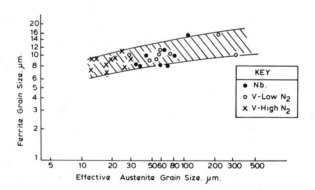

Figure 12   Relationship between austenite and
ferrite grain sizes for partially and
unrecrystallised austenite.

394

i) Austenite grain boundaries: These were the predominant nucleation
   sites, grain edges being preferred. Ferrite nucleation appeared to
   occur regularly along austenite grain boundaries, probably where
   local misorientations give rise to poor atomic matching and therefore,
   higher energy sites. More frequent ferrite nucleation occurred on
   deformed austenite grain boundaries, but it has been reported that
   this effect does not occur at relatively small rolling reductions in
   Nb steels (6). Deformed austenite grain boundaries thus appear to
   exhibit greater lattice mismatch, possibly because of greater lattice
   distortion and a higher dislocation density near deformed grain boundaries. Roberts (3) has shown that deformation leads to the formation
   of grain boundary serrations or bulges, and suggests that these act
   as nuclei in addition to increasing the total grain boundary surface
   area. In the present investigation, however, grain boundary bulging
   was only observed after 20% reduction or after holding following 50%
   reduction, when strain induced boundary migration would be expected
   to cause local bulging. In many cases, bulging was not observed and
   the increased rate of grain boundary nucleation of ferrite must be
   due to some other cause, i.e. local lattice mismatch.

ii) Deformation bands: Copious intra-granular nucleation of ferrite occurred in unrecrystallised austenite, Fig.6(c), and many examples of
   ferrite nucleation on deformation bands were observed, e.g. Figs.4(c)
   and 6(d). It has been shown (5) that deformation bands are potent
   ferrite nuclei, as also indicated by Fig.13. However, not all deformation bands are effective ferrite nuclei, possibly due to variable
   energies associated with them. Higher energy bands and consequent
   ferrite nucleation was usually associated with the larger rolling reductions. It has been suggested that a delay between the formation of
   the deformation band and subsequent transformation gives less ferrite
   nucleation on the band (32) whilst if transformation occurs during
   deformation, ferrite forms on the deformation bands in preference to
   other intra-granular sites. On the other hand slow cooling after deformation inhibited nucleation at deformation bands compared with
   other sites (32). In the present work, fast cooling of deformed austenite caused ferrite nucleation on deformation bands even with low
   rolling reductions, although the effect varied from grain to grain,
   possibly due to variations in the stored energy. It is suggested
   that if recovery of the dislocation structure is allowed to occur
   within the deformation band, the potential for ferrite nucleation is
   greatly decreased. A consistent observation was that holding at 750°C,
   which may be regarded as representing slow cooling, did not give intragranular ferrite in deformed austenite. It is suggested that this
   was because the larger driving force needed to activate nuclei in the
   recovered austenite was not available.

iii) Second phase particles: Nucleation of ferrite on undissolved carbide/
   nitrides has often been suggested, and the present work supports this
   mechanism, Figs.4(c) and 6(b), as does other recent work (7). It is
   suggested that ferrite nucleation may also occur on precipitated car-
   bides/nitrides, although these probably behave differently from undissolved particles, as previously discussed.

iv) Sub-grain boundaries: It is difficult to obtain conclusive microsc-
   opical evidence for ferrite nucleation on sub-grain boundaries produced by recovery, but the copious intra-granular ferrite nucleation
   in the absence of second phase particles or deformation bands, Fig.6(c)

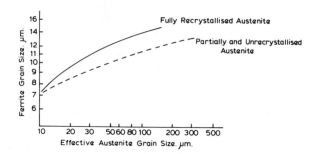

Figure 13   Effect of state of austenite recrystallisation
on relationship between austenite and ferrite
grain sizes.

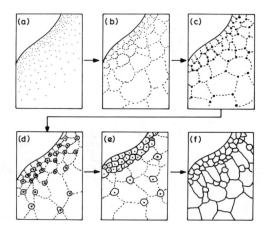

Figure 14   Schematic representation of multiple
nucleation of ferrite at deformed austenite
grain boundaries.

may be interpreted in terms of this effect. It has been postulated that the deformation sub-structure should be unrecovered (32) as recovery minimises the ferrite nucleation potential. Some confirmation for this is provided in the present work by the observation that, with small deformations,holding at 950°C for 1000s allowed the sub-structure to recover and mixed ferrite grain structures to be formed because of the absence of intra-granular nucleation. A shorter holding time of 100s, allowed less recovery, gave more intra-granular nucleation, and resulted in a uniform ferrite grain size. This effect was less in evidence in Nb steels than in V steels, possibly due to the Nb or NbC retarding recovery, and thus maintaining a high sub-boundary nucleation potential. However, multiple nucleation of ferrite at and near to deformed prior austenite boundaries, sometimes called cascade nucleation (33), fig.4(d), may have resulted from a type of sub-structure nucleation. The mechanism suggested is schematically illustrated in Fig.14. During deformation and recovery, a sub-grain boundary network is developed, which is finer close to the boundary because of the strain gradient developed, Figs.14(a) and (b). Precipitates then form on this network, Fig.14(c), and serve as ferrite nucleation sites, Fig.14(d). The ferrite grains thus formed impinge to give a zone of very fine ferrite, Fig.14(e), whilst the grains at the edge of this band grow into the austenite grain, until halted by impingement with other growing ferrite grains nucleated at the sub-structure or precipitates within the austenite grain, Fig.14(f). These ferrite grains are often elongated, as it is unlikely that sub-grain boundaries or their associated precipitates can pin migrating α/γ boundaries (34).

(v) Twin boundaries: Due to their low energy, the coherent twin boundaries are unlikely to act as ferrite nucleation sites. Ferrite was regularly observed at the austenite twin boundaries, however, probably nucleated at the higher energy non-coherent ledges.

4.5 The Formation of Mixed Ferrite Grain Sizes

A mixed ferrite grain structure is reported to be deleterious to the toughness of HSLA steels. From the work which has been described, the following factors may be responsible for mixed ferrite grain sizes:-

(i) A variable and mixed austenite grain structure prior to rolling, which results in unrecrystallised or partially recrystallised austenite of varying grain size after rolling, tends to produce a variable ferrite grain size. Because fine grains recrystallise preferentially to coarse ones, the coarser grains resulting from a variable initial grain size remain unrecrystallised. This situation can be perpetuated and exaggerated in sequential rolling passes. Depending on the size of the recrystallised grains, the size and elongation of the unrecrystallised grains, the hardenability of the austenite and the rate of cooling, either type of grain may form the coarser ferrite grains on transformation. If, however, the local rates of ferrite nucleation are similar in both the recrystallised and the unrecrystallised austenite, a uniform ferrite grain size can be produced.

(ii) Even in a fully recrystallised austenite, if the grain size is very variable, a mixed ferrite grain structure will be formed. Variable fully recrystallised austenite grain sizes may result from variable initial austenite grain sizes prior to rolling, from heterogeneity of composition and particularly carbide/nitride segregation, from inhomogeneous deformation during rolling and from a variable temperature distribution, Fig.6(c).

(iii) Multiple nucleation of ferrite, Fig.4(d), which results in a region
      of very fine ferrite grains at deformed austenite grain boundaries,
      and coarse ferrite in the interior of the austenite grains. This
      type of structure has already been discussed, but it should not be
      confused with the morphologically similar structure resulting from
      deformation and holding within the critical range in V-low N steels.

(v)   Rolling and holding within the critical range, which causes ferrite
      to precipitate at the austenite grain boundaries. This ferrite is
      deformed, and with holding can recrystallise to a very fine grain
      size, particularly in V-low N steels. The remaining coarser un-
      recrystallised austenite then transforms to coarse ferrite, Figs.4(b)
      and 7(b). The Nb and V-high N steels show no recrystallisation of
      the ferrite (35) during 1000s at 750°C.

     An interesting feature which has been observed in partially recryst-
allised austenite is greater heterogenity of ferrite grain structure when
the recrystallised austenite grains were very fine. In the absence of
grain growth inhibition, which allows the recrystallised austenite grains
to grow, the subsequent ferrite grain structure will be more uniform, al-
beit rather coarser.

     It has often been suggested that to maximise ferrite grain uniformity,
the rolling practice should be controlled to produce either very fine recrystallised austenite, or heavily elongated unrecrystallised austenite.
These methods both require large rolling reduction and close control over
the rolling parameters. Other possible means of alleviating the problem
of mixed ferrite grain structures may be considered. What is required is
to increase the nucleation rate for ferrite, by activating more nucleation
sites. This is done in practice by increasing the rolling deformation, and
in the present work it was always observed that increasing the rolling reduction decreased the occurrence of mixed ferrite grain structures by activating more intra-granular nuclei. Other steps, associated with influencing the defect structure in the austenite and the rate of ferrite nucleation by control of the transformation temperature, may also be considered.
In order to allow more intra-granular ferrite nucleation, it has been suggested that recovery of the unrecrystallised austenite should be prevented
(32). This can be done by minimising holding periods between passes at
temperatures where recovery is rapid, and by cooling as rapidly as possible
after the finish of rolling in structures where there is unrecrystallised
austenite. Moreover, holding within the critical range should be controlled so as to produce little or no transformation or substantially complete
transformation to ferrite. Partial transformation to ferrite invariably
leads to mixed ferrite grain sizes, and the crucial objective is to prevent ferrite forming in two very different temperature ranges.

     Accelerated cooling through the transformation range can be benefic-
ial providing no bainite is formed, because it lowers the transformation
temperature and thereby refines the ferrite grain size. Also, decreasing
the transformation temperature increases the driving force for ferrite
formation, thereby activating more intra-granular nuclei. This can be
beneficial where the coarse unrecrystallised austenite would tend, at
slower cooling rates, to transform to coarse ferrite grains.

     Finally, recent work has shown just how important strain and temper-
ature heterogeneity can be in producing variable and mixed austenite grain
sizes and morphologies (36), with a consequent mixed ferrite grain structure. Such heterogeneity will tend to be more pronounced in thicker sect-

ion sizes and can best be overcome by attention to the mechanics of the rolling process.

5. Summary and Conclusions

Some effects of the complex interactions between dissolved solute concentration, austenite grain size prior to rolling, undissolved and strain induced precipitates, and defect structure in deformed austenite grains have been investigated in terms of the morphology of the thermo-mechanically processed austenite and its effect on ferrite formation during the γ/α transformation in Nb and V steels. It has been shown that:-

(i) Nb increased the hardenability more than did V in terms of atomic concentration. Increased N enhanced the hardenability in V steels.

(ii) Even at 1250°C, Nb in solution retarded austenite recrystallisation, whilst NbC precipitates had a similar influence at 950°C. Decreasing the dissolved Nb by lowering the reheating temperature reduced the hardenability, and gave more but coarser ferrite. Undissolved NbC increased the amount of austenite recrystallisation.

(iii) Dissolved V was less effective than dissolved Nb in retarding recrystallisation of austenite. Dissolved N also seemed to have little effect, but strain induced VN precipitation did retard austenite recrystallisation at lower rolling temperatures, the effect being increased by N. Any undissolved or freshly precipitated VC or VN lowered the hardenability, producing more and coarser ferrite.

(iv) During rolling within the critical range ferrite occurred and was deformed. On holding, this deformed ferrite recrystallised to very fine ferrite grains in the V-low N steels, but did not recrystallise in the Nb and V-high N steels.

(v) In general, unrecrystallised austenite produced more ferrite, but the grain size of this ferrite depended upon the hardenability of the austenite, and on the amount of recovery which had occurred in the austenite.

(vi) There was a definite relationship between the austenite grain size and the ferrite grain size formed from it. Unrecrystallised austenite produced a much refined ferrite grain size, due to intragranular ferrite nucleation.

(vii) It has been shown that ferrite nucleates at the boundaries of recrystallised or unrecrystallised austenite, at undissolved micro-alloy carbides/nitrides, at deformation bands and sub-grain boundaries in unrecrystallised austenite. A mechanism for the formation of multiple ferrite nucleation at deformed austenite grain boundaries has been suggested.

(viii) More intra-granular ferrite nucleation, with a resulting finer and more uniform ferrite grain size, occurs if the degree of recovery is restricted in the deformed austenite and if the cooling rate is faster, so giving a lower transformation temperature.

(ix) The conditions for the formation of mixed ferrite grain structures have been identified as a heterogeneous initial austenite grain size, a heterogeneous recrystallised austenite grain size, partial recrystallisation either before subsequent deformation or before transformation, holding to allow partial transformation before producing further transformation on cooling, multiple ferrite nucleation at deformed austenite grain boundaries, and rolling and holding within the critical range to allow ferrite to be deformed and recrystallised.

(x) Suggestions have been made, based on the results reported, for producing uniform ferrite grain structures by heavy rolling reductions, ensuring homogeneity of rolling strain and temperature, activating more intra-granular ferrite nuclei in unrecrystallised austenite by decreasing holding and recovery effects, and by the increasing cooling rate. Finally it is essential that ferrite formation does not occur over a wide range of temperature.

Acknowledgements

Acknowledgement is made to Union Carbide Corporation for providing a bursary to enable this work to be carried out, and to Mr M Korchynsky for many stimulating discussions. Our thanks are due to Dr A W D Hills, Head of Department of Metallurgy, Sheffield City Polytechnic, for providing facilities to carry out the work.

References

1. M Fakuda, T Hashimoto and K Kunishige, "An Investigation on Controlled Rolling of Low Carbon Killed Steels", Sumitomo Search, 9 (1973) pp 8-23.

2. A Brownrigg and R Boelen, "The Effect of Nb on Transformation and Strength of Hot Rolled Medium Carbon Steels", Paper V15 in Phase Transformations, The Institution of Metallurgists, London, 1979.

3. W Roberts, H Lidefelt and A Sandberg, "Mechanism of Enhanced Ferrite Nucleation from Deformed Austenite in Micro-alloyed Steels", pp.38-42 in Hot Working and Forming Processes, The Metals Society, London, 1979.

4. C Ouchi, T Sampei, T Okita and I Kozasu, "Microstructural Changes of Austenite during Hot Rolling and their Effects on Transformation Kinetics", pp.316-340 in The Hot Deformation of Austenite, John B Ballance, ed.; A.I.M.E., New York, 1977.

5. I Kozasu, C Ouchi, T Sampei and T Okita, "Hot Rolling as a High Temperature Thermo-Mechanical Process", pp.120-135 in Micro-alloying '75, Union Carbide Corporation, New York, 1977.

6. R Priestner, "The Origin of Fine-grained Ferrite in Steels Rolled Under Controlled Conditions". Rev de Met. 72 (1975) pp 285-292.

7. D J Walker and R W K Honeycombe, "Effect of Deformation on the Decomposition of Austenite; Part 1 - The Ferrite Reaction" Metal Science Journal, 12 (10) (1978) pp.445-452.

8. I L Dillamore, R F Dewsnap and M G Frost, "Metallurgical Aspects of Steel Rolling Technology", Report No CDL/MT/2, 1974, British Steel Corporation.

9.   J M Chilton and M J Roberts, "Microalloying Effects in Hot Rolled
     Low Carbon Steels Finished at High Temperatures", Met. Trans. A,
     11A (1980), pp.1711-1721.

10. R K Amin and F B Pickering, Unpublished Work 1980.

11. D Webster and J H Woodhead, "Effect of 0.03%Nb on the Ferrite Grain
     Size of Mild Steel", J.I.S.I. 202 (1964) pp.987-994.

12. J M Gray, "Effect of Niobium on Transformation and Precipitation
     Processes in High Strength Low Alloy Steel", pp.19-28 in Heat Treat-
     ment '73, The Metals Society, London, 1973.

13. R K Amin, G Butterworth and F B Pickering, "Effects of Rolling Var-
     iables and Stoichiometry on Strain Induced Precipitation of Nb(CN)
     in C-Mn-Nb Steels", pp.27-31 in Hot Working and Forming Processes,
     The Metals Society, London, 1979.

14. R K Amin, M Korchynsky and F B Pickering, "Effect of Rolling Variabl-
     es on Precipitation Strengthening in High Strength Low Alloy Steels
     Containing Vanadium and Nitrogen", Metals Technology, 8 (7) (1980)
     pp 250-261.

15. J D Jones and A B Rothwell, "Controlled Rolling of Low Carbon, Niob-
     ium-treated Mild Steels", pp.78-82 in Deformation under Hot-Working
     Conditions, Report No.108, The Iron and Steel Institute, London,
     1967.

16. J J Irani, D Burton, J D Jones and A B Rothwell, "Beneficial Effects
     of Controlled Rolling in the Processing of Structural Steels", pp.
     110-122 in Strong, Tough Structural Steels, Report No 104, The Iron
     and Steel Institute, London, 1967.

17. I Kozasu, T Shimizu and K Tsukada, "Further Observations on Micro-
     structural Changes of Structural Steels". Trans I S I Japan, 12
     (1972) pp 305-313.

18. M Fakuda, T Hashimoto and K Kunishige, "Effects of Controlled Roll-
     ing and Microalloying on Properties of Strips and Plates", pp.136-
     150 in Microalloying '75, Union Carbide Corporation, New York, 1977.

19. K J Irvine, F B Pickering and T Gladman, "Grain Refined C-Mn Steels",
     J I S I, 205 (1967) pp 161-182.

20. K Bunghardt, K Kind and W Oelsen, "Solubility of Vanadium Carbide
     in Austenite" Archiv Eisen. 27 (1956) pp 61-66.

21. A D Vassiliou and F B Pickering, Unpublished Work, 1981.

22. L Meyer, F Heisterkamp and W Mueschenborn, "Columbium, Titanium and
     Vanadium in Normalised, Thermo-mechanically Treated and Cold Worked
     Steels", pp 153-167 in Micro-alloying '75, Union Carbide Corporation,
     New York, 1977.

23. G L Fisher and R H Geils, "The Effect of Columbium on the Alpha-Gamma
     Transformation in a Low Alloy Ni-Cu Steel". Trans AIME 245 (1969)
     pp 2405-2412.

24. J S Kirkaldy, Discussion in Hardenability Concepts with Applications to Steels, D V Doane and J S Kirkaldy eds, AIME (1978), p 414.

25. P Maitrepierre. Ibid, p 414.

26. G Begin and R Simoneau, "The Difficulties in Attempting to Quantify the Precipitation of Microalloying Elements", pp 85-87 in Microalloying '75, Union Carbide Corporation, New York, 1977.

27. M Durbin and P R Krahe, "Controlled Rolling and the Properties of Very Low Carbon, High Manganese Steels Containing Strong Carbide Formers", pp 109-132 in Processing and Properties of Low Carbon Steels, J M Gray ed. AIME, New York, 1973.

28. F de Kazinsky, A Axnos and P Pachleitner, "Some Properties of Niobium Treated Mild Steels", Jernkontorets Annalar, 147 (1963) pp.408 et seq.

29. R A Grange, "Estimating the Hardenability of Carbon Steels", Met Trans A, 4A (1973) pp 2231-2244.

30. W C Leslie, "Nitrogen in Ferritic Steels - A Critical Survey of the Literature" American Iron and Steel Institute, (1959), pp 53 et seq.

31. K W Andrews, "Empirical Formulae for the Calculation of some Transformation Temperatures" JISI 203 (1965) pp 721-727.

32. R Priestner and E de los Rios, "Ferrite Grain Refinement by Controlled Rolling of Low Carbon Microalloyed Steel" Metals Technology 7 (1980) pp 309-316.

33. A Le Bon, J Rofes-Vernis and C Rossard, "Strain Induced Precipitation, Austenite Recrystallisation and Transformation", paper No.6 presented at British Steel Corporation Conference Controlled Processing of H S L A Steel, York, 1976.

34. W Roberts, Discussion in Hot Working and Forming Processes, The Metals Society, London, (1979) p.45.

35. R E Hook and H Nyo, "Recrystallisation of Deep Drawing Columbium (Nb) Treated Interstitial-Free Sheet Steels" Met. Trans A, 6A (1975) pp 1443-1451.

36. Private communication, Godwin O Ekebuisi, Sheffield City Polytechnic, Aug. 1981.

DISCUSSION

Comment: The relationship between the austenite grain size and
the resulting ferrite grain size was discussed at length yesterday.
(Lagne borg) showed the ratio of austenite to ferrite grain sizes to be
somewhere between two and three, whereas in the discussion ratio values
as high as 8 and 12 were reported. I believe Figure 13 shows that both
points of view are correct. For austenite grain size between 80 - 100
microns, the resulting ratio, or refinement of ferrite, is close to 10,
while for finer grain size of 20 - 30 microns the ratio is between 2
and 3. A second point is the commonly held belief that for optimum
ferrite grain refinement, the austenite must be unrecrystallized. Again,
the data in Figure 13 show that if the austenite grain size is small
enough, say 20 - 30 microns, it doesn't matter very much whether transformation takes place from recrystallized or nonrecrystallized austenite, and
in both instances very fine ferritic grain size can be achieved.

Comment: In discussing the austenite-to-ferrite grain size ratio we
must remember that in the case of Figure 13 the rolling reduction was
fixed at 50 percent. What can then be deduced from Figure 13 is that this
amount of pancaking doesn't have much effect when the austenite grains
are small. However, additional pancaking would still be expected to
produce additional grain refinement.

Comment: Regarding the difference between transformation from
pancaked or polygonal recrystallized austenite, a major point is that the
pancaked austenite grains are stable for thousands of seconds whereas the
fine recrystallized grains may grow before transformation begins.

Comment: This conclusion may not be valid, because a pancake
structure may recrystallize just as fine grains may coarsen.

Comment: When the austenite is in a recrystallized condition, the
addition of niobium lowers the transformation temperature, but if the
austenite is not recrystallized, niobium raises the transformation temperature.

THE INFLUENCE OF THERMOMECHANICAL TREATMENT ON THE CONTINUOUS-

COOLING TRANSFORMATION OF AUSTENITE IN MICROALLOYED STEELS

Alf Sandberg and William Roberts

Swedish Institute for Metals Research

Drottning Kristinas väg 48, 114 28 STOCKHOLM, Sweden

Kinetic data pertaining to the $\gamma \rightarrow \alpha$ transformation, from both elongated and equiaxed austenite grains during continuous cooling, is presented for a number of microalloyed steels. The nature of the various nucleation sites for polygonal ferrite during the $\gamma \rightarrow \alpha$ transition is discussed and the difference in behaviour when transformation proceeds from deformed as opposed to recrystallized austenite grains rationalized. In particular, it is demonstrated that ferrite nucleation takes place over the entire surface of elongated austenite grains (controlled rolling) whereas grain corners are the principal nucleation sites when transformation proceeds from equiaxed austenite (normal rolling, normalizing). The standpoints presented in Cahn's classical approach to the kinetics of grain-boundary reactions are applied to the situation of ferrite transforming from thermomechanically - processed austenite; some degree of prediction of the experimentally-observed behaviour is possible.

Introduction

The deformation of microalloyed austenite below its recrystallization temperature is conducive to the development of
fine ferrite grains. Thus, for a given steel composition and fixed
cooling conditions, the grain refinement derived from controlled
rolling is usually at least as effective as that accruing from a
normalizing treatment. The density of potential nucleation sites
for ferrite is proliferated by controlled rolling. This derives,
on the one hand, from austenite grain refinement as a result of
recrystallization during predeformation, and, on the other, from
grain elongation and transition band generation in association
with finishing. For a given steel, the effective ferrite nucleation
area/unit volume (S_v, which includes contributions from both grain
boundaries and transition bands) increases as the temperature at
the termination of predeformation is lowered (smaller γ grains
prior to finishing) and as the total reduction during finishing
(ψ) is increased.

In early work, the enhanced ferrite grain refinement, derived
from transformation of deformed as opposed to recrystallized
austenite grains, was ascribed in part to the increase in S_v during
finishing, and in part to the spatial limitations on ferrite growth
imposed by the elongated austenite grains. However, three pieces
of evidence indicate that the above rationale can only account for
a fraction of the observed effect, and that an additional factor
must be involved:

(i) for a given S_v, the ferrite grain size resulting from reaction
of equiaxed austenite is larger than when the grains of the
latter are deformed (1);

(ii) the geometrical limitation on the growth of ferrite, imposed
by the elongation of the austenite grains, is not supported
by direct metallographic study of partly-reacted specimens;
and

(iii) for a given degree of undercooling, the nucleation frequency
of ferrite per unit area of austenite grain boundary is enhanced as a result of deformation below the recrystallization
temperature, the rate of nucleation increasing with the total
degree of reduction during finishing (2).

In this paper, the continuous-cooling transformation to
ferrite from elongated and equiaxed austenite grains is compared
for a number of microalloyed steels. It will be demonstrated that
the transformations from the two types of austenite are characterized by quite dissimilar CCT-kinetics. These variations are
rationalized in terms of (i) the different types of nucleation site
which dominate during the ferrite reaction, and (ii) spatial
effects engendered as a result of the different grain shapes.

Experimental Procedure

The results reported in this article refer to four Al-treated HSLA steels, three microalloyed with Nb and one containing vanadium. The compositions are given below.

Table 1.Compositions of steels

Designation	C	Mn	Si	S	P	Al	Nb	V	N
Nb(low Mn)	0,11	1,00	0,27	0,017	0,005	0,016	0,024	–	0,009
Nb(high Mn)	0,11	1,45	0,27	0,017	0,006	0,018	0,024	–	0,009
Nb(PR)[x]	0,08	1,50	0,27	0,019	0,006	0,022	0,024	–	0,009
V	0,12	1,34	0,30	0,008	0,003	0,018	–	0,091	0,013

x) PR = pearlite-reduced

Rolling simulations were performed via uniaxial hot compression ($\dot{\varepsilon}=2s^{-1}$) of cylindrical specimens; naturally, the deformation schemes adopted were considerably simplified compared to large-scale rolling. Austenitization was for 200s at 1230°C (1150°C for the V steel), which is sufficient to dissolve Nb/V carbonitrides for the relevant microalloying levels. The controlled rolling simulation (CR) comprised a single "predeformation" (25%) followed by a series of "finishing" steps effected between 900 and 800°C. The grain size prior to finishing was varied by altering the predeformation temperature; during finishing, the deformation per step (10-25%) and the number of steps were varied according to the required ψ. Simulation of normal rolling (NR) was accomplished by a series of four deformation steps (17-25%) effected at regular intervals between 1120 and 950°C (finishing temperature). Cooling both during and subsequent to hot compression was controlled in such a way as to simulate the behaviour of 40 mm plate, 12 mm plate or 10 mm ϕ bar in air (average cooling rate (750→550°C) = 0,25, 0,9 and 5°C.s^{-1} respectively). The austenite microstructure at various stages of hot working could be examined, or the progress of the $\gamma \rightarrow \alpha$ transformation monitored by "quenching out" specimens at any preselected point in the thermomechanical schedule.

The transformation behaviour of hot-worked material has been compared with that after normalizing (N). The latter was effected at 920°C/15 min. Cooling rates were identical to those adopted for the hot-deformation experiments.

Measusrements of austenite grain size and shape, density of de-

formation bands in austenite, and ferrite grain size were performed using standard optical metallographic techniques. The stereo-
logical relationships used in the evaluation of S_v for a given
austenite microstructure (size and shape of grains, deformation
band density) were taken from Underwood (3). The extent of the
$\gamma \rightarrow \alpha + Fe_3C$ transformation was estimated by point counting on specimens
quenched from various temperatures.

Results

This paper is limited to a presentation of some important
results from a more extensive study (4), which encompasses
additional steels and more comprehensive process-parameter
variations than are discussed here. The interested reader can glean
more detailed information from the original report (supplied on
request).

Austenite microstructure

The γ grain size after austenitization was 250±30 µm in all Nb
variants; the value for the V-steel was 150 µm. For CR simulation,
the predeformation step (25%) was performed at either 1120°C or
1020°C; ψ was varied between 34 and 67%. The total reduction applied
during NR simulation was 67% (25% at 1120, 1050, 1000, 950°C).
Results pertaining to the quantitative characterization of the
austenite microstructure, for processing by CR, NR and N, are listed
below.

Table 2. Microstructural characteristics of austenite
prior to transformation

Steel designation	Processing	Cooling rate	Predef'n. temp.(°C)	$\psi(\%)$	$D^\gamma(\mu m)$ 900°C	$D^\gamma(\mu m)$ 800°C	Def'n.band density (mm^{-1})	S_v (mm^{-1})
Nb	CR	40mm pl.	1120	51	79	95x31	15,7	54
(low Mn)	CR	12mm pl.	1120	51	80	95x34	14,0	58
	CR	10mm bar	1120	51	55	76x29	15,4	69
	CR	12mm pl.	1020	51	48	62x22	13,0	75
Nb	CR	40mm pl.	1120	51	67	99x34	19,5	59
(high Mn)	CR	12mm pl.	1120	51	71	114x36	15,3	52
	CR	10mm bar	1120	51	57	66x27	15,4	68
	CR	12mm pl.	1020	51	50	62x25	13,9	70
	CR	40mm pl.	1120	34	67	82x42	6,8	43
	CR	40mm pl.	1120	57	67	101x28	21,4	67
	NR	40mm pl.	–	–	–	53x38	$\approx$ 0	47
	N	40mm pl.	–	–	–	12	0	167

Steel designation	Processing	Cooling rate	Predef'n. temp.(°C)	ψ(%)	D^{γ}(μm) 900°C	D^{γ}(μm) 800°C	Def'n.band density (mm^{-1})	S_v (mm^{-1})
Nb	N	1.2mm pl.	-	-	-	12	0	167
(high Mn)	N	10mm bar	-	-	-	12	0	167
Nb(PR)	CR	40mm pl.	1120	67	74	98x26	33,0	82
	CR	12mm pl.	1120	67	72	101x28	32,0	79
	CR	10mm bar	1120	67	48	68x25	29,1	84
	CR	12mm pl.	1020	51	47	57x22	13,4	76
	N	40mm pl.	-	-	-	13	0	154
	N	12mm pl.	-	-	-	13	0	154
	N	10mm bar	-	-	-	13	0	154
V	CR	40mm pl.	1100	53	not measured	13$^{x)}$	~ 0	154
	CR	12mm pl.	1100	53	"	12,5$^{x)}$	~ 0	160
	CR	10mm bar	1100	53	"	11$^{x)}$	~ 0	182
	NR	40mm pl.	-	-	-	25	0	80
	NR	12mm pl.	-	-	-	21,5	0	93
	NR	10mm bar	-	-	-	18,5	0	108
	N	40mm pl.	-	-	-	14	0	143
	N	12mm pl.	-	-	-	14	0	143
	N	10mm bar	-	-	-	14	0	143

x) Grains somewhat elongated; size of equivalent equiaxed grains is given.

Fig.1 shows the effective ferrite nucleation area per unit volume of austenite (sum of contributions from grain boundaries and deformation bands) as a function of ψ (CR; Nb steels). The component derived from deformation bands increases, slowly at first but then more dramatically, with increasing reduction during finishing. The increase in S_v due to grain elongation can be evaluated theoretically assuming an unconstrained shape change of initially cubical grains during uniaxial compression. The "theoretical" curves in Fig.1 are obtained by adding S_v(g.b.) so calculated to the experimentally-measured S_v(d.b.) values (Table 2). The agreement with the experimental points for the total S_v is quite good for two values of D^{γ} prior to finishing (50 and 75 μm); better accord can probably be obtained if a more realistic grain shape is assumed from the beginning. Note that the increase in S_v associated with CR is quite modest; for example ψ-values >60% are necessary in order that S_v be doubled. In no case does the S_v-value for controlled-rolled Nb-austenite approach that characterizing normalized material (Table 2).

The microstructure engendered in the V-steel as a result of controlled rolling merits a separate discussion. The grain sizes

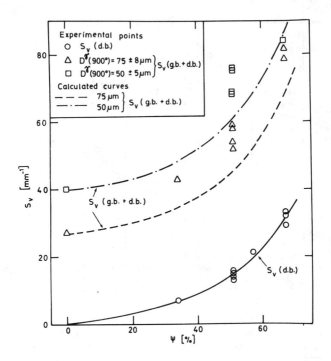

Fig.1 Data for ferrite nucleation area per unit volume of austenite (S_v) as a function of the total reduction during finishing (ψ). The dotted curves are calculated by adding the experimentally-measured contribution from deformation bands (S_v(d.b)) to the component derived from grain boundaries (S_v(g.b.)), the ψ-dependence of which was evaluated assuming unconstrained compression of cubical grains.

obtained are much smaller (and the S_v-values correspondingly greater) than in identically processed Nb-steels; the grains are only slightly elongated. It is now well established that strain-induced Nb(C,N) precipitation is effective in retarding re-crystallization of austenite under 950°C; on the other hand, at normal microalloying levels, V(C,N) starts to be effective only under ~ 850°C (5,6). Hence, while recrystallization is obviated in Nb steels during the entire finishing sequence, V-bearing qualities will recrystallize to a fine grain size during the first finishing passes and then undergo grain elongation only under the final passes. Microstructure development in association with thermo-mechanical treatment of the vanadium grade of interest here, and other V-steels, is taken up in another contribution to this conference (7).

The $\gamma \rightarrow \alpha$ + Fe$_3$C transition
_____

Data relating to the $\gamma \rightarrow \alpha$ + Fe$_3$C reaction, corresponding to the combinations of steel chemistry and processing route listed in Table 2, are presented below.

Table 3. Details pertaining to $\gamma \rightarrow \alpha$ + Fe$_3$C

Steel designation	Process-ing	S_v (mm^{-1})	T($^{\circ}$C) for X$^{\alpha}$ =				T($^{\circ}$C)for X^P=0,03	Complete transf.			
			0,05	0,10	0,25	0,50		D$^{\alpha}$(μm)	X$^{\alpha}$	X^P	
Nb (low Mn)	CR 40 pl.	54	822	806	783	762	639	8,9	0,88	0,12	
	CR 12 pl.	58	805	780	761	729	618	8,3	0,88	0,12	
	CR 10 bar	69	776	750	720	695	577	5,7	0,89	0,11	
	CR 12 pl.	75	811	790	760	733	630	7,9	0,88	0,12	
Nb (high Mn)	CR 40 pl.	59	778	761	722	703	620	7,9	0,85	0,15	
	CR 12 pl.	52	770	747	714	675	590	7,5	0,85	0,15	*
	CR 10 bar	68	735	711	676	640	513	5,3	0,83	0,08	
	CR 12 pl.	70	772	752	723	682	595	7,0	0,85	0,15	
	CR 40 pl.	43	762	745	719	698	–	8,9	0,86	0,14	
	CR 40 pl.	67	782	765	731	707	–	7,1	0,86	0,14	
	NR 40 pl.	47	732	723	711	698	608	10,9	0,86	0,14	
	N 40 pl.	167	764	758	749	738	620	8,9	0,86	0,14	
	N 12 pl.	167	759	752	741	730	605	8,1	0,86	0,14	**
	N 10 bar	167	731	723	711	698	570	7,9	0,79	0,15	
Nb(PR)	CR 40 pl.	82	782	766	745	726	603	7,2	0,90	0,10	
	CR 12 pl.	79	777	756	731	711	564	6,7	0,90	0,10	
	CR 10 bar	84	746	727	698	678	523	5,2	0,82	0,10	***
	CR 12 pl.	76	776	755	728	708	574	8,1	0,90	0,10	

*0,07B

**0,06B

***0,08B

Steel designation	Process-ing	S_v (mm^{-1})	T($^{\circ}$C) for X^α =				T($^{\circ}$C) for X^P=0,03	Complete transf.		
			0,05	0,10	0,25	0,50		D^α(μm)	X^α	X^P
Nb(PR)	N 40 pl.	154	770	764	753	745	610	7,2	0,90	0,10
	N 12 pl.	154	764	757	748	737	590	7,0	0,90	0,10
	N 10 bar	154	736	730	720	706	530	6,3	0,88	0,10(0,02B)
V	CR 40 pl.	154						8,2		
	CR 12 pl.	160	778	758	723	703	-	7,2	-	-
	CR 10 bar	182						5,9		
	NR 40 pl.	80						10,2		
	NR 12 pl.	93	761	742	709	688	-	8,6	-	-
	NR 10 bar	108						7,7		
	N 40 pl.	143						8,0		
	N 12 pl.	143						7,3		
	N 10 bar	143						6,0		

The salient points from the above tabulation are:

(i)   For Nb steels, the Ar_3 temperature is higher after processing
      by CR than with N, in spite of the fact that the S_v value is
      considerably greater in the latter case.

(ii) The $\gamma \to \alpha$ transition in controlled-rolled Nb steels starts off
      at a higher temperature than in normalized material, but proceeds
      at a slower rate. Hence, for a given steel chemistry and cooling
      rate, 50% ferrite is attained at a higher temperature in the
      normalizing case, despite the lower Ar_3 (plots of X^α vs. tempera-
      ture for CR and N intersect - see Fig. 2). These differences are
      very readily appreciated via a comparison of CCT diagrams for
      transformation of a given steel after CR and N (Fig.3).

(iii) Alteration of austenite grain size prior to finishing, or the
      total reduction during finishing (ψ) affects the transformation
      kinetics after CR, partly because S_v is changed and partly as
      a result of more prolific ferrite nucleation (increased ψ). For
      the same ψ, the effect of (albeit small) alterations in D^γ
      (900°C) is quite modest. If, on the other hand, the initial S_v
      prior to finishing is fixed and ψ is increased, the temperature
      range over which X^α changes from 0,05 to 0,25 is raised appreci-
      ably although the effect derived from changes in ψ seems to
      saturate for $X^\alpha>0,5$ (data for normal rolling of Nb(high Mn)
      provide a good indication of the situation when $\psi\sim0$). The
      dependence of Ar_3 on ψ observed in this work, ie. a steep
      augmentation with increased ψ when the latter is small followed
      by a less dramatic rise for $\psi>50\%$, agrees well with the behavi-
      our observed by Kozasu et al.(1) for a similar Nb steel

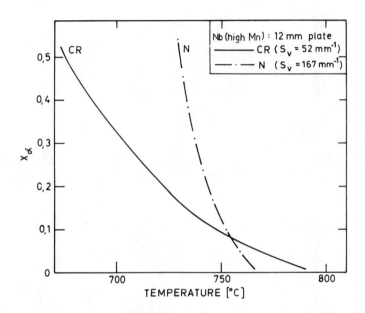

Fig.2 Volume fraction ferrite vs. temperature for transformation of a niobium steel processed by controlled rolling (ψ= 51%) and normalizing.

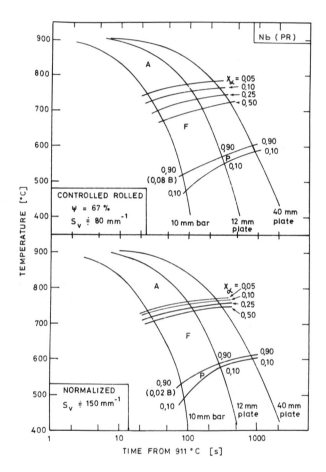

Fig.3 Comparison of CCT diagrams for Nb(PR) steel after normalizing (equiaxed austenite) and controlled rolling with ψ=67% (elongated austenite).

processed by CR. The fact that ψ influences $T(0,05)$ to a greater extent than $T(0,5)$ is illustrated by Fig.4.

(iv) It is difficult to compare the transformation characteristics of the Nb- and V qualities directly because these micro-additions seem to affect the $\gamma \to \alpha + Fe_3C$ transition in different ways. If the V steel after CR is compared with the normalized Nb (high Mn) variant (S_v about equal), then the transformation starts earlier but proceeds at a slower rate in the former case. Comparing these two steels processed identically (CR: 12 mm plate), we see that the transformations start at about the same temperature but $T(X^\alpha=0,5)$ is appreciably lower for the Nb variant. Further work is required in order to elucidate these different behaviours.

It is instructive to plot the mean ferrite grain size as a function of S_v for a given cooling rate. An example is presented in Fig.5 (12 mm plate) where data for a number of additional V-steels are included (4,7). The curves given in the figure are taken from Kozasu et al.(1) and refer to elongated unrecrystallized austenite grains (CR) and equiaxed grains (NR). The present results on Nb steels confirm convincingly the standpoint, first proposed by the Japanese workers, that transformation from elongated austenite engenders a finer mean D^α than does reaction of equivalent (same S_v) equiaxed grains. Thus, for example, controlled rolling of Nb steels promotes ferrite grain sizes as fine or even finer than those attainable via normalizing, in spite of the fact that S_v is at least twice as great in the latter instance.

The behaviour of the vanadium steels in the context of Fig.5 is novel. All points fall on the curve relating to the transformation of equiaxed Nb-austenite, even when the V steel has been processed by CR where the austenite grains prior to transformation are certainly not equiaxed (see previous section). This suggests that the positive effect accruing via transformation from deformed austenite is gradually eliminated as the size of the grains which are elongated diminishes, i.e. the $D^\alpha(S_v)$ curves for equiaxed and elongated grains tend to merge at high values of S_v. A possible explanation for this behaviour is forwarded in the discussion section.

Fully-transformed microstructures

For the range of steels and processing schemes investigated, the fully transformed microstructures comprised, for the most part, only polygonal ferrite plus pearlite. For cooling at the fastest rate (10 mm.bar) and with Mn levels ~1,5%, non-polygonal products are formed after processing by both CR and N. The microstructures found in the controlled-rolled Nb steels are invariably mixed; faster cooling tends to reduce the grain size in the regions with larger

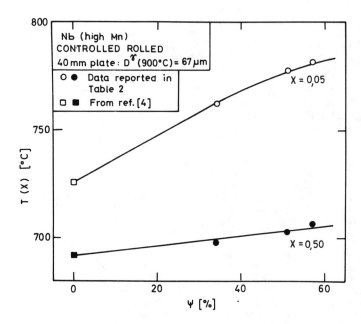

Fig.4 Influence of reduction below the recrystallization temperature (ψ) on $T(X^\alpha = 0,05)$ and $T(X^\alpha = 0,50)$ for transformation of controlled-rolled Nb (high Mn) steel cooled at a rate $\equiv$ 40 mm plate in air.

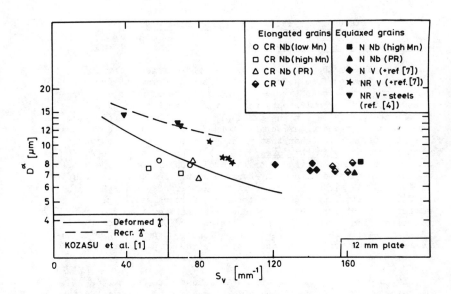

Fig.5 Relationship between ferrite grain size (D^α) and specific
ferrite nucleation area of austenite (S_v). Points pertaining to
transformation of both equiaxed (N,NR) and elongated (CR) γ grains
are included. Corresponding curves, reported by Kozasu et al.(1),
are included for comparison.

grains whilst not affecting the size of the smaller ones (Fig.6a,b). The grain size of Nb variants which have been normalized is much more uniform (Fig.6c), although in the examples shown the average grain size is actually greater than that engendered by CR. The grain size distribution in controlled rolled V steels is also considerably narrower than for equivalently-processed Nb variants (Fig.6d).

Discussion: Theoretical Considerations

Nucleation sites for ferrite

As stated in the introduction, the enhancement of ferrite nucleation derived from deforming Nb-austenite below its recrystallization temperature cannot solely be ascribed to the associated increase in S_v. The results obtained in the present study clearly corroborate this view. In the first place, the ferrite grain size obtained after transformation of deformed, elongated austenite grains, is distinctly smaller than that produced by reaction of equivalent equiaxed grains, with the same S_v (Fig.5). Furthermore, the Ar_3 temperature characterizing controlled-rolled Nb-microalloyed steel is higher than for equivalently-treated normalized material, despite the fact that S_v is much greater in the latter case (Figs.2 and 3); Ar_3(CR) exhibits a pronounced increase as ψ is raised (Fig.4). Another piece of evidence, which supports the view that the grain refinement stemming from CR is not derived exclusively from the increase in S_v, is that the areal nucleation frequency of ferrite is augmented dramatically with increasing degrees of reduction of austenite below the recrystallization temperature (2).

It is clear that deformation of the austenite during the finishing phase of CR generates effective grain boundary sites for ferrite nucleation, i.e. the sites are associated with a low activation energy (ΔG^x) and a high density per unit volume. Now, the driving force for the $\gamma \rightarrow \alpha$ transition is low. For dilute solutions of carbon in both α and γ (Henry's law), it is easy to show that

$$\Delta G^{\gamma \rightarrow \alpha} = RT \ln \frac{1-X}{1-X^\gamma}$$

where X, X^γ are the atom fractions of carbon in the steel and in austenite (equilibrium with α at T). For a 0,11%C, 1,45%Mn steel at 750°C, $\Delta G^{\gamma \rightarrow \alpha}$ = -110 J.mol^{-1} (-1,55.10^7 J.m^{-3}). This is quite a small driving force and nucleation will require particularly potent sites (low ΔG^x). In this context, it is a common misconception that the stored energy of deformation can enhance the driving force for the $\gamma \rightarrow \alpha$ transition and so proliferate ferrite nucleation from unrecrystallized γ. The stored energy (ΔG_d) is about $-\tau \rho_0$ where τ is the line energy of a dislocation ($\sim \mu b^2$) and ρ_0 the dislocation density.

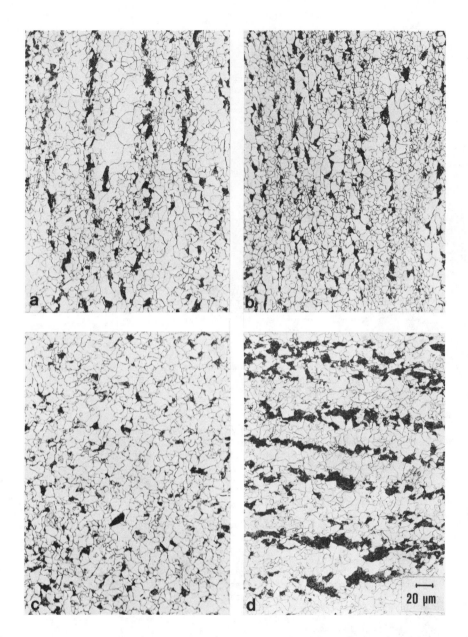

Figure 6 - Microstructures of fully-transformed steels. a,b) Nb(PR) controlled rolled and cooled a rate ≡ 40 mm. plate (a) and 12 mm plate (b) in
air; c) Nb(PR) after normalizing (cooling ≡ 12 mm plate); and d) V steel
after controlled rolling (40 mm plate).

With $\tau=2,3.10^{-9}$J.m^{-1} and $\rho_0=10^{15}$m^{-2} (almost certainly an over-estimate), we find that $\Delta G_d = -2.10^6$ J/m^3, or only 10% of $\Delta G^{\gamma \to \alpha}$. Thus, the stored energy of deformation can hardly account for the proliferation of ferrite nucleation when transformation proceeds from deformed γ.

Grain boundary nucleation in an equiaxed microstructure can take place either on grain surfaces or at grain edges or corners. The relative activation energies for nucleation at such sites were first evaluated by Clemm and Fisher (8), who showed that the nucleation potency decreases in the order corners→edges→surfaces. The problem is that the density of sites increases in the above order, i.e. in all but small grained materials, the most potent sites (corners) will not contribute much to the overall transformation rate because they are too few. The question of site density is taken up further in the next section.

Turning to elongated austenite, it is clear that some other type of site, with about the same nucleation potency as corners or edges, must be generated during deformation below the recrystallization temperature. Two alternatives have been proposed:

(i) Ferrite nucleates on regions of the grain boundaries with locally sharp curvature (bulges)(2). Bulges are formed via strain-induced grain-boundary migration and are a precursor to static and dynamic recrystallization in association with hot working. The contention is that bulges are formed during the finishing phase of CR, even though recrystallization might be suppressed.

(ii) Ferrite nucleates on the austenite grain boundaries and also on subgrain boundaries in the vicinity of the former (9). The growth of the ferrite is limited by strain-induced Nb(C,N) thus permitting further nucleation. The extent of subgrain nucleation is purported to diminish rapidly with increasing distance from the original grain boundary because of the high strain gradient. Hence, the nucleation in the grain-boundary region is initially prolific but the rate decreases rapidly once the original austenite grain boundaries are decorated with ferrite.

With regard to alternative (i), one can envisage several alternative configurations for bulge nucleation, which are depicted schematically in Fig.7; the evaluation of ΔG^x for the various geo-metries is rather involved and only the principal results will be presented here. Nucleus development into only one of the adjacent grains is the simplest case to treat quantitatively (Fig.7a). For a nucleus of critical size, the pressure due to curvature is always balanced by the available driving force and

$$r^x = r^x_H = \frac{2\sigma^{\alpha\gamma} V^\alpha}{-\Delta G^{\gamma \to \alpha}}$$

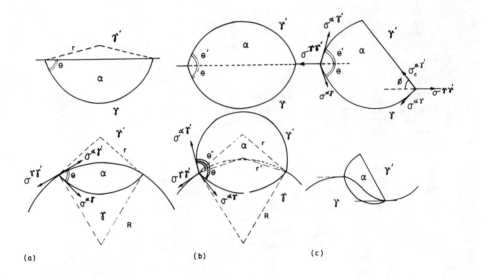

Fig.7 Configurations pertaining to nucleation of ferrite on straight and strongly curved γ-grain boundaries.
a) Nucleus development on one side of boundary only;
b) development on both sides of the boundary;
c) nucleus with low-energy facet illustrating "puckering" - the necessary boundary displacement on either side of the nucleus might already exist in deformed austenite (bulges).

where V^α is the molar volume of α, and r^x_H is the critical nucleus
size for homogeneous nucleation. By substituting for r^x_H in the expression for the total energy of the nucleus, it is possible to
evaluate ΔG^x which can be compared to ΔG^x_H, the activation energy
for homogeneous nucleation. $\Delta G^x/\Delta G^x_H$ depends on r^x_H/R (R is the
radius of curvature of the bulge) and θ which is defined by the
balance of surface energies (Fig.7a). A complete analysis (2) shows
that nucleation at bulges is considerably more favourable than at a
straight boundary, i.e. $\Delta G^x/\Delta G^x_H$ is lower, if $r^x_H/R > 0,1$; for
$r^x_H/R > 0,5$, then the activation energy associated with bulge sites
is comparable to that characterizing grain edges or even corners.
Unfortunately, the situation depicted in Fig.7a cannot exist in
reality because there is no balance of radial tensions; it will,
however, be a reasonable approximation if $\sigma^{\gamma\gamma} \approx \sigma^{\alpha\gamma'} \gg \sigma^{\alpha\gamma}$, but it is
difficult to say under what circumstances this might be realistic.

For the more tenable situation of nucleus development into both
grains (Fig.7b), a tractable solution can be found if one assumes
$\sigma^{\alpha\gamma'} = \sigma^{\alpha\gamma}$, which might often be the case. A full analysis of this
configuration demonstrates that a sharply-curved grain boundary
(bulge) offers no particular energetic advantage over a straight
one, even when the curvature is large (small R). Clearly, Figs.7a
and 7b (with $\sigma^{\alpha\gamma'} = \sigma^{\alpha\gamma}$)represent two extremes.

One might guess that, when the nucleus develops into both grains,
nucleation proceeds more easily at bulges than at a straight boundary
only if $\sigma^{\alpha\gamma'} > \sigma^{\alpha\gamma}$; unfortunately, the mathematical development when
$\sigma^{\alpha\gamma'} \neq \sigma^{\alpha\gamma}$ is rather complicated and we have not attempted it.

It is well known that α nodules growing in austenite are often
faceted, because the energy-misorientation relationship for α/γ interfaces contains cusps corresponding to Kurdjumov-Sachs orientations.
The influence of the presence of facets on the equilibrium shape of
nuclei at grain boundaries has been treated by Lee and Aaronson (10).
These workers showed that a faceted nucleus cannot exist if the
boundary is straight because the surface tensions are not balanced.
In order to achieve a balance of forces, the grain boundary must
"pucker" (their term) such that one side is displaced relative to the
other (Fig.7c). The necessary displacement of the boundary might be
difficult to accomplish if the latter is straight from the beginning.
On the other hand, the configuration of Fig.7c should be much easier
to achieve if the austenite grain boundaries are characterised by
irregular bulges; the required displacements will be smaller and in
some cases might already exist (see Fig.7c). The radius of curvature
of the curved surfaces of a critical faceted nucleus must,of course,
be r^x_H but ΔG^x is obviously much smaller than for the corresponding
unfaceted nucleus because of the smaller volume and surface area in

the former case. Taking $\sigma^{\gamma\gamma}/\sigma^{\alpha\gamma}=1,05$ $(\cos\theta=0,525)$, $\sigma_C^{\alpha\gamma}/\sigma^{\alpha\gamma} = 0,3$ and $\phi=45°$ (for definitions, see Fig.7c), then $\Delta G^x/\Delta G_H^x$ is 0,13 which is appreciably less than the value for a corresponding non-faceted nucleus (0,28); for a more complete discussion, see (10).

The conclusion from the aforegoing discussion is that deformed austenite, with grain boundaries characterized by distinct local alterations in curvature (bulges), can provide very favourable nucleation sites for ferrite; apart from having a low activation energy, the density of such sites per unit volume is high, since they cover the entire grain surfaces (see next section).

The second alternative ((ii) above), which has been proposed in order to explain the enhanced areal nucleation frequency in deformed austenite, is more difficult to deal with quantitatively since the mechanism whereby subgrain boundaries, with low interfacial energy, are able to effectively nucleate ferrite is not particularly clear. However, even this nucleation mode, if feasible, will provide for a high site density.

Effect of density of sites

As has been mentioned several times, the criterion that an effective nucleation site should have low ΔG^x is necessary but not sufficient; it is also important that the appropriate site exists in adequate numbers (high site density). This question has been taken up by Cahn (11) in a classical treatment of the kinetics of grain boundary reactions.

The contribution of a given site to the overall nucleation rate in a transformation can be written

$$\dot{N}_j \propto \left(\frac{\delta}{D}\right)^{3-j} \exp\left(\frac{-\Delta G_j^x + \Delta G_A}{RT}\right)$$

where j is the dimensionality of the sites (0 for corners, 1 for edges, 2 for surfaces and 3 for homogeneous nucleation), D is the grain size, δ the effective grain-boundary thickness and ΔG_A is the activation energy controlling continued growth of a nucleus (usually that for diffusion of some appropriate species). ΔG_j^x, as we have seen, depends on ΔG_H^x (which can be evaluated from the free energy driving the reaction), on the type of boundary site and on the balance of surface tensions around the nucleus (θ in Fig.7). If we form the ratio of nucleation rates for sites of dimensionality i and j, then

$$\ln\frac{\dot{N}_i}{\dot{N}_j} = (i-j)\ln\frac{D}{\delta} - (A_i - A_j)\frac{\Delta G_H^x}{RT}$$

where $A_j = \dfrac{\Delta G^x_j}{\Delta G^x_H}$. Hence, sites of dimensionality i will contribute more to the overall nucleation rate than those of dimensionality j if

$$(j-i) \; Z < A_j - A_i$$

where Z is the dimensionless quantity

$$Z = \frac{\ln{}^D/\delta}{\Delta G^x_H/RT}$$

For example, nucleation at corners will dominate a grain-boundary nucleated transformation if

$$\frac{\ln{}^D/\delta}{\Delta G^x_H/RT} < A_E - A_C \; ,$$

i.e. corner nucleation is favoured for small grain sizes and low driving forces (high ΔG^x_H).

Since A_i will depend on θ, it is possible to construct a Z–θ diagram which maps the conditions under which different types of nucleation site dominate the transformation. Such diagrams are shown in Fig.8 and refer to nuclei developing on one side of a straight boundary and a curved one ($r^x_{H/R} = 0,5$); the appropriate activation energies are taken from references (2) and (8). What these figures illustrate is the greatly extended field of domination for grain surface nucleation when bulges are present on the grain boundary. This is because ΔG^x_S is reduced for a curved boundary and surface nucleation can thus dominate over that at edges or corners down to lower values of Z. Such diagrams tell us nothing about the absolute nucleation rate, of course; this depends on the individual values of ΔG^x_H and D. However, if these variables plus θ are fixed, then it is clear from the basic equation for nucleation rate that N_j increases with j, because of the term D^{j-3}. Hence, the presence of bulges, which engenders domination of sites with higher dimensionality (grain surfaces as opposed to corners or edges for corresponding straight boundaries), will result in a higher nucleation rate for a given undercooling ($\Delta G^{\gamma \to \alpha}$) and grain size (or S_v). The very high Ar_3 temperatures and prolific grain-boundary nucleation, which character-ize the initial stages of $\gamma \to \alpha$ in controlled-rolled Nb steels, are clearly in line with the above predictions. It should also be pointed out that the conclusions arrived at in this section are valid inde-pendent of the mechanism by which the ferrite nucleation efficiency of deformed austenite is augmented; the bulge argument has been used here because some degree of quantification is possible (Fig.8).

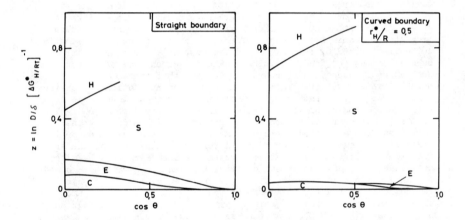

Fig.8 Plots of Z vs cosθ (for explanation, see text) for
straight and curved (r$^{x}_{H}$/R = 0,5; nucleus development on one
side only) grain boundaries showing regions of domination of
various types of nucleation site (H = homogeneous, S = grain
surface, E = edge, C = corner).

Discussion: Experimental Observations

The fact, that deformation of austenite below its recrystallization temperature engenders highly efficient nucleation sites for
ferrite which cover the entire grain surfaces, would seem to be in-
disputable. Thus, for Nb steels processed by CR, the nucleation rate
is high even for low degrees of undercooling in spite of the rather
low S_v (often less than half of the corresponding value for normalized
material). The Ar_3 temperature is concomitantly high and a substantial
degree of transformation occurs at temperatures where a corresponding
normalized steel, cooled at the same rate, is still austenitic. However, once the favourable grain-boundary sites are exhausted (site
saturation), the nucleation rate drops drastically; the volume density
of intragranular sites is low. Metallographic observations indicate
that the interior of the austenite grains transforms via preferred
growth of a certain grain-boundary sites or by intragranular nucleation at favourable sites (e.g. deformation bands, which are evidently
not characterized by the same nucleation potential as true grain-
boundary sites (4), and inclusions). The consequence of this drop in
nucleation frequency, combined with the fact that the volume of
material which must transform via intragranular nucleation is considerable (low S_v), is a rather low transformation rate at lower tempera-
tures (much slower than in normalized material) even though $\Delta G^{\gamma \to \alpha}$ increases. Another ramification of the rather special mode of trans-
formation from elongated γ grains (CR) is that the original austenite
boundaries become delineated by small ferrite grains whereas the grain
interiors react to give relatively coarse-grained ferrite; the microstructure is mixed. The spread of ferrite grain sizes will be reduced
as the γ grain size prior to finishing is refined or the finishing
reduction is increased; both effects tending to reduce the volume
which transforms to coarse grains. Fast cooling promotes intragranular
nucleation and reduces the average size of the coarser ferrite grains.

The rationale outlined above is clearly in substantial agreement
with the experimental observations pertaining to the transformation
of controlled-rolled Nb steels. A problem which remains is to explain
why the grain-boundary nucleated ferrite crystals do not continue to
grow rather than the transformation proceeding via development of
unfavourable intragranular sites, which leads to large ferrite grains
in the middle of the original austenite grains. This will not be
discussed further in this paper.

Turning to normalized steels, it would appear that, in this case,
grain surfaces are not particularly favourable sites and that the
transformation proceeds entirely via nuclei which have developed at
edges or corners. The difference in nucleation behaviour between
material processed by CR and N is convincingly demonstrated via examination of partially-transformed material (Fig.9); in the case of CR,
the entire grain surface is outlined by ferrite nodules while

a

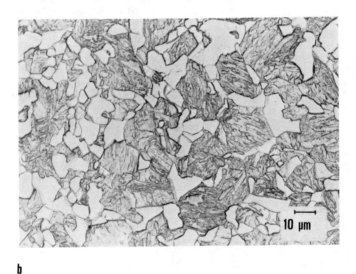

b

Fig.9 Optical micrographs of a) controlled rolled (Nb(high Mn))
and b) normalized (Nb(PR)) niobium steels after about 30% transformation to ferrite. In a) the regions adjacent to the original
γ-grain boundaries are completely transformed, whereas b) is
characterized by nucleation having occurred at selected points on
the boundaries (corners, edges), and segments of the original γ
boundaries are still visible.

427

segments of the original γ boundaries are still visible in a normalized specimen after approximately the same degree of transformation. As discussed in the previous section, the density of sites effect implies that the nucleation rate for a given undercooling will be less for normalized austenite (edge, corner nucleation) than for controlled-rolled material (surface nucleation); Ar_3 is lower in the former case, as found experimentally. However, the small γ grain size after normalizing (high S_v) is associated with concomitantly small growth distances and the transformation can proceed virtually to completion via development of the original grain-boundary nuclei, which precludes the necessity of intragranular nucleation. Hence, the transformation, once initiated, proceeds rapidly to completion, which again is in accord with experimental observation.

Finally, we should discuss the novel results pertaining to microstructure development in V-microalloyed steels which are reported in this work (see Fig.5). For a given S_v and cooling rate, processing by CR and N effects about the same degree of ferrite grain refinement. The austenite microstructure after CR comprises small (10-15 μm) austenite grains which are somewhat elongated. The rather modest degree of deformation below the recrystallization temperature in this case, implies that the density of favourable grain-surface sites will be relatively low, with the result that corner and edge sites (which have a high density) are likely to contribute as much to the overall transformation rate as surface nucleation. This combined with the fine γ grain size means that the progress of the transformation from, on the one hand, controlled rolled and, on the other, normalized austenite should be rather similar for the V steel, at least in the present experiments. An interesting point for future study is the effect of increasing the degree of deformation below the recrystallization temperature on the transformation characteristics of V-bearing qualities, which are characterized by small γ grain sizes prior to grain elongation.

Conclusions

i)   Controlled rolling of Nb steels effects only a modest increase in S_v due to grain elongation and generation of deformation bands during finishing. The positive effect from controlled rolling for ferrite grain refinement stems principally from an increased nucleation efficiency per unit area of austenite grain boundary.

ii) Controlled-rolled V steels are characterized by a very high S_v (equivalent to that in normalized material) and only slightly elongated austenite grains. This derives from effective γ grain refinement as a result of recrystallization during the first finishing passes.

iii) A Nb steel processed by controlled rolling exhibits a higher Ar_3

temperature than after normalizing (fixed cooling conditions); the Ar_3 of controlled-rolled austenite increases as the grain size prior to finishing is reduced and the finishing reduction raised.

iv) The transformation of normalized material proceeds very quickly to completion once initiated; the transformation temperature interval is much wider in the case of controlled rolling.

v) All other things being equal, the tendency for mixed ferrite grain size and the temperature interval over which transformation proceeds in controlled-rolled Nb steels both decrease with decreasing grain size prior to finishing, higher total finishing reduction and increased cooling rate.

vi) For a given $S_v \lesssim 100$ mm^{-1}, transformation from deformed, elongated γ grains engenders a finer average ferrite grain size than when $\gamma \rightarrow \alpha$ proceeds from equiaxed austenite. When S_v is larger (~ 150 mm^{-1}, typical for normalized γ), it would appear that grain elongation has little significance as regards ferrite grain size (further work required).

vii) Acceptable qualitative prediction of the observed differences in transformation behaviour, for the steels and processing procedures studied, is provided by Cahn's approach to the kinetics of grain boundary reactions. In particular, the initial ferrite nucleation in controlled-rolled Nb steels occurs at favourable sites dispersed over the entire grain surfaces; grain corners and edges constitute the principal nucleation sites during transformation of normalized austenite.

viii) Further work is required in order to establish unequivocally the nature of the efficient ferrite nucleation sites in austenite which has been deformed below its recrystallization temperature.

Acknowledgements

This work has been supported financially by the following Scandinavian steel companies; Halmstads Järnverks AB, Norsk Jernverk A/S, SKF Steel, Smedjebackens Valsverks AB and Svenskt Stål AB. The authors gratefully acknowledge this support and valuable discussions with representatives of the above organizations. The experimental steels were melted and fabricated at Swedish Steel (Oxelösund).

References

1. I.Kozasu, C. Ouchi, T. Sampei and T. Okita, "Hot rolling as a high-temperature thermomechanical process, in Microalloying '75, Ed.M.Korchynsky (Union Carbide Corp., New York), vol.1, pp.100-114 (1976).

2. W.Roberts, H.Lidefelt and A.Sandberg, "The mechanism of enhanced ferrite nucleation from deformed austenite in microalloyed steels". in Proc. Sheffield Conf. on Hot Working and Forming Processes, Eds. C.M.Sellars & G.J.Davies (Metals Society, London), pp. 38-42 (1980).

3. E.E. Underwood, "Quantitative microscopy" (McGraw-Hill, New York), p. 77 (1968).

4. A.Sandberg and W. Roberts,"Inverkan av termomekanisk behandling på austenitens fasomvandling i mikrolegerade stål". Swedish Institute for Metals Research Report No. 1439 (1980). (in Swedish).

5. W. Roberts, "Hot deformation studies on a vanadium microalloyed steel". Swedish Institute for Metals Research Report No. 1333 (1978).

6. R.K. Amin, M. Korchynsky and F.B. Pickering, "The effect of rolling variables on precipitation strengthening in HSLA steels containing vanadium and nitrogen". Report from Sheffield City Polytechnic, to be published.

7. T. Siwecki, A. Sandberg, W. Roberts and R. Lagneborg, "The influence of processing route and nitrogen content on microstructure development and precipitation hardening in V-microalloyed steels". These proceedings.

8. P.J. Clemm and J.C. Fisher, "The influence of grain boundaries on the nucleation of secondary phases". Acta Met. $\underline{3}$ (1955),70-78.

9. R.K. Amin and F.B. Pickering, Discussion of ref.(2) pp. 44-45.

10. J.K. Lee and H.I. Aaronson, "Influence of faceting on the equilibrium shape of nuclei at grain boundaries". Acta Met. $\underline{23}$ (1975), 799-808 and 809-820.

11. J.W. Cahn, "The kinetics of grain-boundary nucleated reactions". Acta Met. $\underline{4}$ (1956), 149-160.

DISCUSSION

Q: When you plotted your results against S_v, did you take into account the small bulges in the austenite boundaries in estimating the value of S_v, or did you assume the austenite boundaries are smooth, and get the S_v from microscopic measurement of the grain size and shape?

A: We think that they are so small that they are not measurable.

Q: We did a similar study with a V steel, and we looked at nucleation at austenite bulges. You mentioned that you found a high density of ferrite nucleation at the bulges, but as you moved away from the prior austenite boundaries to the interior of the grain, the density was lower. How large was this difference in ferrite nucleation?

A: Our work indicates that the ratio might be something like 2 or 3.

CONTROLLED ROLLING SIMULATION IN MICROALLOYED STEELS

G.A. Crosbie and A.J. Baker
Department of Metallurgy
The University of Leeds
Leeds, England

The controlled rolling of C-Mn steels, with and without Ti and V additions, has been studied by single pass rolling to simulate low temperature roll finishing. Rolling caused progressive ferrite grain refinement and microalloy additions increased this refinement. Ferrite was found to nucleate in unrecrystallised austenite at grain boundaries, at the austenite-ferrite interface and at intergranular slip band sites within the austenite grains. Combinations of these nucleation mechanisms were responsible for the creation of microstructures containing duplex ferrite grain sizes. Observations of hot deformation in the steels have also been carried out "in-situ" using the Photo-Emission Electron Microscope with a hot straining stage attachment. Recrystallisation of the C-Mn steels during hot deformation was observed to be inhibited by the microalloy additions. Deformation at lower temperatures created deformation bands in the austenite and these influenced subsequent ferrite formation in the unrecrystallised structure.

Introduction

Over recent years the increasing use of welding as a fabrication technique, along with demand for better formability and fracture toughness, has led to the progressive lowering of carbon content in high strength structural steels, and new methods of strengthening have had to be found. Grain refining during processing, by controlled rolling and addition of microalloying elements (Nb, V, Ti), has been the key development in the production of low carbon, high strength structural steels, and carbon levels down to 0.03% are now being used (1, 2, 3). At these carbon levels, steels are "pearlite reduced", and the absence of the lamellar carbide is an important factor in promoting good toughness and resistance to lamellar tearing. Microalloy additions also make it possible to produce fine grained steels of a "balanced" type, rather than the earlier Al-refined "killed" variety, and this may have considerable economic advantages in terms of ingot yield.

Grain refinement is one of the few methods available for strengthening a steel and improving its impact properties at the same time. Consequently, controlled rolled steels can have very attractive combinations of mechanical properties. The properties of controlled rolled steels are closely related to their transformed microstructures (4) and these in turn are dependent upon the condition of the austenite immediately prior to transformation. The most important factors in determining the mechanical properties of controlled steels are grain size, state of precipitation and dislocation density (5).

This paper describes the results of experiments in which the controlled rolling of C-Mn steels, with and without Vanadium and Titanium additions, has been studied by single pass rolling to simulate low temperature roll finishing. The ferrite grain structures developed by varying types of rolling have been examined metallographically and the ferrite nucleation processes in deformed austenite have been characterised. Observations of hot deformation in the steels have also been made directly, using a Photo-Emission Electron Microscope, and the relationship between these observations and the results from the rolling experiments is discussed.

Experimental

Controlled Rolling

The steels used in this investigation were supplied by the British Steel Corporation, Sheffield division. The analyses of these steels are given in the table below.

Compositions of Steel (wt%)

	C	Mn	Si	Ti	V
Plain Carbon-Mn	0.20	0.45	0.07	<0.005	<0.005
C Microalloy	0.094	2.16	0.3	–	0.15
Ti Microalloy	0.092	2.03	0.34	0.15	–

Specimens used for rolling were 10 x 10 mm in cross section, and 50mm long. They were made by cold rolling 50% from an initial 20mm thickness, annealing for 30 minutes at 625°C and then cutting to the correct size. For single pass reduction of 50% or more the fronts of the specimens were tapered to a wedge shape, to decrease the initial angle of bite. Rolling was carried out on a two-high mill at speeds of 2-13 m/min.

So that accurate temperature measurements could be made during rolling, a hole 8mm deep and 2mm diameter was drilled into the back of each specimen, into which was cemented a thermocouple, using high temperature adhesive.

The output signal from the thermocouple was recorded at one second intervals in digital form on punch tape. The tape produced was then processed to provide temperature - time curves for the complete rolling sequence.

Specimens were solution treated for 10 mins at 1000°C (plain C-Mn) or 1100°C (microalloy steels), prior to rolling, and the recording equipment was then switched on, as they were removed from the furnace. They were placed on the roll input table and allowed to aircool to the rolling temperature. Rolling was carried out in the temperature range 600-900°C, with single pass reductions of 20-85% and rolling speeds of 2, 7, and 13m min^{-1}. After rolling, specimens were either quenched into iced water or allowed to air cool on the exit table. Thermocouple monitoring was continued throughout the treatment, until the specimens reached ambient temperature.

After rolling, specimen thickness was measured with a micrometer and from these % reductions were calculated. Longitudinal and transverse sections were prepared from each specimen by standard metallographic methods and photographed.

Photo-Emission Electron Microscopy

Detailed descriptions of the Photo-Emission Electron Microscope (P.E.E.M.) and its operation are available in the literature (6, 7) and will not be dealt with here. For the work reported in this paper a special straining stage was utilised. With this stage small sheet tensile specimens, 0.2 to 1.0mm thick, could be heated or cooled at rates upto 100°C/sec and strained by up to 100% of their original gauge length, while being observed in the microscope. The specimens were heated by means of electron beam bombardment and temperatures up to 1000°C could be achieved in the specimen gauge area. The temperature was monitored by means of a thermocouple welded to the edge of the specimen.

Using this equipment, direct dynamic observations of the microstructural changes occurring during hot or cold deformation of the tensile specimens were obtained.

Results

Plain C-Mn Steel

The cooling curve for the C-Mn Steel, air cooled from 1000°C, showed arrests at 750°C and 640°C. Quenching experiments confirmed that these temperatures corresponded to A_3, when ferrite grains first nucleated, and A_1, the eutectoid temperature. The air cooled structure consisted of a coarse mixture of grain boundary and Widmanstatten ferrite, with a pearlite infilling.

The effect of Rolling Temperature

Specimens were aircooled from 1000°C and underwent 50% reduction, with 7 m/min rolling speed, at various temperatures, followed by water quenching.

Rolling at 950°C produced an equiaxed austenite structure, which was finer than in a specimen quenched directly from 1000°C, indicating that the austenite had recrystallised during rolling. Rolling at 885°C again produced a finer equiaxed austenite structure, which had partially transformed to bainite at the grain boundaries. Rolling at 800°C produced a structure consisting of 60% large, unrecrystallised austenite grains and 40% of a fine equiaxed ferrite and bainite mixture. The austenite had partially recrystallised, and ferrite and bainite had formed in these areas of reduced austenite grain size, during cooling.

In the specimen rolled at 760°C (i.e. just above the A_3 temperature) austenite grains were elongated and unrecrystallised and had a grain boundary network of equiaxed ferrite grains in them, 5-8 μm in size. A few Widmanstatten ferrite needles were seen, and some of the ferrite had been nucleated at manganese sulphide precipitates within the austenite grains. At 740°C, just below the A_3 temperature, austenite grains were again deformed, but there was very little evidence of deformed pro-eutectoid ferrite. The equiaxed ferrite formed mainly in austenite grain boundaries, often in multiple rows, although some nucleated within the austenite grains. The structure was very banded. At 700°C, a continuous network of ferrite appeared in the grain boundaries of the deformed austenite, and this showed a heavily worked substructure, sub-grain size 2 μm. A number of equiaxed ferrite grains (5 μm grain size) also appeared, in rows within the austenite grains.
    Rolling at 650°C, just above the A_1 temperature, produced a structure similar to that at 700°C (Figure 1). More ferrite was present, both in the deformed state (subgrain size 1-2 μm) and as parallel rows of equiaxed grains (3-5 μm in size), nucleated inside deformed austenite grains. These rows were oriented at 15° to the rolling plane.

The specimen rolled at 600°C was below A_1, prior to rolling, and had a structure of deformed grain boundary and Widmanstatten ferrite in deformed pearlite (Figure 2). This was a 'cold worked' modification of the normalised structure.

The Effect of Cooling Rate

    To investigate the effect of cooling rate upon microstructure, specimens were given the same treatment as in the previous section, but were air cooled after leaving the rolls. Rolling at 800°C, followed by air cooling, produced a banded ferrite and pearlite structure with a ferrite grain size of 12-20 μm. No Widmanstatten ferrite was seen. Rolling at 750°C produced a structure which was finer and more banded than at 800°C with a ferrite grain size of 10 μm.

    At 700°C, the specimen was partially transformed before rolling, and this produced a very banded normalised structure with two distinct ferrite morphologies (Figure 3). Outlining the austenite grain boundaries were regions of fine equiaxed ferrite (6 μm grain size), which contained no eutectoid product, and inside these were much coarser, elongated ferrite grains (10-20 μm grain size), lying parallel to each other, at orientations of 30°-50° to the rolling plane. Rolling at 650°C (just above A_1) produced a microstructure consisting of deformed, unrecrystallised ferrite, containing no eutectoid products, and elongated ferrite grains formed inside the deformed prior austenite grains (Figure 4).

The Effect of Amount of Deformation

    To investigate the effect of amount of reduction upon microstructure, specimens were treated as previously, but using reductions of 20% and 70%.

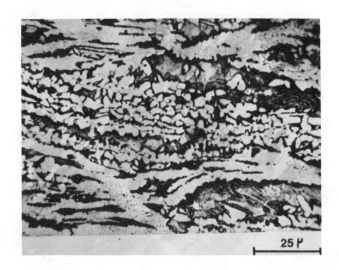

Figure 1.   C-M n steel, rolled 50% at 650°C and water quenched. Deformed
             and recovered proeutectoid ferrite around austenite grain bound-
             aries, and rows of equiaxed intragranular ferrite grains formed
             in deformed austenite.

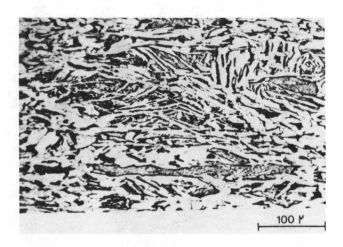

Figure 2.   C-Mn steel, rolled 50% at 600°C and water quenched. Deformed
             ferrite and pearlite structure.

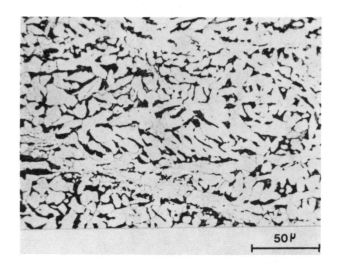

Figure 3.   C-Mn steel, rolled 50% at 700°C and air cooled. Duplex ferrite structure of fine equiaxed grain boundary ferrite and coarser, elongated intragranular ferrite.

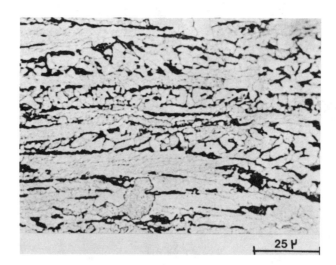

Figure 4.   C-Mn steel, rolled 50% at 650°C and air cooled. Deformed and unrecrystallised proeutectoid ferrite with equiaxed intragranular ferrite in prior austenite grains.

(i) 20% Reduction. Generally, 20% rolling reduction had little effect upon microstructures, which were only slightly modified from undeformed structures. At 800°C, an equiaxed austenite structure was obtained, with some ferrite and bainite at the austenite grain boundaries. At 735°C, the austenite was again equiaxed, with small colonies of proeutectoid ferrite in the grain boundaries; these showed little sign of deformation. Some Widmanstatten ferrite was also formed. When the rolling temperature was reduced to 700°C, the austenite grains became more elongated, and the proeutectoid ferrite showed some raggedness at the austenite-ferrite interface. The amount of Widmanstatten ferrite was greater than at 735°C. Rolling at 650°C produced a structure which contained coarse Widmanstatten ferrite needles, in deformed austenite. These were formed in parallel rows, with the carbide trapped between them. Others showed jaggedness at the austenite-ferrite interface.

(ii) 70% Reduction. All of the structures rolled 70% and water quenched showed a great deal of structural modification after rolling. There was little tendency for Widmanstatten ferrite formation, even at 650°C, and structural deformation in both austenite and ferrite (at lower rolling temperatures) was very marked. The microstructure had transformed 70% during, or immediately after rolling to fine exquiaxed ferrite grains (2-5 μm grain size) in heavily deformed austenite. Figure 5 shows this banded structure, in an area where very fine ferrite has formed; the regularity of the rows of ferrite is very pronounced.

At 700°C, the structure had again transformed 70% before quenching. At this temperature, there was 15% heavily deformed ferrite, which lay in flat bands, marking the austenite grain boundaries. The rest of the ferrite appeared within the austenite as rows of equiaxed grains. After rolling at 650°C, the ferrite was nearly all of the deformed proeutectoid ferrite type (Figure 6), and there was a very marked contrast between this structure and that of the specimen deformed 20% at the same temperature.

Steels Containing Vanadium and Titanium Additions

A series of controlled rolling experiments was carried out on low carbon steels, microalloyed with either 0.15% vanadium or titanium, using 70% reductions at rolling speeds of 2 m/min, to optimise the effect of single pass controlled rolling. Specimens were solution treated at 1100°C, to facilitate carbide dissolution, because alloy carbides have higher solution temperatures than cementite.

Vanadium Microalloyed Steel

(i) Rolling at 900°C. The specimen rolled at 900°C and water quenched had a structure of very elongated unrecrystallised austenite grains, with small ferrite colonies nucleated in the austenite grain boundaries, and in rows within the austenite grains. The ferrite was formed at 685°C, between rolling and water quenching. A specimen rolled at 900°C and water quenched after holding for five seconds, showed a similar deformed structure, with more equiaxed ferrite present. Again this specimen had cooled to 675°C prior to quenching.

(ii) Rolling at 740°C. Rolling at 740°C produced a roll cooling of 100°C, to below the A_3 temperature. The quenched specimen had a structure which had transformed from deformed austenite, containing 15% equiaxed ferrite grains (1-2 μm grain size) which were nucleated at grain boundaries and in bands within the austenite (Figure 7). Air cooling produced a banded structure of fine equiaxed ferrite (2-4 μm grain size) and broken bands of a light etching product. These bands were 2-3 μm wide, and had a 10-12 μm spacing.

439

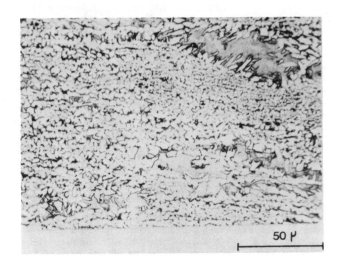

Figure 5.   C-Mn steel, rolled 70% at 750°C and water quenched. Very fine equiaxed ferrite grains nucleated around deformed prior austenite grain boundaries.

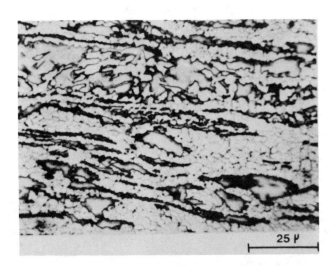

Figure 6.   C-Mn steel, rolled 70% at 650°C and water quenched. Bands of deformed and recovered ferrite formed in austenite.

When this air cooled structure was tempered for 20 minutes at 400°C, the bands etched very darkly, due to carbide precipitation, and it was therefore concluded that they were a bainitic structure.

(iii) Rolling at 700°C. The microstructures were similar to those obtained at 740°C. In the water quenched specimen, more equiaxed ferrite was present, and this formed as many parallel rows (1 μm grain size), within the austenite grains and at grain boundaries. The air cooled structure (Figure 8) also had a finer ferrite grain size (1-2 μm) and the bainite bands were more continuous and closely spaced (8 μm).

(iv) Rolling at 650°C. The microstructure of this specimen was different from that formed at 750°C, as there was 30% deformed ferrite present, and many rows of equiaxed intragranular ferrite had also formed. A particularly striking example of this intragranular ferrite formation is seen in Figure 9. The air cooled specimen differed from Figure 8 in that the spacing of the bainite rows had become less uniform, and their continuous banded structure was much broken up, the bands tending to deviate considerably from the rolling plane. This was because the ferrite formed around equiaxed austenite grains, prior to deformation had interfered with subsequent ferrite nucleation.

Titanium Microalloy Steel

(i) Rolling at 900°C. After rolling at 900°C and water quenching, a structure of deformed austenite grains was seen with 5% fine equiaxed ferrite grains (1 μm grain size) delineating the flattened austenite grain boundaries.

(ii) Rolling at 750°C. The quenched microstructure was deformed austenite with 1 μm equiaxed ferrite grains nucleated at austenite grain boundaries and within austenite grains. Many fine precipitates (less than $\frac{1}{2}$ μm) were seen throughout the specimen. Air cooling produced a mixture of fine grained ferrite, and small colonies of a light etching phase (2-4 μm in size), which showed slight alignment in the rolling direction, although this was very much less pronounced than that seen in the vanadium microalloy steel, after similar treatment. Tempering this structure for 30 minutes at 400°C, again caused the second phase to etch much more darkly, indicating that a bainite structure had been formed.

(iii) Rolling at 700°C. At 700°C, the microstructures obtained were very similar to those at 750°C. In the water quenched specimen (Figure 10) more ferrite was seen nucleated inside the austenite grains, both in bands at 30° to the rolling plane, and at precipitates. Air cooling produced a finer bainite colony size than at 750°C.

(iv) Rolling at 650°C. Again, there was little change in structure from those rolled at 750°C. In the water quenched specimen, equiaxed ferrite was present in bands within the austenite grains. The air cooled specimen was more banded than those rolled at higher temperatures, but the bainite colonies were still very discreet compared with those seen in the vanadium steel.

Photo-Emission Electron Microscopy

Austenite Deformation

Plain C-Mn steel specimens were deformed in the austenite region at 920°C, 825°C and 740°C. Specimens were initially fast heated to 920°C and solution treated for ten minutes, to allow some austenite grain growth, such that the initial austenite grain size was around 100 μm. They were then fast cooled to deformation temperatures and deformed. After deformation, the specimens were

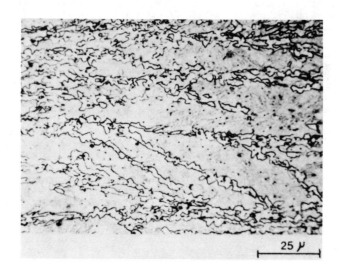

Figure 7.   Vanadium microalloy steel, rolled 70% at 740°C, 2m/minute rolling speed and water quenched. Deformed austenite grains with ferrite nucleated at grain boundaries and in intragranular bands.

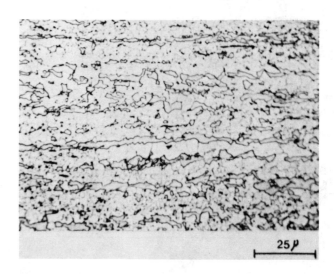

Figure 8.   Vanadium microalloy steel, rolled at 70% at 700°C, 2m/minute rolling speed and air cooled. Fine equiaxed ferrite grains, with continuous bands of bainite.

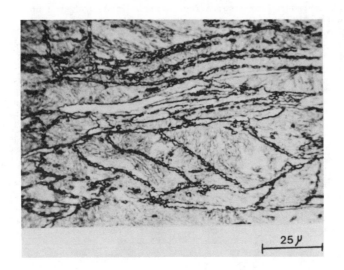

Figure 9.   Vanadium microalloy steel, rolled 70% at 675°C, 2m/minute
             rolling speed and water quenched. Deformed and recovered
             grain boundary ferrite and fine equiaxed intragranular
             ferrite grains in transformed austenite.

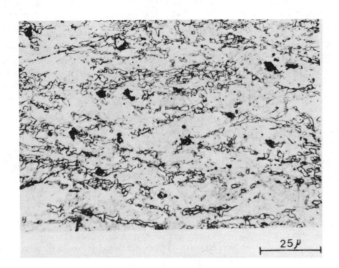

Figure 10.   Titanium microalloyed steel, rolled 70% at 700°C, 2m/minute
             rolling speed and water quenched. Deformed austenite structure
             containing fine equiaxed ferrite grains formed at grain bound-
             aries and in intragranular bands.

held at temperature, to observe recrystallisation, and then cooled to below 700°C, where the ferrite transformation was observed.

(i) Austenite Deformation at 920°C. The deformation of austenite (γ_1) at 920°C resulted in austenite recrystallisation. After initial solution treatment at 920°C, which caused deep thermal grooving of austenite grain boundaries, a 5% strain produced austenite grain rotation, and the disruption of the specimen surface. Holding for ten minutes resulted in the appearance of a fine grooved substructure within the austenite grains. Regions separated by these boundaries showed no orientation contrast, and it was therefore concluded that they were probably austenite subgrains. A further 2% deformation resulted in complete austenite recrystallisation within 90 seconds. The recrystallised austenite grains (γ_2) were very large (>150 µm) and did not alter during further holding.

Cooling to 660°C caused ferrite nucleation within two minutes. Ferrite was nucleated to γ_2 grain boundaries and it was seen to grow by bowing out of the austenite-ferrite interface into the austenite grains. Growth was not inhibited by the initial (γ_1) boundaries. Migration of some interfaces was discontinuous, and this resulted in surface grooves, which formed where the interface had been stationary. The final ferrite grain size, after cooling to 100°C, was approximately 100 µm.

Partial austenite recrystallisation was also seen at 920°C (Figure 11). After 5% deformation, a rapid 25% stress relaxation was accompanied by γ_1 grain boundary migration, and the formation of new, recrystallised (γ_2) grains (b). Both processes were discontinuous, which again produced surface grooving. A further 2% deformation (c) resulted in more grain boundary sliding at γ_1, migrated γ_1, and γ_1/γ_2 boundaries. Surface grooves, formed by γ_1 grain boundaries, which were inside the new twinned γ_2 grain, were unaffected by this deformation. These observations clearly showed that changes in contrast observed after high temperature austenite deformation were due to the formation of new recrystallised austenite grains.

(ii) Austenite Deformation at 825°C. Austenite deformation at 825°C did not result in recrystallisation. A 5% deformation at this temperature caused heavy grain boundary separation but holding for 20 minutes failed to produce any large scale changes in the austenite structure. However, some localised grain coarsening was observed. Little stress relaxation was recorded, the load drop-off rate being about 1% per minute. Cooling to 680°C resulted in the start of the ferrite transformation within 30 seconds. The final ferrite structure was finer and more equiaxed than those seen after deformation at 920°C, with a grain size of around 20 µm.

(iii) Austenite Deformation at 740°C. Figure 12 (a) shows an austenite structure, cooled to 740°C after solution treatment at 920°C. The white precipitate on the specimen surface is graphite, formed during cooling. A 4% deformation failed to cause recrystallisation during a 20 minute holding period but the structure (b) suffered severe austenite grain boundary rotation, and also had parallel surface relief markings, with a regular 3 µm spacing, in most of the grains. The markings did not cross austenite grain boundaries and they had different orientations in different grains. They were therefore thought to be austenite slip bands.

Cooling to 700°C caused the start of ferrite transformation in less than one minute. The austenite-ferrite interface movement was slow at this temperature, but during the ten second plate exposure time, it had moved 8-10 µm, giving the diffuse boundary seen in Figure 12 (c). After six minutes holding, and gradual cooling to 650°C, the transformation was still incomplete, the

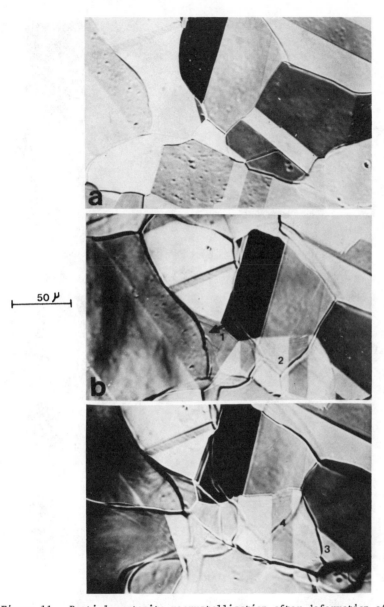

50 μ

Figure 11. Partial austenite recrystallisation after deformation of C–Mn steel at 920°C.
a) Austenite structure (γ_1) after solution treatment at 920°C.
b) 5% deformation results in γ_1 boundary migration (1) and the formation of recrystallised (γ_2) grain (2).
c) Further 2% deformation causes grain boundary decohesion at γ_1/γ_2 interface (3) and gradual disappearance of γ_1 thermal grooves (4).

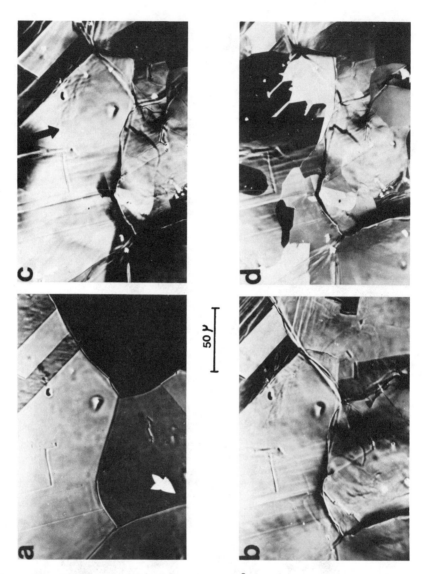

Figure 12. Deformation of C-Mn steel at 740°C.
a) Austenite at 740°C, after solution treatment at 920°C.
b) 4% deformation and 20 minutes holding at 740°C. Austenite slip
band formation, and no recrystallisation or recovery.
c) Cooled to 700°C. Ferrite forming in austenite.
d) After six minutes transformation and cooling to 650°C, the
transformation is still incomplete. Ferrite growth front is
modified by slip bands.

ferrite growth front being modified and made very angular, by the austenite slip bands (d). The final structure consisted of nearly equiaxed ferrite with a grain size of approximately 30 μm. Some of the ferrite grains remained very angular.

Austenite Deformation in Vanadium Microalloy Steel

Deformation of the vanadium microalloyed steel at 930°C did not result in austenite recrystallisation, and the subsequent specimen behaviour was considerably different from that seen in the plain C-Mn steel under similar conditions. Figure 13 (a) shows the austenite structure, grain size approximately 50 μm, after solution treatment above 1050°C and cooling to 890°C. An 8% deformation at 930°C, resulted in severe grain boundary decohesion, and the formation of parallel slip bands within the austenite grains (b). Holding resulted in the disappearance of some slip bands, concurrent with the formation of subgrain boundaries (c). No recrystallisation was seen after 25 minutes holding.

The specimen was cooled, and rapid transformation occurred at about 640°C. Heavy surface precipitation, which occurred during the austenite to ferrite transformation, obscured ferrite formation, but surface cleaning revealed a structure with two distinct ferrite morphologies (d). Ferrite formed in areas which had recovered was uniform and equiaxed, with no surface markings, but ferrite formed in slipped areas was non-equiaxed, with jagged grain boundaries, and heavy surface markings, which were coincident with slip lines in the pre-transformation austenite structure. Thus, the state of the austenite prior to transformation had a pronounced effect upon the final ferrite morphology.

The surface precipitation during transformation was observed in all of the experiments carried out on this steel, and in some cases it was seen as fine curved rows of precipitates, with approximately 0.2 μm spacing, which went across individual ferrite grains. This was clear evidence that interface precipitation occurred in this alloy during ferrite formation.

(ii) Austenite Deformation in Titanium Microalloy Steel. Deformation of the titanium microalloyed steel at 900°C did not result in austenite recrystallisation. The austenite structure had a 30-40 μm grain size, after eight minutes solution treatment at above 1050°C, followed by cooling to 900°C The structure was very pitted because of the presence of high temperature titanium precipitates, and it suffered rapid thermal grooving during solution treatment. These factors severely affected the image quality. After initial rapid austenite grain growth, the grain structure remained stable, and no observable grain growth occurred, despite marked grain size heterogeneity.

Five percent deformation at 900°C resulted in grain boundary decohesion and subsequent precipitation at austenite grain boundaries. A further 4% deformation, followed by two minutes holding, resulted in more decohesion and a subgrain boundary network started to develop within the austenite grains. Cooling to 750°C and holding for five minutes resulted in the austenite-ferrite transformation starting. The austenite subgrain structure became more pronounced at this temperature and fine precipitation had occurred at some subgrain boundaries. The austenite grains showed thermal grooving. Ferrite grain contrast in the fully transformed structure was very poor, so it was not possible to obtain detailed information about the effect of deformation upon ferrite formation.

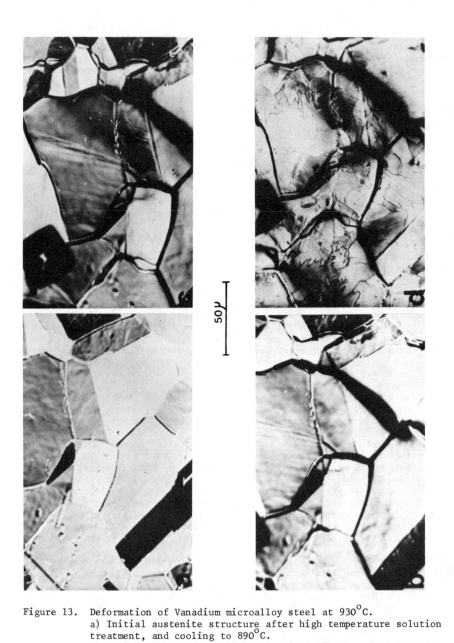

Figure 13. Deformation of Vanadium microalloy steel at 930°C.
a) Initial austenite structure after high temperature solution
treatment, and cooling to 890°C.
b) Grain boundary decohesion and slip band formation after 8%
deformation at 930°C.
c) Austenite subgrain formation, after holding at 890°C.
d) After transformation at 640°C two distinct ferrite morphologies
occur:

(i) Equiaxed ferrite, formed in areas of recovered austenite.
(ii) Jagged ferrite, formed in areas of non-recovered austenite.

Discussion

Ferrite Formation in Austenite

The results of this and other studies indicate that ferrite formation in controlled rolling is considerably more complex than transformation in undeformed steels, particularly when large reductions (with their accompanying roll chilling effects) at low rolling temperatures are used. In underformed austenite, ferrite nucleation occurs entirely at austenite grain boundaries, but after deformation, ferrite has also been found to form at the austenite-ferrite interface and at intragranular sites.

(i) Ferrite Formation at Austenite Grain Boundaries. In undeformed or recrystallised austenites, ferrite forms in equiaxed austenite grain boundaries at temperatures below A_3. In air cooled specimens, the ferrite takes on a more Widmanstatten type morphology, as the specimen transforms, due to undercooling. If a partially transformed specimen is rolled, then both austenite and ferrite are deformed, and a structure consisting of unrecrystallised single austenite grains, with a network of deformed and recovered grain boundary ferrite around them, is formed (Figure 6). If this is water quenched or finished below 680°C, the dynamically recovered ferrite substructure is retained, as there is no time for static recrystallisation to occur (Figure 4). Air cooling, from above 680°C allows enough time for statically recrystallised ferrite (4 μm grain size) to form. These grains are characteristic as they form in bands which are completely free of eutectoid products (Figure 3).

In deformed austenite, grain boundary ferrite forms as single rows of small equiaxed grains (Figure 7). If the austenite is highly deformed, these rows will be close together and parallel to the rolling plane, giving a banded appearance upon cooling (Figure 8). It has been suggested that this mode of ferrite nucleation is responsible for banding in controlled rolled structures which are finished below their recrystallisation temperatures (8,9).This would explain why semi-continuous bands of bainite products are seen in the air-cooled vanadium steel (Figure 8). The bands are the outlines of carbon rich austenite grains, trapped between parallel rows of grain boundary nucleated ferrite, which have transformed to bainite at lower temperatures.

(ii) Ferrite Formation at the Austenite-Ferrite Interface. Ferrite nucleation at the austenite-ferrite interface is seen in steels rolled both above and below the A_3 temperature. Large deformations provide the combined driving forces of both deformed structure and undercooling, which are needed to form ferrite in this way. Figure 5 shows a striking example of "cascade"(10) nucleation of fine grained equiaxed ferrite in a steel deformed just above A_3. After deformation the specimen was chilled by the rolls to below A_3, causing rapid saturation of all grain boundary nucleation sites. The undercooling provided enough driving force for further rows of ferrite to be successively nucleated at the austenite-ferrite interface, producing a very fine, equiaxed structure. The ferrite became finer further away from the original austenite grain boundary, indicating that they were sequentially nucleated at progressively lower temperatures. Similar structures were seen in specimens quenched after large deformations below A_3, the equiaxed grains being nucleated at the interface between deformed ferrite and austenite. Air cooled specimens showed similar rows of equiaxed ferrite grains around the austenite grain boundaries and these had undergone some grain growth.

(iii) Ferrite Nucleation at Intragranular Sites. Intragranular nucleation of ferrite grains was seen in steels which did not recrystallise between rolling and transformation; the tendency for nucleation at these sites was increased at lower temperatures, because of specimen undercooling.

Intragranular ferrite formed as characteristic rows of fine equiaxed ferrite grains, usually at $30^{\circ}-50^{\circ}$ to the rolling plane (Figure 1, 7, 10). Sekine and Maruyama (11) reported similar rows of ferrite to lie in the same orientations as austenite deformation bands, and therefore concluded that the bands acted as nucleation sites for ferrite grains. There is no direct evidence in this study to confirm these conclusions, but the fact that no similar ferrite rows were seen in undeformed or recrystallised austenite is a strong indicator that austenite deformation bands act as ferrite nucleation sites.

When specimens were air cooled, many of the intragranular ferrite grains were elongated, and lay parallel to each other at $30^{\circ}-50^{\circ}$ to the rolling plane (Figure 3). This was probably due to coalescence of the small equiaxed grains during cooling. These small grains formed from the same austenite grains, and thus inherited similar crystallographic orientations to the austenite matrix(12). During cooling, ferrite grain growth occurred, and the grains coalesced to form larger, elongated grains.

It should be noted that ferrite formation, at sites other than the austenite grain boundaries, was severely inhibited or completely prevented if roll cooling caused the start of the eutectoid reaction during rolling. Structures produced consisted of coarse mixtures of deformed ferrite and degenerate pearlite, with little or no fine equiaxed ferrite present. To avoid these undesirable structures, the rolling finishing temperature must be above A_1.

(iv) Duplex Ferrite Structures. Microstructures consisting of two distinct ferrite grain sizes were seen in some of the rolled structures (Figure 3) and it has been shown by Jones and Rothwell (13) that these are formed after partial austenite recrystallisation has occurred, prior to transformation, subsequent to rolling low in the austenite range.

This study has shown that the formation of duplex structures may be explained by three different mechanisms, depending upon the rolling temperature. Firstly, rolling above A_3, and finishing at a temperature high enough for partial recrystallisation to occur before transformation, will produce a mixed austenite grain size which will subsequently air cool to a duplex ferrite structure (13). Secondly, large reductions above A_3, resulting in cooling into the two-phase region, will cause rapid "cascade" nucleation (10) around unrecrystallised austenite grain boundaries, and intragranular ferrite nucleation at austenite slip bands. Since the concentration of intragranular nucleation sites is low compared with areas around the grain boundaries (11), air cooling will produce intragranular grain growth and a resultant duplex structure. Thirdly, duplex ferrite may be formed by large reductions below A_3. This again will result in fine ferrite grains, formed from deformed and recrystallised ferrite, around the austenite grain boundaries, with intragranular ferrite forming and growing at deformation band sites.

Photo-Emission Electron Microscopy

In general, the results obtained using hot deformation in the P.E.E.M. were, within the limitations of the technique, in good agreement with observations made in this and other hot deformation studies.

(i) High Temperature Austenite Deformation. Controlled rolling at high temperatures (>825°C in the austenite region, at strains less than the critical strain to produce dynamic recrystallisation, resulted in static recovery and recrystallisation. In the P.E.E.M., deforming and holding austenite resulted in subgrain formation, followed by a sudden change in phase contrast, as strain free recrystallised austenite grains (γ_2) were formed. This phase contrast

450

was a real effect, because in partially recrystallised austenites the ghosts of γ_1 grain boundaries disappeared from within recrystallised grains and γ_2 grain boundaries suffered decohesion on subsequent deformation (see Figure 11).

Recrystallised (γ_2) grains in controlled rolled steels are finer than the initial (γ_1) austenite grains, but in the P.E.E.M. experiments it was found that the γ_2 grain size was always coarser. This was because only limited amounts of deformation could be used in the P.E.E.M. experiments, so that straining was to just above the critical strain for recrystallisation nucleation. In this situation, very few recrystallisation nuclei were formed, and rapid growth occurred, resulting in coarse γ_2 grain sizes. In the controlled rolling expdriments however, deformations of greater than 50% were used, and these resulted in the formation of many recrystallisation nuclei, which gave a refined γ_2 structure.

The coarse recrystallised austenites formed in the P.E.E.M. transformed to coarse ferrite structures. This observation clearly shows that recrystallised austenite grain size has an important effect upon subsequent ferrite grain size, because the number of ferrite nuclei per unit area of austenite grain boundary is constant (11, 14).

(ii) Low Temperature Austenite Deformation. Low temperature austenite deformation may be defined as deformation in a temperature range where no austenite recrystallisation is observed (15, 16). In the P.E.E.M., deforming 5% at 825°C and 4% at 740°C produced no recrystallisation. Disruption of specimen surfaces increased at lower austenite temperatures, so the maximum amount of deformation possible was also decreased. Generally, at lower deformation temperatures, the effect of working upon the structures became more severe, with an increasing tendency to slip band formation as the thermally activated recovery and recrystallisation processes became more difficult.

The unrecrystallised austenite structures produced finer ferrite grain sizes than recrystallised austenite, illustrating that transformation from deformed austenite promotes ferrite grain refinement (11). Results from deformation at 740°C (Figure 12d) indicate that ferrite grain growth is influenced by slip bands in deformed austenite, but difficulties arise in obtaining information on ferrite nucleation, because specimen free surfaces act as a preferential nucleation site for ferrite grains.

(iii) Deformation of Microalloy Steels. Deformation of microalloy steels in the P.E.E.M. showed that at 900°C, both titanium and vanadium had a delaying effect upon austenite recovery and recrystallisation, although the effect of vanadium was more pronounced than that of titanium.

In the vanadium treated steel, no recrystallisation was observed, and the structure only partially recovered. Unrecovered areas clearly showed slip band formation (Figure 13), which was seen in the C-Mn steel only below 800°C. This indicated that vanadium additions stabilised the "low temperature" austenite structure, by delaying softening processes. In the titanium treated steel, again no recrystallisation was observed, but large scale recovery had occurred, and there was no evidence of slip band formation.

The presence of two distinct structures, after deformation and holding of the V microalloy steel, was due to high temperature vanadium nitride precipitation. Vanadium carbide is soluble in austenite above 720°C (17), so at 900°C VC precipitates could not affect softening kinetics. Recovered structures were stable in areas where recovery had occurred prior to precipitation, but in unrecovered areas, VN precipitation had stabilised the "low temperature" slip band structure. Strain induced VN precipitation prevented sub-grain

451

formation by inhibiting dislocation climb (18).

Subsequent transformation produced equiaxed ferrite from areas of recovered austenite, and jagged, non equiaxed ferrite from slipped areas. This
shows that in hot worked microalloy steels, the presence of unrecrystallised
austenite affected subsequent ferrite formation. During transformation, there
was an interaction between the ferrite grain boundaries, and the deformed
structure, with grain boundary pinning occurring along slip band lines (Figure 13d). The pinning, which was associated with precipitates formed on the
slip bands, inhibited grain boundary movement, causing the jagged nature of
some ferrite boundaries. The surface markings in the low temperature structure, were due to V(C, N) precipitation on slip band relief steps during
transformation. Subsequent ion beam cleaning of the specimen removed the
precipitates, and heavily etched the surface in the unrecovered region.

In the titanium microalloy steel, precipitates were not dissolved during
initial solution treatment, so high temperature austenite grain growth was
prevented. The heterogeneious austenite grain structure should have provided
a large driving force for grain growth (9), but Ti(C, N) precipitates prevented this by pinning austenite grain boundaries. Titanium microalloy additions
inhibited austenite softening processes, although the effect was not as strong
as that of vanadium. Because Ti(C, N) particles were not dissolved, the fine
scale reprecipitation necessary to inhibit recovery was not possible, so a
well developed subgrain structure was formed. However, recrystallisation by
high angle boundary migration can be prevented by precipitates much larger
than those needed to inhibit recovery (19), so the retained high temperature
precipitates were probably responsible for the recrystallisation inhibition.

Conclusions

The main experimental findings and conclusions that are drawn from this
work are summarised below:

1. Controlled rolling of plain C-Mn steels, with or without small additions of titanium or vanadium, refines ferrite grain size. As rolling temperature is decreased, ferrite grain size is progressively refined, but there
is a greater tendency to form structures which are banded, or contain deformed
and unrecrystallised ferrite. Increased rolling reduction also increases
grain refinement, but it may cause specimen cooling to below A_1, so that cementite is nucleated before rolling is completed. Microalloying additions
cause additional grain refinement and the production of more homogeneous
structures than are found in controlled rolled C-Mn steels, and structures
produced are less sensitive to variations in rolling conditions. At high Mn
levels, microalloy additions encourage bainite formation in controlled rolled
and air cooled specimens, due to modified ferrite nucleation mechanisms.

2. After controlled rolling, ferrite may be nucleated in unrecrystallised austenite:

    (i)   at austenite grain boundaries,
    (ii) by a "cascade nucleation" mechanism at the austenite-ferrite inter-
        face,
  (iii) at intragranular slip band sites within the austenite grains.

3. After controlled rolling, duplex ferrite structures may be formed in
air cooled specimens:

    (i)   by transformation from partially recrystallised austenite structures,
    (ii) by transformation from unrecrystallised austenite rolled above A_3,

(iii)　by transformation from structures rolled in the austenite + ferrite region.

4. Titanium and vanadium microalloy additions to C-Mn steels inhibit austenite recrystallisation, in hot deformed structures, below 1000°C. In the P.E.E.M. vanadium was observed to be more effective than titanium in preventing softening, as it partially inhibited austenite recovery, whereas titanium prevented recrystallisation only, and allowed a fully recovered structure to develop. Titanium was, however, found to be a very effective austenite grain growth inhibitor.

References

1. "Microalloying '75'" International Conference Proceedings, Washington, U.S.A., October 1975. Union Carbide Corporation.

2. "Controlled Processing of High Strength Low Alloy Steels" Conference Proceedings, York, England. September 1976. British Steel Corporation.

3. "Low Carbon Structural Steels for the Eighties" Conference Proceedings, Plymouth, England. March 1977. Institution of Metallurgists.

4. I. Kozasw et al, Ref. 1, p.120.

5. L. Meyer et al, Ref. 1, p.153.

6. L. Wegmann, Prakt. Metallog. 5 (1968) p.241.

7. K.E. Parker, B. Burnett and A.J. Baker. 1st International Conference on Emission Electron Microscopy, September 1979, Tubingen, Germany, p.123.

8. D.J. Naylor, Ph.D Thesis, 1970, Leeds University.

9. T. Gladman and D. Dulieu, Metal Science 6 1974, p.441.

10. A. Le Bon et al, Ref. 2, Paper No. 6.

11. H. Sekine and T. Maruyama, Trans. ISI Japan, 16, 1976, p.427.

12. A. Jones and B. Walker, Metal Science, 8, 1974, p.397.

13. J.D. Jones and A.D. Rothwell, ISI Special Publication No. 108, 1968, p.78.

14. R. Priestner et al, J.I.S.I., 206, 1968, p1252.

15. A. Le Bon and Saint-Martin, Ref. 1, p.90.

16. T. Tanaka et al, Ref. 1, p.107.

17. A.M. Sage, Ref. 3, II p.1.

18. W. Roberts, Swedish Inst. for Metals Research, Report No. 1M-1211, 1977.

19. B.L. Phillipe and F.A.A. Crane, J.I.S.I., 211, 1973, p.653.

DISCUSSION

Q: I was interested in your intercritical rolling experiments. We have done similar experiments where you are deforming, and apparently recrystallizing the ferrite, to that extremely fine structure that you observed. But, as we looked at the development, it wasn't quite clear whether this was really recrystallization, or the development of a very well characterized substructure which started from very low misorientations and then increased. In trying to check this with the transmission microscope, we concluded that these were, in fact, just highly developed subgrains with misorientations of 2 or 3 degrees that etched in nital just like high angle grain boundaries. Do you have any comment on this?

A: Yes, that would be the case. I believe that the angle of misorientation would be quite high, even if it is a subgrain, or not. They are very fine by normal grain size standards. I think the other question is, whether they nucleate within an existing mass of ferrite, or whether they have formed progressively at the edges of the ferrite.

Q: In the specimen which you have been studying with an emission microscope, you have shown that the grain size of the ferrite is the same as the austenite. Have you checked the inside of the specimen after the tests, that is, have you cut the specimen and looked at the ferrite grain size inside the specimen or not?

A: Yes, we normally do take sections. The grain size in our case, using slow deformate rates, was similar, and there wasn't any degree of preferential cusping at the surface.

STRAIN-INDUCED γ→α TRANSFORMATION IN THE ROLL GAP IN

CARBON AND MICROALLOYED STEEL

R. Priestner
Joint University of
Manchester/UMIST
Metallurgy Department,
Grosvenor Street,
Manchester M1 7HS, U.K.

In two microalloyed steels, strain-induced transformation to ferrite
(SIT) during rolling nucleated on elongating γ grain boundaries and deformation bands. In a C-steel, initial SIT stopped dynamic recrystallization
and allowed more SIT on elongating γ grain boundaries and deformation bands.
Ferrite grain refinement by SIT in the roll gap was thus similarly efficient
in both types of steel. SIT in the roll gap started very rapidly and then
slowed dramatically, and yielded two characteristic ferrite morphologies:
(a) fine, impinged α grains, and (b) extremely large, heavily substructured
single crystals of ferrite occupying large areas of γ grain boundary. It is
suggested that these features of SIT may be explained by relative crystallographic rotations of α allotriomorphs in the γ matrix caused by plastic
deformation after nucleation.

Introduction

The research reported here is an attempt to obtain direct insight into the way austenite transforms to ferrite in the roll gap during continuing plastic deformation. Transformation within a few seconds of exit from the roll gap is also considered. It is easy enough to achieve an experimental situation in which transformation begins at some arbitrary point in the roll gap. Systematic control of the temperature and position in the roll gap at which transformation begins is rather more difficult to achieve, and requires prior knowledge of the steels under investigation. The general background to the experimental method will, therefore, be reviewed briefly before coming to the details of the present experiments.

If a steel slab containing an embedded thermocouple is subjected to a hot working pass, a temperature record like that in Figure 1(a) would indicate that transformation began after its exit from the roll gap, one like Figure 1(b) would indicate that transformation began before exit from the roll gap. By varying the reduction in the pass and the cooling rate after exit, for situations corresponding to Fig. 1(a) (rolling "above" transformation), a partial CCT diagram can be constructed. Fig. 2 is such a CCT diagram for one of the microalloyed steels used in the research. Deformations upto 70% (given in one or two passes) progressively raised the temperature of the start of transformation and moved the start to shorter times. For a particular cooling rate the transformation-start temperature may vary with rolling reduction as illustrated (schematically) in Figure 3. If the slab temperature falls during rolling at that same cooling rate, according to the second line in Figure 3, then transformation should be expected to start at that place in the roll gap where the transformation-start temperature was raised to meet the slab temperature, at (R',T'). With sufficient information, R' and T' can be varied systematically, as in the series of experiments to be reported.

Quantitative information about the effect of rolling on the transformation can be obtained from samples that have been either quenched-out directly on exit from the roll gap or allowed to cool to yield a ferritic structure. For slowly cooled, fully ferritic specimens it is useful to define the "Fractional Refinement" of ferrite grain size as:

$$\text{F.R.} = \frac{\bar{d}_{\alpha 0} - \bar{d}_{\alpha R}}{\bar{d}_{\alpha 0}} \qquad (1)$$

where $\bar{d}_{\alpha R}$ is the mean-linear-intercept ferrite grain size after a rolling reduction R, and $\bar{d}_{\alpha 0}$ is the ferrite grain size of a dummy sample given exactly the same thermal treatment but without being rolled (rolling reduction = 0). F.R. is not only a measure of the grain refining efficiency of a controlled-rolling operation; it also has some fundamental connotations (1). Suppose the mechanisms of transformation (e.g., nucleation rate, growth rate, etc.) remain constant, so that the ratio of austenite to ferrite grain sizes remains unaltered:

$$\bar{d}_{\gamma} = K\bar{d}_{\alpha} \propto \frac{1}{S_{\gamma}} \qquad (2)$$

where $\bar{d}_{\gamma}$ is the austenite grain size and S_{γ} is the austenite grain boundary area per unit volume. Then

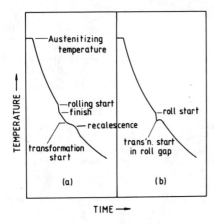

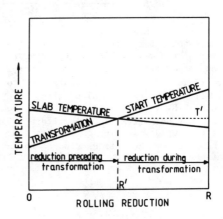

Fig.1. Schematic cooling curves during rolling: (a) transformation starts after finish of roll pass; (b) transformation starts during rolling.

Fig.3. Method of estimating temperature T', and reduction, R', at which transformation starts in the roll gap.

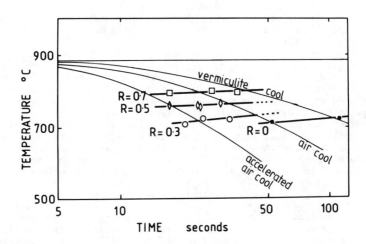

Fig.2. Effect of prior deformation by rolling on the start of transformation in the V-steel during continuous cooling.

$$\text{F.R.} = \frac{\bar{d}_{\alpha0} - \bar{d}_{\alpha R}}{\bar{d}_{\alpha0}} = \frac{S_{\gamma R} - S_{\gamma 0}}{S_{\gamma R}} = \Delta S_{\gamma}$$

(3)

where the subscripts 0 and R refer the austenite grain boundary area to rolling reduction. ΔS_{γ} is the fraction difference between austenite grain boundary areas in the rolled and unrolled states. S_{γ} may be altered by rolling, due to grain elongation or to recrystallization, and ΔS_{γ} may be easily determined by calculation or by measurement. In experiments for which F.R. is numerically equal to ΔS_{γ}, the mechanisms of transformation have not been significantly altered by the rolling process and K in Eq. (2) is unchanged: if F.R. is greater than ΔS_{γ} then K in equation (2) has been increased by the introduction of new nucleation sites and a greater rate of nucleation. Metallographic examination of quenched-out samples then yield information about such extra nucleation sites.

Experimental Details

The composition of the steels were:

Table I. Compositions of steels.

	C	Mn	Nb	V	N	Si	P	S	(wt%)
C-steel	0.2	1.2	-	-	-	0.2	0.02	0.02	
Nb-steel	0.093	1.22	0.06	-	0.009	0.26	0.014	0.016	
V-steel	0.085	1.24	-	0.17	0.012	0.26	0.014	0.016	

Slabs approximately 3.5 cm wide were machined from the original hot-rolled plates to a series of thicknesses such that, after rolling to various reductions, the finishing thickness was constant for each steel. This allowed consistent cooling rates to be obtained after rolling. For the C-steel the final thickness was 5 mm, for the microalloyed steels it was 3.8 mm. A chromel-alumel thermocouple was embedded on the centre plane of each slab, and a handling rod attached to one end. The C-steel was austenitized at 1020°C and the Nb- and V-steels at 1150°C for 30 minutes. After removal from the furnace, the slabs cooled in air until a predetermined roll-start temperature was reached, and then rolled at 10 m/minute. Deformations up to 50% were given in one pass, higher deformations in two. The roll-start temperatures were selected to allow transformation to start at pre-selected positions in the roll gap (as described in the introduction) or, in further experiments with the microalloyed steels, within a few seconds of emergence from the rolls. Immediately after rolling the slabs were plunged into a bed of vermiculite to simulate slow cooling of thicker plates, or cooled in still air, or in an air blast, or quenched in cold water or brine. Thermocouple output was recorded continuously except during rolling, when spurious emfs caused by contact with the rolls degraded the signal. Start- and finish-rolling temperatures were recorded, and a linear variation of temperature between these extremes was assumed for the purpose of constructing figures like Fig. 3, in order to determine R'. For the further experiments with the microalloyed steels the thermo-couple output recordings also yielded estimates of the time elapsed between the end of rolling and the start of transformation.

Volume fractions of ferrite present in quenched-out samples, and grain

sizes, were determined using a Q.T.M. 720 image analysing computer after
standard metallographic preparation, at the centre-planes of the rolled-
slabs. Aspect ratios of austenite grains, and austenite grain sizes, were
determined by manual counting, using quenched-out specimens in which
austenite grain boundaries were outlined by ferrite (the remainder of the
structure being dark-etching quench products). Micrographs were prepared
using optical and scanning electron microscopes.

Results and Discussion

When transformation started at R' in the roll gap, the amount of
ferrite present at the end of the total rolling reduction, R, may have been
a function of the amount of rolling that followed the start of transformation, (R - R'),or of the amount that preceded the start, or both. In
Figure 4 the amount of ferrite present in the Nb-steel at exit from the roll
gap is plotted against (R - R'), experiments in which the total reduction
was 0.3, 0.5 and 0.7 being distinguished by graph symbol. All the data fall
near a straight line intercepting the ordinate at approximately 11% ferrite.
Of course, no ferrite can be present for (R - R') = 0, as indicated by the
dashed extension of the curve through the origin. The simplest interpretation of these data is that transformation began extremely rapidly at
R' and then slowed down, and was independent of the amount of deformation
that preceded the start of transformation. In the V-steel, Figure 5, the
amount of ferrite present on exit increased also as the deformation preceding R' increased. It would appear that the extra nucleation sites, on
which very rapid initial ferrite formation could take place, were introduced
very early in the deformation process in the Nb-steel, and reached a
saturation level, but were more gradually introduced in the V-steel, and
continued to increase with prior deformation. In the C-steel only passes
of 50% reduction were made, and, again, the amount of ferrite present on
exit from the roll gap increased as the amount of deformation following the
start of transformation increased, Figure 6. Again, it is suggested that
transformation was most rapid at the initiation of transformation.

The most prolific source of extra ferrite nucleation sites observed in
quenched-out samples appeared to be deformation bands traversing austenite
grains, an example of which is shown in Figure 7. Ferrite nucleation on
such bands has been observed by others (2,3,4), and Priestner and de los
Rios (1) quantitatively identified the degree of grain refinement in slowly
cooled specimens with the density of such bands. Fractional Refinements of
ferrite grain size are plotted in Figures 8 and 9, for the microalloyed
steels, as a function of the rolling reduction which was applied before the
slabs were cooled in vermiculite. In each figure the dashed line represents
the variation of ΔS_γ with R calculated from the austenite grain elongation:
it is known (1) that F.R. follows exactly the variation of ΔS_γ if no extra
intragranular nucleation sites are introduced. In each figure the upper
curve represents the F.R. of ferrite grain size that was achieved when
transformation began in the roll gap: the degree of grain refinement F.R.,
was greater than ΔS_γ due to nucleation on deformation bands. The numbers
against open data points are the number of seconds that elapsed between the
exit from the roll gap and the start of transformation during subsequent
cooling in vermiculite, in experiments conducted according to the schematic
Figure 1(a). It is clear that within 35 seconds much of the grain refining
potential of the deformation bands present at exit from the rolls was
annealed out. This recovery of substructure seemed to be more rapid in the
Nb-steel than in the V-steel, and faster after lower reductions.

From the data available, the ferrite grain refinement in slowly cooled
microalloyed steels seemed to be independent of how early in the roll pass

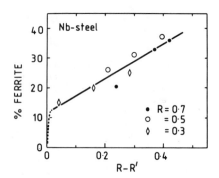

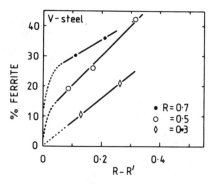

Fig.4. Ferrite content on exit
from roll gap as function
of amount of rolling
reduction that followed
start of transformation
Nb-steel.

Fig.5. Ferrite content on exit
from roll gap as function
of amount of rolling
reduction that followed
start of transformation
V-steel.

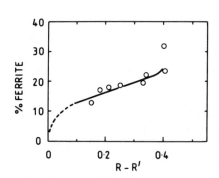

Fig.6. Ferrite content on exit
from roll gap as function
of amount of rolling
reduction that followed
start of transformation
C-steel

Fig.7. V-steel quenched out on
exit from roll gap. R=0.5,
R'=.184, T'=719°C

460

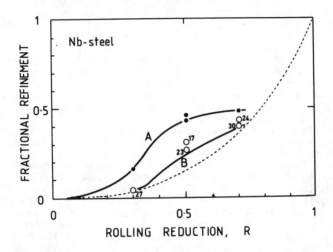

Fig.8. Effect of rolling reduction, R, on fractional refinement of ferrite grain size. Nb-steel cooled in vermiculite after rolling. Solid points, transformation started in roll gap; open points, transformation started the number of seconds indicated after exit from roll gap.

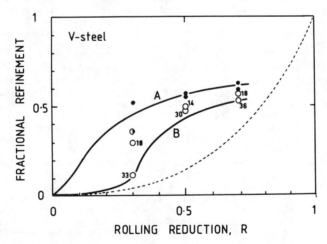

Fig.9. Effect of rolling reduction, R, on fractional refinement of ferrite grain size. V-steel cooled in vermiculite after rolling. Solid points, transformation started in roll gap; open points, transformation started the number of seconds indicated after exit from roll gap.

strain-induced transformation started. This was not the case in the C-steel. In Figure 10 the Fractional Refinement of vermiculite cooled slabs for which transformation started in the roll gap is plotted versus the amount of deformation, R-R', that followed the onset of transformation. Examination of quenched-out samples showed that the earlier transformation started in the roll gap, the coarser and more elongated were the austenite grains. It was deduced that recrystallization of austenite occurred up to R', and was then stopped by the precipitation of ferrite on grain boundaries. During the remainder of the rolling strain the austenite grains then elongated. Austenite grain boundary areas per unit volume, S_γ, were determined, and the variation of ΔS_γ with R-R' is shown as a dashed curve in Figure 10. (N.B. Grain elongation is a very inefficient way of increasing grain boundary area, relative to recrystallization to a finer grain size, for reductions below about 70%.) Clearly, some extra agency promoted nucleation of ferrite, with increasing power as the amount of deformation following initial ferrite formation increased. This extra agency was identified in quenched-out samples as nucleation on transgranular deformation bands in the austenite, Figure 11. It seems that when transformation started in the roll gap recrystallization of austenite was inhibited; continuing deformation then produced deformation bands in increasing numbers as deformation continued. In the microalloyed steel it seemed that all the deformation bands that participated in ferrite nucleation formed during the deformation that preceded the onset of ferrite nucleation.

The morphologies of ferrite in quenched-out samples fell into two groups. First, as shown in Figure 7 and 11, were a range of "normal" ideomorphs which, had the samples not been quenched, would have developed into the roughly equiaxed grain structure observed in specimens slowly cooled after rolling. Secondly, very long ferrite grains, often stretching for long distances along austenite grain boundaries and deformation bands, and containing much substructure, were also observed in specimens in which transformation began in the roll gap. Figure 12 illustrates such a ferrite grain of moderate size; even so it is over 100 μm long and only 8 to 10 μm across. Such ferrite grains were often more extensive, and gave rise to the incorporation of large, substructured grains in slowly cooled samples, Figure 13.

In samples in which transformation had progressed somewhat further, the first morphology developed into many small grains in the vicinity of prior austenite grain boundaries, e.g., Figure 14. That is, after impingement, the first grains that formed along a boundary neither coalesced nor grew further, but transformation preferred to proceed by the renucleation of ferrite to form a band of very fine ferrite grains. In Figure 14, up to ten ferrite grains can be counted across a band only 15 μm thick. In the second morphology, transformation continued by the renucleation of ferrite grains in austenite adjacent to the large plates of ferrite, as seen in Figure 12. Figure 15 is a further illustration, in which it appears that new grains formed adjacent to a large, old one are being incorporated by some coalescence mechanism into the original grain. The coalescence seems to be progressing from the upper, left corner of the micrograph, at the expense of incorporating considerable substructure into the enlarged grain. Similar ferritic constituents have been observed by others in controlled-rolled steel (2,3,4).

In attempting to explain the diverse ferrite morphologies, it is emphasized that ferrite formed in the roll gap, and deformation continued as transformation proceeded. Thus, each ferrite grain was itself subjected to plastic deformation in compression from the moment it appeared. The continuing plastic deformation would, therefore, cause the crystallographic orientation of each ferrite grain to rotate, in a manner determined

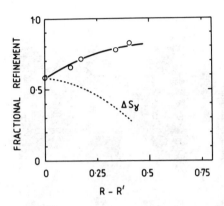

Fig.10. C-steel. Dependence of fractional refinement of ferrite grain size on R-R'. Vermiculite-cooled after rolling reduction R=0.5.

Fig.11. C-steel quenched-out on exit from roll gap SEM micrograph, ferrite shows dark. R=0.5.

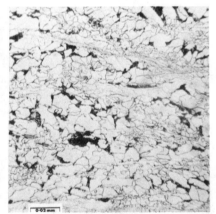

Fig.12. V-steel quenched-out from roll gap. R=0.5. R'=0.184. T'=719°C. Long ferrite grain probably formed by coalescence of many small ones.

Fig.13. Nb-steel, vermiculite-cooled. Long grains like that in Fig.12, inherited in fully-transformed structure.

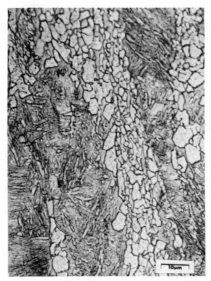

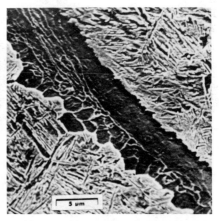

Fig.14. V-steel, quenched-out on
exit from roll gap.
R=0.5, R'=0.333,
T'=729°C.

Fig.15. V-steel, quenched-out on
exit from roll gap.
R=0.5, R'=0.184.
T'=719°C. SEM micrograph,
ferrite is dark.

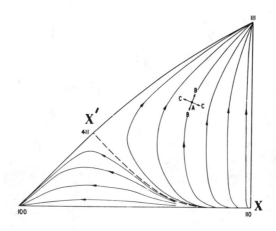

Fig.16. Orientation in single crystal b.c.c. grains
during compression. Compression axes move as arrows
(After Dillamore and Kato, ref. 5).

principally by its initial crystallographic orientation with respect to the compression axis. It is held that such rotations may, on the one hand, be such as to bring a number of adjacent grains to a common orientation, so increasing the probability of coalescence, or, on the other hand, may increase the orientation difference between grains, thus leading to a more stable, polycrystalline aggregate. It is, of course, by such crystallographic rotations in individual grains in an aggregate that rolling textures are formed. As individuals, the orientation of ferrite grains would tend to rotate according to Figure 16, taken from Dillamore and Kato (5). Grains to the right of the boundary X-X' would rotate so their <111> directions moved towards the compression axis, grains to the left of X-X' would move towards a <100> orientation. If two adjacent grains had orientations which straddled X-X', their orientations during compression would always diverge, one going eventually to <111>, the other to <100>. If both were on the same side of X-X', then their orientations could either diverge or converge, initially, during compression. For example, two orientations near the <100>-<110> boundary would initially diverge, two near the <100>-<111> boundary would converge.

The rotations described are, of course, subject to various restraints, because the ferrite grains are each part of a complex, polycrystalline aggregate of ferrite and austenite. Such a constraint on rotation of crystallographic structure is likely to be the nature of the interface between a ferrite grain and the austenite matrix. The nature of the interface is determined by the relative orientation of the two crystal structures and by the crystallographic orientation of the boundary in the structures. If it is a high energy interface, or contains many ledges, at which rapid ferrite growth can occur, then an extra driving force for rotation exists, if the rotation is such as to reduce the energy of the boundary. Conversely, if rotation achieved a low energy interface, the structure would be stabilized against further rotation, and at the same time, the growth rate of the ferrite grain should become very slow. Locally, continued transformation may then proceed by re-nucleation of ferrite instead of continued growth, and this argument is independent of whether the rotations had caused coalescence of ferrite grains or a greater divergency of orientation among neighbouring grains. The rotation mechanism, operating until a low energy $\gamma-\alpha$ interface is achieved, is thus capable of explaining why transformation in the roll gap occurs by multiple re-nucleation (called "cascade" transformation by Maîtrepierre (6)) in preference to continued growth, as well as the presence, simultaneously, of long ferrite grains formed by coalescence, and goes far towards explaining why initially rapid transformation changes to relatively slow transformation.

Summary and Conclusions

When the $\gamma \rightarrow \alpha$ transformation occurred in the roll gap, it generally started rapidly, and then slowed down. In the microalloyed steels deformation bands, formed before transformation started, provided the extra nucleation to explain the initially rapid, strain-induced transformation. In the Nb-steel, the deformation bands appeared to reach a constant ability to nucleate ferrite early in the deformation process, whereas in the V-steel their ability to nucleate ferrite appeared to depend directly on the amount of deformation that preceded the start of transformation. In the C-steel, austenite recrystallized during rolling preceding transformation, but recrystallization was stopped by the onset of transformation: continuing deformation then caused austenite grain elongation and deformation bands, which increased initial transformation rate. Nucleation on deformation bands thus led to substantial ferrite grain refinement in both types of steel, when transformation started in the roll gap.

465

When transformation was delayed beyond exit from the roll gap, recovery processes in the austenite reduced the ability of deformation bands to nucleate ferrite, a substantial loss of grain refinement taking place within 35 seconds. Recovery appeared to be somewhat slower in the V-steel than in the Nb-steel, and slower after heavier deformations.

Strain-induced transformation in the roll gap resulted in two kinds of ferrite morphology. In one, ferrite grains nucleated on elongating austenite grain boundaries and deformation bands, grew to impingement, and then remained as separately identifiable grains. Transformation continued by the nucleation of new ferrite grains close to the original ones, instead of by the growth of those that formed first. This led to bands of extremely fine-grained ferrite associated with austenite grain boundaries. In the second morphology, impinging ferrite grains appeared to coalesce, leading to extremely long, heavily substructured grains along austenite grain boundaries and deformation bands. Transformation then also continued by the nucleation of fresh grains close to old ones, and these in turn could coalesce with the old grain. These morphologies may be explained, at least qualitatively, by consideration of crystallographic rotations imposed on newly formed ferrite grains by continuing plastic deformation.

Acknowledgements

Research at the University of Manchester/UMIST Joint Metallurgy Department on Strain Induced Transformation in the Roll Gap During Controlled Rolling is supported by the Science Research Council.

References

1. R. Priestner and E.de los Rios: <u>Metals Tech.J.</u> 1980, vol. 7, pp.309-316.

2. C. Ouchi, T. Sanpei, T. Okifa and I. Kozasu: "The Hot Deformation of Austenite", AIME, 1977, p.316; also in "Micro-Alloying '75", ASM, 1977, p. 120.

3. G.R. Speich and D.S. Dabkowski: "The Hot Deformation of Austenite", AIME, 1977, p.557.

4. R. Irani and E.A. Almond: "Structural Steels for the Eighties", AIME, 1977, p.557.

5. I.L. Dillamore and H. Kato: <u>Met.Sci.J.</u> (1974), vol. 8, p.73.

6. P.H. Maîtrepierre and J. Rofes-Vernis: "Heat Treatment '76", The Metals Society, London, 1976, Written discussion.

DISCUSSION

Comment: We have also observed large single grains similar to those that Mr. Priestner showed. These grains were observed in materials that transformed below the last deformation step. We deformed at 800°C, and the transformation began at 860°C. We think that those large single crystals are formed at deformation bands or twin boundaries that occur within the austenite matrix, and that they will then develop to a single ferrite grain because there are so few nucleation sites for other ferrite crystals.

BEHAVIOR OF AUSTENITE DURING THERMOMECHANICAL PROCESSING WITH

REGARD TO THE FINAL PROPERTIES*

B. Engl
Forschung und Qualitäts-
kontrolle
ESTEL Hüttenwerke Dortmund AG
Eberhardstr. 12

4600 Dortmund 1
Fed. Rep. of Germany

K. Kaup
Forschung und Qualitäts-
kontrolle
ESTEL Hüttenwerke Dortmund AG
Eberhardstr. 12

4600 Dortmund 1
Fed. Rep. of Germany

The main object of the thermomechanical treatment is to obtain a fine
transformed structure, because both strength and toughness are
enhanced. To investigate the mechanisms of grain refinement semi large
scale rolling experiments were carried out on microalloyed steels.
It is shown that the refinement of ferrite can directly be attributed
to the refinement of the austenite structure. Rolling with recrystallization decreases the austenite grain size progressively until a
limiting value is obtained. Rolling without recrystallization leads
to a reduced grain dimension normal to the rolling plane. The influence
of various combinations of rolling reductions in the recrystallizing
and the not recrystallizing range on the ferrite structure and the
resulting mechanical properties is shown.

* Research partly sponsored by the Bundesministerium für Forschung
und Technologie of the Fed. Rep. of Germany

Introduction

A better understanding of the changes in the austenite structure during hot deformation is necessary in order to get more benefits from the controlled rolling process. The most essential object is to get a fine transformed structure, because both strength and toughness are enhanced (1-4). To investigate the mechanisms of grain refinement, semi large scale experiments were carried out on microalloyed steels. The aim of the study was to enlarge the knowledge of the structural changes and to get hints for a more efficient use of the controlled rolling process.

Experimental

The compositions of the steels used in this study are shown in the table 1.

Table 1. Chemical Composition of Steels

Steel	C	Si	Mn	P	S	Cr	Mo	V	Nb	Ti
A	.06	.39	1.60	.010	.002	.28	-	.09	.05	-
B	.04	.08	1.67	.011	.006	-	.34	-	.06	-
C	.10	.43	1.64	.017	.009	-	-	.07	.04	-
D	.09	.39	1.44	.008	.009	-	-	-	-	.20
E	.09	.20	1.39	.016	.011	-	-	-	-	-

The microalloyed steels A, B, C, D contain Ti, Nb, V single or in combi-

Fig. 1 - The hot rolling laboratory mill at ESTEL Hüttenwerke Dortmund AG (Roll diameter 550 mm, max. rolling force 6,6 MN, length of the barrel of the roll 600 mm, max. slab thickness 200 mm, max. rolling speed 0,7 m/sec)

nation. Further important alloying elements are Cr or Mo. Some of these steels are already produced since several years with great success, or are new developed especially for the main applications of thermo-mechanically treated steels: large diameter line pipes and cold formable steels for the automotive industry.

The rolling experiments were carried out in a laboratory two high mill (Fig. 1). The average strain rate is 6 per second. This hot rolling mill allows, due to its size and facilities, rollings with conditions similar to mill plant rolling processes. Temperatures and other rolling data are controlled with special measurement equipments. Strip simulation can be performed by using spray or laminar flow cooling on the run out table and an additional controlled furnace cooling. Slabs of various thicknesses up to 200 mm were rolled to plates with a constant thickness of 18 mm using rolling schedules similar to mill plant processes. In different stages of the process plates were quenched in iced water within five seconds. The austenitic structures were revealed by etching in saturated aqueous picric-acid solution containing a wetting agent. The average grain dimensions of the austenite and of the ferrite structures were determined parallel and normal to the rolling plane using a semi automatic image analyzer. The grain aspect ratio was taken as indicator of the degree of recrystallization.

Results and Discussion

From the results of many investigations it is evident that a fine recrystallized austenite can be transformed into fine grained ferrite (2, 5, 6). This is confirmed by the results of the present study (Fig. 2). The relationship between austenite and ferrite grain size shows a constant ratio. It is shown in Fig. 2 that the ferrite grain sizes after transformation from elongated unrecrystallized austenite grains (dark symbols) are as fine as those from the recrystallized austenite grains having the same diameter as the elongated grains dimension normal to the rolling plane (open symbols). The same correlation for the equiaxed (recrystallized) as for the elongated (non-recrystallized) austenite structure and the ferrite grains seems to be valid up to at least 75 % rolling reduction producing elongated austenite grains. More details will be discussed later on.

The question is how to get on the most effective way a fine austenite structure by recrystallization or by avoiding recrystallization or by a combination of both mechanisms in order to get a fine ferrite structure. As the first steps in this trial the influence of reheating on the austenite grain size and the recrystallizing behavior were investigated.

The knowledge of the grain sizes after reheating is essential because they act as the initial grain sizes for the following recrystallization process. In Fig. 3 the influence of the reheating temperature on the austenite grain size is shown. The addition of the microalloying elements reduces the austenite grain size. This is explained by the growth and solution of particles. The small grain size of steel A is caused by a most effective dispersion of particles stable up to a reheating temperature of $1300^{\circ}C$.

○ steel A, austenite recrystallized
● " " not recrystallized
(rolling reduction 75%, grain
dimension ⊥ rolling plane)
▼ steel E, austenite recrystallized

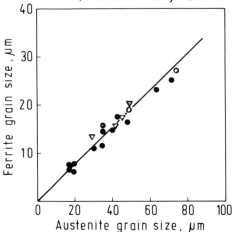

Fig. 2 - Relationship between the austenite
and the ferrite grain size

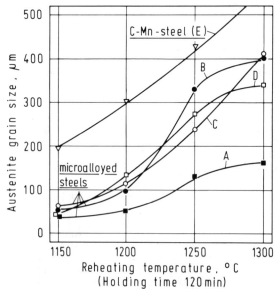

Fig. 3 - Influence of the reheating temperature
on the austenite grain size

The results of the recrystallization examinations are summarized in Fig. 4. In the present investigation static and not dynamic recrystallization seems to be dominant. This is concluded from the data of other investigations for similar test conditions (7-9). It is to be seen from Fig. 4 a) that the recrystallization behavior is mainly influenced by the temperature. For the comparison of the three investigated steels the initial grain sizes and the total reduction are kept constant. The transition from full recrystallization to non-recrystallization occurs in a narrow range between about 1050°C and about 925°C for these steels. There is also a slight influence of the reduction per pass. Increasing pass reductions lower the temperature limit for full recrystallization.

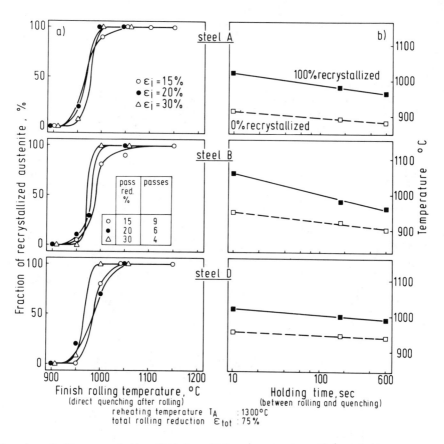

Fig. 4 - Influence of the finish-rolling temperature (a) and the holding time after rolling (b) on the recrystallization

As shown by the results in Fig. 4 b) the recrystallization is not so much influenced by the isothermal holding time after rolling. The results of the recrystallization experiments allow to divide the hot rolling process in a fully, partially and non-recrystallizing stage.

Rolling with recrystallization decreases the austenite grain size progressively until a limiting value is obtained (Fig. 5 a). This observation is principally in agreement with data reported in the literature (10). It is to be seen that the limiting grain size is the same for the three steels, being reheated at the same temperatures but showing different initial grain sizes. Fig. 5 b) shows that grain growth during holding after rolling is not quite important especially for the steel A.

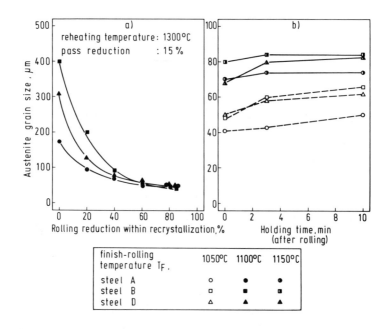

Fig. 5 - Influence of the rolling reduction (a) and the holding time after rolling (b) on the recrystallized grain size

In Fig. 6 the other significant influences on the limiting fineness of the recrystallized austenite grain size are shown. From Fig. 6 a) it is to be seen that it will become smaller with decreasing rolling temperatures. Fig. 6 b) demonstrates that the limiting grain size will also become finer if the pass reduction is increased.

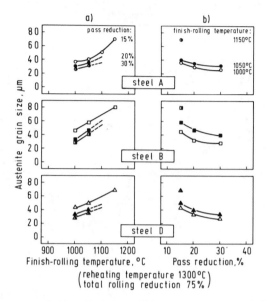

Fig. 6 - Influence of the finish-rolling temperature (a) and the
pass reduction (b) on the recrystallized grain size

In Fig. 7 the influence of deformation with (a, b) and without
(c, d) recrystallization on the austenite structure is shown.
Rolling with recrystallization refines the structure from different
initial grain sizes to a constant value (Fig. 7 a). The curves
representing the corresponding grain refinement are given in Fig. 7 b).
The total refinement is lowered with a decrease of the initial grain
size. If this was already as fine as the limiting value, it remained
unchanged during rolling. The increase in refinement becomes smaller
with increasing rolling reductions and it is very small if the
rolling reduction exceeds about 60 %.

Rolling without recrystallization reduces the austenite grain
dimension normal to the rolling plane (Fig. 7 c). The grain dimension
reductions are in a good agreement with the values calculated from
the thickness reductions of the rolled material. The data of the
ferrite grain sizes show the close relationship to the austenite grain
dimensions as could be expected from the correlation shown in Fig. 2.
The grain refinement of the austenite (dimension normal to the
rolling plane) and of the corresponding ferrite structure is shown in
Fig. 7 d). Both follow the same relationship.

From these data it is evident that the austenite grain boundaries
are the dominant nucleation sites. It is shown in Fig. 8 and Fig. 9
that the nucleation of ferrite apparently occurs at the austenite
grain boundaries. They are the nucleation sites in the coarse (Fig. 8
a) as well as in the fine elongated structure (Fig. 9 a). The coarse
ferrite grains in Fig. 8 b) are thought to originate from the coarse
elongated austenite grains. The lack of efficient ferrite nucleation
in the interior of these grains is obvious. This is confirmed by

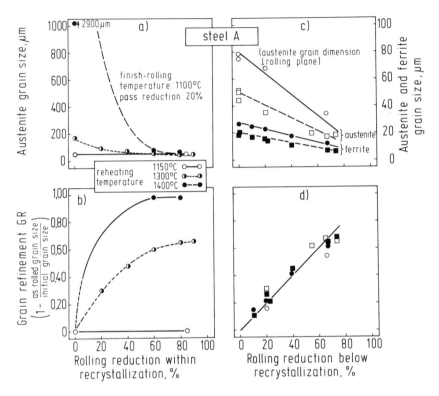

Fig. 7 - Comparison of the influence of deformation within (a, b)
and below (c, d) the recrystallizing temperature range on
the grain size and the grain refinement.

the result that only quite few deformation bands which could act as
nucleation sites were observed in the investigated structures.
Summarizing the results of these investigations there is no evidence
of any important intragranular ferrite nucleation potential in the
not recrystallized austenite deformed up to 75 %.

From the data in Fig. 7 the efficiency of refinement of the
austenite structure by recrystallization (Fig. 7 b) compared with
that by grain elongation (Fig. 7 d) can be evaluated. From the slope
of the grain refinement curves it is to be seen that the increase
of the refinement by recrystallization is generally stronger with
lower rolling reductions, and that the refinement becomes more
efficient by austenite grain elongation with higher reductions. Due
to the close relationship between the austenite and the corresponding
ferrite grain size (Fig. 2) the efficiency of the ferrite grain
refinement can also be estimated. Exceeding a certain rolling reduction,
ferrite grain refinement caused by austenite deformation below
recrystallization seems to be more efficient than that caused by
recrystallization of austenite. This is proved by the experimental
results. If at a constant total deformation the rolling reduction
below recrystallization is increased (Fig. 10, 11), the austenite

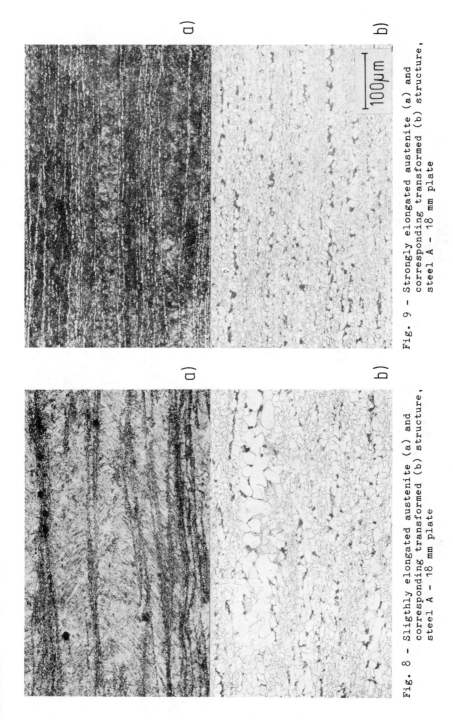

a)

b)

Fig. 9 – Strongly elongated austenite (a) and
corresponding transformed (b) structure,
steel A – 18 mm plate

a)

b)

Fig. 8 – Sligthly elongated austenite (a) and
corresponding transformed (b) structure,
steel A – 18 mm plate

100μm

475

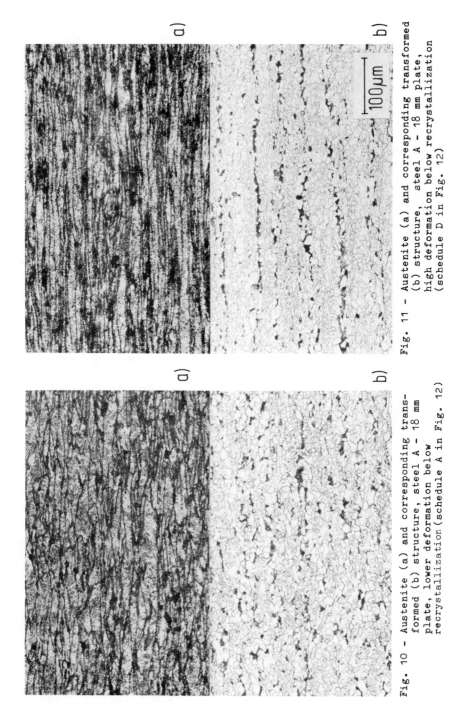

Fig. 11 – Austenite (a) and corresponding transformed (b) structure, steel A – 18 mm plate, high deformation below recrystallization (schedule D in Fig. 12)

Fig. 10 – Austenite (a) and corresponding transformed (b) structure, steel A – 18 mm plate, lower deformation below recrystallization (schedule A in Fig. 12)

476

steel A

Rolling schedule		Ferrit grain size μm	ΔY.S.* N/mm²	Y.S. N/mm²	50% Shear FATT °C
A		7,6	86	452	- 55
B →C Reduced pre-rolling temperature →C		B C 6,7 5,7	B C 98 74	B C 480 468	B C -85 -105
D		6,0	96	487	- 100
			* fraction of yield strength by precipitation and dislocation strengthening		

Increased deformation below recrystallization (vertical label, left margin)

Fig. 12 - Influence of increased deformation below recrystallization and reduced pre-rolling temperatures on the ferrite grain size and the mechanical properties (constant total deformation)

structure becomes more elongated and the corresponding ferrite grain size more refined from 7.6 to 6 μm. The resulting mechanical properties (Fig. 12) show that the increase of deformation below recrystallization leads to higher yield strengths and to lower FAT-temperatures. These changes are believed to be mainly caused by the grain refinement, though some additional effect on FATT might arise from separations in the fractured specimens.

477

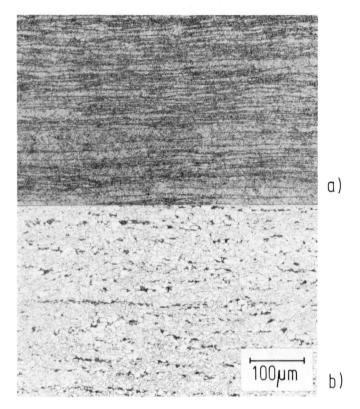

Fig. 13 - Austenite (a) and corresponding transformed
(b) structure, steel A - 18 mm plate
(schedule C in Fig. 12)

From the results of the investigations of the limiting
recrystallized austenite grain size it is expected that an additional
effective method to obtain a further ferrite refinement is reducing
the recrystallized austenite grain size by lowering the pre-rolling
temperatures, for example by the rolling schedule C (in Fig. 12).
The resulting structures are presented in Fig. 13. The ferrite grain
is refined from 6.7 to (B) 5.7 μm (C). With the decreased pre-rolling
temperature range the yield strength is slightly lowered, but the
FATT is strongly improved. The fraction of yield strength caused by
precipitation and dislocation strengthening was lowered by this
process, whilst there was no marked influence on these strengthening
mechanisms by increasing the rolling reductions below recrystallization
(sched. A. B. D).

The described results of ferrite grain size are all received
from experiments cooling the deformed austenite with a constant rate
(air-cooled 18 mm plate). A further ferrite grain refinement is

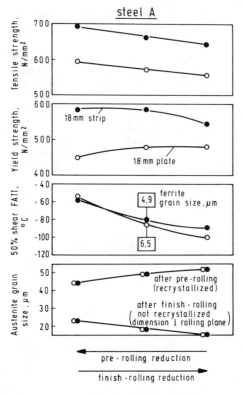

Fig. 14 - Comparison of the structural and mechanical
properties between strips and plates

expected if the deformed austenite undergoes an accelerated cooling.
This is confirmed by the results in Fig. 14. The accelerated cooling
was performed by a strip simulation with a cooling rate of about
10°C/sec. The austenite structures after pre-rolling and after finish-
rolling were constant for the plates and the strip due to the identical
schedules (A, B, D in Fig. 12). It is shown that the yield and the
tensile strengths for the strips are significantly higher than for the
plates, the FATT being nearly unchanged. This behavior is explained
in terms of the finer grain size in combination with a more effective
precipitation and dislocation hardening in the strips.

Conclusions

From the results given in this report some conclusions for the improvement of the thermomechanical treatment of austenite are drawn.

The main object of this process is to yield a fine ferrite grain size, because both strength and toughness are enhanced. Due to the close relationship between the austenite and the ferrite grain size the austenite structure has to be refined as much as possible.

For this it is generally advantageous to start the process with a grain size as fine as possible. This can be obtained by using steels with a particle distribution sufficiently stable at reheating temperatures or by reducing the reheating temperatures. In the latter case one has to care for a sufficient solution of microalloying elements to obtain the required precipitation hardening.

With regard to the properties of thermomechanically rolled steels it is proposed to apply a reduction just sufficient to refine the austenite close to its limiting grain size by recrystallization in the higher temperature range and impose the remaining reduction in the temperature range in which the austenite grains elongate. Smaller limiting recrystallized austenite grain sizes are obtained by higher pass reductions and lower rolling temperatures in the recrystallizing range. Applying the proposed procedure in the mill plant, an increase of rolling forces is to be taken into account.

Finally, the application of an accelerated cooling of the deformed austenite is always a suitable method to obtain a further improvement of the properties.

The different effects occuring during the thermomechanical processing on strength and toughness of the transformed structure can be interpreted in terms of the ferrite grain refinement and the concurrent changes in precipitation and dislocation strengthening.

References

(1) R. Phillips and I. A. Chapman,
"Influence of finish rolling temperature on the mechanical
properties of some commercial steels rolled to 13/16 in diameter
bars", Journal of the Iron and Steel Inst., June 1966 pp 615-622

(2) I. Kozasu, Ch. Ouchi, T. Sampei and T. Okita,
"Hot Rolling as a High-Temperature-Thermo-Mechanical Process",
pp. 120-135 in Microalloying 75, Union Carbide Corporation,
New York, N. Y., 1977

(3) G. R. Speich and D. S. Dabkowski,
"Effect of deformation in the austenite and austenite-ferrite
regions on the strength and fracture behavior of C, C-Mn, C-Mn-Cb,
and C-Mn-Mo-Cb steels", pp. 557-598 in The Hot Deformation of
Austenite, John B. Ballance, ed.; AIME, New York, N.Y., 1977

(4) K. Kaup and W. Zimnik,
"Perlitarme und perlitfreie Baustähle", pp. 278-291 in Werkstoff-
kunde der gebräuchlichen Stähle, Teil 1, Verlag Stahleisen m.b.H.
Düsseldorf, 1977

(5) H. Sekine and T. Maruyama,
"Retardation of Recrystallization of Austenite during Hot-rolling
in Nb-containing low-carbon Steels", Transaction ISIJ, 16 (1976)
pp. 427-436

(6) R. Priestner and E. de los Rios,
"Ferrite grain refinement by controlled rolling of low-carbon and
microalloyed steel", Metals Technology, August 1980, pp 309-316

(7) C. M. Sellars and I. A. Whiteman,
"A Look at the Metallurgy of the Hot Rolling of Steel", The
Metallurgist and Materials Technologist, October 1974 pp. 441-447

(8) M. J. Luton, R. Dorvel and R. A. Petkovic,
"Interaction Between Deformation, Recrystallization and
Precipitation in Niobium Steel", Metallurg. Trans. A, 11 A (1980)

(9) R. Kaspar, L. Peichl and O. Pawelski,
"Metallkundliche Vorgänge beim Vorwärmen und Vorwalzen von mikro-
legierten Baustählen", Stahl und Eisen, 101 (12) 1981 pp. 17-21

(10) R. Priestner, C. C. Early and J. H. Rendall
"Observations on the Behavior of Austenite during the Hot
Working of some Low-carbon Steels", JISI, December 1968
pp. 1252-1262

DISCUSSION

Q: I was pleased to see this paper because it agrees so well with our results. I would like to add that the grain refinement you showed with an increase in finishing reduction (or increase in flatening the austenite which seemed to be linear) has a tendency to saturate if you carry this out to higher reductions (80-90%). There is some limiting grain size and there seems to be some recovery process that occurs. So that, while you get further and further flatening, the austenite grain size and the ferrite that results seems to saturate somewhere around 4 or 5 microns. It does not seem to get much finer with that straight kind of rolling without accelerated cooling and it is at a fixed cooling rate.

A: This is our experience too, but we are not very sure we have obtained such results. The scatter in the data was too wide in our tests, and it is extremely difficult to evaluate substructures and to make the relationship with these higher reductions; therefore, we are not very sure how to interpret these results.

Q: You have shown that the minimum austenite grain size developed in the recrystallization range is coarser than one which is developed after applying deformation below recrystallization. Would you have any information, or speculation, about what would happen, if after attaining the minimum austenitic grain size in the recrystallization range, you would apply accelerated cooling? Could we reach a comparable grain size, as you have done, with deformation below recrystallization?

A: To reach a fine recrystallized austenite, there is a limit, due to the lower limit of the recrystallizing range in the microalloyed steels. It is possible to reach finer austenite grain size if the limit of the recrystallizing range is lower. We obtained some results of this, but we have not performed the investigation with accelerated cooling.

Q: With such a fine grain size, I am curious whether you have explored the superplasticity aspect of this material? If you did, what kind of difficulties have you encountered thus far?

A: We did not investigate the effect of superplasticity, and I wonder if the structures we obtained are really so fine to have this superplasticity effect.

Q: I am curious about your tensile tests. What kind of strain-rate sensitivity do you get, and what is the value of M?

A: We performed normal tensile tests and didn't vary the strain rate. We conducted all other experiments with rolling.

Q: You did not explore intermediate temperatures?

A: No, only at room temperature.

EFFECT OF HOT WORKING ON DEFORMATION AND RECRYSTALLIZATION BEHAVIOR OF AUSTENITE AND ITS TRANSFORMATION STRUCTURE IN A LOW-CARBON Mn-Nb STEEL

Xiang Deyuan, Yie Zongfa, Li Shuchuang, and Luan Fengying
Central Iron and Steel Research Institute, Beijing

Shen Ruzhuang, Shen Bide, Lin Youzuo, and Xu Wenzhuang
Shanghai First Iron and Steel Works, Shanghai, China

The effect of hot working on the deformation and recrystallization behavior of austenite and its transformation structure has been studied. The main topics discussed in this paper are:

. Critical deformation of austenite recrystallization
. Temperature of austenite recrystallization
. Behavior of austenite recrystallization
. Initial grain size of austenite
. Percentage of recrystallization of austenite
. Grain size of recrystallized austenite
. Rolling reduction below recrystallization temperature of austenite
. Strain-induced grain-boundary migration of austenite
. Rolling in two-phase zone ($\gamma + \alpha$)
. Grain size of ferrite
. Transformation structure

Based on the experimental results, a chemical composition and set of improved controlled rolling specifications are recommended.

Introduction

With regard to controlled rolling of low-carbon Mn–Nb Steel, many investigations have been carried out abroad in the last ten years or so (1-7), but in our country the development of this study only started recently. With the aim of comprehensively using the abundant niobium resources of our country and developing the techniques of controlled rolling, for the past two years the Central Iron and Steel Research Institute of Beijing (Peking) and Shanghai First Iron and Steel Works have collaborated in an investigation of 09MnNb steel, using native niobium iron for microalloying. The present paper describes the effect of hot working on deformation and recrystallization behavior of austenite and its transformation structure in the case of a low-carbon Mn–Nb steel.

Experimental Method

Steel used in the experiment was melted in a 30t top-blown oxygen converter, conti-cast to 150x1050mm slab, and rolled in a 2-high rolling mill to 55mm thick slab. The chemical composition is shown in Table I.

Table I. Chemical Composition of Steel for Experiment						
Element	C	Mn	Si	S	P	Nb
Content %	0.09	1.16	0.33	0.017	0.024	0.036

Batches of samples from the slab were heated to required temperatures (1150°C, 1200°C, 1250°C), soaked 30 minutes, and air cooled outside of the furnace until the surface temperature reached the predetermined temperature (1150°-750°C, at intervals of 50°C, and extra two temperatures of 980°C and 920°C). The samples were then transferred to a furnace whose temperature corresponded to the rolling temperature, and held there for 5 minutes (tolerance of furnace temperature $\pm$ 5°C). After holding, the samples were rolled one pass with reduction of 0-60%.

In order to study the effect of the post-rolling hold period and the cooling rate on recrystallization behavior and transformation structure of austenite, three samples were rolled at each temperature, and after rolling the samples were either air cooled continuously to room temperature or quenched 3 seconds or 15 seconds after the start of air cooling. From the rolled materials test specimens were taken for metallographic inspection and grain-size measurement on the cross-section of the plate.

Experimental Results

1. Deformation and Recrystallization Behavior of Austenite
1.1. Critical deformation of austenite recrystallization
The critical deformation of austenite recrystallization (hereafter simply called critical deformation) denotes the minimum deformation causing complete recrystallization of austenite. It is affected by rolling temperature, delay time after rolling, and reheating temperature, etc. The critical deformation will be larger the lower the rolling temperature, the shorter the

delay time, or the higher the reheating temperature. As shown in Fig. 1 and Table II, when the reheating temperature is 1200°C and the delay time after rolling is 3 seconds, the critical deformation will increase from 30% to 60% if the rolling temperature is decreased from 1050° to 950°C. Likewise when rolled at 950°C, if the hold period after rolling is increased from 3 seconds to 15 seconds, the critical deformation will decrease from 60% to 50%. When the rolling temperature is 950°C and the holding after rolling is 15 seconds, it can be seen from Table II that the critical deformation increases from 50% to 60% due to increase of reheating temperature from 1200°C to 1250°C.

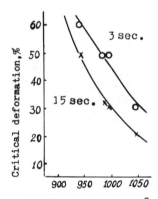

Fig. 1 - Effect of rolling temperature
and hold period after rolling
on critical deformation (reheated at 1200°C).

Table II. Effect of Rolling Temperature, Holding after Rolling
and Reheating Temperature on Critical Deformation

Reheating Temperature °C	Rolling Temperature °C	Critical Deformation, %	
		Holding 15 sec.	Holding 3 sec.
1200	1050	20	30
1200	1000	30	>30
1200	980	30	>30
1200	950	50	60
1200	920	60	>60
1150	980	30	>30
1150	950	>30	50
1150	920	60	60
1250	980	>30	50
1250	950	60	>60
1250	920	>60	>60

1.2. Recrystallization temperature of austenite

Here the recrystallization temperature of austenite denotes that deformation temperature at which holding for less than 3 seconds after rolling
will cause 3-5% recrystallization. As shown in Fig. 2, recrystallization
temperature decreases sharply with the increase of deformation, but the dependence is diminished at large strains. When the deformation is increased
over 60%, the recrystallization temperature cannot drop below 850ºC because
850ºC is already below the absolute recrystallization temperature of this
particular steel.

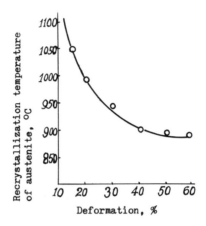

Fig. 2 - Effect of deformation on recrystallization
temperature (reheated at 1200ºC, quenched
at 3 seconds, holding after rolling).

1.3. The recrystallization behavior of austenite

When steel is deformed above the recrystallization temperature of austenite, recrystallization will take place in austenite under appropriate conditions. Whether the austenite recrystallizes completely or partially and
whether the speed of recrystallization is quick or slow is determined by the
deformation amount, deformation temperature, and holding period after rolling.
If rolling temperature is high and deformation amount large, dynamic recrys-
tallation may take place during rolling. For example, in the present
experiment, when the rolling temperature is above 1100ºC and the reduction is
larger than 20%, recrystallization is complete before the quench after rolling, which suggests that dynamic recrystallization has happened. The effect
of dynamic recrystallization on the refining of austenite is not good; when
rolling at 1150ºC even deformations as large as 60% only produce an austenite
grain size of ASTM No. 5.5. When the rolling temperature is not very high
and the deformation also not large, static recrystallization follows the deformation. In the present experiment the recrystallization which took place at
a rolling temperature of 1050ºC and deformations not larger than 30% is static
recrystallization. The static recrystallization has a good effect on refining
of austenite grain structure. For example, with 50% deformation at 950ºC, the
austenite grains can be refined to ASTM No. 6.5-7.0. When the rolling

486

temperature was between 1050–920°C, the austenite can recrystallize completely or partially or does not recrystallize at all, according to the amount of deformation and the duration of holding after rolling, as shown in Photo 1. The percentage of austenite recrystallization increases with increasing rolling temperature and the amount of deformation, as shown in Fig. 3. The effect of hold period on percentage of austenite recrystallization is shown in Fig. 4. The effect of reheating temperature on recrystallization behavior of austenite will be described later.

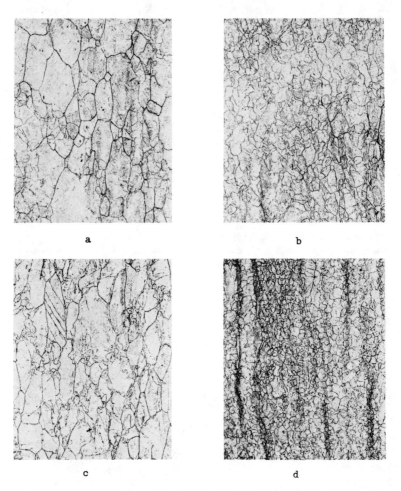

Photo 1 – Effect of deformation and rolling temperature on austenite recrystallization (reheated at 1200°C, quenched at 15 sec. holding after rolling)

        a. Rolling at 980°C, Reduction 20%;
        b. Rolling at 980°C, Reduction 30%;
        c. Rolling at 920°C, Reduction 30%;   ├─────┤
        d. Rolling at 920°C, Reduction 60%.   100 μm

Rolling Temperature

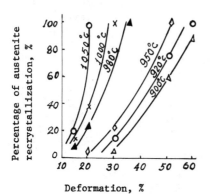

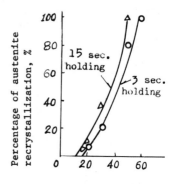

Fig. 3 – Effect of rolling temperature and deformation on
percentage of austenite recrystallization (reheated
at 1200°C, quenched at 15
sec. holding after rolling).

Fig. 4 – Effect of holding after rolling on percentage of austen-
ite recrystallization
(reheated at 1200°C, rolled
at 950°C).

1.4. Rolling between recrystallization temperature and Ar₃

When steel is rolled in this temperature range the austenite does not
recrystallize. The austenite deforms continuously, as shown in Photo 2.
As deformation increases in the range 15 to 60 percent, the austenite grains
gradually elongate and deformation bands form within the grain.

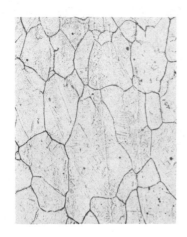

Photo 2 a   Photo 2 b

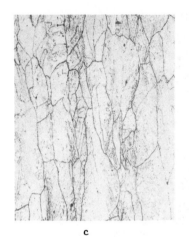 

c   d

Photo 2 – When rolled in zone of temperature below recrystallization
temperature to Ar_3, the effect of deformation on the form
of austenite grain (reheated at 1200°C, rolled at 850°)
a. Reduction 15%; b. Reduction 30%;
c. Reduction 50%; d. Reduction 60%.

⊢————⊣
100 μm

1.5. Rolling in the 2-phase zone ($\gamma + \alpha$)

The critical points of the experimental steel are Ar_3 = 792°C, Ar_1 =
612°C. Therefore, when rolled at 750°C, not only untransformed austenite
deforms but also the precipitated ferrite. With the increase of deformation
or delay time, the deformed ferrite begins to recrystallize, as shown in
Photo 3. The deformation in Photos 3a and 3b is 30%, but the delay time
after rolling is different; in 3a it is 3 seconds and the ferrite has not
recrystallized, in 3b it is 15 seconds and a part of the deformed ferrite
recrystallized. Although in 3c the holding time is only 3 seconds, the deformation is larger (60%) and a part of the deformed ferrite recrystallized.

 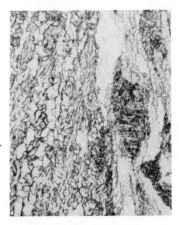

Photo 3 a   Photo 3 b

20 μm

c

Photo 3　When rolled in 2-phases zone (γ + α), the effect of
deformation and holding after rolling on the deformation and recrystallization of ferrite (reheated at
1200°C, rolled at 750°C)
    a. Reduction 30%, holding 3 sec.;
    b. Reduction 30%, holding 15 sec.;
    c. Reduction 60%, holding 3 sec.

1.6. Rolling at small reduction
    In order to study the effect of small rolling reductions, the samples
with 7% reduction were metallographically observed. It was found that in samples rolled in the temperature range of 1000° – 850°C some austenite grains
were very large, even bigger than that of undeformed initial austenite grains,
as shown in Photo 4. This is the strain-induced grain-boundary migration of
austenite discussed in References 5 and 6.

2. Effect of Reheating Temperature
    The effect of reheating temperature on the recrystallization behavior of
austenite and its transformation structure is brought about through the influence on the initial grain size of austenite.

2.1. Effect of reheating temperature on initial grain size of austenite
    The initial grain size of austenite increases with increase of reheating
temperature, as shown in Fig. 5. When the reheating temperature is raised
from 1050° to 1250°C, the initial grain size of austenite is coarsened by 3
grain size numbers. But the speed of growth of the initial austenite grains
is not the same at different reheating temperatures. It is slow below 1200°C,
and increases suddenly when over 1200°C. When reheated at 1250°C for 30 min.,
the growth is uneven but some grains have grown to ASTM No. –1 to –1.5.

2.2 Effect of reheating temperature on recrystallization behavior of austenite
    When the reheating temperature is high, the initial austenite is coarse
grained and is therefore difficult to recrystallize. For example, experimental results, shown in Fig. 6, point out that when the initial austenite is
coarse, the critical deformation will be larger.

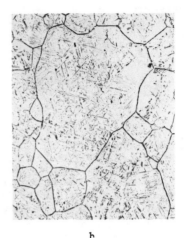

a   b

Photo 4 - Effect of small reduction rolling on the
growth of austenite grain (reheated at
1200°C, rolled at 950°C).
a. Reduction 0%;   b. Reduction 7%.

⊢————⊣
100 μm

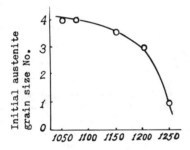

Reheating temperature, °C

Initial austenite grain
size number.

Fig. 5 - Effect of reheating temperature on critical
austenite grain size.

Fig. 6 - Effect of initial austenite
grain size on critical deformation for recrystallization (rolled at 980°C,
holding after rolling for
15 sec.).

2.3. Effect of reheating temperature on the grain size of recrystallized
austenite

Under same deformation condition (temperature, reduction) the recrystallized austenite grain is coarser when the slab is reheated to a higher
temperature, as shown in Table III. The grain size of recrystallized
austenite at 1200°C reheating temperature is finer than at 1250°C by 0.5-0.8
grade.

Table III   Effect of Reheating Temperature on Grain Size of Recrystallization Austenite (Reduction 60%)

Reheating Temp. °C	Initial Austenite Grain Size Number	Rolling Temperature and Grain Size of Recrystallization Austenite		
		980°C	950°C	920°C
1200	3	7.0	7.5	7.8
1250	1	6.5	6.8	7.0

3. Effect of Deformation Condition on Austenite Grain Size before Transformation and on Transformed Structure

3.1. Effect of deformation condition on austenite grain size

Experimental results are shown in Fig. 7 and Table IV. The general rule is that the austenite grain size before transformation is refined with an increase of deformation and decrease of rolling temperature. For example, at the same rolling temperature, no matter whether recrystallization is complete or partial, the austenite grain size before transformation is finer with increase of deformation. But for the same deformation, the effect of rolling temperature is more complicated. As niobium raises the recrystallization temperature of steel by 100-200°C (4, 8-10), it may cause steel to recrystallize completely or partially or not at all in the course of rolling. Under the condition of complete recrystallization, the austenite grain size before transformation decreases with lowering of rolling temperature (see the right half of curve in Fig. 7). But when rolling temperature decreases to the region where partial recrystallization takes place, the reverse happens and the austenite grain size before transformation increases with decrease of rolling temperature (see left half of curve in Fig. 7). As the rolling temperature decreases, the number of unrecrystallized austenite grains increase; therefore the mean austenite grain size before transformation will increase and the grain size distribution is very uneven. When the rolling temperature decreases to the region where recrystallization does not take place, then austenite grain size before transformation will be independent of temperature. The temperatures to which the peaks correspond are therefore the lowest rolling temperatures at which austenite can completely recrystallize under that deformation, and it means that these temperatures are the ones at which austenite can be refined the most for the given deformation.

Table IV   Effect of Deformation Condition on Austenite Grain Size Before
Transformation (Reheated at 1200°C, holding 15 sec. after Rolling)

Rolling Temp.°C	1150			1100			1050			1000			950			900			850		
Deformation %	30	50	60	30	50	60	30	50	60	30	50	60	30	50	60	30	50	60	30	50	60
Austenite Grain Size No.	4.7	4.2	5.5	5.0	5.5	5.8	5.5	6.0	6.3	6.0	6.7	7.0	5.8	5.9	7.5	3.0	5.0	7.0	3.0	3.0	3.0

3.2. Dependence of ferrite grain size on the size and form of the austenite grain structure

Experimental results are shown in Fig. 8. The finer the austenite grain before transformation, the finer the ferrite grain will be. But with the austenite grain becoming finer, the refining effect of austenite on ferrite will be smaller, as is shown in Fig. 9. As is stated in literature (5), even with deformations greater than 70% in high temperature zone, the austenite grain cannot be made smaller than 20 μm (about ASTM No. 8.5). In the present experiment, under the condition of giving 60% deformation at 1000°C, austenite grain size of only 36 μm (corresponding to No. 7) is obtained. Therefore, by means of refining only austenite grain before transformation, it is difficult

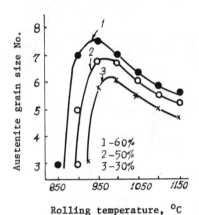

Rolling temperature, °C

Fig. 7 - Effect of rolling temperature and deformation on austenite grain size (initial austenite grain of Grade 3, holding 15 sec. after rolling).

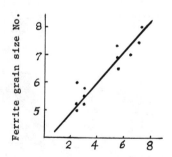

Austenite grain size No.

Fig. 8 - Effect of austenite grain size before transformation on ferrite grain size after transformation (rolled above recrystallization temperature).

to obtain ferrite grain size $\leqslant$ ASTM No. 9. But if the austenite grain size cannot be refined in the high temperature region, it is also difficult to refine ferrite grains only by deformation in the low temperature region by increasing the grain boundary and deformation band surface areas. In the present experiment, even when the deformation in the low temperature region is increased to 60%, only ferrite grains of ASTM No. 7-8 are obtained. Furthermore, the grain structure is not even; there exist relatively more big cake grains. The structure is composed of deformed ferrite and very fine recrystallized ferrite ($\leqslant$ ASTM No. 10-12), as shown in Photo 5. This structure enhances the anisotropy of steel, and this processing method is not recommended. When adopting the process which combines the refining of austenite grain in the high temperature region with giving sufficient large deformation in the low temperature region, it is possible to obtain uniformly small ($\leqslant$ ASTM No. 10) ferrite grains, see Photo 6.

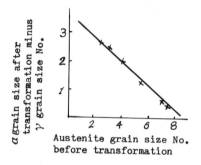

Fig. 9 - Refining effect of austenite grain size on ferrite grain size.

3.3. Effect of deformation condition on transformation structure

The transformation structure after hot working of steel, no matter whether it is air cooled or quenched, is related to the austenite grain size and the condition of hot working. For the air-cooled structures obtained in the present experiment, the relation between ferrite (F), pearlite (P), granular bainite (B_G), and deformation is shown in Fig. 10.

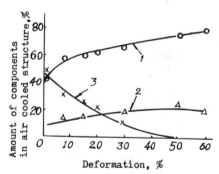

Fig. 10 - Effect of deformation on amount of components in air-cooled structure (reheated at 1200°C, rolled at 980°C).

1. Ferrite;   2. Pearlite;   3. Granular bainite

It can be seen from Fig. 10 that the elimination of granular bainite necessitates sufficiently large deformation. When rolled below 1000°C, it is necessary to increase the deformation to over 50%. The decrease of granular bainite with the increase of deformation is shown in Photo 7. From the viewpoint of austenite grain size, it is necessary to refine the pretransformation grain structure to ASTM No. 6.5 or finer.

With regard to the effect of rolling temperature, it can be said that in general the higher the temperature, the more will be the amount of granular bainite and the coarser the structure. But when rolled below 1000°C, the effect of temperature on the amount of granular bainite is not so apparent.

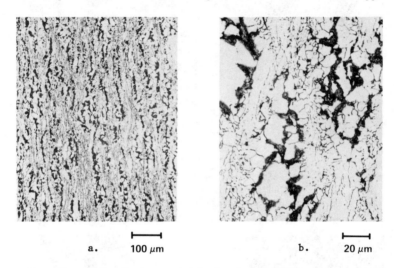

a. ├────┤ 100 µm   b. ├────┤ 20 µm

Photo 5 - Grain structure obtained only by deformation in low
temperature region (reheated at 1200°C, rolled at
800°C, reduction 60%).

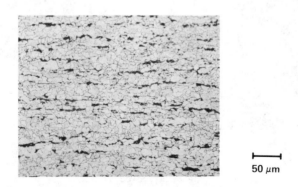

├────┤ 50 µm

Photo 6 - Grain structure obtained by optimal rolling process
(reheated at ≤ 1200°C, deformed 67% at 1150°-1030°C,
66% at 890°-820°C).

The structure obtained by quench after rolling contains lath martensite (M) and granular bainite (B_G). When the deformation increases, the amount of M_L decreases and that of B_G increases, as shown in Fig. 11 and Photo 8. This phenomenon is possibly induced by the decrease of hardenability which accompanies the increase of deformation.

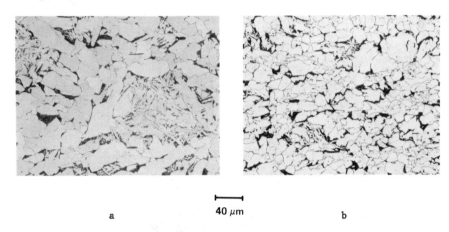

a   40 μm   b

Photo 7 - Effect of deformation on amount of granular bainite in air-cooled structure (reheated at 1200°C, rolled at 980°C, air cooled after rolling).
a. Reduction 15%;   b. Reduction 30%

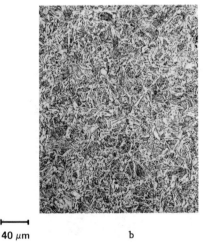

a   40 μm   b

Photo 8 - Effect of deformation on amount of M_L and B_G in quenched structure (reheated at 1250°C, rolled at 980°C, holding 3 sec. after rolling).
a. Reduction 30%;   b. Reduction 60%

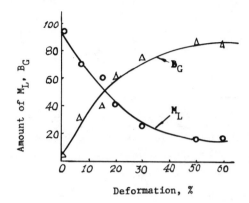

Fig. 11 - Effect of deformation on M_L and B_G in
quenched structure (reheated at 1200°C,
rolled at 1000°C).

The effect of rolling temperature on quenched structure is somewhat similar to that on air-cooled structure. Generally speaking, the grain structure is coarser when the rolling temperature is high. But when above 1000°C, the effect on the amount of M_L and B_G is not apparent, and when under 950°C the amount of M_L decreases a little with the decrease of rolling temperature.

Based on the above-mentioned results and through the investigation of the relation between chemical composition, structure, properties of steel and technology, a proper chemical composition (see Table V) and a set of more appropriate technical specifications for controlled rolling (see Table VI) are recommended.

Table V Adequate Chemical Composition of Steel

Element	C	Mn	Si	S	P	Nb
Content %	0.08–0.12	1.10–1.50	0.02–0.04	≤0.025	≤0.035	0.02–0.05

Controlled-rolled steel plates produced by using this chemical composition and technology have a good combination of high strength and high toughness, excellent weldability, and formability. The plates were successfully used to build a 3000t scoop-type suction dredger. In addition, the steel plates were used for making the chassis of 20t capacity mining trucks.

497

Table VI Proper Technology of Controlled Rolling

Plate	Slab Reheating		Rolling on 2-High Stand	
Thickness,	Reheating	Reheating	Finish Rolling	Total Reduction, %
mm	Temp. °C	Time, hrs.	Temp. °C	
150	1170–1200	1.5–2.0	≥ 1000	≥ 60

	Rolling on 4-High Stand			Thickness
Plate	Start	Finish Rolling		of Final
Thickness,	Temperature	Temperature	Total Reduction	Products,
mm	°C	°C	%	mm
150	< 920	830–790	> 60	8–20

Conclusions

1. The condition of hot working has a large effect on the critical deformation for austenite recrystallization and on the recrystallization temperature. Other conditions being equal, when the rolling temperature is lowered from 1050°C to 950°C, the critical deformation increases from 30% to 60%. Likewise, when the deformation is increased from 15% to 30%, the recrystallization temperature decreases from 1050°C to 940°C. When the reheating temperature is raised from 1200°C to 1250°C, the critical deformation is increased about 10%.

2. When rolled above the recrystallization temperature, the recrystallized austenite grain size decreases with the increase of deformation and lowering of rolling temperature. When the reheating temperature is raised, the austenite grains will be coarser. Among these factors the effect of deformation plays the main role, followed by the rolling temperature, and least effective is the reheating temperature.

3. The finer the austenite grain structure before transformation, the finer is the ferrite grain structure after transformation. But with the increased refinement of austenite its effect on refining the ferrite structure weakens. As the extent of refining of austenite is limited, it is necessary to give the austenite, which has already been refined in high temperature region, a deformation of over 60% in low temperature region. Rolled according to this technology, steel plate contains a uniform and fine structure in which the ferrite grain size can reach ASTM No. 9-12.

4. The condition of hot working has a large effect on the transformed structure. When air cooled after rolling, the amount of granular bainite gradually decreases with the increase of deformation and lowering of rolling temperature, and the grain structure also becomes finer. With the rolling temperature lowered to 950°C and the deformation increased to over 50%, the

amount of granular bainite can be reduced to less than 5%, or even eliminated. In the structure produced by quenching after rolling, the amount of lath martensite decreases and granular bainite increased with an increase of deformation and a lowering of the rolling temperature, and the structure is refined.

5. Based on the study of austenite deformation and recrystallization behavior, and through the investigation of the relation between chemical composition, structure, properties of the steel and technology, a proper chemical composition and a set of more adequate and applicable controlled rolling specifications has been suggested for industrial practice.

References

1. Guan Gen-Kuan, Iron Making Research, No. 289 (1976), pp. 43-61.

2. Guan Gen-Kuan, Tetsu-to-Hagané (10) (1972), pp. 1624-1637.

3. H. Kobayashi, Japan Inst. Metals (1976), pp. 828-833.

4. V. I. Pogorzhelskyi, Y. J. Matrosov, and A. G. Nasibov, Microalloying 75, New York, NY, 1977, pp. 100-106.

5. T. Tanaka, Tech. Rep. Kawasaki Steel (10), 1974, pp. 34-49.

6. T. Tanaka, N. Tabato, T. Hatomura, and C. Shiga, Microalloying 75, New York, NY, 1977, pp. 107-119.

7. I. Kozasu, O. Ouchi, T. Sampei, and T. Okita, ibid, pp. 120-135.

8. M. Fukuda, T. Hashimoto, and K. Kunishige, ibid, pp. 106-152.

9. I. Kozasu, et al., The Various Properties of Plate, Japan, 1976, pp. 31-50.

10. Central Inst. Iron Steel Research, Controlled Rolling Group, Iron and Steel (1978), pp. 38-46.

DISCUSSION

Q: Referring to Figure 7, why does the austenite grain size increase for the condition of rolling at 850 to 900°C?

A: Because in this region there is only partial recrystallization, and some of the grains still have the initial grain size. (ASTM No3)

CONTROLLED ROLLING PRACTICE OF HSLA STEEL

AT EXTREMELY LOW TEMPERATURE FINISHING

Tamotsu Hashimoto*
Hiroo Ohtani*
Tsuneaki Kobayashi**
Nobuo Hatano***
Shu Suzuki***

Sumitomo Metal Industries., Ltd.
 * Central Research Laboratories
   3-1 Nishinagasu Hondori, Amagasaki Japan 660
 ** Wakayama Steel Works
*** Kashima Steel Works

In the aim of strengthening the steel plate without deterioration of toughness, the extremely low temperature finishing process was successfully developed in austenite and ferrite region. Good toughness can be obtained by the rolling at the temperature range where deformed ferrites recover or recrystallize. The prior hot rolling just above Ar_3 temperature is emphasized to improve the strength and toughness more over. This produces many deformation bands acting as fine ferrite nucleation sites and promotes the strain enhanced transformation and development of {100} texture. Based on these results, the new technique of extremely low temperature finishing in the 680-650°C after heavy reduction of austenite no-recrystallization range was performed in plate mill. Steel plates with less alloy content with excellent combination of strength and toughness can be newly manufactured for line pipe of grade X70-X80 and low carbon equivalent high tensile steel for ship building.

Introduction

The strengthening mechanisms utilized in controlled rolled steel now a days are coupled of various factors such as grain refining, precipitation hardening, dislocation hardening and transformation hardening. These factors are interacted so strong that one of metallurgical fields has been built up on controlled rolling, as shown in Fig. 1.

The dislocation hardening in these mechanisms is the most economical method to increase strength, so that this has been noticed many years ago. As shown by Duckworth in 1966[1] and Irvine in 1971[2], although the rolling of austenite and ferrite two phase was expected to strengthen steel by dislocation hardening but toughness deterioration still remained. They showed that the best to finish rolling is at the temperature of 850 $\sim$ 750°C just above Ar$_3$ transformation in niobium bearing steel or low carbon steel. The controlled rolling technique utilized finishing temperature of 700°C was introduced for line pipe steel with good toughness in 1969 by Sumitomo Metal. The result exhibited that separation in broken Charpy specimen was observed often[3]. It will be understood that the light rolling of austenite and ferrite two phase region was practiced already for such line pipe steel rolling (as shown by Tanaka in 1975[4]). The attempt to utilize progressively this rolling method in plate mill was done last few years by Melloy[5], Speich[6], Gohda[7] and Hashimoto[8].

In rolling of austenite and ferrite region, it is most important problem to strengthen steel with less deterioration of toughness. This research was aimed to cralify that the relation between toughness changes and some conditions such as prerolling in austenite region, rolling in austenite and ferrite region, and microalloying element addition were investigated. Then

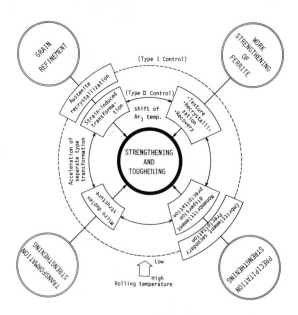

Fig. 1 Effect of controlled rolling on strengthening
mechanism and its corelation

the morphology of deformed ferrite and texture development were discussed. As a result, the rolling method of extremely low temperature finishing of 680 ∿ 650°C has been developed for line pipe steel and other high strength steel with good toughness in Wakayama and Kashima Steel Works.

Preliminary Discussion of Controlled Rolling

Improvement of mechanical properties by controlled rolling

The effect of rolling temperature of 50% reduction on mechanical properties was investigated in the range from one phase of austenite to austenite and ferrite two phase. Because the reheating temperature is 930°C, the initial grain size of austenite is very fine. The result is shown in Fig. 2. The change of strength will be divided to three ranges: i.e. recrystallized, unrecrystallized austenite and two phase of austenite and ferrite. Increasing the strength is the largest in austenite and ferrite region but the smallest in recrystallized austenite. It is only in two phase rolling where change of strength shows strongly temperature dependence. On the other hand, both rolling methods of unrecrystallized austenite and two phase are useful for improvement of toughness. This suggests that mechanical properties will

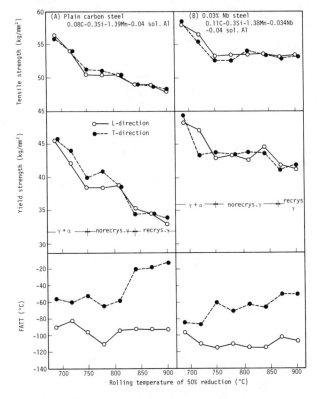

Fig. 2   Effect of rolling temperature with 50% reduction on mechanical properties. Steels were reheated at 930°C for rolling.

be affected with not only grain refining but also dislocation hardening, texture and so on. It is interesting to note that the improvement of toughness in T-direction is larger than in L-direction. The question how anisotoropy of both direction in nature will be reduced by some effect on deformation such as grain size, metal flow, separation is still open.

Effect of prerolling in austenite region on two phase rolling

The deformation in austenite region affects the morphology of austenite and ferrite, ferrite grain size, and also enhances transformation from austenite to ferrite. How are effective these phenomena on two phase rolling? Fig. 3 shows the effect of prerolling on mechanical properties of Nb steel rolled in austenite and ferrite two phase. It appears that 33% reduction in austenite range is effective to improve strength and toughness of rolled steel at 750°C or 720°C with reduction of 50% in two phase region. Fig. 4 shows the results on 1.8%Mn-0.3%Mo-0.04%Nb, so called acicular ferrite steel. The prerolling refines ferrite grain size and reduces acicularity in Mo steel.

In Fig. 3 and 4, prerolling enhances nucleation of separations in Charpy specimen, which contributes to improve toughness. This is due to increasing

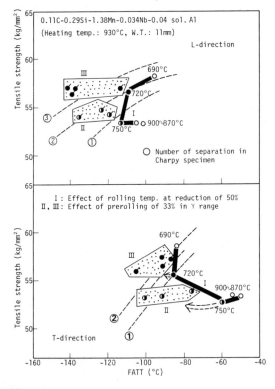

Fig. 3 Effect of rolling temperature with 50% reduction on strength and toughness of 0.03%Nb steel and effect of prerolling in austenite range before (γ + α) two phase rolling.

of fine deformed ferrite by strain enhanced transformation and inheritance of texture by rolling in austenite.

It turns out that the prerolling of heavy reduction in unrecrystallized austenite before two phase rolling is effective to improve the toughness.

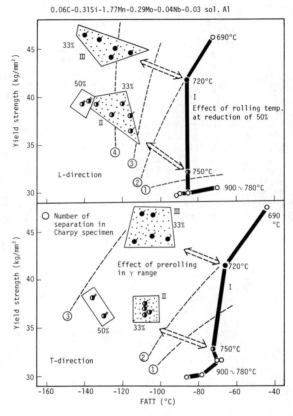

0.06C-0.31Si-1.77Mn-0.29Mo-0.04Nb-0.03 sol. Al

Fig. 4 Effect of rolling temperature with 50% reduction on strength and toughness of 1.8%Mn-0.3%Mo-0.04%Nb steel and effect of prerolling in austenite range before (γ + α) two phase rolling.

Strength and Toughness Rolled in Austenite and Ferrite Region

Effect of rolling condition on mechanical properties

The relation between reduction ratio, rolling temperature in austenite and ferrite two phase region and mechanical properties was researched in the steels with the various contents of carbon, niobium, vanadium and titanium. Fig. 5 shows that results of plain carbon and Nb steel in condition of reheated 1100°C and 12 mm in wall thickness. The strength is raised by lowering rolling temperature and maximum strength was obtained at 50% reduction. The FATT in 2V Charpy test is improved by high reduction ratio. In this figure, Nb steel has different behavior from plain carbon steel in change of mechanical properties by rolling conditions, at the temperature of 675 °C and 600°C. This will consist in difference of volume fraction of deformed ferrite. Nb steel transforms to ferrite at 675°C more than plain carbon steel, but less at 600°C. Table I shows strength change by lowering finishing temperature of 10 °C. The high strengthening rate was obtained at temperature range 700 to 650°C, because of rapid increase of deformed ferrite in this range.

In ordinary controlled rolling with finishing temperature at 720 ～ 700°C, tensile strength will be 50 kg/mm^2 for plain carbon steel and 55 kg/mm^2 for Nb steel. So, it can be concluded that heavy reduction at 675 ～ 650°C in austenite and ferrite region is effective to strengthen steel without loss of toughness.

Change of microstructure

The optical microstructure is shown in Fig. 6, corresponding to result of Fig. 5. The rolled steel showing maximum strength with 50% reduction consists of large amount of deformed ferrite in Photos. b and e. The change of strength in 33 and 50% reduction is supposed to be due to this difference of volume fraction of deformed ferrite. At 67% reduction, most of deformed ferrite recrystallized at 675°C (Photo. c) and unrecrystallized but recovered at 600°C. This is reason why strength decreases along with increase of reduction ratio from 50% to 67%.

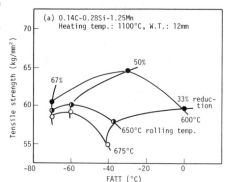

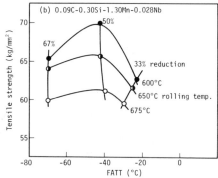

Fig. 5 Relation between tensile strength and ductile-brittle transition temperature in 2V Charpy test with variation of reduction ratio and temperature in (γ + α) region.

The electron microscopic observations of deformed ferrite are shown in Fig. 7 for 650°C rolling. In Fig. 7, the increase of reduction ratio corresponds to the increase of density of deformed structure in ferrite grains as shown in Photos. a and b. At 67% reduction, the subgrains were formed in deformed ferrite (Photo. c). Directly observation in Fig. 8 shows the change of dislocation density and its re-arrangement. The rolled steel

Table I. Increase of strength by lowering 10°C of finishing roll temperature

Finishing temp.	700 ～ 650°C	650 ～ 600°C	600 ～ 500°C
Δ Yield strength	1.5 (kg/mm^2)	0.5 (kg/mm^2)	0.8 (kg/mm^2)
Δ Tensile strength	1.5	0.3	0.6

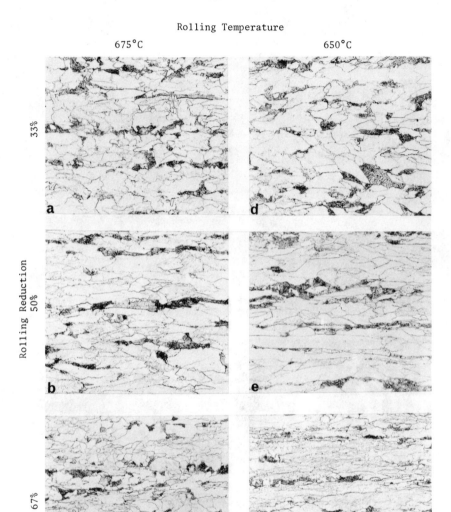

Fig. 6 Changes of microstructure by rolling temperature and reduction at (γ + α) region in 0.09%C-1.30%Mn-0.03%Nb steel.
Reheating temperature is 1100°C.

with reduction of 67% in Photo. b shows dynamically recovered structure with high density and rearrangement of dislocation.

These morphology of substructure in deformed ferrite have good relation with replica method and directly observation. It is understood that the change of strength and toughness of rolled steel in austenite and ferrite region is affected with such a tendency of recovery and cell size in deformed ferrite.

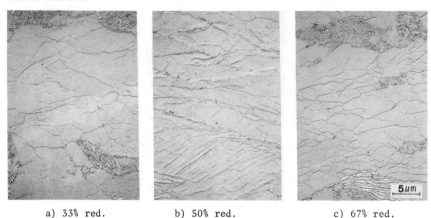

| a) 33% red. | b) 50% red. | c) 67% red. |

Fig. 7 Electron microscopic observation by replica method showing the difference of recovery of deformed ferrite in Nb steel hot rolled at 650°C.

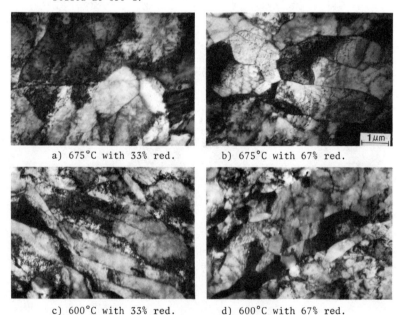

| a) 675°C with 33% red. | b) 675°C with 67% red. |

| c) 600°C with 33% red. | d) 600°C with 67% red. |

Fig. 8 Directly observed electron microscopic structure of deformed ferrite in Nb steel rolled at 675°C or 600°C.

Effect of microalloying element

Fig. 9 shows effect of microalloying elements such as niobium, vanadium and titanium on the strength and the toughness rolled at austenite and ferrite two phase region. The addition of vanadium or titanium raised strength but lower toughness compared with plain carbon steel. This tendency is clear at low temperature of rolling. On the other hand, niobium steel shows high strength and good toughness in examined rolling temperature. What is the reason of this difference?

Nitrides of vanadium or titanium act strongly as nucleation sites of polygonal ferrite and enhances transformation from austenite to ferrite. But niobium in solution seems to supress the ferrite transformation than other steels. At temperature of 650°C and below, the recovery of deformed ferrite will be retarded strongly. The increase of unrecovered ferrite increases the strength but it is not preferable for toughness. On such a point of view, niobium addition is able to retard ferrite transformation and heavy reduction of small amount of fine ferrite can be performed, which recover easily at high temperature. Fig. 10 shows that the increase of Nb content is preferable for improvement of mechanical properties. Niobium addition bring out good combination in austenite and ferrite two phase rolling. It is believed due to increase of niobium in solid solution and suppress austenite-ferrite transformation.

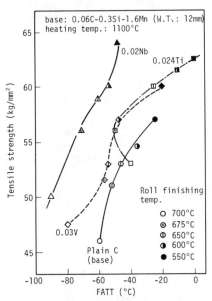
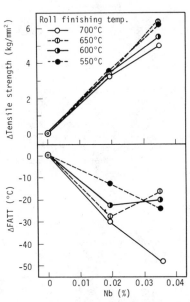

Fig. 9 Comparison of effect of microalloying elements such as Nb,
V and Ti on strength and 2V
Charpy FATT with hot rolling
in (γ + α) region.

Fig. 10 Effect of Nb content on
strength and 2V Charpy FATT
with variation of hot rolling
in (γ + α) region after reheating at 1100°C.

Two phase rolling of transformation hardened steel

Ferrite and pearlite steel may have limit for its available strength. Transformation hardening such as fine bainite and or martensite is considered one method of obtaining higher strength for X70 or X80 grade. One of these kind of steel is acicular ferrite steel, which was good toughness at high finishing temperature at near 800°C. However, the transformation behavior of steel with high hardenability is not studied in detail when heavy reduction is given in the vicinity of Ar_3 temperature.

Fig. 11 shows CCT diagrams of deformed and undeformed steel of 0.08%C-1.52%Mn-Cu-Cr-Ni-V-Nb after reheating at 1100°C. In practical rolling, the cooling time from 800°C to 500°C takes about between 300 and 1500 seconds. In no deformation, ferrite and coarse bainite will be produced in this range. At deformed steel in austenite region, bainitic transformation curve shifts to short time side and martensitic transformation curve is expanded to long time side. As the results, the mixed microstructure of large amount of polygonal ferrite and fine martensitic structure can be produced. This is due to occurance of concentration of solute atom in austenite during progress of austenite to ferrite transformation and, then, refinement and stabilization of austenite grain also proceed. It is, therefore tend to become fine bainite and or martensite. This is called "a phenomena to enhance separate transformation by controlled rolling" (Hashimoto et al. in 1979[9]).

Fig. 12 shows electronmicroscopic observation of 0.05%C-1.9%Mn-0.4%Mo-0.06%Nb steel rolled in two kinds of method. These are of micro duplex structure, mainly of fine ferrite with small amount of martensite or bainite scattered. Furthermore, as austenite to ferrite transformation is progressed in CR-B, because of fine austenite grain size by low reheating temperature, deformed ferrite is contained more in CR-B than CR-A.

The effect of microalloying element such as C, Mn, Mo, Si and V on separate type transformation is shown in Fig. 13. The volume fraction of each microstructure was measured by linear analysis method. When separate type transformation is enhanced, following characteristics are obtained;
1 Amount of ferrite (Sum of deformed ferrite F_D and undeformed ferrite F_P) increases, while volume fraction of low temperature transformation structure (bainite and martensite) decreases.
2 Transformation temperature of rest of austenite lowers and invert it into harder structure, so average hardness and strength increase.
3 Good toughness is expected because of large amount of fine polygonal ferrite.

Fig. 13 shows that increasing of austenite former elements such as C and Mn retards separate type transformation as it suppresses ferrite but increases of bainite and martensite. On ferrite former elements, increase of Mo and V also have the same tendency. Only Si is effective for enhancement of separate type transformation, because it has small affinity of carbide, and may change diffusion rate or solubility of C.

The combination of transformation hardening with micro duplex structure and dislocation hardening with deformed ferrite was adopted in practical manufacture of X-65 to X-70 grade line pipe with thick wall, higher strength line pipe of X-80 grade and structural steel with 80 kg/mm^2 tensile strength.

510

(a) No reduction

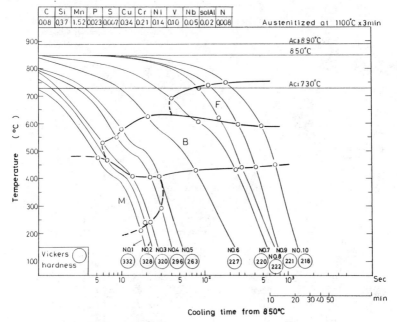

(b) Strain at 1000°C and 850°C

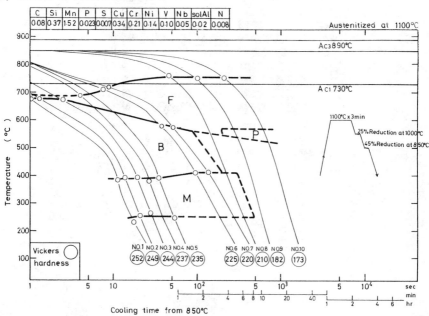

Fig. 11   Comparison of CCT diagram of no reduction and strained at austenite range.

511

 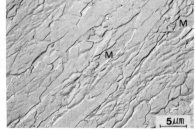

a) CR-A Heating temp. 1100°C
   Finishing temp. 700°C

b) CR-B Heating temp. 920°C
   Finishing temp. 700°C

Fig. 12 Electron micrographs showing duplex structure of deformed, undeformed ferrite with bainite or martensite on 0.05%C-0.25%Si-1.87%Mn-0.38%Mo-0.06%Nb steel.

		CR-A 50 / 100 (%)	Hv*	T.S (kg/mm²)	CR-B 50 / 100 (%)	Hv*	T.S (kg/mm²)
C (%)	0.03		200	63		205	65
	0.05		210	67		235	71
	0.08		235	75		230	73
Mn (%)	0.93		(210)	64		(205)	62
	1.95		210	67		235	71
	4.10		300	101		285	98
Mo (%)	0		(190)	60		(225)	63
	0.25		210	67		235	71
	0.40		235	72		245	72
Si (%)	0.05		200	66		185	63
	0.25		210	67		235	71
	0.55		235	73		260	72
V (%)	0		210	67		235	71
	0.08		220	73		250	75

▨ Deformed ferrite (F_D)   ☐ Polygonal ferrite (F_P)
■ Bainite+Martensite (B+M)   ▨ Pearlite
※ Hardness of (B+M)

Fig. 13 Effect of alloying element on micro duplex structure and strength, hardness (base: 0.05%C -0.25%Si-1.95%Mn-0.25%Mo-0.06%Nb steel, CR-A: heating temp. of 1100°C and finishing temp. of 700°C, CR-B: heating temp. of 920°C and finishing temp. of 700°C.

Discussion

Strengthening mechanism

In austenite and ferrite two phase rolling, the lower rolling temperature and the higher reduction ratio, it increases strengthening. This strengthening is considered mainly due to the following factors, as: ①

grain refining by subgrains and sell structure. ② dislocation hardening. ③ enhancement of precipitation hardening of V or Nb.

Fig. 14 shows relationship of volume fraction of deformed ferrite and strength, when rolling temperature of 50% reduction is continuously changed. They have a good correlation to all kinds of steels with strengthening of $3 \sim 3.5$ kg/mm^2 by 10 % volume fraction of deformed ferrite. This can be explained by Y.S increasing rate owing to subgrain size refinement obtained by Tanaka[4], Mangonon[10], et al. The authors consider equally important the strengthening by rise in dislocation density. But it seems to difficult to separate these factors as the higher the dislocation density, the smaller the cell size becomes.

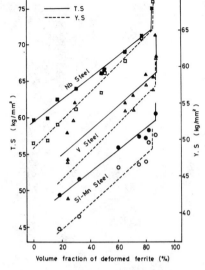

Fig. 14 Relation between volume fraction of deformed ferrite and tensile strength in 0.10%C-1.3%Mn steel (Heating temp. 1200°C, Rolling temp. is varied from 725°C to 550°C)

Toughness

The toughness in two phase rolling correlates with 4 factors as: ① grain size of ferrite ② degree of recovery of deformed ferrite ③ volume fraction of deformed ferrite and ④ texture (separation, anisotoropy). Grain refining is most effective to improve toughness and fine ferrite grain size make it easy to recrystallize at two phase rolling. But the most important factors are degree of recovery of deformed ferrite and its volume fraction.

The correlation of toughness and degree of recovery in deformed ferrite was examined in details by electron microscopic observation using the steel showing no separation. As the result, substructure of deformed ferrite and change in toughness shows good correspondence and good toughness can be maintained when deformed ferrite recrystallize or recovered as shown in Fig. 7-C or Fig. 8-b. When as rolled structure is remained such as straightly deformed or incomplete recovery of deformed ferrite, toughness can not be expected.

It could be probable that the introduction of subgrains and cell structure into deformed ferrite serves to improve toughness as grain refining effect. However, this was denied by the results of unit size of fracture facet in Charpy test[11] which was hardly changed in substructure as shown in Fig. 15 and Table Ⅱ. These subgrains consist of low angle grain boundary with least misorientation between them and they are considered insufficient to compose one unit size of fracture facet.

Texture plays important role in austenite and ferrite two phase rolling. Ogawa et al. showed in 1980[12] that the fiber structure of R.D. // <110> consisted of {001}<110>, {113}<110> and {112}<110> is developed in controlled rolled steel. Fig. 16 shows the change in texture with rolling temperature

513

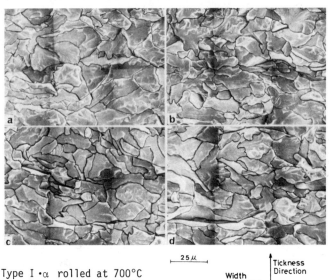

25μ

Width Direction

Tickness Direction

a: Type I·α rolled at 700°C
b: Type I·α rolled at 600°C
c: Type II·α rolled at 700°C
d: Type II·α rolled at 600°C

Fig. 15 Brittle fracture appearances of 0.10%C-1.3%Mn steel rolled in (γ + α) region. (Heating temp.: 1200°C, Type I·α: prerolling in γ recrystallizing range; Type II·α: prerolling in γ norecrystallizing range)

Table II Facet size of brittle fracture appearance in Fig. 15.

| Type | Rolling temp. (°C) | Facet size (μ) | | ℓ_W/ℓ_T |
		Thickness direction (ℓ_T)	Width direction (ℓ_W)	
Type I	700	8.9	11.8	1.33
	600	8.2	12.3	1.50
Type II	700	6.1	9.1	1.49
	600	7.4	13.1	1.77

in two phase region at two kinds of rolling conditions. Type Iα means of prerolling at recrystallized austenite region and Type IIα at unrecrystallized austenite region. {001} texture was developed strongly in rolling at Type II α. {001}<110> texture develops easily in controlled rolled steel and this causes to nucleate separation at broken specimen in Charpy test. It is well understood that separation is to lower FATT. Thus, it may be considered that the mechanism whereby good toughness is demonstrated when separation masks the embrittlement even though it is made of unrecovered and brittle structure.

To ascertain this mechanism, the experiment of cold rolling was carried out where deformed ferrite hardly recover dynamically. The result is shown in Fig. 17. In the finished rolled steel at 780°C and normalized steel show no separation as received, but separations appear in Charpy specimen if cold rolling reduction exceeds a certain critical reduction and FATT is improved at over this reduction. Otherwise, finished rolled steel at 700°C nucleates separations as received, and this steel was not raised FATT by cold rolling.

The prerolling at norecrystallized austenite region enhances not only fine ferrite nucleation but also {100} texture and these contribute to keep good toughness at austenite and ferrite two phase rolling.

The main factors effecting on toughness of two phase rolling are summarized as follows.

    Improvement : • grain refining of ferrite
                    (recrystallization of deformed ferrite and fine ferrite
                    produced in deformed austenite)
                  • separation
    Deterioration : unrecovered ferrite and its volume fraction
    Harmlessness : recovered ferrite with subgrains

The toughness of rolled steel at austenite and ferrite region will be determined by the summation of each factor whether they can keep good toughness or not.

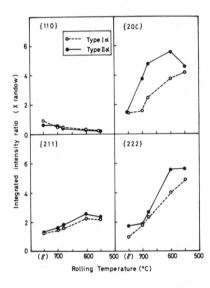

Fig. 16 Change in X-ray reflaction
intensity with 50% rolling
temperature in 0.10%C-1.3Mn
steel. (Type Iα: prerolling
in γ recrystallizing range,
Type IIα: prerolling in γ norecrystallizing range)

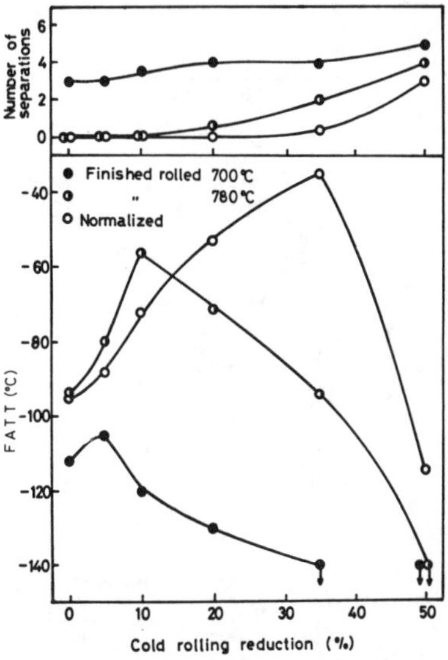

Fig. 17   Change in Charpy FATT and
number of separations with
cold rolling reduction in
0.10%C-1.3%Mn-0.06%V steel
subjected to three different
treatment.

Rolling practice

For application of austenite and ferrite two phase rolling, the rolling processes as shown in Fig. 18 practiced to make clear which is the best way in actual plate mill. Fig. 19 shows the test results and process B exhibits superior to combination of strength and toughness than process A and C. It can be said that good strength and toughness are obtained by heavy rolling in short temperature range from 700 to 650°C, where recrystallization and recovery of deformed ferrite easely occurred. The following requirement

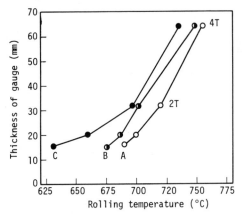

Rolling process in inter-critical range

A: high finishing temp. with small reduction

B: high reduction in narrow temp. range

C: high reduction in wide temp. range

Fig. 18   Three kinds of rolling process in (γ + α) region tested in plate rolling mill

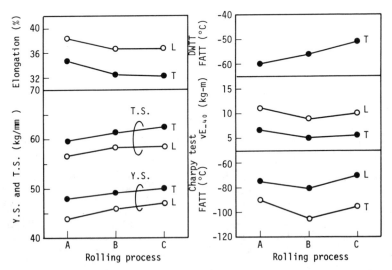

Fig. 19   Test results of rolling process in Fig. 18.   (0.15%C-0.23%Si -1.45%Mn-0.04% sol.Al, Heating temp.: 1120°C, W.T.: 15.7 mm)

must be met for this purpose. Especially, stage ① and ② are important.

① The grain refinement of initial austenite grain by low temperature reheating

② The heavy reduction at just above Ar₃ temperature in unrecrystallized austenite region. This produces many deformation bands acting as fine ferrite nucleation sites and promotes the strain enhanced transformation.

③ Two phase rolling with reduction more than 50% and below the surface temperature of 700°C.

④ Precise control of temperature and reduction of each pass near the final stage of rolling. This leads to the minimum scatter of the strength.

Thus, new technique of extremely low temperature finishing in the 680   650 °C range consists of the above multi stage control.

Application

Steel plates with less alloy content with excellent combination of strength and toughness can be newly manufactured for the following applications.
Steel plate for line pipe.
•Alloy saving for X60 ∿ X70 Grades line pipe
•As rolled X80 Grade or T.S. 65 ∿ 70 kg/mm^2 line pipe
Steel plate for ship building
•Y.P. 36 kg/mm^2 Grade

Some examples for line pipe steel are shown in Table Ⅲ.

Table Ⅲ   X-70 and X-80 grade line pipes with thickwall manufactured
by controlled rolling

| Steel | Chemical composition (wt %) | | | | | | | | | | | | Ceq* | PCM** | Steel making*** |
	C	Si	Mn	P	S	Cu	Ni	Cr	Mo	Nb	V	sol.Al			
A	0.06	0.13	1.52	0.018	0.006	—	—	0.05	0.13	0.042	0.09	0.029	0.367	0.156	CC, Ca treated
B	0.04	0.20	1.59	0.017	0.002	0.29	0.14	0.10	—	0.041	0.08	0.032	0.370	0.156	CC, Ca treated
C	0.07	0.23	1.65	0.017	0.003	—	0.15	0.16	0.27	0.031	—	0.026	0.443	0.190	CC
D	0.06	0.41	1.52	0.021	0.004	0.32	0.16	0.11	0.09	0.038	0.09	0.026	0.404	0.189	CC

$$* \ Ceq = C + \frac{Mn}{6} + \frac{Cr + Mo + V}{20} + \frac{Cu + Ni}{15}$$

$$** \ PCM = C + \frac{Si}{30} + \frac{Mn + Cu + Cr}{20} + \frac{Ni}{60} + \frac{Mo}{15} + \frac{V}{10} + 5B$$

*** Continuous casting

| Pipe size‡‡ | Steel | Base metal | | | | | | | | Welded Seam | |
| | | Tensile test† | | | | Impact test†† | | DWTT test | | Weld | HAZ |
(mm)		Y.S (kg/mm²)	T.S (kg/mm²)	El (%)	Y.R (%)	vE₋₂₀ (kg·m)	FATT (°C)	50%Sa (°C)	SA(-20°C) (%)	vE₋₂₀ (kg·m)	vE₋₂₀ (kg·m)
609.5 x 31.8	A	55.1	62.4	39.8	88	14.2	-82	-40	100	13.1	10.3
914.4 x 34.9	B	54.7	61.5	55.8	89	29.5	-75	—	100	17.3	22.9
914.4 x 38.1		54.7	63.1	57.1	87	26.7	-52	—	100	18.9	15.2
1422 x 25.9	C	59.7	73.5	38.2	81	23.7	-140	-76	100	10.4	4.2
1422 x 25.9	D	59.7	71.8	32.6	83	11.5	=140	-56	100	8.5	6.5

‡‡ Diameter x wall thickness

†   specimen ; API, Direction ; Transverse

†† 2mm V-notch (10 x 10 x 55 mm)

Conclusion

(1) If the deformed ferrite grains partially recrystallize or recover to
such an extent that subgrains are formed by rearrangement of dislocations, the low temperature toughness does not deteriorate by two phase
rolling at the temperature range from Ar₃ to 620°C.

(2) The unrecovered structure of ferrite obtained by two phase rolling after
the prerolling in recrystallized austenite grain raises the Charpy
transition temperature. However, it is not so harmfull on toughness, if
unrecovered ferrite included by two phase rolling subsequent to the prerolling in unrecrystallized austenite region which produces many fine
ferrite grains by strain enhanced transformation.

(3) The texture of deformed ferrite is another important factor affecting
toughness of rolled steel. The prerolling in unrecrystallized austenite
region develops the {100} texture and this causes the nucleation of many
separation (splitting) at fracture facet and lower ductile-brittle
transition temperature in Charpy test. The mechanisms of increasing
separation and masking effect of brittleness of unrecovered structure
were discussed.

(4) The toughness of rolled steel at austenite and ferrite region may be determined by the summation of factors such as grain size, volume fraction
of unrecovered ferrite and separation whether they can keep good toughness or not.

(5) The new technique of extremely low temperature finishing in 680 ∿ 650°C range were developed. This may be usefull to produce high strength steel required for good arrestability of brittle fracture or good weld-ability with low carbon equivalent steel.

References

(1) W.E. Duckworth, "Thermomechanical Treatment of Metals", Journal of Metals, 18 (8) (1966) pp. 915-922.

(2) K.J. Irvine, T. Glodman, J. Orr. and F.B. Pickering, "Controlled Rolling of Structural Steels", Journal of The Iron and Institute, 208 (8) (1970), pp. 717-726.

(3) E. Miyoshi, M. Fukuda, H. Iwanaga, T. Okazawa, "The Effect of Separation on the Propagating Shear Fracture", paper presented at Meeting on Crack Propagation in Pipelines, Newcastle Upon Tyne, England, March, 1974.

(4) T. Tanaka, N. Tabata, T. Hatomura and C. Shiga, "Three Stage of the Controlled-Rolling Process", pp. 107-118 in Micro Alloying 75, Washington D.C., October, 1975.

(5) G.F. Melloy and J.O. Dennis, "Continuum Rolling -- a Unique Thermo-mechanical Treatment for Plain-Carbon and Low-Alloy Steels", pp. 60-64 in The Strength of Metals and Alloys, 1973.

(6) G.R. Speich, and D.S. Dabkowski, "Effect of Deformation in the Austenite and Austenite-Ferrite Regions on the Strength and Fracture Behavior of C, C-Mn, C-Mn-Cb, and C-Mn-Mo-Cb", pp. 557-597 in The Hot Deformation of Austenite, J.B. Ballance, ed.; AIME, Cleveland, Ohio, 1976.

(7) S. Gohda, K. Watanabe, and Y. Hashimoto, "Effect of Intercritical Roll-ing on Structure and Properties of Low Carbon Steel", Tetsu-to-Hagane, Journal of the Iron and Steel Institute of Japan, 65 (9) (1979) pp. 1400-1409.

(8) T. Hashimoto, T. Sawamura, and H. Ohtani, "Mechanical Properties of High Strength Low Alloy Steel Controlled Rolled at Austenite and Ferrite Two Phase Regions", ibid, 65 (9) (1979) pp. 1425-1433.

(9) T. Hashimoto, T. Sawamura, and H. Ohtani, "Transformation and Mechanical Properties of Low Carbon-High Mn-Mo Steel by Controlled Rolling", ibid, 65 (10) pp. 1589-1597.

(10) P.L. Mangonon, Jr., and W. Heitmann, "Subgrain and Precipitation-Strengthening Effect in Hot-Rolled, Columbium-Bearing Steels", pp. 59-70 in Micro Alloying 75, Washington D.C., October, 1975.

(11) H. Ohtani, F. Terasaki, and T. Kunitake, "The Microstructure and Tough-ness of High Tensile Strength Steel", Transactions of Iron and Steel Institute of Japan, 12 (1972) pp. 118-127.

(12) R. Ogawa and T. Yotori, "Texture and Trough Thickness Toughness of Controlled Rolled Steel Plate", Kobe Steel Engineering Report, 31 (1) (1981), pp. 57-61.

DISCUSSION

Q: In your first slide you indicated that accelerated cooling would, among other things, decrease costs. However, I didn't find any other comments in your talk about the effects of accelerated cooling. Could you possibly just indicate what things it may do to the structure and properties.

A: Accelerated cooling can lead to extra grain refinement, increased precipitation hardening and an increased amount of non-polygonal ferrite.

Q: Have you any general idea what sort of change of chemistry should be done to promote or to allow recrystallization of ferrite, or to widen the temperature range for this recrystallization?

A: In my experience, the ferrite recrystallization stop temperature is 675° or 680°C. As you said, V- or Nb would shift this temperature a little compared to plain carbon steel.

Q: Would it help to add elements such as silicon to increase the gamma to alpha transformation temperature to give a greater chance to roll ferrite at high temperatures. That is, add elements to increase the transformation temperatures or lower hardenability so that you have ferrite available at higher temperatures. Would that be helpful?

A: Yes.

MATHEMATICAL MODEL TO CALCULATE STRUCTURE DEVELOPMENT AND MECHANICAL PROPERTIES OF HOT-ROLLED PLATES AND STRIPS

S. Lička, J. Wozniak, M. Košar, T. Prnka

Iron and Steel Research Institute (Výzkumný ústav hutnictví železa), 739 51 Dobrá, Czechoslovakia

A mathematical model of the effect of the technological conditions of hot rolling on the material structure development and final mechanical properties is presented.

The model has been applied for scheduling technological procedures in order to obtain optimum product properties. An example of its utilization during rolling strips made of ferritic-pearlitic V-N steel is discussed.

INTRODUCTION

A considerable amount of research work /1-3/ has been in progress with the aim of proposing at least a probable mathematical model permitting a better analysis of the technological procedure of rolling than procedures based only on practical experience have made possible so far. The paper outlines the structure of a mathematical model used for the analysis of technological conditions which apply during hot rolling and cooling.

1. Model Structure

The model provides the possibility of evaluating the suitability of the chosen technological procedure as well as prediction of the structure and mechanical properties of the rolled product for the requested reduction schedule and chosen rolling conditions (temperature, rolling speed, cooling mode) at the specified mill train. The optimum theoretical solution on which the rolling experiment or technological project is based can be found by repeated calculations for the changed input data. The model consists of three parts, a brief description of which is given below.

1.1 Rolling Model

This part of the computer program provides conventional engineering calculations of the rolling stock temperature and mill stand loading. The rolling stock temperature is determined by the numerical solution of the Fourier equation of heat conduction; common formulas are used for calculations of rolling forces, moments and power.

1.2 Material Structure Development Model

The algorithm of the phase volume fraction calculations during anisothermal decomposition of austenite with an arbitrary course of material cooling is given in Appendix A and Fig. 1. The course of cooling the material is approximated by a gradual curve (Fig. 1), the kinetics of the austenite decomposition during individual time intervals is calculated as isothermal transformation at corresponding temperature. The needed kinetics data are given by the time-temperature-transformation (TTT) diagram which can be digitized, e.g. by means of the procedure described in /2/. The calculation method can also account for the austenitic grain size and the effect of prior deformation on the kinetics of phase deformations.

Empirical relationships permitting an approximate determination of the fraction of the statically or dynamically recrystallized grains and of the austenitic grain size /3,4/ represent additional parts of the model.

1.3 Model of Mechanical Properties

Various empirical formulas taken over from literature /1/ and modified in conformity with our own experimental results have been utilized for the prediction of the mechanical properties of the rolled product. Examples are shown in Table I.

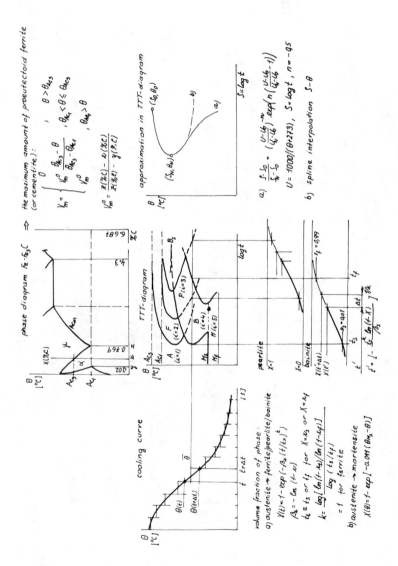

Fig. 1 Diagrammatic presentation of the structural portion calculation during anisothermal decomposition of austenite.

TABLE I
Empirical formulas for the calculation of the
mechanical properties of ferritic-pearlitic steels

Yield Stress:

$$YS = X_F^{1/3} [35 + 58Mn + 17.4 \; d_\alpha^{-1/2}] + (1-X_F^{1/3}) \; x$$
$$x \; [178 + 3.8 \; s^{-1/2}] + 63Si + 3535 \sqrt{V \cdot N} \qquad [MPa]$$

Tensile Strength:

$$TS = X_F^{1/3} [246 + 18.2 \; d_\alpha^{-1/2}] + (1-X_F^{1/3}) \; [720 + 3.5 \; s^{-1/2}] +$$
$$+ 97Si + 1047V + 2294 N \qquad [MPa]$$

Impact Transition Temperature:

$$ITT = -(19 + 11.5 \; d_\alpha^{-1/2}) + 44Si + 2.2 \; (\% \; pearlite) +$$
$$+ 919 \sqrt{V.N} \qquad [^oC]$$

X_F = transformed volume fraction

C, Mn, Si, ... wt. %

d_α [mm] ferritic grain size

s [mm] mean interlamellar

spacing of pearlite

2. Example of Model Application

Table II illustrates the resulting proposal of the optimum thermomechanical schedule for rolling strips 6 mm thick made of C-Mn steel microalloyed with vanadium. The proposal is based on the technological possibilities
of the analyzed mill train whereby the resulting product should meet the
requirements of the A.P.I. Standard for materials intended for the production of pipelines.

With regard to the short intervals between deformations taking place
in the finishing rolling line, the material comes out of the mill train
with dynamically recrystallized grain. As reported earlier /3/ austenitic
grain size d_γ = 7.6 μm (11 ASTM) conforms to finish rolling temperature of
900°C and to the deformation rate 83 /s^{-1}. This grain size as well as the
proposed cooling conditions are in conformity with the course of anisothermal decomposition given in Table III and calculated by using the modified
model of the TTT diagrams described in /2/.

Austenite transformation is completed at the temperature of 575°C, i.e.
prior to strip coiling and the material contains 69.7% ferrite and 29.8%
pearlite.

Further formulas provide the ferrite grain size 12 ASTM and inter-lamellar spacing of pearlite s=7.3 .10^{-5}mm. The values of the mechanical properties calculated for these structural parameters are shown in Table II.

Theoretical proposal of the optimum rolling process is in good agreement with the technological practice which justifies a wider application of the model in engineering practice.

CONCLUSION

Mathematical modeling is a useful means for analyzing complex tech-nological procedures and permits controlled scheduling of the rolling experiment. In the stage of analyzing the mill train, mathematical modeling is practically the only way of getting an idea of the technological possibilities of the designed facilities.

TABLE II
Example of the technological procedure for
rolling C-Mn steel microalloyed with V

Steel: 0.14 C, 1.45 Mn, 0.38 Si, 0.030 Al, 0.10 V, 0.015 N

Slab: 115 x 410 x 3000 mm

Temperature: 1240°C

Rolling conditions:

I. Universal two-high reversing mill (rolls of 760 mm dia.)

i	H /mm/	ε	$\dot{\varepsilon}/s^{-1}/$	°C	Note
1	95	0.19	2.0		Intervals between deformations in the range of 5 up to 16 s.
2	72	0.12	2.7		
3	52	0.32	3.4		
4	37	0.34	4.0		
5	27	0.32	4.6	1100	

II. Four-high rolling mill (rolls of 440 mm dia.)

6	22	0.21	8.6	1070/1020	

III. Finishing line (rolls of 430 mm dia.); interval 45 s.

7	15.6	0.35	14.1		Intervals between deformations $\leqslant$1.5 s.
8	11.1	0.34	23.3		
9	8.5	0.27	28.3		
10	7.2	0.16	38.7		
11	6.0	0.17	83.5		900 Austenitic grain size 11 (ASTM)

IV. Controlled cooling

900/650°C water spray 8.5 s (30°C/s)
650/550°C equalization of temperature in the air 14 s. (7°C/s)
550/500°C coiling
500°C   cooling the coil in the air (aging)

V. Resulting properties/calculated/

69.7% ferrite + 29.8% pearlite + 0.5% bainite
YS = 536 MPa, TS = 733 MPa, ITT = -51°C
Ferritic grain size: 12 (ASTM)

TABLE III

Course of anisothermal decomposition of austenite during
controlled cooling of strip 6 mm thick
(See Table II)

°C	ferrite	pearlite	Volume fraction of: bainite	martensite	austenite
900	.000	.000	.000	.000	1.000
825	.000	.000	.000	.000	1.000
775	.022	.000	.000	.000	.880
725	.178	.000	.000	.000	.822
675	.498	.025	.000	.000	.477
625	.593	.180	.000	.000	.227
575	.697	.298	.005	.000	.000

B_S = 661°C, M_S = 435°C

Algorithm of the structural portion calculations

1. Temperatures of the transformation starts: θ_{Ac3}, θ_{Ac1}, θ_{Bs}, θ_{Ms};

2. $t \leftarrow 0$; $\theta(0) \leftarrow \theta_0$; $V_1(0) \leftarrow 1$; $V_i(0) \leftarrow 0$, $i = 2,3,4,5$;

   2.1 mean temperature in the interval $\langle t, t + \Delta t \rangle$

   $\bar{\theta} = (\theta(t) + \theta(t+\Delta t))/2$;

   if $\bar{\theta} \leq \theta_{Ms}$ then go to 3;

   if $\bar{\theta} \leq \min_{j} (\theta_{js})$ then $n \leftarrow j$;

   2.2 for $i = 2,\ldots, n$ do:

   − calculation of the transformable fraction amount of austenite

   $V_{mi}(t)$

   (for $i = 2$ see Fig. 1; for $i > 2$ $V_{mi} = 1$);

   − calculation of transformation start and completion t_{si}, t_{fi}

   and exponent k_i (Fig. 1) for $\theta /^{\circ}C/$ (for ferrite $k_2 = 1$);

   − fictitious volume fraction of the so far transformed volume

   fraction $X_i = V_i(t)/\left[(V_1(t) + V_i(t)) \cdot V_{mi}(t)\right]$;

   − fictitious time related to the previously transformed volume

   fraction X_i

   $t'_i = (- t_{si}^{ki} \cdot \ln(1-Xi) / B_s)^{1/ki}$;

   − fictitious volume fraction at time $t'_i + \Delta t$

   $X_i(t'_i + \Delta t) = 1 - \exp\left[-B_s((t'_i + \Delta t)/t_{si})^{ki}\right]$;

   − volume fraction of the structural component at time $t + \Delta t$

   $V_i(t+\Delta t) = X_i(t'_i + \Delta t)\left[V_1(t) + V_i(t)\right] \cdot V_{mi}(t)$;

   2.3 new value of the residual austenite

   $V_1(t + \Delta t) = 1 - \sum_{i=2}^{n} Vi(t + \Delta t)$;

   If $V_1(t + \Delta t) \leq 0$ then end of transformation.

   2.4 $t \leftarrow t + \Delta t$; go to 2.1.;

3. Martensite transformation for $\theta_{Mf} \leq \theta \leq \theta_{Ms}$

   $V_5(\theta) = (1-V_2-V_3-V_4) \cdot \left[1-\exp(-0.011(\theta_{ms} - \theta))\right]$;

4. Residual austenite with $\theta \leq \theta_{Mf}$

   $V_1 = 1 - \sum_{i=2}^{5} Vi$;

Note: i = 1 − austenite
      2 − ferrite
      3 − pearlite
      4 − bainite
      5 − martensite

Ferritic grain size:
$GS_{ferrite}$ (ASTM) $= 8 + 0.35 d_\gamma^{-0.5}$
(d_γ [mm] austenitic grain size)

Interlamellar spacing of pearlite:
$s = .018 /(996-\bar{\theta})$ [mm]
($\bar{\theta}$ [$^{\circ}C$] mean temperature during pearlitic transformation)

REFERENCES

/1/   D. V. Doane, J. S. Kirkaldy (Ed.):   Hardenability Concepts with
       Applications to Steel,Metallurgical Society of AIME, New York,
       1977.

/2/   M. Ito et al.:   Tetsu-To-Hagane, 64 (11), 352, S 806 (1978); 65
       (8), 69-72, A 185 - A 188 (1979).

/3/   C. M. Sellars: in: Hot Working and Forming Processes,(Proc. Int.
       Conf. Sheffield, July 1979), Metals Society, London 1980, p. 3-15.

/4/   C. M. Sellars, J. A. Whiteman: Met. Sci., 13 (3/4), 187-194 (1979).

THE RESISTANCE OF HSLA STEELS TO DEFORMATION DURING HOT ROLLING

J. A. DiCello
Raychem Corporation
&
D. Aichbhaumik
National Steel Corporation

Eleven Cb-bearing HSLA steels were hot rolled to compare their deformation behavior to that of a base C-Mn steel. The C content, Cb content, and Si content in these steels varied from 0.05% to 0.21%, from 0.004% to 0.075% and from 0.003% to 0.72%, respectively. The deformation resistance of these steels was characterized by mean flow strength-temperature curves. The shape of these curves depended on the Cb content. Low and high Cb contents yielded linear curves, while medium Cb contents resulted in two linear segments of different slopes. Both the Cb content and the product of the concentration of C, Cb, and Si (i.e., C% x Cb% x Si%) influenced the magnitude of the flow strength. Steels with low-Cb content and a Si content above 0.5% had deformation resistance equal to or greater than steels with high Cb content (>0.05%). The high-Si steels (Si>0.5%) and the high-Cb steels were 20% to 30% stronger than the base C-Mn steel in the temperature range of 1600°F to 1700°F (871°C to 927°C).

Introduction

The use of HSLA steels by the automotive, agricultural and other industries is based on being able to reduce component thickness, taking advantage of the higher strength to achieve effective weight reductions. The HSLA steels are designed to take advantage of the solid solution hardening, precipitation strengthening and grain refinement characteristics produced by small additions of columbium, vanadium or titanium. Although the alloy content is minimal, the HSLA steels are significantly stronger than C-Mn steels during and after hot rolling.

Production experience has shown that hot rolling of HSLA steels result in substantially higher rolling loads than those observed with C-Mn steels. The alloy additions used in these steels have been shown to retard the rate of recovery and recrystallization of austenite.[1] Heavily strained unrecrystallized austenite is developed as the steel moves through the finishing train. Consequently, the flow strength increases substantially from stand to stand generating excessively high rolling loads relative to C-Mn steels with the same rolling schedule. The high deformation resistance of the HSLA austenite causes considerable difficulty in meeting the widths and thicknesses specified by the customer. Conventional hot mills were designed to roll C-Mn steels and, therefore, do not necessarily have the capacity needed to accommodate the higher rolling loads imposed by hot rolling of HSLA steels at the conventional widths and thicknesses. In addition, HSLA steels are often "controlled-rolled" where large reductions are taken at lower temperatures to ensure a fine ferrite grain size.

This investigation was initiated to determine the flow strength-temperature characteristics of C-Mn and C-Mn-Cb steels to establish the temperature ranges where these steels exhibit excess resistance to deformation. The influence of various alloy additions was also studied.

Materials

The chemical analyses of the steels rolled in this study are shown in Table I. These steels are all columbium bearing steels except the base steel, which is a C-Mn steel. The columbium range covered was from 0.004 percent to 0.075 percent. Five steels were treated with Zr for inclusion shape control. Two heats of steel H were made and designated H-1 and H-2. The major difference in the two heats was the silicon content where H-1 contained 0.58 percent and H-2 contained 0.72 percent.

Procedure

Apparatus

The experimental work was conducted with a two-high laboratory hot rolling mill with rolls 8.5 inches (21.6 cm) in diameter by 10 inches in length. Roll loads were measured with 120,000 pound (54,500 Kg) capacity strain gauge load cells inserted between the screwdown assemblies and their bearing blocks. Slab temperatures were measured with 0.125 inch (0.32 cm) diameter chromel-alumel thermocouples which were placed in a 0.25 inch (0.63 cm) diameter hole, 3.0 inches (7.6 cm) deep into one end of each slab. The thermocouple wires were insulated (and encased in an Inconel sheath) with only the thermocouple ends exposed to the hot slab. The roll loads and slab temperatures were continuously recorded with a Visicorder and an X-Y recorder, respectively.

Table I. Chemical Composition of the Hot Rolled Steels (Wt.%)*

Steel	C	Mn	Si	Al	Cr	Cb	Zr
Base	0.05	0.36	0.003	0.005	0.02	–	–
A	0.08	0.45	0.025	0.078	0.02	0.010	0.096
B	0.15	0.60	0.035	0.008	0.04	0.009	–
C	0.20	0.76	0.028	0.005	0.02	0.010	–
D	0.20	0.96	0.010	0.007	0.02	0.023	–
E	0.09	0.45	0.017	0.100	0.02	0.021	0.124
F	0.20	1.06	0.031	0.003	0.06	0.053	–
G	0.17	1.08	0.035	0.001	0.05	0.075	0.080
H-1	0.14	0.72	0.580	0.010	0.02	0.004	0.066
H-2	0.15	0.85	0.720	0.012	0.09	0.005	0.140
I	0.08	0.82	0.590	0.025	0.61	0.027	–
J	0.20	1.03	0.023	0.016	0.06	0.032	–

*Phosphorous and sulfur were within acceptable limits for the above steels. All steels were production heats cropped after the final roughing stand prior to entering the finishing train. The slabs were 1.25 inches (3.2 cm) to 1.5 inches (3.8 cm) thick by coil width. Each slab was subsequently torch cut and machined to 6 inches (15.2 cm) x 10 inches (25.4 cm) x thickness.

Rolling Procedure

Each slab was reheated to 2350°F (1290°C) for 1.5 hours to solution treat the slab and generate a sufficiently thick scale that could be easily removed by a skin pass on the hot mill. Complete scale removal is very important for obtaining accurate thickness measurements before and after each rolling pass and also to ensure a uniform force distribution across the slab. After scale removal, the slab was placed in a furnace set at the rolling temperature, 1600°F to 2000°F (871°C to 1093°C), to homogenize the slab temperature. Previous studies[2,3] have shown that no significant precipitation of columbium carbide or carbo-nitride [Cb(C,N)] occurs between these temperatures at short times in the absence of prior strain.

After temperature homogenization, a slab was removed from the furnace and a thermocouple was inserted into the drilled end prior to rolling. The rolling sequence consisted of two passes. The first pass reduction was either 25 or 35 percent to break up the reheated structure, to set the thermocouple and to introduce sufficient strain to allow any possible strain-assisted strengthening mechanisms to operate in the second pass. Second pass reductions were held near 30 percent. The roll velocity was a constant 10.5 inches per second (26.6 cm per second) which resulted in strain rates from 2.5 to 4 per second. The interpass time was kept near 15 seconds. The temperature of the second pass (T_2) was kept within 50°F (28°C) of the first pass temperature (T_1). The roll load and temperature were measured during each pass along with the entry and exit thickness.

Metallographic Procedure

Recrystallization kinetics of selected steels were followed by metallographic examination of steels quenched immediately after the first pass.

These specimens were almost completely martensitic after rolling and quenching. An etchant,[4] consisting of 100 ml saturated aqueous solution of
picric acid, 2 grams of Teepol (a wetting agent), 2 grams of ammonium per-
sulfate and six drops of hydrogen peroxide, was used to reveal the prior
austenite grain boundaries. The samples were mounted and polished by standard metallographic procedures with extra care taken to remove all traces of
flowed metal. The most important feature of the etchant is to retard the
development of the martensite matrix and, therefore, enhance the prior
austenite grain boundaries.

Results

Flow Strength

The resistance to deformation of the rolled steels was compared to that
of the base C-Mn steel using their individual flow strengths between $1600^{\circ}F$
($871^{\circ}C$) and $2000^{\circ}F$ ($1093^{\circ}C$). The flow strength is calculated from the roll
load, entry and exit thickness and roll diameter in a manner as follows:[5]

$$P = K_m \cdot L \cdot W \cdot Q$$

where P = roll load or separation force (lbs)
K_m = mean flow strength (psi)
L = projected arc length of contact between
roll and slab (in.)
Q = geometrical function
W = slab width (in.)

The projected arc length of contact is defined as follows:

$$L = \left(\frac{RCP}{W} + Rrh_o \right)^{1/2} \quad \text{from Hitchcock}[6]$$

where R = roll radius (in.)
C = roll constant (in. ^{2}/lb)
$= \dfrac{16 (1-\nu^2)}{\pi E}$ E = Young's Modulus
  ν = Poisson's Ratio
r = slab reduction = $(h_o - h_f)/h_o$
h_o, h_f = entry and exit slab thickness, and
Q = $1.57 + L/(h_o + h_f)$ as defined by Ford and Alexander.[7]

Solving for the mean flow strength gives

$$K_m = P/(L \cdot W \cdot Q).$$

Mean Flow Strength - Temperature Curves

Mean flow strength-temperature curves were plotted using the second
pass flow strength and temperature data. The reason for plotting second
pass data is that the structure rolled in the second pass more closely
approaches the structure and flow strength of the steel as it moves through
the finishing train of a hot mill. The first pass was used to refine the
reheated grain structure and to generate sufficient strain to trigger
accelerated precipitation or other strain induced strengthening mechanisms.

The mean flow strength-temperature curves are shown in Figures 1 to 11.
Figures 2 to 11 contain a dashed line for the strength of the base steel for
comparison. The base steel shows a linear relationship between flow strength
and temperature (Figure 1). The columbium-bearing steels, on the other hand,

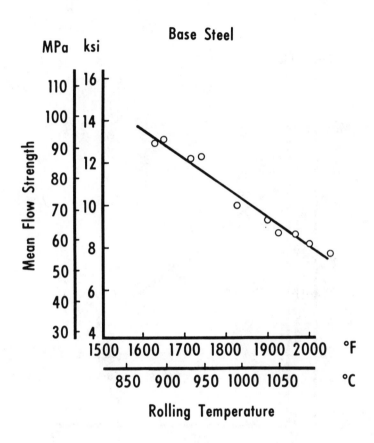

Figure 1.   MEAN FLOW STRENGTH CURVE FOR SECOND PASS

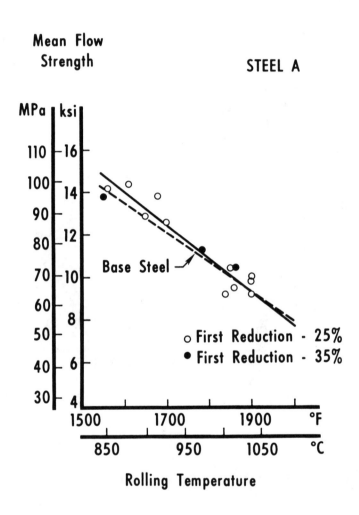

Figure 2.   MEAN FLOW STRENGTH CURVE FOR SECOND PASS

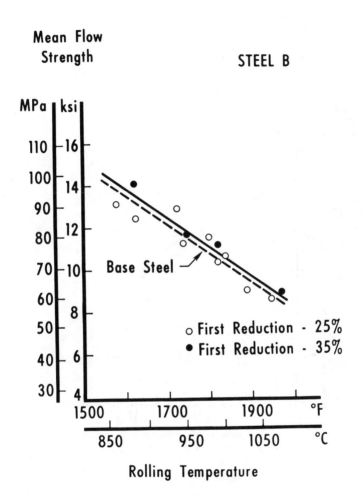

Figure 3. MEAN FLOW STRENGTH CURVE FOR SECOND PASS

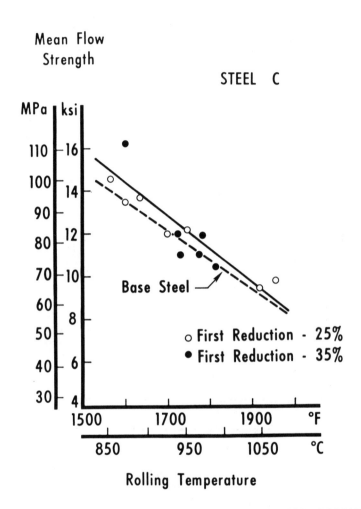

Figure 4.   MEAN FLOW STRENGTH CURVE FOR SECOND PASS

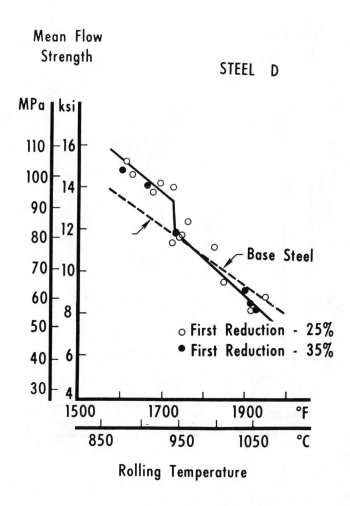

Figure 5. MEAN FLOW STRENGTH CURVE FOR SECOND PASS

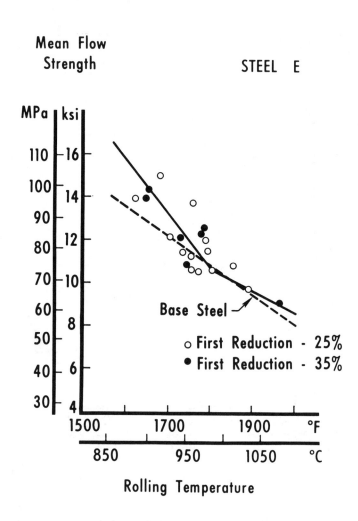

Figure 6.   MEAN FLOW STRENGTH CURVE FOR SECOND PASS

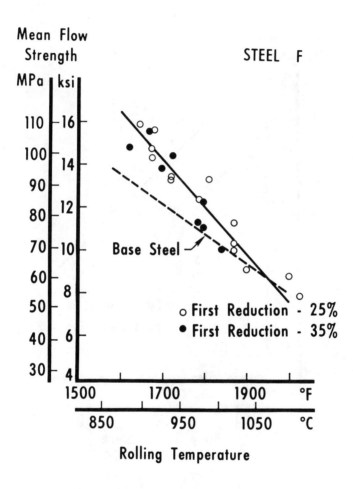

Figure 7.   MEAN FLOW STRENGTH CURVE FOR SECOND PASS

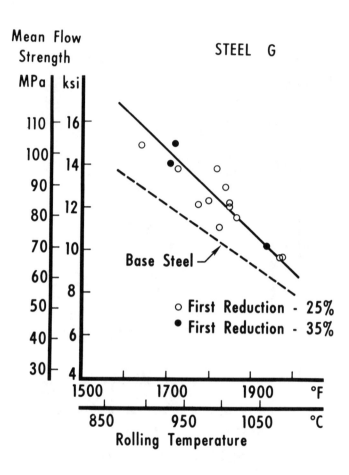

Figure 8. MEAN FLOW STRENGTH CURVE FOR SECOND PASS

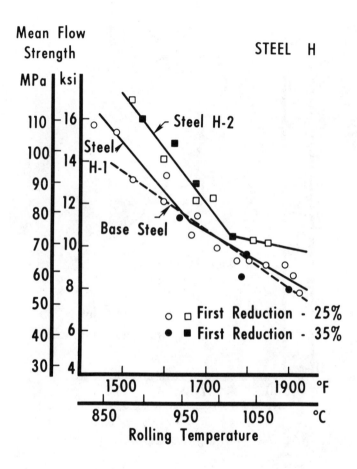

Figure 9. MEAN FLOW STRENGTH CURVE FOR SECOND PASS

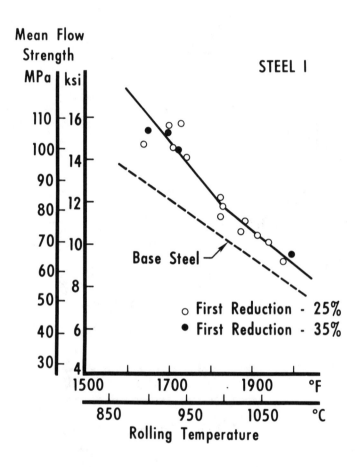

Figure 10.   MEAN FLOW STRENGTH CURVE FOR SECOND PASS

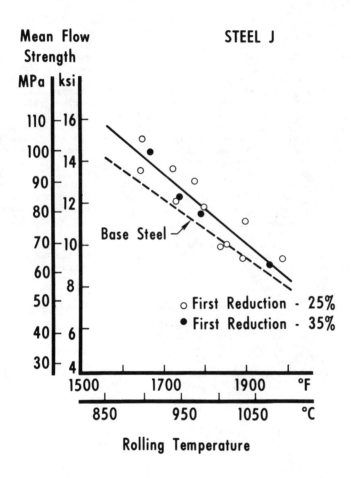

Figure 11. MEAN FLOW STRENGTH CURVE FOR SECOND PASS

are not all linear (refer to Figures 5, 6, 9, and 10). The flow strength-temperature data points in each of these figures fit two straight line segments of different slopes. Although each straight line segment is a linear regression curve fit, the temperature at which the slope change occurs was established graphically (i.e., the intersection of these two linear segments).

These figures point out that at high temperatures most of the Cb-bearing steels have nearly the same flow strengths as the base steel. On the other hand, at lower temperatures there is a significant difference in flow strengths, with the Cb-bearing steels becoming much stronger than the base steel. The shape of the flow strength-temperature curves and the flow strength levels of the high strength steels seem to be dependent on the Cb content. Steels with Cb levels less than 0.015 percent act very similarly to the base steel. They have a linear flow strength-temperature relationship and have nearly the same strength in the hot rolling temperature range (refer to Figures 2, 3, and 4), except for steels H-1 and H-2 which will be discussed later. With Cb contents between 0.015 percent and 0.030 percent, the slope of the flow strength-temperature curves shows a change between 1700°F and 1825°F (927°C and 996°C) (refer to Figures 5, 6, and 10). At temperatures below the slope change temperature, the Cb-bearing steels become much stronger than the base steel. When the Cb content is above 0.030 percent, the steel flow strengths are substantially higher than the base steel at all rolling temperatures; they show linear flow strength-temperature relationships with no slope change (refer to Figures 7, 8, and 11).

The flow strength of the base steel was used to determine the relative flow strengths of the Cb-bearing steels. Table II presents the data on the mean flow strength ratio, K_m (steel)/K_m (base), for temperatures between 1600°F and 1900°F (871°C and 1038°C).

Table II. Mean Flow Strength Ratios, K_m(Steel)/K_m(Base)

Steel	Rolling Temperature °F (°C)						
	1600 (871)	1650 (899)	1700 (927)	1750 (954)	1800 (982)	1850 (1010)	1900 (1038)
Base	1.00	1.00	1.00	1.00	1.00	1.00	1.00
A	1.03	1.02	1.01	1.00	0.99	0.99	0.98
B	1.02	1.02	1.02	1.02	1.02	1.02	1.03
C	1.07	1.07	1.06	1.05	1.05	1.04	1.03
D	1.14	1.13	1.13	1.00	0.98	0.97	0.94
E	1.16	1.12	1.08	1.03	0.98	1.00	1.02
F	1.22	1.19	1.17	1.15	1.12	1.08	1.05
G	1.22	1.21	1.20	1.18	1.17	1.15	1.13
H-1	1.06	1.04	1.01	0.99	0.96	1.00	1.03
H-2	1.28	1.26	1.23	1.20	1.16	1.14	1.14
I	1.28	1.26	1.23	1.20	1.16	1.14	1.14
J	1.11	1.11	1.10	1.10	1.09	1.08	1.05

This table shows that the C-Mn-Cb and C-Mn-Si-Cb steels can be 20 to 30 percent stronger than the base steel at 1600°F (871°C) depending on the Cb and Si contents.

Microstructural Characteristics

Recrystallization studies[2,8] have shown that C-Mn steels require very short times to recrystallize during hot rolling. Similar studies[2,4,9] show that columbium steels require substantially longer times to recrystallize as the hot rolling temperature is reduced. Since the recrystallization kinetics of the two types of steels are much different, they may be traced by their microstructures. The structure desired for observation is the austenite grain morphology just after the first pass. Quenching after rolling is necessary to "freeze" the structure for future study. As a result of the quenching, the microstructure will be predominantly martensite with substantially no austenite grain remaining. Therefore, a suitable etching technique has to be used to reveal the prior austenite grains.

Samples of the base steel and steel D were reheated to 2350°F (1290°C) for 1/2 hour, held at the rolling temperature between 1700°F (927°C) and 1900°F (1038°C) for 10 minutes, given a 40 percent reduction and either immediately quenched in water or held 15 seconds before quenching. The etching technique, described earlier, was employed to reveal the prior austenite grains. Figures 12 and 13 are the resulting microstructures for the base steel and steel D, respectively. The general microstructure, that can be seen from these figures, is a white matrix of unetched martensite and a cell-like structure of etched impurity-rich boundaries. The size and shape of this network reveal the prior austenite grain structure.

The base steel has a nearly equiaxed recrystallized structure at both hold times from 1900°F (1038°C) to 1750°F (954°C). At 1700°F (927°C) the microstructure is partially equiaxed and elongated at the 3-second hold. This indicates that immediately after rolling at 1700°F (927°C) the structure is only partially recrystallized. However, after 15 seconds the structure becomes completely recrystallized. In a production situation C-Mn steels will recrystallize between passes in the finishing train except possibly between the last two stands where the time is short and the temperature is low. Shortly after leaving the last stand the C-Mn steels will be completely recrystallized.

The C-Mn-Cb steel, on the other hand, is partially recrystallized at 1900°F (1038°C) immediately after rolling; even 15 seconds after rolling there are some remnants of unrecrystallized grains. At 1800°F (982°C) and below no recrystallization is evident, the grains remain elongated, and deformation bands are evident in some grains. This type of steel may recrystallize between the first few stands since the temperature is high and the time between stands is relatively long. No recrystallization will be evident after the first few stands and may not occur at all before the austenite-to-ferrite transformation.

Discussion

Second Pass Significance

The mean flow strength-temperature curves (Figures 1 to 11) were all generated using the second pass data. It is particularly important that the data be taken during the second pass. All the steels rolled, except the base steel, contained Cb. In such Cb-bearing steels, precipitation of $Cb(C,N)$ takes place during hot rolling. The $Cb(C,N)$ precipitates strengthen the austenite and the presence of Cb either in precipitates or solid solution causes recovery and recrystallization to become sluggish, tending to strengthen the austenite. The purpose of the first pass is to break up the large grained reheated structure, which is soft, while introducing sufficient strain to activate the precipitation of $Cb(C,N)$ that would normally be encountered during hot rolling. After the first pass, the slab should then more closely represent the elongated unrecrystallized grain structure and strength of the austenite rolled in the finishing stands of a hot mill.

BASE STEEL

Hold Time

3 Sec.   15 Sec.

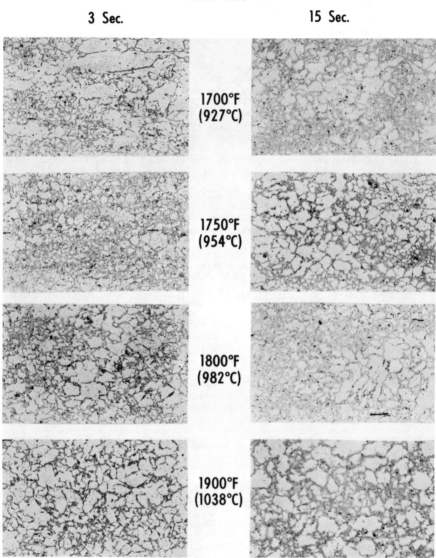

1700°F
(927°C)

1750°F
(954°C)

1800°F
(982°C)

1900°F
(1038°C)

Figure 12. Austenite morphology after hot rolling (40% reduction)
at the temperature shown.

⊢ .01″ ⊣

STEEL D

Hold Time

3 Sec.   15 Sec.

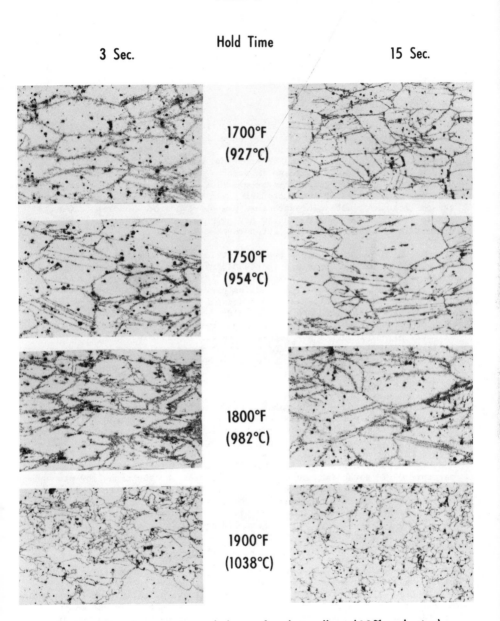

1700°F
(927°C)

1750°F
(954°C)

1800°F
(982°C)

1900°F
(1038°C)

Figure 13. Austenite morphology after hot rolling (40% reduction)
at the temperature shown.

.01"

Effect of First Pass Reduction

First pass reductions were either 25 or 35 percent to determine whether the deformation resistance during the second pass was a function of the degree of prior strain. Data presented in Figures 1 to 11 show no significant difference between 25 and 35 percent prior reductions. Although some change may be expected, the 10 percent difference in first pass reduction may not be sufficient to alter the second pass flow strength.

Austenite Morphology

The microstructural data, as well as the flow strength-temperature curves, do indicate a correlation between the deformation resistance and the morphology of the austenite entering the roll gap. If the austenite continues to recrystallize during hot rolling, which is characteristic of C-Mn steels (Figure 12), any increase in deformation resistance is due primarily to the reduction in rolling temperature plus an incremental flow strength increase as a result of the austenite grain refinement. This equiaxed recrystallized austenite microstructure was observed for a number of C-Mn steels and resulted in linear flow strength-temperature curves[3] very similar to the base steel flow strength-temperature curve.

As Cb is added to steel, the recrystallization kinetics of the steel are altered. As Cb levels are increased, unrecrystallized austenite may be found at higher hot rolling temperatures. Figure 13 shows the characteristic, elongated, unrecrystallized grains near 1900°F (1038°C) at short hold times. The deformation resistance continues to increase since these grains and other recrystallized grains are being deformed at lower and lower temperatures. The flow strength-temperature curves for columbium steels are slightly higher than the base steel curve at temperatures near 1900°F (1038°C) but become substantially higher at lower temperatures as the volume of elongated unrecrystallized grains increases.

Chemistry Effects

The flow strength ratio and shape of the flow strength-temperature curve are directly related to the Cb content in C-Mn-Cb steels. It is well documented[2,10] that the lack of recrystallization is the primary reason for the flow strength increase in C-Mn-Cb steels over C-Mn steels. An increase in Cb content causes recrystallization to become delayed at higher and higher temperatures resulting in flow strength increases at those temperatures. To establish the effect of Cb content on the mean flow strength of the C-Mn-Cb steels rolled in this study, the mean flow strength-ratio vs. Cb content curves were plotted for steels having nearly the same levels of C (0.17-0.2%) and Mn (0.8-1.0%), no Si, and Cb contents ranging from 0.010 to 0.075 percent. These curves, shown in Figure 14, clearly indicate that the mean flow strength ratio increases with increasing columbium content and decreasing temperature.

The shape of the flow strength-temperature curves seems to be a function of the Cb content. As mentioned earlier, low Cb levels produce a linear flow strength-temperature curve equal to or slightly higher than the base steel. Medium Cb levels show a slope change in the flow strength-temperature curve. At medium Cb levels, the flow strength at high temperatures is much like the base steel; however, at lower temperatures (below the slope change temperature) the flow strength increases substantially over the base steel. Steels with higher Cb content have linear flow strength-temperature curves and are substantially higher in flow strength at all temperatures compared to the base steel.

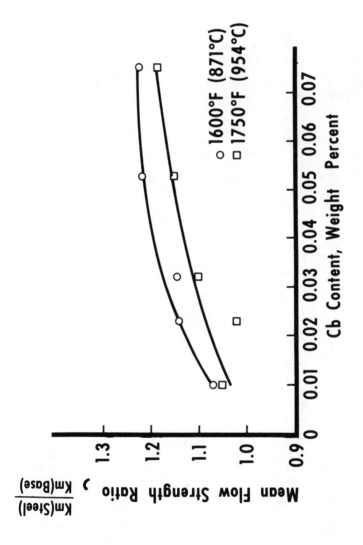

Figure 14. EFFECT OF Cb CONTENT ON THE MEAN FLOW STRENGTH RATIO

The flow strength of low Cb-C-Mn steels is near the base steel since, at these Cb levels, the delay time for recovery and recrystallization is short allowing for softening throughout the hot rolling range.

The slope change in the medium level Cb steels may be due to the combination of the volume of precipitate and amount of recovery and recrystallization occurring at any given temperature. Above 1850°F (1010°C), the precipitation rate of Cb(C,N) is relatively slow and thus the precipitate volume is low. As a result, the hot rolled structure observed at about 1850°F (1010°C) shows a partially or totally recrystallized structure. Those grains not recrystallized probably have enough energy at these temperatures to soften somewhat by recovery. Therefore, the flow strength of the C-Mn-Cb steels, partially or totally recrystallized and recovered, is nearly equal to the flow strength of the base steel at temperatures above 1850°F (1010°C). As the rolling temperatures are reduced, the precipitation rate of Cb(C,N) increases to a maximum between 1700°F and 1850°F (927°C and 1010°C).[2,11] The increased precipitation rate causes the hot rolled microstructure to become unrecrystallized and delays the recovery rate. This combination causes the flow strength to increase at a faster rate than it does above 1850°F (1010°C). It is speculated that this is the main reason for a slope change in the flow strength-temperature curve. At temperatures below 1700°F (927°C), the precipitation rate drops off with no recovery or recrystallization evident as the temperature is reduced. This results in the continuation of the curve maintaining the same slope.

In high Cb-bearing steels, the precipitation of Cb(C,N) in sufficient quantities occurs at all rolling temperatures, thereby effectively eliminating substantial recovery and recrystallization. As a result, the austenite becomes very strong throughout the hot rolling range. Steels with high Cb content require much higher strain to initiate recrystallization even at high rolling temperatures. Therefore, there is no slope change in the flow strength-temperature curves. These steels have higher flow strength than the base steel throughout the rolling range.

In the present study, silicon, when specific conditions exist, has also been found to increase significantly the flow strength during hot rolling. Si (over 0.04 percent) in these steels was present with Cb levels between 0.004 and 0.027 percent. There were no steels with high levels of both Si and Cb. The flow strength data for steels H-2 and I, for example, were as high as those for steels F and G at low temperatures even though the latter grades had at least twice as much Cb. Another interesting fact is shown in Figure 9. The two H steels shown have a Si variation and nearly the same Cb content but the higher Si steel is significantly stronger.

The role of Si in strengthening the austenite in these Cb steels is not completely clear. Si is a substitutional element in iron which would indicate some solid solution strengthening effect. This would explain the difference in flow strength between C-Mn and C-Mn-Si steels. However, when small amounts of Cb are added to a C-Mn-Si steel, the flow strength change is a drastic one, much greater than either the Si or Cb effect alone. It has been reported by Koyama et.al.[12] that Si reduces the solubility of Cb(C,N) in austenite. Therefore, the Si effect may be the result of an increase in the volume of Cb(C,N) above what would normally be produced during hot rolling. This would increase the sluggishness of recovery and recrystallization comparable to higher Cb steels. It may also explain why the Si effect is not as great with higher Cb steels where sufficient precipitation is available to delay recrystallization effectively at higher temperatures.

To determine the effect of composition on the flow strength at various

temperatures, the variation of the mean flow strength ratio (at three rolling temperatures) with the product of the percent carbon, percent columbium
and percent silicon [i.e., (C%) x (Cb%) x (Si%)] is shown in Figure 15.
This figure shows that the mean flow strength ratio increases with the increasing magnitude of the factor [(C%) x (Cb%) x (Si%)] until approximately
a value of 70×10^{-5} is reached and beyond this, the mean flow strength ratio
levels off. These curves together with the flow strength-temperature curves
can be used to predict the relative flow strength of steels with C, Mn, Si
and Cb level in the range reported in this study. Moreover, these curves
do indicate some unusual characteristics that one must be aware of to model
these steels effectively for computer controlled hot rolling.

Conclusions

1. The mean flow strength of C-Mn-Cb steels increases with the columbium
   level.
2. The shape of the mean flow strength-temperature curves is a function of
   the columbium content. Low and high columbium contents yield linear
   curves, while a medium columbium content generally results in two
   linear segments of different slopes.
3. The mean flow strength ratio, K_m(steel)/K_m(base), increases with the
   product of the percent carbon, percent columbium and percent silicon
   [i.e., (C%) x (Cb%) x (Si%)].
4. Silicon is a very effective austenite strengthener in the presence of
   small amounts of columbium.
5. Some of the steels, C-Mn-Cb or C-Mn-Cb-Si, are 20% to 30% stronger than
   the base steel between 1600°F and 1700°F (871°C and 927°C).

References

1. J. H. Cordea and R. E. Hook, Met. Trans., Vol. 1, 1970, p. 111.

2. A. Lebon, J. Rofes-Vernis and C. Rossard, Metal Science, Vol. 9, 1975,
   pp. 36-40.

3. J. A. DiCello, National Steel Corporation, Weirton, WV, unpublished
   research, January 1978.

4. A. T. Davenport, R. E. Miner and R. A. Kot, Proceedings of the Conference on The Hot Deformation of Austenite, Cincinnati, OH, 1975, TMS-
   AIME, 1977, p. 186.

5. G. E. Dieter, Mechanical Metallurgy, McGraw-Hill, 1961.

6. J. H. Hitchcock, ASME Res. Pub. App. 1, 1930.

7. H. Ford and J. M. Alexander, JISI, Vol. 92, 1963-64.

8. M. J. Stewart, "Hot Deformation of C-Mn Steels with Constant True Strain
   Rates," Canada Centre for Mineral and Energy Technology, August 1975.

9. M. J. Stewart, "The Effects of Niobium and Vanadium on the Softening of
   Austenite During Hot Working," Canada Centre for Mineral and Energy
   Technology, August 1975.

10. L. J. Cuddy, Proceedings of the Conference on The Hot Deformation of
    Austenite, Cincinnati, OH, 1975, TMS-AIME, 1977, p. 169.

11. R. Coladas, J. Masounave and J. Bailon, Proceedings of the Conference on
    The Hot Deformation of Austenite, Cincinnati, OH, 1975, TMS-AIME, 1977,
    p. 140.

12. S. Koyama, T. Ishii and K. Narita, Kinzoku Gakkai-shi, Vol. 35, No. 11,
    1971, p. 1089.

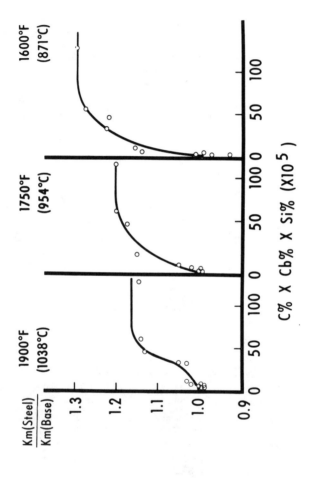

Figure 15. EFFECT OF ALLOY CONTENT ON THE MEAN FLOW STRENGTH RATIO AT VARIOUS TEMPERATURES

DISCUSSION

Q: In some of the steels you added up to 0.15% zirconium. Did you find any change in mean strength and also what was the influence of 0.15% zirconium on sulfide inclusions?

A: Each showed good sulfide shape control. But, with regard to the effect on the flow strain at the rolling temperature, we did not see any effect.

Q: Why did you apply a constant reduction for the first pass?

A: The first pass was to break the cast structure so that we can get similar material that we encounter in the finishing train of the rolling mill.

Q: Were all the specimens as-cast?

A: No, they were bloomed, but we still found some remnants of as-cast structures present.

Q: With regard to the correlation of Cb, C, and Si, that product with higher flow stresses would appear, or might be interpreted as a precipitation effect, in as much as silicon appears to lower the solubility of Cb carbide in austenite. Did you make any observations of fine particles?

A: No, we did not, but we presume the reason for the effect of silicon on the flow strain is because of the point you just raised. Silicon is known to reduce the solubility of Cb-carbonitride in steels. In the presence of the silicon, then we get more precipitation of Cb-carbonitride, and that is the reason for their higher strength.

Q: I have almost the same question as the previous one. It concerns the increase in flow stress due to the additions of C, Cb, and Si. Beyond some critical amount, resistance to stress increases very rapidly. So, below this critical amount, this effect is due to solute-dragging effect, and above this amount, it is due to precipitation.

A: I can't answer that, but I think that might be the mechanism that is working.

Comment: We have been doing similar studies to investigate hot deformation. One comment is that we find that silicon has a noticable hardening effect, whether the Cb is present or not. We generally find in the course of our mill experience with high-silicon steels, in the absence of microalloying elements, that the hot strength goes up. The other point I'd like to bring up is that the increase in flow strength that you see is associated with the nonrecrystallization effect. Our indications are that the influence of nonrecrystallization effects, when you first see them, that is at high temperatures, increases the flow strength by the amounts that you presented into something in the order of 10% or maybe 20%. But the significant increases of flow strength occur at lower temperatures, which our data would indicate, is associated with a burst of precipitation. The conclusions that we have drawn is that the increase of rolling load in the pancaking range can be quite modest if you do your pancaking deformation at high temperatures above the precipitate burst, and you can get excellent levels of ferrite grain refinement due to the deformation of the austenite. Whereas, at lower temperatures if you get a precipitation burst, not only do the hot loads go up, but if your grain refinement doesn't increase and your strength can come down because you lose precipitation in ferrite. So the point I am raising in this discussion is the interplay between precipitation and of deformed austenite structure, which I think have two distinctly different contributions to hot strength.

Comment: We keep seeming to re-invent these same interactions. I'd like to point out that some of the work on low-carbon, pearlite-reduced steels at BISRA in 1966-67, in the accelerated cooling program, steels with up to 1% silicon were examined. There was a consistant effect of silicon in improving the fracture-appearance transition temperature of these low-carbon niobium steels. Today, I think we can rationalize that in terms of silicon increasing the pancaking temperature for these materials. Then when we use accelerated cooling, we gather genefit of the microstrucutre on the nucleation of ferrite from that increased accumulated strain in the austenite.

THERMOMECHANICAL TREATMENT OF Ti- AND Nb-Mo-MICROALLOYED STEELS

IN HOT STRIP ROLLING

R. Kaspar, A. Streißelberger, and O. Pawelski
Department of Metal Forming
Max-Planck-Institute for Iron Research
Düsseldorf, W.Germany

Simulating hot strip rolling with the plane strain compression test two low carbon micro-alloyed steels containing either titanium or niobium and molybdenum were investigated. In a basic study nonisothermal precipitation in austenite and in upper ferrite was evaluated. Gamma-to-alpha transformation kinetics and changes in the fraction and morphology of transformation products were determined. Special attention was paid to the effect of the deformation of austenite. Simplified working schedules were designed to obtain either completely recrystallized fine-grained or unrecrystallized and dislocation hardened austenite entering gamma-to-alpha transformation. After deformation both simulated coiling and water-quench processing were applied. The relations between microstructural features and mechanical properties are discussed.

1. Introduction

In the development of HSLA steels both advantages of chemistry and modifications of the hot-rolling process such as controlled rolling and controlled cooling are effective in producing an optimum balance of mechanical properties and production costs. The benefits of combining a low carbon content with a fine grain size and strengthening by fine precipitations or by introducing a dislocation substructure were recognized earlier (1, 2). In the microalloying strategy two typical steel groups have been developed in the last ten years:
(i) Pearlite-reduced steels with a polygonal ferrite matrix and a small amount of pearlite,
(ii) bainitic-ferritic steels with a complex microstructure containing a mixture of polygonal ferrite, acicular ferrite, bainite, martensite and retained austenite.

In the present work some metallurgical processes occurring during thermomechanical treatment have been investigated for each of both microalloyed steel types. Simulation of typical rolling and cooling schedules has been applied to both experimental steels. The relations between microstructural features and mechanical properties are discussed.

2. Experimental Procedure

Both the Ti-bearing pearlite-reduced steel and the Nb-Mo-bearing bainitic-ferritic steel were received as commercial hot-rolled plate having the composition listed in Table I. A low-manganese content of the Ti-steel was aimed to provide improved weldability and better inclusion control. The molybdenum and the high manganese content in the Nb-Mo-steel are used to increase the hardenability to suppress the formation of polygonal ferrite and pearlite, favoring acicular-ferrite or bainitic structures. Silicon promotes the formation of coarse polygonal ferrite and is therefore held at a low level (3) in this steel.

Using a computer-controlled servohydraulic hot deformation simulator (WUMSI) the experimental steels have been investigated by simulating hot rolling in the plane strain compression test. The tool set-up allowed controlled heating, deformation and cooling, details of which are given elsewhere (4). Samples (100x140x20 mm^3) were large enough to permit complete measurement of mechanical properties. For the detection of the phase-transformation temperatures during cooling of plastically deformed austenite a measuring apparatus for a thermal analysis has been built and adapted to the testing machine. All specimens were soaked at 1200°C for 30 min. For the simulation of simplified hot strip rolling schedules specimens were predeformed at 1100°C in one deformation step of logarithmic strain φ = 0.3. The finishing passes were simulated in three deformation steps each of φ = 0.3 at temperatures 1000° to 650°C. In most cases after finishing the specimens were cooled at a rate of $\dot{v}_{FC}$ = 4°C/s to 600°C, then held at this temperature for 2 h (simulating coiling) and air cooled to room temperature. In some tests cooling rate $\dot{v}_{FC}$ was varied in the range of 1 to 12°C/s.

Table I. Chemical composition of experimental steels, wt %

Steel Type	C	Si	Mn	P	S	Mo	Nb	Ti	N
Ti-bearing	0.075	0.26	0.59	0.012	0.008	–	–	0.14	0.009
Nb-Mo-bearing	0.038	0.08	1.80	0.010	0.007	0.28	0.07	–	0.009

3. Metallurgical Processes after Deformation

Thermomechanical treatment of steel generally implies
strain hardening of the austenite by deformation that takes place
prior to or during the gamma-to-alpha transformation. The condition of the austenite immediately preceding transformation exerts
a strong effect on resulting microstructure and mechanical properties. According to chemistry and processing history microalloying elements can be soluted or partially precipitated as carbides,
nitrides or carbonitrides in the austenite. The size and shape
of austenite grain and the defect structure resulting from hot
working can be either preserved or changed due to recovery and
recrystallization. The processes like recrystallization, precipitation and gamma-to-alpha transformation interfere to each other
and run mostly simultaneously within a broad range of temperatures. For the resulting microstructure and mechanical properties
not only the condition of austenite prior to the start of transformation must be considered but also further changes occurring
within the austenite-ferrite phase field should be paid attention.

Anisothermal Precipitation in Austenite and Upper Ferrite

The niobium and titanium that are in solid solution before
hot working precipitate partially in austenite and further during
the gamma-to-alpha transformation and thereafter. Suppressing
precipitation in austenite is desirable to preserve more microalloying potential for the precipitation in the ferrite, which
is dispersed finely and more effective for strengthening. Therefore precipitation is a decisive process in rolling and cooling
schedules and should be controlled. The temperature range of the
most rapid precipitation results from the balance between the
supersaturation of the alloying elements as a driving force and
diffusion as a feasibility of motion during formation of precipitates. Consequently C-curves with a "nose" arise in a time-
temperature-precipitation diagram (5). Deformation of austenite
largely accelerates the precipitation but does not change the
shape of curves generally. In the bainitic-ferritic Nb-Mo-steel
the transformation is delayed and so the existence of metastable
austenite is extended to lower temperatures. Therefore the critical temperature range of rapid precipitation is expected to be
in the one-phase austenite field. Corresponding to production

conditions an anisothermal method of test has been applied to investigate the precipitation in austenite. After deformation below the recrystallization temperature in the range from 910° to 820°C the specimens were cooled at various cooling rates $\dot{\vartheta}_A$ to a temperature close to Ar_3, then quenched and age-hardened. The results for two values of $\dot{\vartheta}_A$ are shown in Fig. 1. The decrease of hardness in the age-hardened condition indicates an extensive precipitation of coarse carbonitride in austenite which does not contribute to precipitation strengthening (6). An increasing strain φ of austenite accelerates the precipitation. Within a critical temperature range (820° to 850°C) even a small value of φ causes a drop of hardness, Fig. 1a. Above this range (880° to 910°C) a higher strain is necessary to produce an extensive precipitation. Nevertheless, because of the longer duration in austenite during the cooling from higher temperatures the drop in hardness is greater in the latter case. Owing to a higher cooling rate $\dot{\vartheta}_A$ the precipitation in austenite was reduced, which corresponds to a smaller loss in hardness, Fig. 1b.

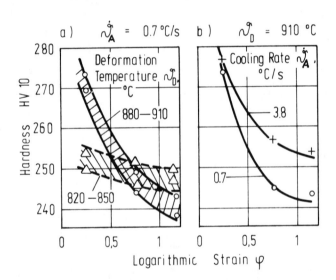

Fig. 1 - On the effect of anisothermal precipitation in the deformed austenite of
the Nb-Mo-steel:
The drop in hardness after age-hardening describing the influence of strain
a) at various deformation temperatures
ϑ_D
b) for two cooling rates $\dot{\vartheta}_A$ prior to
quenching.

In the pearlite-reduced Ti-steel with a relatively low manganese content the ferrite transformation range is extended to higher temperatures, especially after deformation of austenite. Thus the range of a rapid precipitation in austenite falls into the two-phase austenite-ferrite field. In the tests described the specimens were deformed at ϑ_D = 900°C, then cooled at two cooling rates ϑ_{AF} = 0.7 and 3.8°C/s to the temperature of 830°C arriving at the austenite-ferrite range for the ϑ_D applied. Finally the specimens were quenched and age-hardened. A markedly greater difference in the drop of hardness for the two cooling rates, Fig. 2a, as compared with the Nb-Mo-steel (Fig. 1b), indicates a cumulative effect of proceeding transformation and precipitation in both austenite and ferrite. As expected the fraction of ferrite increases due to strain of austenite and is higher after lower cooling rate, Fig. 2b. Whilst the microhardness of age-hardened bainite is relatively insensitive to austenite history, that of ferrite decreases with φ and is lower for the case of a longer stay in the upper ferrite range during the lower cooling at a rate of ϑ_{AF} = 0.7°C/s. This loss in precipitation strengthening is related to some fraction of coarse precipitation in ferrite, the field of which is shifted to higher temperatures. Thus an early formation of proeutectoid ferrite of the low hardenable pearlite-reduced Ti-steel after low finishing temperature brings about an additional detrimental component to the resulting yield strength due to a loss in precipitation strengthening.

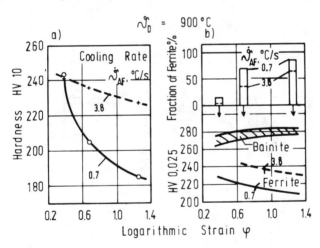

Fig. 2 - On the effect of anisothermal precipitation in austenite and upper ferrite of the Ti-steel:
a) Influence of strain and subsequent cooling rate ϑ_{AF} on the hardness after age-hardening
b) Influence of strain on the fraction of ferrite and microhardness of ferrite and bainite.

Transformation gamma-to-alpha

Little systematic basic work has been done on studying the
continuous cooling transformation kinetics of thermomechanically
worked austenite. The major factor is the prior hot working
strain below the recrystallization temperature resulting in formation of elongated "pancake" grains. Consequently the specific
austenite grain boundary area is increased providing more favored sites for nucleating ferrite during transformation (7). Additionally, increasing of density of potent ferrite nucleation
sites per unit area of austenite grain boundary is expected as
a consequence of extending strain (8). Generation of deformation
bands of high dislocation density creates a further source of
nucleation sites. Fig. 3 shows the relation between the resulting mean ferrite grain size and the pre-existing mean grain size
of austenite with or without prior deformation. The strong influence of the austenite grain size after transformation of an
undeformed austenite decreases with increasing strain. A high
strain of austenite gives rise to a fine ferrite grain nearly
unaffected by the original austenite grain size.

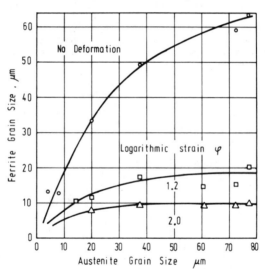

Fig. 3 - Influence of the austenite grain size and
deformation of austenite on the ferrite
grain size of the Ti-steel.

Continuous cooling transformation diagrams of recrystallized and unrecrystallized austenite of Ti-steel are presented in
Fig. 4a and 4b, respectively. For methodical reasons the deformation temperature ϑ_D had to be sufficiently higher than the
start of transformation. Some preliminary tests showed that
ϑ_D = 930°C, which met this limitation, was still low enough to
suppress recrystallization for this steel. The influence of deformation is mostly reflected in terms of an extended ferrite
range. An accelerated formation of ferrite at higher temperatures causes a strain-induced transport of carbon and perhaps
alloying elements from ferrite into a coexisting austenite. The

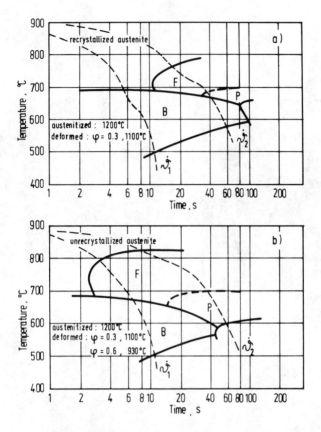

Fig. 4 - Continuous cooling transformation diagram
for the Ti-steel cooled from 900°C
a) Deformed and recrystallized austenite
b) Additional deformed (φ= 0.6) unrecrystallized austenite.

latter one becomes stabilized and consequently the bainite-start
is shifted to lower temperatures. Applying lower cooling rates
on deformed austenite the formation of diffusional transformation products was promoted giving rise to a larger amount of
pearlite, which had a nonlamellar morphology.

In order to compare microstructural features of transformation products originating from both austenite conditions
the resulting microstructures are presented in Fig. 5, for cor-
responding cooling curves $\dot{\vartheta}_1$ and $\dot{\vartheta}_2$ see Fig. 4. After cooling
at the rate $\dot{\vartheta}_1$ the high ferrite fraction observed in the unrecrystallized specimens (c) exhibits the most substantial difference in comparison with the pure bainite in (a). A grain refining effect together with the promotion of pearlite is shown
in (d) if compared to (b).

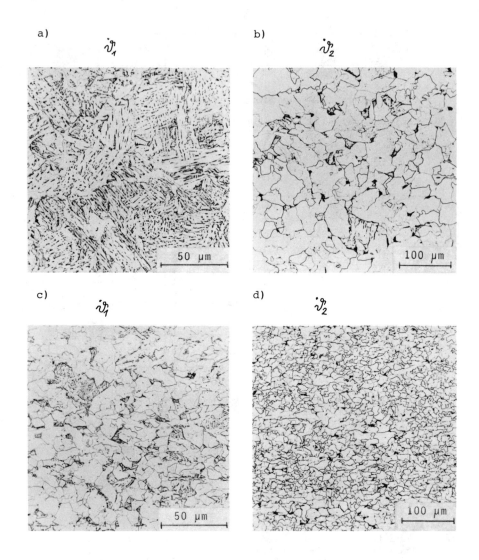

Fig. 5 - Effect of deformation of austenite on the
microstructure of the Ti-steel after two
cooling rates $\dot{\vartheta}_1$ and $\dot{\vartheta}_2$ (see Fig. 4)
a)b) Transformed from a recrystallized
austenite
c)d) Transformed from an unrecrystallized
austenite.

CCT-diagrams of recrystallized and unrecrystallized Nb-Mo-steel, Fig. 6, have a flat top so that the transformation start temperature does not vary substantially with cooling rate. Without regard to the range of polygonal ferrite (F) no noticeable differences in the transformation temperatures were recorded. It should be emphasized that the deviations on the differentiated cooling curves corresponding to ferrite-start and ferrite-finish were rather week so that the detection of these ranges was subject to a greater error. The bainitic-ferritic range (BF) includes formation of a combined structure covering a pure bainite, a mixture of acicular ferrite and carbides making the terminology difficult (3). Subdividing of this field by means of thermal analysis was impossible.

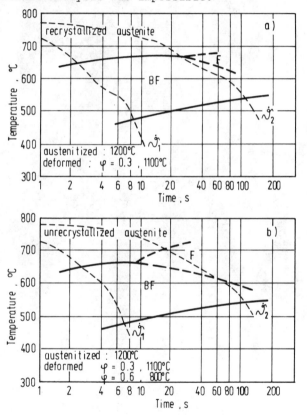

Fig. 6 - Continuous cooling transformation diagram for
the Nb-Mo-steel cooled from 780°C
a) Deformed and recrystallized austenite
b) Additional deformed (φ = 0.6) unrecrystallized austenite.

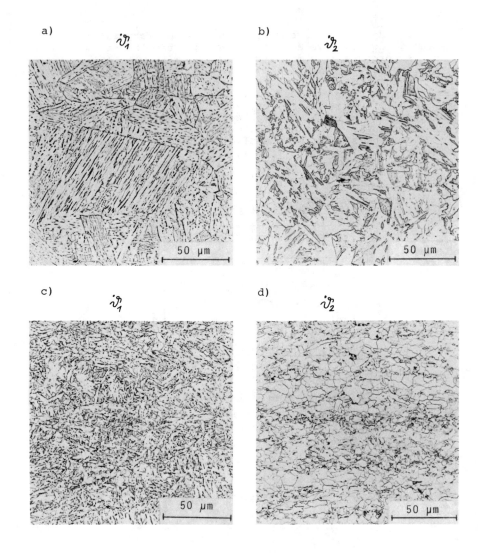

Fig. 7 - Effect of deformation of austenite on the
microstructure of the Nb-Mo-steel after
two cooling rates $\dot{v}_1$ and $\dot{v}_2$ (see Fig. 6)
a)b) Transformed from a recrystallized
austenite
c)d) Transformed from an unrecrystallized austenite.

a)

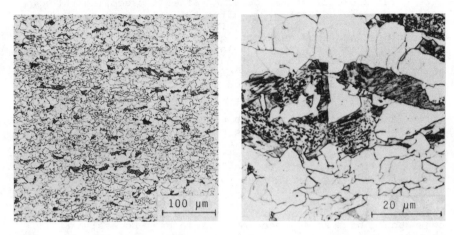

b)

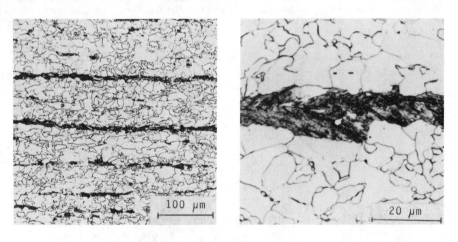

Fig. 8 - Partitioning features in the microstructure
of the Nb-Mo-steel
a) martensitic-bainitic islands (finishing
temperature ϑ_F = 700°C, cooling rate between finishing and coiling ϑ_{FC} = 0.02°C/s,
coiling temperature ϑ_C = 550°C)
b) martensitic-bainitic layers (ϑ_F = 900°C,
ϑ_{FC} = 0.02°C/s, ϑ_C = 550°C).

The microstructures obtained after two cooling rates $\dot{\vartheta}_1$ and $\dot{\vartheta}_2$ are depicted in Fig. 7. After the rapid cooling $\dot{\vartheta}_1$ the bainite transformation from deformed austenite (c) is much finer without distinguished austenite grain boundaries in comparison with the bainite (a) showing a large bainitic ferrite lath size. As a transformation product of unrecrystallized austenite after cooling rate $\dot{\vartheta}_2$ a fine mixture of polygonal and acicular ferrite together with dispersed enriched constituents (d) replaces a coarse bainite (b) produced from recrystallized austenite.

A slow cooling within an upper temperature range of transformation promotes partitioning processes in structure (9). During transformation diminishing regions of austenite are enriched in carbon predominantly in the vicinity of an early polygonal ferrite and become so stable that they will transform in martensite even if the cooling to room temperature is slow. These regions of martensite, bainite and a certain amount of retained austenite (10) can be distributed either uniformly as islands, Fig. 8a, or as parallel oriented layers, Fig. 8b, as a consequence of a more advanced partitioning process which in this latter case results in a higher fraction of martensite and retained austenite at the expence of bainite in these regions. In production of hot strip the conditions for formation of such enriched regions in bainitic-ferritic Nb-Mo-steel are given if the cooling between finishing and coiling is slow and/or if the coiling temperature is high. The result of a simple classification of such partitioning phenomena in the microstructure of Nb-Mo-steel after simulated rolling and coiling schedules is given in Fig. 9. For two low cooling rates $\dot{\vartheta}_{FC}$ the effect of coiling

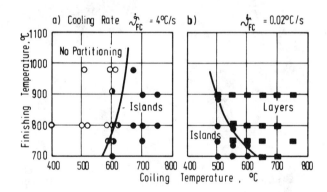

Fig. 9 - Occurrence and morphology of the martensitic-bainitic regions of the Nb-Mo-steel for various processing conditions.

and finishing temperature on the partitioning morphology is presented. While layers can hardly be considered as a beneficial feature, the structure with uniformly distributed martensite-austenite islands exhibits a high ductility and formability attributed to the dual-phase microstructure (11). However, the best combinations of yield strength and impact transition temperature have been achieved on the specimens without distinct islands in the structure of this steel.

4. Simulated rolling schedules

Production of hot strip was simulated as an application of the studied phenomena and thereafter microstructure and mechanical properties of both experimental steels were investigated.

Two groups of processing schedules with final deformation in the austenite field were designed

i. to obtain completely recrystallized fine-grained austenite (schedule A1) or
ii. to obtain unrecrystallized and dislocation hardened austenite (schedule AØ) before entering gamma-to-alpha transformation.

In a further group of schedules final deformation was extended to temperatures within the austenite-ferrite phase (schedule AF). Finally, two additional treatments were applied subsequent to the schedule AØ:

i. direct water quenching ($\dot{\mathcal{V}}$= 100°C/s) from the finishing temperature of processing AØ (schedule WQ)
ii. normalizing after AØ at 950°C (schedule N).

Ti-bearing steel

The summary of tensile and impact properties for the schedules applied on this steel are presented in Fig. 10. Each discrete symbol within the AØ and A1 fields corresponds to the mean value of properties measured for a definite cooling rate $\dot{\mathcal{V}}_{FC}$ between finishing and coiling increasing in the arrow direction (1, 4 and 12°C/s).

Surprisingly, no significant difference in yield strength has been determined for AØ and A1. A much finer ferrite grain size has been observed after deforming austenite below recrystallization temperature (AØ). However, this strengthening effect was almost compensated by the loss in precipitation potential due to coarse precipitation in the strain-induced upper ferrite. The rather poor impact properties of specimens processed according to schedule A1 may be ascribed to embrittlement caused by a greater extent of precipitation hardening and a lack in beneficial grain refining. In both schedule groups AØ and A1a faster cooling increases the yield strength but leads to some loss in toughness. These results reflect a detrimental effect of precipitation strengthening on transition temperature. A steeper slope of yield strength versus impact transition temperature at lower cooling rates corresponds to a higher contribution of grain refinement observed in microstructure.

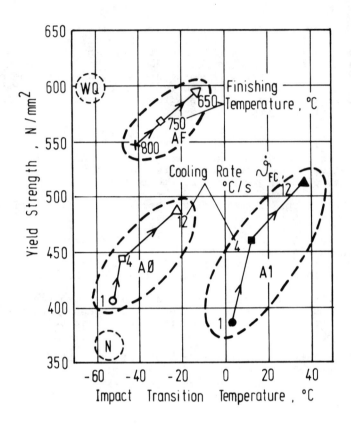

Fig. 10 - Mechanical properties of the Ti-
steel resulting from characteristical schedules of thermomechanical
treatment
(A1 - recrystallized austenite,
finishing temperature ϑ_F = 1000°C;
AØ - unrecrystallized austenite,
ϑ_F = 900°C; AF - finishing in the
austenite-ferrite range; WQ - water
quenched after AØ; N - normalized
after AØ).

Applying deformation temperatures in the austenite-ferrite two-phase region (schedule AF) provides a substantial strengthening contribution in comparison with AØ or A1. In this steel with a wide temperature range for such intercritical deformation, which is extended up to high temperatures, a rather large volume fraction can be transformed before final deformation resulting in a considerable increase in the yield strength. Lowering the temperatures of intercritical deformation (see arrow direction in AF) deteriorates impact properties as the dislocation strengthening component increases. No recrystallization tendency of the strained ferrite could be detected in this steel. The best properties have been obtained accordant to schedule WQ on quenching the specimen immediately after the processing AØ. The bainite-martensite structure ensures both the high yield strength and the lowest impact transition temperature compared to all schedules applied, because of the fine microstructure units.

A subsequent tempering brings about an additional increase in the yield strength due to full exploitation of the precipitation strengthening. This will be accompanied by a loss in impact properties.

Additional normalizing treatment (N) exhibits a loss in yield strength values achieved after preceding schedule AØ permitting only a small improvement in the impact transition temperature.

Nb-Mo-bearing steel (Fig. 11)

Rather different results of simulated processing schedules have been achieved for the bainitic-ferritic Nb-Mo-steel. Especially impact properties were markedly better giving a low impact transition temperature largely unsensitive to changes in finishing conditions. Hence the significance of finishing temperature in austenite (AØ compared to A1) is considered only concerning the yield strength, prefering finishing below the recrystallization temperature (AØ). Within AØ and A1 schedules a certain cooling rate is desirable to avoid the formation of coarse polygonal ferrite. This seems to be a reason for a considerable higher yield strength measured for the faster cooling rates ϑ_{FC} = 4 and 12°C/s.

For deformation in the austenite-ferrite two phase region this bainitic-ferritic steel offers much smaller amount of pre-existing ferrite to be deformed. The temperature range for such processing is rather narrow. In the intercritical schedules (AF) the ferrite transformed prior to final deformation was obtained in two ways: Conventionally by lowering the finishing temperature to 650°C or by a holding period of 2 min at 700°C prior to final deformation. According to structural investigation a supplementary strengthening component, resulting from the structure refinement due to dynamic recovery or recrystallization of deformed ferrite (12), may come into question in this steel. Only a negligible rise of impact transition temperature seems to support this consideration.

The improvement of properties by direct quenching compared to the simulated coiling can be derived from microstructure in this steel containing fine bainite and a low carbon martensite. A fine acicular structure and suppressed precipitation are supposed to be the reason for such an excellent low impact transition temperature. Comparable to the Ti-steel a subsequent aging further increases the yield strength but not without impairing

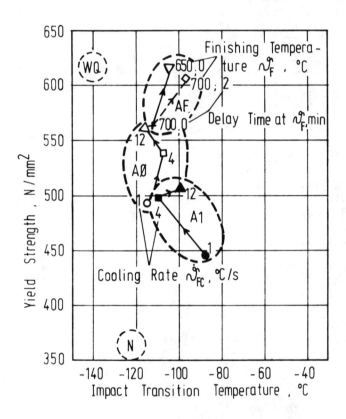

Fig. 11 - Mechanical properties of the Nb-Mo-steel resulting from characteristical schedules of thermomechanical treatment (see key to Fig. 10 but ϑ_F = 980°C for A1 and ϑ_F = 800°C for AØ).

impact transition temperature.

A considerable drop of yield strength after normalizing (N) indicates a higher precipitation strengthening component which has been lost during normalizing if compared with the Ti-steel.

5. Summary

As representatives of two families of low carbon micro-alloyed steels the behaviour of a pearlite-reduced Ti-bearing and a bainitic-ferritic Nb-Mo-bearing steel has been investigated. The essential results of some basic studies as well as of simulated thermomechanical treatments of hot strip can be summarized as follows:

1. Coarse precipitation after hot working can occur both in strained austenite and in ferrite that is strain induced transformed at high temperatures. A faster cooling within the critical temperature range suppressed this process.

2. The grain size of ferrite transformed from a heavily strained austenite was independent of the pre-austenite grain size in a wide range.

3. In the CCT-diagram obtained on the strained austenite of pearlite-reduced steel the ferrite field was widened to higher temperatures as well as to the range of bainite. A similar effect on the polygonal ferrite field of the bainitic-ferritic steel was much weaker.

4. The partitioning structure phenomena giving rise to martensitic-bainitic islands or layers strongly depended on processing conditions.

5. Improvement in properties by finishing in the austenite-ferrite range was effective especially on the pearlite-reduced steel with a large amount of deformed ferrite.

6. As for the impact properties the bainitic-ferritic steel with a fine acicular structure was superior to the pearlite-reduced steel examined.

7. Thermomechanical schedules using direct quenching (cooling rate ≈ $10^2 \, °C/s$) after hot deformation produced excellent mechanical properties for both experimental steels.

6. Acknowledgement

Financial support from the German Ministry of Research and Technology and from the European Commission for Steel and Coal is gratefully acknowledged.

References

(1) F. B. Pickering, "High-Strength, Low-Alloy Steels - A Decade of Progress", paper presented at 'Microalloying 75', Washington, Oct. 1975.

(2) T. Gladman, I. D. McIvor, and D. Dulieu, "Structural Property Relationships in Micro-Alloyed Steels", ibid.

(3) A. P. Coldren, and J. L. Mihelich, "Acicular Ferrite HSLA Steels For Line Pipe", pp. 14-28 in Molybdenum Containing Steels for Gas and Oil Industry Application, Climax Molybdenum rep., 1978.

(4) R. Kaspar, and O. Pawelski, "A Computer-Controlled Simulation of Hot Working by the Flat Compression Test on a High Speed Servo-Hydraulic Testing Machine", paper presented to 19th Int. MTDR Conf., Manchester, Sept. 1978.

(5) H. Watanabe, Y. E. Smith, and R. D. Pehlke, "Precipitation Kinetics of Niobium Carbonitride in Austenite of High-Strength Low-Alloy Steels", John B. Ballance, ed.: AIME, New York, N. Y., 1977.

(6) A. le Bon, J. Rofes-Vernis, and C. Rossard, "Recrystallization and Precipitation during Hot Working of a Nb-Bearing HSLA Steel", Metal Science 9 (1975) pp. 36-40.

(7) I. Kozasu, C. Ouchi, T. Sampei, and T. Okita, "Hot Rolling as a High-Temperature Thermo-Mechanical Process", paper presented at 'Microalloying 75', Washington, Oct. 1975.

(8) W. Roberts, H. Lidefelt, and A. Sandberg, "Mechanisms of Enhanced Ferrite Nucleation from Deformed Austenite in Micro-alloyed Steels", pp. 38-42 in Hot Working and Forming Processes, C. M. Sellars and G. J. Davies ed.; The Metal Society, London, 1980.

(9) V. Biss, and R. L. Crydeman, "Martensite and Retained Austenite in Hot-Rolled, Low-Carbon Bainitic Steels", Metallurgical Trans., 2 (8) (1971) pp. 2267-2276.

(10) T. Gold, and B. Garbarz, "The Structure of Martensite-Austenite Islands in Granular Bainite", Praktische Metallographie 17 (1980) pp. 338-393.

(11) A. P. Coldren, G. Tither, A. Cornford, and J. R. Hiam, "Development and Mill Trial of As-Rolled Dual-Phase Steel", in Formable HSLA and Dual-Phase Steels, A. T. Davenport ed.; AIME, New York, N. Y., 1979.

(12) S. Gohda, K. Watanabe, and Y. Hashimoto, "Effect at Intercritical Rolling on Structure and Properties of Low Carbon Steel", Trans. ISIJ 21 (1981) pp. 6-15.

DISCUSSION

Q: What kind of structure do you have in your normalized steel? I ask this question because the transition temperatures in normalized conditions is very typical. Because of the occurrance of splitting, transition temperature in normalized steels is much higher than in controlled-rolled steels, particularly when material is rolled in the alpha/gamma phase region.

A: You have a fine structure in both steels after normalizing. What may be the first reason for such low transition temperatures in the Ti-bearing steel was a small amount of ferrite with a very fine grain size in the Nb-Mo steel.

Q: I have two questions on the effect of deformation on the transformation temperature. Did you apply the straining above or below the A_{C3} temperature?

A: The exact temperatures are in the paper, but this temperature was the lowest temperature which was possible for this technique.

Q: What temperature was that?

A: 930°C was the temperature for the Ti-steel and something about 800°C for the Nb-Mo steel.

Q: My second question then has to do with time base for the CCT diagram for the deformed condition. In your diagrams, in the underformed condition, the ferrite formation start time is around 10 seconds in one of the steels. In the deformed case, for another steel, the ferrite-start time is around 2 seconds. Therefore, I wonder what your time base was in the second case. How did you choose the zero time? Was it at the time beginning to be counted from the moment of deformation?

A: It was from the moment of the end of deformation.

Q: The point of my question is that if the deformation is done below the transformation temperature (below the A_{C3} temperature), then you could argue that you should start the stop watch. You should accumulate the time when you first go below the transformation temperature, or at least take into account, the time that the steel is below the transformation temperature before the deformation is applied. Otherwise, the two time bases are not actually comparable between the lower and the upper diagrams.

A: Temperature for the start of time basis, in most cases was the same. I would have to check the exact temperatures in the paper. It was not exactly the A_{C3} temperature because it is difficult to determine this temperature, but we have gone with the underformed temperature to reach the exact same temperature in the underformed specimens, and this was the start of the time scale.

Q: In last two figures, what temperatures was the simulated coiling temperature?

A: The temperature was 600°C for this steel and the other steels.

Q: Also, in the Nb-containing steel and the other steels?

A: Yes.

Comment: This comment is a supplement to the previous one. We have had to look at this problem because we measured (1) the start of transformation after rolling, (2) when the rolling is finished, and (3) before transformation starts, in order to get the sufficient data to do our experiments. Now we always stop the stop clock at the time when the cooling goes through the A3 temperature. Between that moment, and the start of transformation, you can put the rolling pass at several points. It doesn't make any difference, because the pass takes only a fraction of a second.

Q: I was concerned with whether the stop watch was started when the steels went through the A_{C3} temperature or not.

A: You must be below the recrystallization temperature because you are close to this A_{C3} temperature. After deformation, you can't measure the cooling curve because of heat effects in the short time after deformation. That is why the exact A_{C3} temperature should be difficult to achieve as a starting point.

Comment: We did some work on previous deformation by tensile testing, and then continuous cooling, to get the start of transformation. We found that if you are cooling at a very slow rate after deformation, you did get recrystallization before the transformation started. That cast a lot of doubt on the results of that part of the continuous-cooling transformation diagram.

Comment: I think the critical temperature which is pertinent to this discussion is the A_{R3} not A_{C3}, since the transformation is occurring during cooling not heating.

COMPARISON OF TEXTURES IN LOW-ALLOYED AUSTENITE AND FERRITE
AFTER HOT ROLLING

Dr. Stanislaw Gorczyca
Professor of Physical Metallurgy
Academy of Mining and Metallurgy
Krakow, Poland

My study was concerned with the kinetics of the recrystallization of
low alloy steel austenite after hot working. The results were first
published in the Polish Metallurgical Journal "HUTNIK" in 1958 (1), and
later in Revue de Metallurgie (2). This was, probably, the first system-
atical study of the relation between the temperature of deformation and the
rate of recrystallization (3).

The photomicrographs in Figures 1, 2, and 3 illustrate (a) how the
nucleation of new grains is related to the temperature of working and (b)
that the nuclei of the new grains are formed either on the high angle grain
boundaries or on the boundaries of recrystallization twins.

The photographs in Figures 4 and 5 represent the structure of a plate
rolled at 1175°K, quenched immediately after rolling and after holding 5
seconds held at the temperature of deformation, to obtain some static
recrystallization. It is evident that the rate of recrystallization is
strongly related to the strain distribution in the cross-section of the
plate. At the surface, static recrystallization is more advanced than in
the mid-thickness of the plate.

During the hot rolling of austenite the deformation texture developed
changes with the progress of static recrystallization. This argument was
used by Brown and DeArdo (4) to distinguish the static recrystallization
from the dynamic recrystallization. In a low alloy steel, the texture of
hot-worked austenite could be transformed to the texture of ferrite, provided
there is a definite orientation relationship between the parent phase and
the product phase, as is the case in both bainitic and martensitic transformations. In other words, austenite with a well developed texture should
transform to ferrite with a well defined texture.

This idea was used by Andrews (5) and Karp (6) to explain the texture of
ferrite in the austenitic stainless steel wire in which the martensite was
formed during the wire drawing operation. Transformation of austenite in
hot-rolled plate to textured ferrite via martensitic transformation was
studied by Kula and Lopata (7) and Borik and Richman (8). In all these
studies, the explanation was based on the assumed rotation of the ideal pole

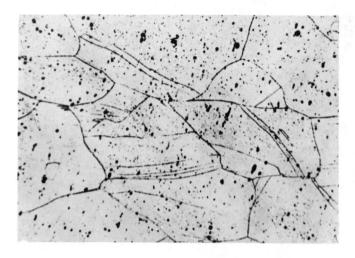

Figure 1 -- Microstructure of steel (C-0.36%; Mn-0.72%; Si-0.29%; Cr-1.43%; Ni-1.46%; Mo-0.23%) after hot-working at 1125°K to 30% strain, quenched in water; sample was etched to reveal the grain boundaries of austenite. 100X.

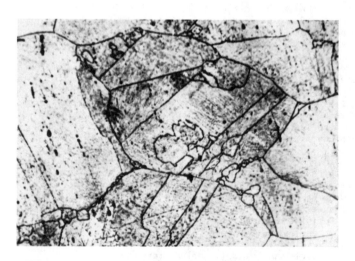

Figure 2 -- Microstructure of steel (composition same as Figure 1) after hot-working at 1125°K to 30% strain, held at temperature of working for 30 sec. and quenched in water. 100X.

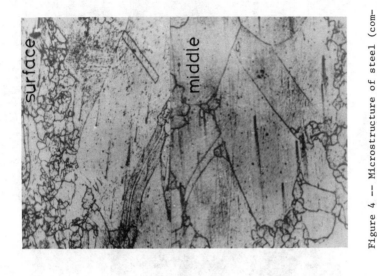

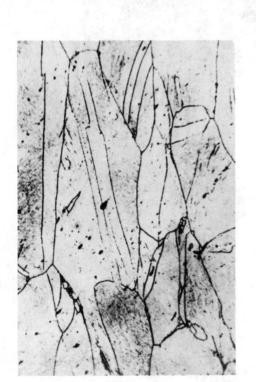

Figure 3 -- Microstructure of steel (composition same as Figure 1) after hot-working at 1225°K to 30% strain, quenched in water. 100X.

Figure 4 -- Microstructure of steel (composition same as Figure 1) after hot-rolling at 1175°K to 30% strain, quenched in water. Top of micrograph represents the surface, and bottom -- the mid-thickness of the hot-rolled plate. 100X.

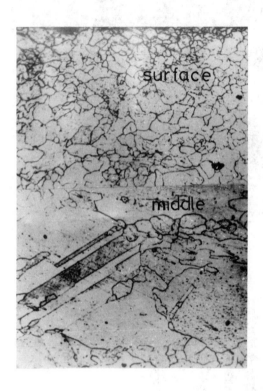

Figure 5 -- Microstructure of steel (composition the same as Figure 1)   after
hot-rolling at 1175°K to 30% strain hold
at the temperature of rolling for 5
seconds and quenched in water.   Top and
bottom of micrograph same as Figure 4.

figure of austenite, according to the Kurdjumov-Sachs orientation relationship for martensitic transformation. In our study on the transformation
texture in bainitic steels, conducted in cooperation with Maciosowski,
Pospiech and Jura (9, 10), we determined the pole figure of retained
austenite. For the bainite constituent, it was possible to determine the
orientation distribution function (ODF), (Figure 6), using the Bunge
method (9).

Using a matrix of all 24 Kurdjumov-Sachs relations, it was possible to
design a computer program which transformed ODF of ferrite to ODF of austenite, or vice versa. Having ODF function of austenite (Figure 7), it is
possible to calculate every pole figure {hkl}. The {200} pole figures of
austenite determined experimentally (Figure 8a), can be compared with those
calculated from the transformation of the ODF of bainite (Figure 8b). The
close agreement between these two sets of data provides strong support for
all the theoretical assumptions.

The basic concepts of this method were first presented at the conference
in Pont a Moussen in 1973 (10), and full account of the results was published in Polish in Acta Met (1976) (9, 12). Similar results were published
by G. J. Davies and his colleague (13, 14), but they were not aware of our
earlier study. To assist the English speaking scientists, more recently we
published our results in English (15).

One could advance arguments that texture, observed in deep drawing steel
after hot-rolling (Goodman and Hsun-Hu 16) and Karp and others (17) is the
result of a texture formed in austenite. However, the following results
suggest that some caution is warranted before accepting the validity of the
aforementioned argument.

In cooperation with Drs. Ratuszek and Jura, we have measured the texture
of a hot-rolled 3.5% silicon steel. Microstructures of the sheet, parallel
to the rolling direction and perpendicular to the rolling plane, are shown
in Figures 9 and 10. At the surface, there is a thin layer of recrystallized
ferrite, whereas in the middle of the sheet grains of ferrite are elongated
and deformed. In the elongated grains well developed subgrains resulting
from the dynamical recovery were revealed by TEM (Figure 11). Using the
Bunge method, the ODF was determined both for the surface layer (Figure 12)
and for the middle of the sheet (Figure 13). The two ODF graphs show a
distinct difference. For the middle of the sheet, ODF is very close to the
texture of cold-rolled ferrite, and at the surface of the sheet, the ODF is
close to the texture of recrystallized ferrite.

In the 3.5% silicon steel, the ferrite is stable from room temperature
to the melting point. However we transformed ODF of this ferrite via
Kurdyumov-Sachs relationship to the ODF of theoretical austenite (Figures
13, 14, 15) and obtained similar results to those reported for the bainitic
steel (Figure 7). These results are interesting, since they suggest that
without a phase transformation, the hot-rolled texture of austenite and the
hot-rolled texture of ferrite are related in a way which could be described
as a reciprocal relation, and the matrix of transformation or mis-orientation
distribution function is described by the Kurdjumov-Sachs relationship. In
other words, in the quoted example of deep drawing steel, after hot-rolling,
the texture could be either the result of deformation of ferrite, or of
forming oriented nuclei of ferrite in deformed, textured, austenite. In the
latter case, it has to be assumed that the oriented nuclei is the main cause
of the transformation texture in a similar way as it was proposed for the
transformation of texture during recrystallization.

Fig.6 Orientation distribution function of ferrite of a bainitic steel (C-0.16; Mn-0.50; Si-0.20; Cr-2.18; Ni-1.20; Mo-0.48; B-0.0017) after hot-rolling of austenite at 975°K to 75% strain; the sample was quenched to bainite.

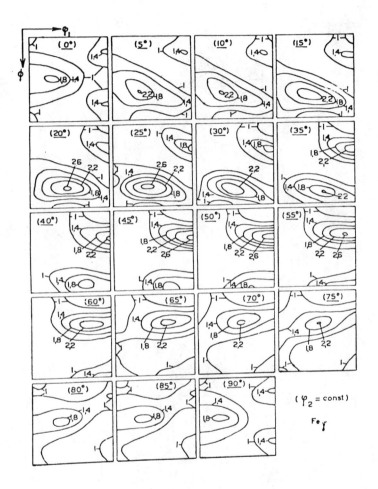

Fig.7 Orientation distribution function of austenite calculated from ODF of ferrite (Figure 6) using Kurdyumov-Sachs orientation relationship.

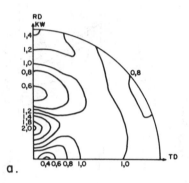

a.

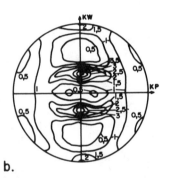

b.

Figure 8 -- {200} pole figure of
austenite
(a)   experimental for bainite steel
(b)   calculated from ODF of austenite
      (Figure 7)

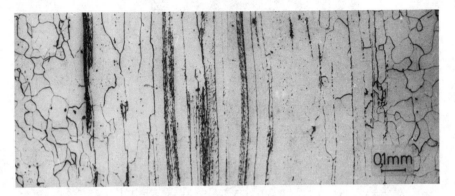

Figure 9 -- Microstructure of 3.5% silicon steel after hot-rolling
to the final rolling temperature of 1175°K. Parallel to the rolling
direction.

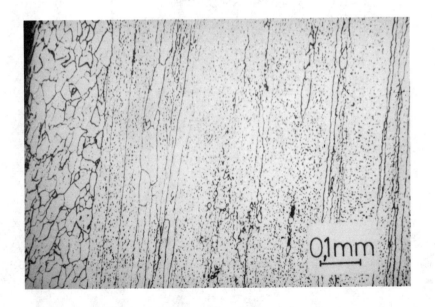

Figure 10 -- Microstructure of 3.5% silicon steel after hot-rolling
to the final rolling temperature of 1175°K. Perpendicular to the
rolling plane.

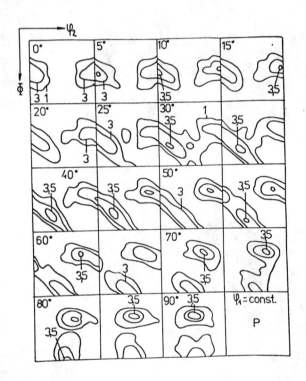

Fig.11 Orientation distribution function of ferrite from the surface layer of the sheet shown in Figure 9.

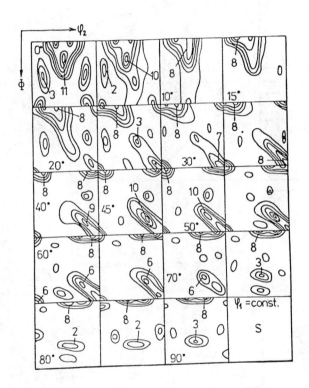

Fig.12 Orientation distribution function of ferrite from the mid-thickness of
the sheets shown on Figure 9.

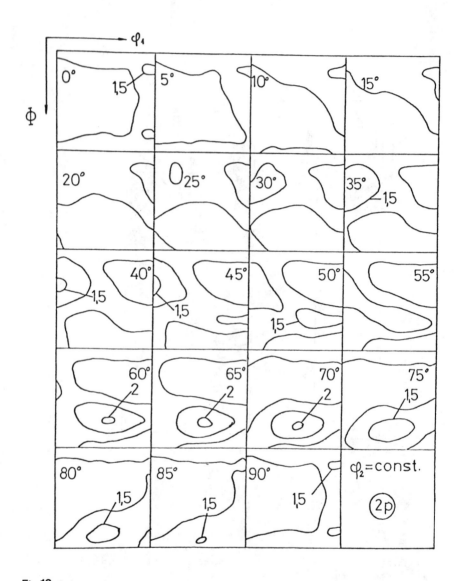

Fig.13 Orientation distribution function for a hypothetical austenite, calculated from Figure 11 via Kurdjumov-Sachs relationship for the surface layer.

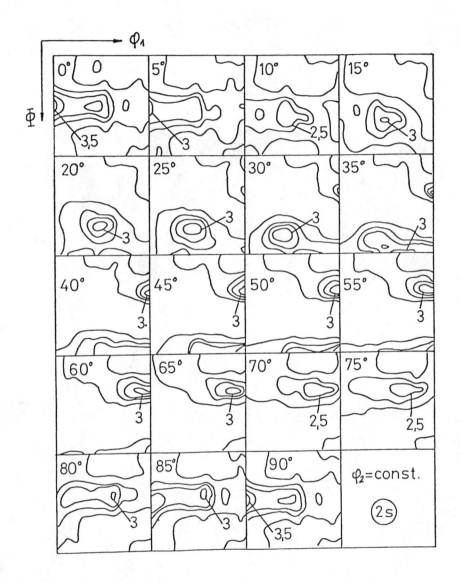

Fig.14 Orientation distribution function of a hypothetical austenite calculated from Figure 12 via Kurdyumov-Sachs orientation relationship.

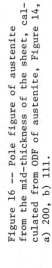

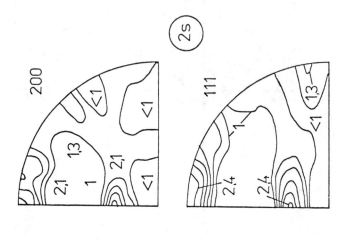

Figure 16 -- Pole figure of austenite from the mid-thickness of the sheet, calculated from ODF of austenite, Figure 14, a) 200, b) 111.

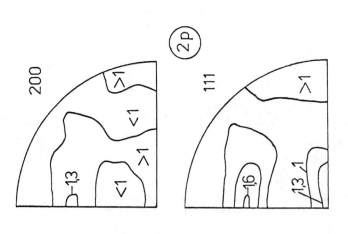

Figure 15 -- Pole figure of austenite from the surface layer, calculated from the ODF of austenite, Figure 13, a) 200, b) 111.

It follows that if during phase transformation the oriented nucleation determines the transformation of the texture, than this observation agrees with the theory regarding the relation between oriented nucleus and transformation of texture during the recrystallization of deformed metals.

REFERENCES

1. S. Gorczyca, Hutnik, vol. XXV, 1958, pp. 269-278.

2. S. Gorczyca, Mem. Scient. Rev. Metall., vol. LVII, 1960, pp. 154-158.

3. J. J. Jones, C. M. Sellars, W. J. McG. Tegart, Metall. Reviews, nr 130, 1969.

4. E. L. Brown, A. J. DeArdo, Metall. Transactions A, vol. 12A, 1981, pp. 39-47.

5. K. W. Andrews, J. Iron Steel Institute, vol. 191, 1956, p. 274.

6. J. Karp, Thesis, AGH 1960, not published.

7. E. B. Kula, S. L. Lopata, Trans. Metall. Soc. AIME, vol. 215, 1959, p. 980.

8. F. Borik, R. H. Richman, Trans. Metall. Soc. AIME, vol. 239, 1967, p. 675.

9. A. Maciosowski, Thesis, AGH, Krakow, 1973, not published.

10. J. Pospiech, J. Jura, A. Maciosowski, Prac. 3^e colloque european suv les textures, p. 117, Pont a Mousson, 1973.

11. J. Bunge, Mathematische Methoden der Texture Analyse, Akademie Verlag, Berlin, 1969.

12. A. Maciosowski, S. Gorczyca, Prace BIII Konf. Metalonznawczej, Gliwice-Wisla, wrzesien, 1974, p. 77.

13. G. J. Davies, J. S. Kallend, P. P. Morris, Acta Met., vol. 24, 1976, p. 159.

14. J. S. Kallend, P. P. Morris, G. J. Davies, Acta Met., vol. 24, 1976, p. 361.

15. A. Maciosowski, S. Gorczyca, J. Pospiech, J. Jura, Archiwum Hutnictwa, vol. 23, 1978, p. 217.

16. S. R. Goodman, Hsun-Hu, Metall. Trans., vol. 1, 1970, p. 1629.

17. J. Karp, J. Lesiecki, O. Nielubowicz, A. Panta, J. Jachowski, Z. Oswiecimski, A. Wegrzyn, Prace Rady Naukowo-Technicznej Huty im. Lenina, Krakow, 1973, p. 146.

18. W. Ratuszek, J. Jura, S. Gorczyca, The Analyses of Texture in Fe-3,5 Steel, not published.

MICROSTRUCTURAL CHANGES DURING
THERMOMECHANICAL PROCESSING OF X80 PIPELINE STEELS

A.P. Coldren, V. Biss, T.G. Oakwood
Climax Molybdenum Company of Michigan
Ann Arbor, Michigan

To produce thick as-rolled plate for large-diameter X80 linepipe, it is necessary to rely upon thermomechanical processing to maximize contributions from grain refinement, substructure, and precipitation strengthening. Weldability and cost limit the extent to which alloy additions can be employed for strengthening purposes; therefore, the thermomechanical processing is a crucial factor in achieving high strength. Results are presented from a study of microalloyed pearlite-reduced steels which were severely controlled-rolled. As a result of substructure development and strain-enhanced precipitation, the strength rises rapidly as hot rolling temperatures are reduced into the austenite plus ferrite region. Finish hot rolling conditions both above and below the Ar3 temperature were examined to provide a comparison of the effects of deformation of austenite and austenite/ferrite mixtures on the final structure and properties.

Introduction

The development of new high-strength low-alloy (HSLA) steels is a continuing effort to find improved combinations of composition and thermomechanical processing that will yield properties superior to those of
steels currently used. Usually, the main objective is to upgrade the
strength so that higher design stresses can be employed. Increased design
stresses permit the use of more economical combinations of wall thickness
and operating pressure in oil and gas transmission pipelines.

In this study, the strength properties of relatively heavy gauge
microalloyed steel plates were evaluated as a function of controlled
rolling practices. The steels studied are being considered as candidates
for X-80 line pipe applications. Specifically, the tensile strength of controlled rolled 19 mm (0.75 in.) laboratory test plates was analyzed in terms
of the contributions from various strengthening mechanisms.

When controlled rolling is employed with a finishing temperature well
below the Ar3 temperature, the contributions to strength from solution
hardening, grain refinement and precipitation hardening are supplemented
by additional strengthening effects associated with ferrite deformation
in the two-phase regime. These additional contributions come from the
dislocation substructure (including dislocation tangles and cell walls),
preferred crystallographic orientation or texture effects, and strain-
enhanced precipitation of microalloy carbonitrides in the ferrite (1-4).
There is considerable evidence to show that a high dislocation density
or a well defined substructure can be developed in hot rolled steels, and
that strength as a result can be significantly increased (1,2). Texture
effects contribute to the anisotropic behavior of strength in steels hot
rolled at low temperatures (2,3). This anisotrophy is of interest in
line pipe steel development because strength is often measured transverse
to the rolling direction, i.e., the direction that bears the hoop stress.
In this direction, the strengthening effects due to preferred orientation
would be most easily seen. Strain enhanced precipitation in ferrite is
of significant interest since precipitation is a potent strengthening mechanism. In the particular case of line pipe steels, high manganese levels,
as well as molybdenum additions, are often employed. These elements lower
austenite-to-ferrite transformation temperature with the result that some
of the precipitation expected during and after transformation may be suppressed. If deformation of ferrite could accelerate this precipitation,
the strengths of these steels could be significantly increased. There is
evidence to suggest that hot rolling below the Ar3 temperature accomplishes
this result; however, the quantitative strengthening effects of such precipitation requires further definition (4). The focus of this work, therefore, was to quantitatively assess the role of these supplemental strengthening mechanisms.

Experimental Approach

Materials

Nine laboratory steels were induction melted in an argon atmosphere
and cast as 30 kg (65 lb) ingots. The compositions of the steels are given
in Table I. The ingots were press forged at 1230 C (2250 F) to 96 mm
(3.75 in.) thick slabs. The slabs were hot rolled to 19 mm (0.75 in.)
plates using eight-pass rolling schedules. The rolling schedules, depicted
schematically in Figure 1, were identical for all plates through the roughing
sequence and holding period. After reheating to 1220 C (2230 F), the slabs
were rough rolled from 96 mm (3.75 in.) to 42 mm (1.65 in.) in three passes

Table I. Compositions of Steels

Steel No.	Climax Heat No.	Element, Wt. %												Carbon Equiv.[a]
		C	Mn	Si	Mo	Nb	V	Ti	Al	N	S	P	Other	
1	P2588	0.079	1.47	0.10	0.30	_[b]	–	0.014	0.052	0.0045	0.015	0.011	0.30Ni	0.38
2	P2457	0.079	1.40	0.10	0.30	0.054	–	–	0.022	0.0046	0.010	0.009	–	0.37
3	P2458	0.080	1.50	0.10	0.30	0.052	0.076	–	0.024	0.0047	0.007	0.008	–	0.41
4	P2459	0.079	1.51	0.10	0.30	0.056	0.078	0.012	0.016	0.0056	0.009	0.008	–	0.41
5	P2589A	0.082	1.61	0.12	0.30	0.056	0.078	0.013	0.036	0.0064	0.009	0.011	–	0.43
6	P2589B	0.081	1.71	0.12	(0.30)[c]	0.055	0.082	0.011	0.025	(0.0064)	0.009	0.012	–	0.44
7	P2592A	0.081	1.74	0.36	0.30	0.057	0.075	0.016	0.042	0.0065	0.009	0.012	–	0.45
8	P2592B	0.081	1.72	0.69	(0.30)	0.058	0.075	0.018	0.036	(0.0065)	0.009	0.013	–	0.44
9	P2467	0.080	1.25	0.13	0.31	0.059	–	0.014	0.009	0.0057	0.008	0.008	0.44Cr	0.44

[a]Carbon Equivalent $= C + \frac{Mn}{6} + \frac{Cr + Mo + V}{5} + \frac{Ni + Cu}{15}$

[b]Dash indicates none added and not analyzed

[c]Parentheses indicate an assumed value based on analysis of similar ingot from same heat.

593

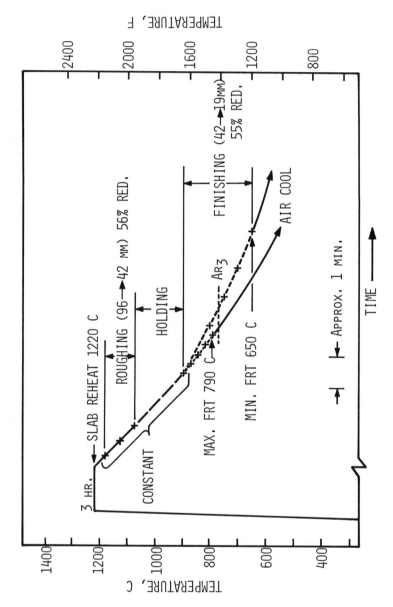

Fig. 1 Schematic Presentation of Rolling Schedules.
FRT = finish rolling temperature.

594

above a temperature of 1065 C (1950 F). The plates were then cooled to 900 C (1650 F) for the finish rolling sequence. The finishing sequence consisted of rolling the plates from 42 mm (1.65 in.) to 19 mm (0.75 in.) in five passes. The temperature of each pass was adjusted so that reductions were taken at uniform temperature intervals between 900 C (1650 F) and the final finishing temperature. Finishing temperatures varied from 650 C (1200 F) to 790 C (1450 F). Total reductions below the Ar_3 temperature ranged from 0-45%.

Tensile Testing

Duplicate round tensile specimens with threaded grips were machined from each of the plates. The reduced sections of the specimens had a diameter of 6.4 mm (0.25 in.) and a gage length of 25 mm (1.0 in.). The axis of tension was parallel to the direction of rolling. Tensile testing was carried out on a hydraulic tensile machine at a strain rate of 1.0×10^{-4} sec^{-1} to 1.5% strain and 6.7×10^{-4} sec^{-1} thereafter to fracture.

Metallographic Evaluation

Optical and Scanning Electron Microscopy. Metallographic specimens were cut from each hot rolled plate. The specimens were cut such that the surfaces to be examined were parallel to the rolling direction and perpendicular to the rolled surfaces. The specimens were mechanically polished and etched in nital for optical metallographic examination. Ferrite grain sizes were measured by a linear intercept method, and volume fractions of pearlite and M-A were measured by point counting. For scanning electron microscopy, selected specimens were electropolished in a perchloric acid-ethanol electrolyte.* The specimens were etched in a reagent containing 4% picric acid plus 1% nitric acid in ethanol and examined in a scanning electron microscope.

Transmission Electron Microscopy. Carbon extraction replicas were obtained from metallographic specimens given the surface preparation described above for scanning electron microscopy. The replicas were separated from the specimen surfaces by etching in a 2% bromine-alcohol solution cooled to -20C (-4 F). Some of the extraction replicas were examined in a scanning-transmission electron microscope (STEM) with energy dispersive X-ray (EDX) analysis capability in order to obtain a qualitative chemical analysis of the precipitates formed during rolling.

Thin foil specimens were prepared by electrolytic thinning, employing a double-jet, hollow cathode technique, for examination in a transmission electron microscope. The electrolyte consisted of a mixture of chromic and acetic acids (25 g CrO_3, 20 ml H_2O and 120 ml glacial acetic acid).

Results

Tensile Test Results

Table II summarizes the mechanical property data of all the steels examined in the investigation. Figure 2 shows the 0.2% offset yield strengths of the various alloys as a function of finish rolling temperature (FRT), and Figure 3 shows the ultimate tensile strengths as a function

---

*120 ml dist. water, 700 ml ethanol, 100 ml butyl cellosolve, 78 ml perchloric acid (70%).

Table II. Summary of Properties[a] of As-Rolled 19 mm (0.75 in.) Test Plates

Steel No.	Climax Heat No.	Nominal Composition, Wt. %[b]	Finish-Rolling Temperature, C (F)[c]	Yield Strength, MPa (ksi) 0.2%[d]	Yield Strength, MPa (ksi) 1.5%	Tensile Strength, MPa (ksi)	Y/T Ratio	El. in 25 mm (1 in.), %	R.A., %
1	P2588	0.08C-1.50Mn -0.10Si-0.30Mo -0.30Ni-0.02Ti	730 (1350) 715 (1315) 675 (1250) 650 (1200)	422 (61.3) 435 (63.1) 475 (69.0) 507 (73.6)	533 (77.3) 538 (78.1) 570 (82.8) 610 (88.6)	624 (90.6) 629 (91.3) 656 (95.2) 688 (99.9)	0.68 0.69 0.72 0.74	26 24 24 23	70 68 67 66
2	P2457	0.08C-1.40Mn -0.10Si-0.30Mo -0.06Nb	790 (1450) 745 (1375) 705 (1300)	442 (64.2)* 490 (71.1)* 530 (76.9)*	440 (63.8) 497 (72.1) 553 (80.2)	555 (80.5) 593 (86.1) 630 (91.5)	0.80 0.83 0.84	32 28 26	75 72 70
3	P2458	0.08C-1.50Mn -0.10Si-0.30Mo -0.06Nb-0.08V	790 (1450) 745 (1375) 705 (1300)	465 (67.5)* 508 (73.7)* 590 (85.7)	490 (71.1) 537 (78.0) 625 (90.7)	595 (86.4) 636 (92.3) 699 (101.4)	0.78 0.80 0.85	30 27 26	70 74 70
4	P2459	0.08C-1.50Mn -0.10Si-0.30Mo -0.06Nb-0.08V -0.02Ti	790 (1450) 745 (1375) 705 (1300)	468 (67.9)* 520 (75.5)* 567 (82.3)*	464 (67.3) 539 (78.3) 589 (85.5)	583 (84.6) 631 (91.6) 662 (96.1)	0.80 0.82 0.86	32 28 25	74 71 69
5	P2589A	0.08C-1.60Mn -0.10Si-0.30Mo -0.06Nb-0.08V -0.02Ti	705 (1300) 675 (1250)	577 (83.7) 602 (87.4)	626 (90.9) 641 (93.1)	692 (100.4) 706 (102.5)	0.83 0.85	26 25	70 71
6	P2589B	0.08C-1.70Mn -0.10Si-0.30Mo -0.06Nb-0.08V -0.02Ti	705 (1300) 675 (1250)	569 (82.6) 639 (92.7)	625 (90.7) 684 (99.3)	701 (101.8) 759 (110.2)	0.81 0.84	25 26	69 70
7	P2592A	0.08C-1.70Mn -0.40Si-0.30Mo -0.06Nb-0.08V -0.02Ti	705 (1300) 675 (1250)	611 (88.7) 624 (90.6)	679 (98.6) 683 (99.2)	789 (114.5) 789 (114.5)	0.77 0.79	26 24	61 63
8	P2592B	0.08C-1.70Mn -0.70Si-0.30Mo -0.06Nb-0.08V -0.02Ti	705 (1300) 675 (1250)	602 (87.4) 646 (93.2)	686 (99.6) 718 (104.2)	807 (117.1) 827 (120.1)	0.75 0.78	25 24	59 60
9	P2467	0.08C-1.30Mn -0.10Si-0.30Mo -0.40Cr-0.06Nb -0.05Ti	790 (1450) 745 (1375) 705 (1300)	402 (58.3) 499 (72.4) 522 (75.7)	449 (65.2) 555 (80.6) 574 (83.3)	575 (83.5) 641 (93.0) 650 (94.3)	0.70 0.78 0.80	30 27 26	73 73 70

[a] Averaged values from duplicate tests on longitudinal specimens.
[b] See Table I for actual analyses.
[c] See Figure 1 for rolling schedules used.
[d] Asterisk indicates that one or both of the duplicate tests showed an upper yield point, the value of which is not tabulated here.

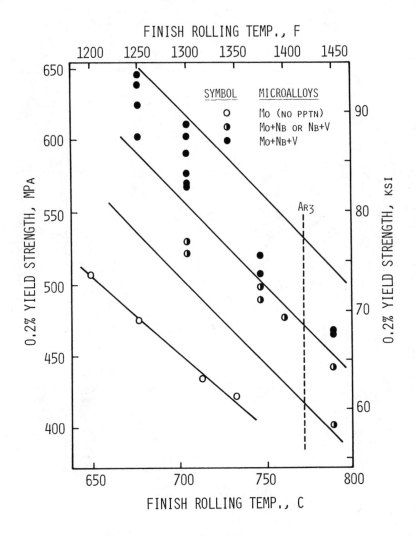

Fig. 2 Effect of Finish Rolling Temperature on the 0.2% Offset
Yield Strength of As-Rolled Plates

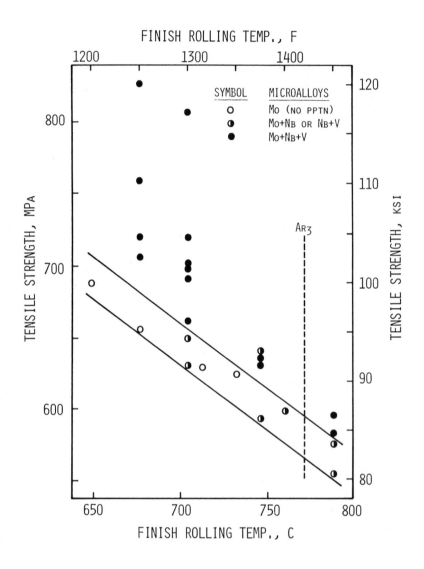

Fig. 3   Effect of Finish Rolling Temperature on the
Tensile Strength of As-Rolled Plates

of the same variable. Included in Figures 2 and 3 are the strengths of Nb-V steels obtained in earlier studies (5). Both yield strength and tesnile strength increase rapidly as the finish rolling temperature decreases.

The yield strengths of the various steels fell within three rather distinct regimes. The steel containing neither niobium nor vanadium microalloy additions (microalloy-free) exhibited the lowest strength at all finish rolling temperatures. Intermediate strength levels were obtained from the Mo-Nb and Nb-V steels. The highest strengths were observed when molybdenum along with niobium and vanadium additions were employed.

A comparison of ultimate tensile strength behavior as a function of finish rolling temperature, illustrated in Figure 3, shows that the microalloy-free steel develops tensile strengths equal to those obtained for the Mo-Nb and Nb-V steels. The highest values were again obtained where triple additions of Mo-Nb-V were used. In two instances where higher silicon levels were present, very high tensile strengths [(>800 MPa (116 ksi)] were observed.

Metallographic Results

The very low hot rolling temperatures employed resulted in microstructures characterized by mixtures of polygonal ferrite, martensite/austenite (M-A), and ferrite/cementite aggregates. The morphology of the ferrite/cementite aggregates varied from that of pearlite to what appeared in some instances to be upper bainite. Typical scanning electron micrographs are shown in Figures 4a and b. For simplicity the ferrite/carbide aggregate is referred to as pearlite throughout this paper.

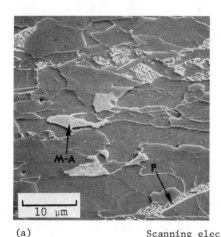 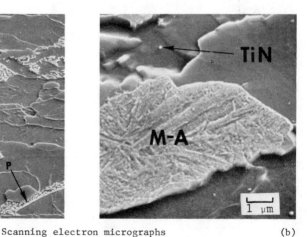

(a)   Scanning electron micrographs   (b)

Fig. 4 Microstructure of Mo-Nb-V Steel No. 5 Finish Rolled
705 C (1300 F)

Table III gives the results obtained from quantitative metallographic analyses. Despite the variability in finish hot rolling temperature, the ferrite grain sizes obtained were surprisingly uniform. The average grain size number for all of the steels and all rolling conditions was ASTM No. 12.0. Variations were no more than ±0.5 from this average. Also shown in Table III are the volume percentages of M-A and pearlite. Steel Nos. 1, 7 and 8 contained relatively high amounts of M-A (6.5-11.1%) while the rest of the steels had M-A contents of 2.3% or lower. The volume percentages of pearlite did not vary significantly except for Steel Nos. 7 and 8: in these two cases, lower amounts of pearlite were observed. There was no consistent variation of M-A content or pearlite content as a function of finish rolling temperature.

Table III. Summary of Metallographic Data

Steel No.	Finish-Rolling Temperature, C (F)	Ferrite Grain Size, ASTM No.	Vol. % M-A	Vol. % Pearlite
1	730 (1350)	11.7	6.9	13.9
	715 (1315)	11.7	6.5	14.3
	675 (1250)	11.7	8.0	14.2
	650 (1200)	11.7	10.9	10.9
2	790 (1450)	12.1	1.1	10.9
	745 (1375)	11.9	0.4	12.7
	705 (1300)	11.8	0.3	11.1
3	790 (1450)	11.6	0.1	17.7
	745 (1375)	12.0	0.0	16.0
	705 (1300)	12.0	0.4	15.0
4	790 (1450)	12.1	0.5	18.6
	745 (1375)	12.2	0.6	15.1
	705 (1300)	12.2	0.8	13.0
5	705 (1300)	11.7	0.9	18.3
	675 (1250)	11.8	0.9	14.3
6	705 (1300)	12.4	2.3	14.5
	675 (1250)	12.5	0.5	15.3
7	705 (1300)	12.1	10.4	10.6
	675 (1250)	12.0	8.4	8.1
8	705 (1300)	11.6	11.1	4.0
	675 (1250)	12.0	9.8	4.0
9	790 (1450)	11.9	1.4	15.0
	745 (1375)	12.3	0.9	18.8
	705 (1300)	12.0	1.1	14.0

Of interest is the observation that Steel No. 1 developed a significant volume fraction of M-A and a relatively high volume fraction of pearlite. Steel Nos. 7 and 8, however, developed a high percentage of M-A but a relatively small amount of pearlite. This indicates that the relatively high tensile strength of these three steels is due in large part to their high volume fractions of M-A.

Figure 5 is a representative extraction replica micrograph of Steel No. 3 which was finish hot rolled at 705 C (1300 F). As can be seen, a profusion of microalloying element carbonitride precipitation in austenite has occurred. This behavior was observed in all of the microalloyed steels studied in this investigation.

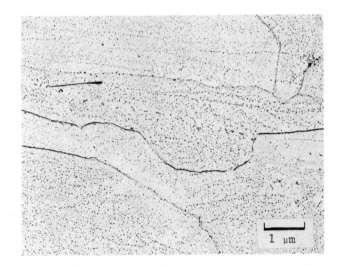

Fig. 5 Microalloy Carbonitride Precipitates in Mo-Nb-V Steel
(No. 3) Finish Rolled at 705 C (1300 F)
Extraction Replica

The dislocation structure underwent distinct changes as finishing temperature was reduced. Figure 6a is a transmission electron micrograph of a thin foil specimen taken from Steel No. 3 finish rolled at 790 C (1450 F). The ferrite is nearly dislocation free. Figure 6b shows the same steel finish rolled at 705 C (1300 F). In this case the low temperature deformation of ferrite resulted in a significantly higher dislocation density and the formation of subgrain boundaries.

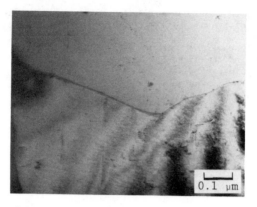

(a)   Finish Rolled at 790 C (1450 F)

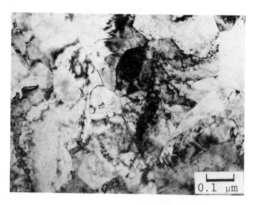

(b)   Finish Rolled at 705 C (1300 F)

Fig. 6   Thin Foil Transmission Micrographs
of Mn-Mo-Nb-V Steel No. 3

Precipitate Analysis

The precipitates observed both on the extraction replicas and in the thin foils were found to fall in roughly three size classifications as follows:

Class	Size Range	Occurrence
I	30 - 1300 nm	Austenite grain boundary and sub-grain boundary precipitation during early, high temperature part of rolling schedule.
II	4 - 15 nm	General random precipitation in austenite during latter part of rolling schedule above Ar_3.
III	1.5 - 3 nm	General precipitation in ferrite (perhaps some interphase) during latter part of rolling schedule below Ar_3 or during cooling after rolling.

Examples of the various precipitate sizes for Steel No. 5 are shown in Figure 7. The very large particles (Class I) were found at or near the prior austenite grain boundaries. These particles were either dense spheroidal particles (IA) or planar particles, relatively transparent to the electron beam (IB). The Class II particles were found along what is believed to have been prior austenite subgrain boundaries as well as being uniformly distributed throughout the austenite grains. The very fine particles (Class III) are believed to have formed either during or after the transformation to ferrite and that these are the precipitates responsible for the precipitation strengthening contribution.

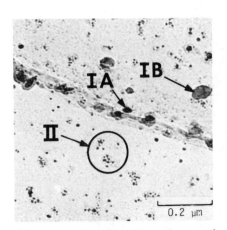
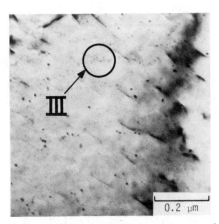

(a) Extraction of replica micrograph   (b) Thin foil transmission micrograph

Fig. 7   Classification of Various Precipitate Sizes in Mo-Nb-V Steel No. 5 Finish Rolled at 705 C (1300 F)

The compositions of some of the largest particles were qualitatively analyzed in the STEM by energy dispersive X-ray analysis technique. These results are shown in Figure 8. The precipitates are complex particles containing varying amounts of niobium, vanadium and molybdenum. (The iron peaks are believed to be spurious peaks resulting from nearby cementite particles in pearlite colonies). This complexity of composition is in contrast to the generally made assumption that most precipitates in microalloyed

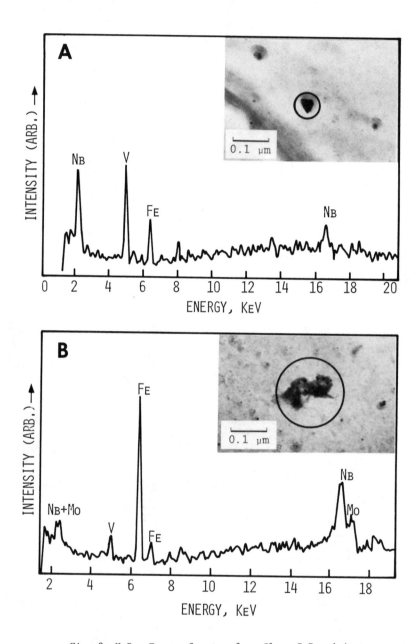

Fig. 8   X-Ray Energy Spectra from Class I Precipitates

steels are carbides, nitrides, or carbonitrides of a single microalloying element. These results are in qualitative agreement, however, with other similar experiments on microalloyed steels (6).

Discussion

It is evident from a comparison of the strength levels, shown in Figures 2 and 3, with the metallographic evidence obtained, that several strengthening mechanisms are operative in the steels investigated in this study. Grain refinement, precipitation hardening, substructure and texture strengthening, and second phase hardening are all contributing to overall strength. Furthermore, the relative contribution of any individual mechanism apparently varies as steel composition or rolling practice is changed.

As was mentioned earlier, the principal aim of this work was to identify the nature of the additional strengthening which could be achieved by hot rolling below the Ar_3 temperature. The fact that strength increased rapidly when finish rolling was extended below the Ar_3 temperature, suggests that strain-enhanced precipitation in the ferrite, along with dislocation substructure and texture, is a significant contributor to strength.

To ascertain at least semiquantitatively the relative contributions of these strengthening mechanisms, an approach utilizing ultimate tensile strength was devised. Tensile strength was selected because, as illustrated in Figure 9, some of the steels exhibited discontinuous yielding whereas others yielded continuously. The ultimate tensile strength of any one of the steels was assumed to consist of the sum of several additive strengthening mechanisms, i.e.:

$$\sigma_{UTS} = \sigma_0 + \sigma_{SOLN} + \sigma_{GS} + \sigma_{M-A} + \sigma_{PEARL} + \sigma_{PPTN} + \sigma_{DISLOC} + \sigma_{TEXT} \quad (1)$$

$$
\begin{aligned}
\text{WHERE, } \sigma_0 \quad &= \text{UTS of Iron Single Crystal} \\
\sigma_{SOLN} \quad &= \text{Solid Solution Effect} \\
\sigma_{GS} \quad &= \text{Ferrite Grain Size Effect} \\
\sigma_{M-A} \quad &= \text{Martensite-Austenite Effect} \\
\sigma_{PEARL} \quad &= \text{Pearlite Effect} \\
\sigma_{PPTN} \quad &= \text{Precipitation Hardening Effect} \\
\sigma_{DISLOC} \quad &= \text{Dislocation Substructure (including subgrain boundary) Effects} \\
\sigma_{TEXT} \quad &= \text{Texture Effect}
\end{aligned}
$$

For simplicity, interactions between mechanisms were ignored. The tensile strengths of all of the steels were normalized to values representing a composition containing 1.6% Mn, 0.1% Si, and 0.3% Mo. Also the strengths were corrected to reflect a microstructure with 15% pearlite, 0% M-A, and an ASTM grain size of 12. The various correction factors applied were as follows:

	Factor	Reference
	16 ksi/1% Si	7
	7 ksi/1% Mn	7
σ_{SOLN}	6.5 ksi/1% Ni	7
	5 ksi/1% Mo	7
	1 ksi/1% Cr	7
σ_{GS}	0.35 ksi/in.$^{-1/2}$ (average)	8,9,10
σ_{PEARL}	0.56 ksi/1% PEARLITE	11
σ_{M-A}	1.5 ksi/1% M-A	10

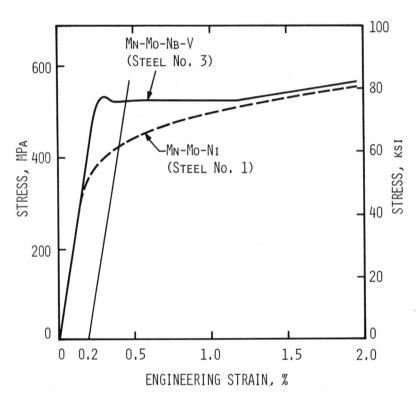

Fig. 9 Comparison of Stress-Strain Curves of Microalloyed
and Microalloy-Free Steels

By applying these assumptions, differences in strength would be due to differences in the relative contributions from precipitation hardening, texture, and dislocation substructure. The normalized (corrected) tensile strengths of the steels were then plotted against finish rolling temperature as shown in Figure 10. A least squares linear regression of corrected tensile strength as a function of finish rolling temperature for the microalloy-free steel was extrapolated to the Ar_3 temperature. This procedure established a base tensile strength representative of the composition and microstructure described above.* Such an extrapolation becomes possible since the only contribution to the tensile strengths of the microalloy-free steel over the base value would be from dislocation substructure and texture. The extrapolation assumes, however, that the contribution to strength from these two mechanisms is negligible when rolling occurs above the Ar_3 temperature. The transmission micrographs shown in Figures 6a and 6b indicate this assumption is reasonable. Also, the base strength determined in Figure 10 is compatible with a value calculated by Irani et al (8).

Figure 10 thus provides a convenient means of determining the contributions to tensile strength from substructure and texture strengthening as well as precipitation hardening. Since the microalloy-free steel was not precipitation hardened, the effects of precipitation hardening alone can be established by substracting the values for this steel from those for the microalloyed steels. No convenient means was found to separate the effects of texture and dislocation substructure. In addition, longitudinal testing was employed which would not reflect the maximum effect of crystallographic orientation. Therefore, the two terms were combined as a single contribution to strength in this analysis.

As Figure 10 shows, the contributions from both precipitation hardening and the combined effects of substructure and texture increase as finish rolling temperature decreases. Furthermore, the former is increasing at a more rapid rate than the latter. Thus, as finish rolling temperature is progressively decreased below the Ar_3, precipitation hardening becomes an increasingly larger contributor to strength. Up to 45 MPa (6.5 ksi) can be obtained due to substructure effects in the deformed ferrite. Furthermore, the 40 MPa (6 ksi) precipitation strengthening increment, obtained by finish rolling above the Ar_3, is increased to 120 MPa (17 ksi) by finish rolling below the Ar_3. This 80 MPa (11 ksi) improvement in precipitation strengthening is ostensibly due to strain-enhanced precipitation in ferrite, in spite of the fact that excessive precipitation in austenite was observed (Figure 5).

The Mo-Nb-V steels which were observed to have the highest as-rolled tensile strengths, Figure 3, also have somewhat higher corrected tensile strengths than the Mo-Nb or Nb-V steels. It is concluded that molybdenum increases the degree of precipitation hardening which can be obtained from the microalloying elements. Other investigators have reported that molybdenum reduces precipitation in austenite (12). This is probably related to an increase in solubility resulting from a decrease in carbon activity effected by molybdenum (13). With less precipitation in austenite, more numerous precipitates could form in ferrite resulting in enhanced strength. The observed presence of molybdenum in the precipitates may also increase their strengthening effectiveness by increasing coherency strains or by increasing volume fraction of precipitate or both.

*The Ar_3 temperatures of the various steels, as measured by a thermal arrest during or after rolling, varied somewhat with composition. The varability was small, however, and a single temperature of 775 C (1427 F) was used for this analysis.

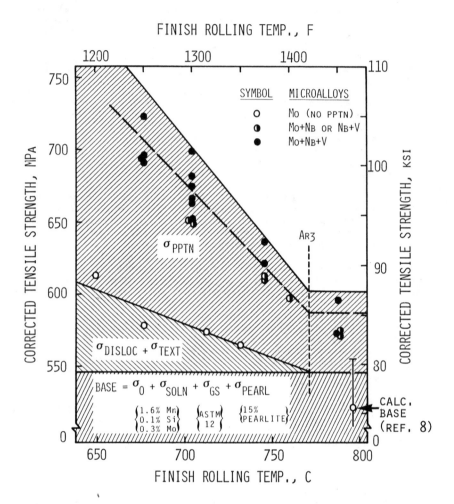

Fig. 10 Corrected Tensile Strength vs. Finish Rolling Temperature for 19 mm (0.75 in.) Plates Controlled Rolled with a 1040 C/900 C (1900 F/1650 F) Hold. Contributions from precipitation strengthening (σpptn) and from dislocation substructure and texture strengthening (σdisloc + σtext) are indicated by shaded areas.

Conclusions

1. When air cooled 19 mm (0.75 in.) plates of Mo-Nb-V ferrite-pearlite steels are severely controlled rolled at temperatures up to 100 C (180 F) below the Ar_3, the following contributions to tensile strength can be obtained:

   *   Up to 45 MPa (6.5 ksi) from the dislocation substructure in the deformed ferrite.

   *   Up to 120 MPa (17 ksi) from precipitation of microalloy carbonitrides, or 80 MPa (11 ksi) more than the 40 MPa (6 ksi) obtained by finish rolling just above the Ar_3.

2. Higher strength levels can be obtained by the triple addition of Mo + Nb + V than by additions of Mo + Nb or Nb + V. The Mo-Nb-V steels exhibit larger contributions from precipitation strengthening than do the double-addition steels.

3. Qualitative analysis of the largest carbonitride particles indicates that some particles contain Mo, Nb and V, while others contain Nb and V.

Acknowledgements

The authors would like to acknowledge the support and help of the Climax Molybdenum Company Research and Development staff in the preparation of this manuscript. Special thanks are due to Dr. J.M. Tartaglia for his help in preparation of some of the electron micrographs.

References

1. P.L. Mangonon and W.E. Heitmann, "Subgrain and Precipitation Hardening Effects in Columbium-Bearing Steels," Microalloying 75, Union Carbide Corporation, 1977, 59-70.

2. B.L. Bramfitt and A.R. Marder, "The Influence of Microstructure and Crystallographic Texture on the Strength and Notch Toughness of a Low Carbon Steel," Processing and Properties of Low Carbon Steel, AIME, 1973, 191-220.

3. B. Mintz, "Low Carbon Steels for Pipeline Applications," Low Carbon Structural Steels for the Eighties, Chameleon Press Ltd., 1977, 47-53.

4. A.P. Coldren, G.T. Eldis, G. Tither, "Structure Property Relationships for Pearlite-Reduced Mo-Nb Steels Finish-Rolled Moderately Below Ar_3," MICON 78, ASTM STP 672, 1979, 126-144.

5. Climax Molybdenum Company, Unpublished Research.

6. D.C. Houghton, J.D. Embury, and G.C. Weatherly, "Characterization of Carbonitrides in Ti Bearing HSLA Steels," Presented before the International Conference on Thermomechanical Processing of Microalloyed Austenite, Pittsburgh, PA, 1981.

7. C.E. Lacy and N. Gensamer, "The Tensile Properties of Alloyed Ferrites," Trans. ASM, Vol. 32 (1944) 88-105.

8. J.J. Irani, D. Burton, J.D. Jones and A.B. Rothwell, "Beneficial Effects of Controlled Rolling in the Processing of Structural Steels," Proceedings of Conference on Strong Tough Structural Steels, Scarborough (April 1967) 110-122.

9. T. Gladman, B. Holmes and F.B. Pickering, "Work Hardening of Low-Carbon Steels," J.I.S.I., Vol. 208 (February 1970) 172-183.

10. J.H. Bucher and E.G. Hamburg, "High Strength Formable Sheet Steel," SAE Paper No. 770164, presented in Detroit (February 1977).

11. K.J. Irvine and F.B. Pickering, "Low-Carbon Steels with Ferrite-Pearlite Structures," J.I.S.I. Vol. 201 (November 1963) 944-959.

12. B. Bacroix, M.G. Akben and J.J. Jonas, "Effect of Mo on Dynamic Precipitation and Recrystallization in Nb- and V-Bearing Steels," presented before the International Conference on Thermomechanical Processing of Microalloying Austenite, Pittsburgh, PA 1981.

13. T. Wada, H. Wada, J.F. Elliott, and J. Chipman, "Activity of Carbon and Solubility of Carbides in the FCC Fe-Mo-C, Fe-Cr-C, and Fe-V-C Alloys," Met. Trans., V. 3 (November 1972) 2865-2872.

DISCUSSION

Q: You pointed out that all the steels contained titanium for austenite grain size control at the beginning of rolling. For steel No. 1, which did not contain any Nb, did you observe any difference in the grain refinement of the austenite during rolling, compared to the other Nb-bearing steels?

A: The final grain size was slightly coarser in steel No. 1.

Q: What was the evidence that the very small Nb carbonitrite particles were coherent or were there.

A: None, that's an assumption.

Q: I presume you've made the assumption that the sub-structure strengthening effect was independent of your alloying elements. I did note that in your silicon-effect series, the high silicon steel seemed to show more heavily retained sub-structure effects. That is one question I have. The second question is: Did you try to check or confirm your precipitation strengthening component by actual analysis of the precipitate size and distribution effects?

A: To answer your last question first. Maybe you misunderstood my comments regarding the high silicon steels. We did not separate the sub-structure effect. That effect can be separated only in the micro-alloying-free steel. So we don't know that high silicon steels have a different sub-structure contribution than the lower silicon steels. In addition, no measurements were made of the precipitates or their distributions.

THE EFFECT OF MICROALLOYING ELEMENTS ON
THE RECOVERY AND RECRYSTALLIZATION IN DEFORMED AUSTENITE

S. Yamamoto, C. Ouchi, and T. Osuka

Technical Research Center
Nippon Kokan K.K.
Kawasaki, JAPAN

The retarding effect of Nb, V, and Ti on recovery and recrystallization behavior in deformed austenite was investigated using the stress-strain curve in double compression test. Particular emphasis was placed on the differentiation of the role of solute atom from that of strain induced precipitate. To meet this purpose, 0.002C steels with 0.002N were prepared by decarburization method so that microalloying atoms remain dissolved in almost all testing conditions. The effect of precipitates was studied in other steels of higher C content with the aid of quantitative analysis of precipitates. Nb, V, and Ti in solution retard static recovery and delay the onset of recrystallization. X-ray diffraction analysis suggests that this retarding effect results from the interaction between dislocation and lattice distorsion by solute atom. Both onset and progress of recrystallization are substantially delayed as soon as strain induced precipitation begins.

Introduction

Controlled rolling of microalloyed steels has been successfully applied
to the mass production of HSLA steels for the last two decades. As many
papers have indicated, microalloying elements such as Nb, V, and Ti play very
important roles in the controlled rolling process and resultant properties
through the suppression of grain growth, the retardation of recovery and
recrystallization in deformed austenite, and precipitation strengthening.
Regarding their role on the retardation of recovery and recrystallization,
two mechanisms have been proposed by several investigators: the pinning
effect by strain induced precipitates, and the solute drag effect by solute
atoms. The pinning effect has been studied by a number of investigators
using the method of hot compression (1 -3), hot tension (4 - 6), hot
torsion (7), and hot rolling (8 - 12), and shown to be considerably dominant in the processing range where strain induced precipitation occurs. The
solute drag effect might exist and could be the principal mechanism at
higher temperatures where no precipitation occurs, but this does not seem to
have been confirmed yet in any systematic research works, because these
microalloying elements form carbide or nitride easily in the presence of
very small amount of C and N. However, if we could control the microstructure of austenite at higher temperature region through this solute drag
effect, it would be very attractive for the improvement of mechanical properties and hopefully mill productivity as well. Therefore it should be made
clear whether and how solute microalloying element can influence recovery and
recrystallization behavior.

In this paper, the retarding effect of Nb, V, and Ti on static recovery
and recrystallization in deformed austenite was investigated by the stress-
strain curve in double compression test. Particular emphasis was placed on
the differentiation of the role of solute atoms from that of strain induced
precipitates. To meet this purpose 0.006C steels with 0.002N content were
decarburized to less than 0.002C level so that microalloying atoms remain
dissolved in almost all testing conditions. Each effect of solute Nb, V, and
Ti atoms was separately examined with these ultra low carbon steels. The
interaction between the progress of recrystallization and precipitation was
also determined in the steels of higher C. The results were discussed on the
basis of lattice distorsion for the effect of solute atom, and with the aid
of quantitative analysis for the effect of precipitate.

Experimental Procedures

Chemical composition

Restoration behavior in deformed austenite was studied on a number of
C-Mn, C-Mn-Nb, C-Mn-V, and C-Mn-Ti steels. The chemical compositions are
given in Table 1. Ultra low carbon steels containing less than 0.002C were
prepared by decarburizing 0.006C - 0.008C steels and used to study the effect
of solute Nb, V, and Ti (steels 1-4, 6-7, and 12-14). Higher carbon steels
were also melt to study the effect of precipitated Nb(C,N), VC, and TiC
(steels 5, 10-11, and 15-16). In the case of V bearing steels, higher N
steels were also prepared to look at the effect of VN.

Table I Chemical compositions of steels

(wt%)

	C	Si	Mn	P	S	Nb	V	Ti	Sol Al	T.N.	
1	0.006 (0.002)	0.30	1.56	0.005	0.006	–	–	–	0.030	0.0020	base steel
2	0.006 (0.002)	0.29	1.54	0.005	0.006	0.050	–	–	0.033	0.0023	
3	0.006 (0.002)	0.25	1.50	0.005	0.006	0.097	–	–	0.033	0.0027	Sol. Nb
4	0.006 (0.002)	0.30	1.55	0.005	0.006	0.171	–	–	0.039	0.0027	
5	0.019	0.25	1.49	0.005	0.006	0.095	–	–	0.032	0.0028	Nb(C,N)
6	0.006 (0.002)	0.24	1.50	0.009	0.001	–	0.047	–	0.033	0.0018	Sol. V
7	0.004 (0.002)	0.25	1.51	0.009	0.001	–	0.092	–	0.039	0.0025	
8	0.006 (0.001)	0.24	1.48	0.008	0.001	–	0.087	–	0.034	0.0050	VN
9	0.007 (0.001)	0.24	1.48	0.009	0.001	–	0.088	–	0.034	0.0134	
10	0.014	0.25	1.50	0.009	0.001	–	0.091	–	0.034	0.0021	VC
11	0.034	0.24	1.50	0.009	0.001	–	0.091	–	0.035	0.0023	
12	0.006 (0.001)	0.27	1.51	0.011	0.004	–	–	0.054	0.039	0.0025	
13	0.006 (0.001)	0.27	1.51	0.011	0.003	–	–	0.095	0.034	0.0030	Sol. Ti
14	0.008 (0.002)	0.29	1.49	0.012	0.0035	–	–	0.230	0.037	0.0033	
15	0.029	0.27	1.51	0.012	0.003	–	–	0.094	0.036	0.0025	TiC
16	0.042	0.27	1.51	0.012	0.003	–	–	0.098	0.040	0.0028	

(   ) after decarburized

Decarburizing treatment

The plates of 12mm thickness were reheated at 1150°C in wet hydrogen atmosphere (Ar: H_2: H_2O = 87 : 10 : 3) for decarburizing treatment. The concentration C at a given position X in plate thickness at time t can be calculated by the following equation assuming a diffusion controlled process in an infinite plate (13).

$$\frac{C - Co}{Cs - Co} = 1 - 2Y\left(\frac{X}{\sqrt{2Dt}}\right) \qquad \left(Y = \frac{1}{\sqrt{2\pi}}\int_0^z \exp\left(-\frac{z^2}{2}\right)dz\right) \qquad (1)$$

Co is initial concentration, Cs is surface concentration, Z is a half of plate thickness, and D is diffusion coefficient of solute atom. For the calculation of this decarburizing treatment Co = 0.006(%), Cs = 0(%), Z = 0.6(cm), and D = 0.49 exp (-36600/RT) (13) were used. By integrating equation (1) through the whole thickness, average carbon concentration Cav was obtained as follows.

$$Cav = \int_0^Y CdX/Y \qquad (2)$$

The result calculated by equation (2) at 1150°C was shown in Fig. 1 in comparison with the experimental data. Good agreement was obtained between calculated values and experimental values. According to these results, the condition of decarburization was determined to be 32 hour treatment at 1150°C for lowering C content less than 0.002%.

Stoichiometric consideration

Most microalloying atoms are expected to remain dissolved in ultra low C steels in the present study, since the amount of stoichiometric precipitates is very low as shown below. Fig. 2(a) shows the stoichiometric amount of Nb as Nb(C,N) and the excessive amount of Nb in solution with total Nb content. In 0.002C steels, stoichiometric Nb(C,N) corresponds to only 0.02Nb, and [total Nb-0.02Nb] remains as solute atom at all temperatures. Moreover, if we take into consideration the solubility product of Nb(C,N) in austenite phase (14) it is presumed almost all Nb remain as solute atom in the temperature range studied in this paper. Fig. 2(b) shows the stoichiometric amount of Ti as TiC and TiN, and the excessive amount of Ti in solution with total Ti content. [Total Ti-0.02Ti] corresponds to be the amount of solute Ti. Similarly, [total V - 0.02V] turns out to be the amount of solute V. Solubility calculation suggests again that almost all Ti and V remain dissolved in the test temperature range in the decarburized steels except Ti precipitated as TiN which is very stable in the whole temperature range. Fig. 3 shows the change of the stoichiometric amount of Nb as Nb(C,N) or Ti as TiC with C content in higher carbon steels with 0.10Nb or 0.10Ti respectively. Nb(C,N) increases with C content increase, and all of 0.10Nb can precipitate stoichiometrically with C content of 0.01%. Based on the same argument, all of 0.10Ti can precipitate stoichiometrically with C content of 0.023%, and all of 0.10V can precipitate in the steels with 0.023C or 0.025N.

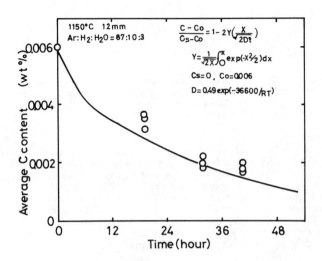

Fig.1 The change of average C content with
decarburizing time.

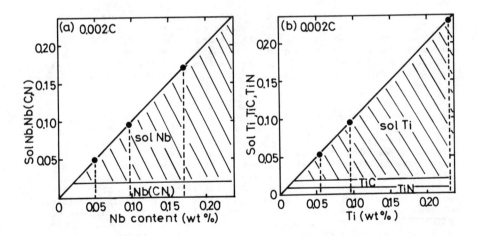

Fig.2(a)

The change of the amount of
solute Nb and the stoichiometric
amount of Nb as Nb(C,N) with the
increase of total Nb in 0.002C
steel.

Fig.2(b)

The change of the amount of
solute Ti and the stoichiometric
amount of Ti as TiC and TiN with
the increase of total Ti in
0.002C steel.

617

Hot deformation test

Static restoration behavior in deformed austenite was investigated using a hot deformation simulator. The simulator is capable of compression test with wide range of strain rate from 4×10^{-4}/sec. to 10/sec. and with three consecutive steps of compression. It is equipped with high frequency induction coil in vacuum and He gas quenching system. Load and displacement can be recorded in digital transient memory. Then true stress and true strain are calculated and plotted by means of a built-in micro computer. Specimens with 6mm in diameter and 10mm in length were taken from hot rolled or additionally decarburized plates. These specimens were reheated to the temperatures between 1120 and 1250°C to have a constant initial austenite grain size of 140μ, and cooled down to the deformation temperature ranging from 850 to 1000°C. Double compression test was carried out with strain rate of 10/sec. and with strain of 0.69 and 0.36 for the first and the second compression respectively. The interpass time was varied from 0.075 to 1,800 sec. The softening ratio X is obtained by the following equation (3).

$$X = \frac{\sigma_{max} - \sigma_y}{\sigma_{max} - \sigma_{yo}} \tag{3}$$

σ_{yo} is the yield stress and σ_{max} is the maximum stress in the first compression, and σ_y is the yield stress in the second compression. When no restoration takes place as in cold deformation. σ_y equals σ_{max} and therefore X=0. On the contrary, when recrystallization is complete, σ_y equals σ_{yo} and X=1. Thus softening ratio X changes continuously from 0 to 1 in accordance with the progress of recovery and recrystallization. Fig. 4 shows an example of true stress-true strain curves for the decarburized 0.002C-0.23Ti steel at 900°C. For the interpass time of 0.075 sec, the yield stress σ_y in the second deformation is nearly equal to the maximum stress σ_{max} in the first deformation, giving a softening ratio X of 2%. σ_y decreases with the increase of interpass time, and therefore X increases continuously.

Microscopic observation of the recrystallization in the decarburized steels was very difficult because of their very low hardenability. Therefore, a correlation between double compression test result and microstructural examination was made in a steel with high hardenability. Photo. 1 shows the correspondence between softening ratio and the microstructure quenched just before the second compression in 0.04C-2.0Mn-0.3Mo-0.10Nb steel. Prior austenite grain is elongated and no recrystallization occurs at the softening ratio of 18%. But at the softening ratio of 28%, recrystallized grain can be observed at elongated grain boundaries especially at triple junctions of grains. The extent of recrystallization increases in accordance with the increase of softening ratio. This result indicates that recrystallization starts at the softening ratio of around 20%. Only recovery proceeds in the region below 20% softening ratio, and recrystallization dominates in the region above 20% softening ratio. In this way it is made possible to distinguish recovery region from recrystallization region in the double compression test.

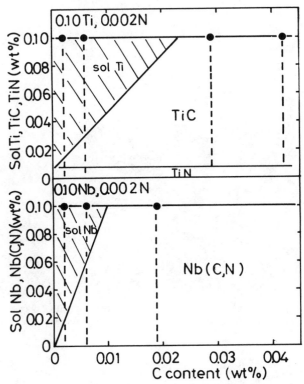

Fig.3 The change of the stoichiometric amount of
Nb as Nb(C,N) and Ti as TiC with C content
in 0.10Nb and 0.10Ti steels respectively.

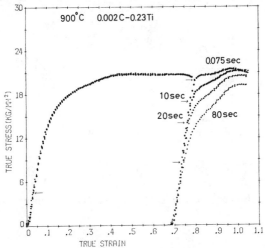

Fig.4 True stress-true strain curves in the
double compression test in 0.002C-0.23Ti
steel at 900°C.

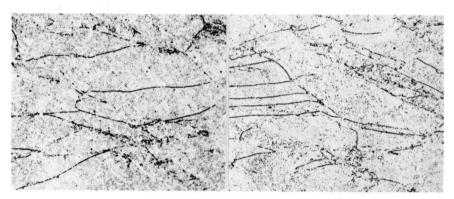

Softening ratio 18%
Recrystallization ratio 0%

Softening ratio 28%
Recrystallization ratio 2%

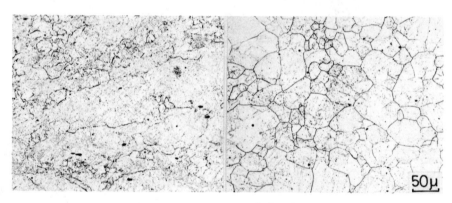

Softening ratio 60%
Recrystallization ratio 35%

Softening ratio 90%
Recrystallization ratio 100%

Photo.1 The relation between softening ratio and recrystallization rate.

Experimental results

The effects of solute atoms

The effect of solute Nb, V, and Ti on static recovery and recrystallization behavior was studied, based on the softening behavior in the double
compression test by using decarburized 0.002C steels. Fig. 5 shows the
change of softening ratio with holding time at 900°C for decarburized Nb
steels. Softening is retarded with the increase of Nb, most of which is
considered to exist as solute Nb. This retarding effect is very remarkable
in the region below 20% softening ratio, and less in the region above 20%
softening ratio. Considering that the region below 20% softening ratio
corresponds to recovery stage as shown before, it can be said that solute Nb
retards recovery and the onset of recrystallization. Fig. 6 shows the
results at 1000°C, 900°C, and 850°C for 0.002C steel and 0.002C-0.097Nb
steel. The recovery stage and the onset of recrystallization are retarded
at all temperatures, although the degree of the retardation becomes smaller
at higher temperatures.

The results for decarburized V and Ti steels are also shown in Fig. 7
and Fig. 8 respectively. Softening is delayed slightly in V steels and
considerably in Ti steels. This delay also results from the retardation of
the recovery stage below 20% softening ratio as in the case of Nb bearing
steels. Thus solute Nb, V, and Ti are all found to have the retarding
effect on the onset of recrystallization. The comparison among the three
types of steels is given in Fig. 9. It reveals that Nb has the strongest
effect in terms of the degree of retardation resulting from a unit addition.
Ti has medium effect, and V has the weakest effect. It should be noticed
again that the difference among three microalloying elements is primarily
appears in the recovery stage and the softening behavior above 20% softening
ratio does not seem to be influenced in the semi-log plot (15).

The effects of strain induced precipitates

The effect of precipitates such as Nb(C,N), VC, VN, and TiC on the
static recovery and recrystallization was studied using higher carbon or
higher nitrogen microalloyed steels. Fig. 10 shows the softening behavior at
900°C in higher C steels with 0.10Nb. Softening behavior does not change
with C content in shorter holding times. But in longer holding times than
around 10 sec., softening is delayed by the increase of C content. It should
be noticed that both $t_x=20\%$ and $t_x=50\%$ which correspond to the onset of
recrystallization and 30% recrystallization respectively were considerably
retarded in higher C steels. Fig. 11 shows the softening behavior of 0.002C-
0.097Nb and 0.019C-0.095Nb steels at 850°C, 900°C, and 1000°C. While two
steels show almost same softening behavior at 1000°C, higher C steel shows
slower softening behavior at 900°C and 850°C than ultra low C steel.
Retardation in higher C steel at lower temperatures is more marked in longer
holding times.

As already shown in Fig. 3, the stoichiometric amount of Nb as Nb(C,N)
increases with C content in higher C steels, although the actual precipitation is influenced by the amount of deformation, deformation temperature, and
holding time. With this point of view, strain induced precipitation behavior
of Nb(C,N) and its correlation with softening behavior in deformed austenite
were investigated using chemical analysis. For this study, specimen was

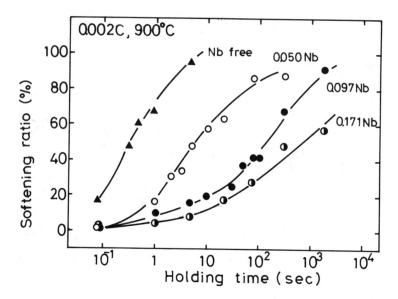

Fig. 5   The effect of Nb on the softening behavior in
0.002C steels.

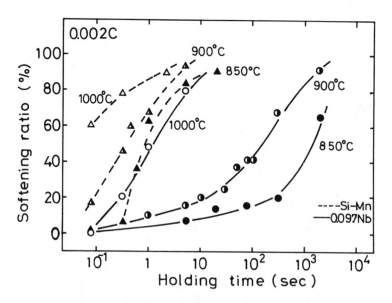

Fig.6   The effect of deformation temperature on the
softening behavior in 0.002C and 0.002C-0.097Nb
steels.

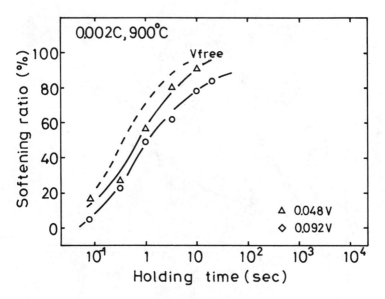

Fig.7　The effect of V on the softening behavior in
0.002C steels.

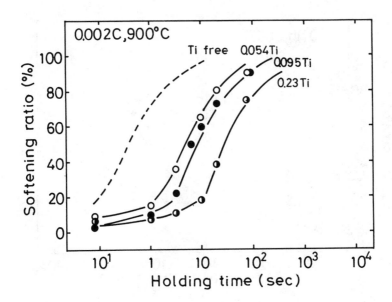

Fig.8　The effect of Ti on the softening behavior in
0.002C steels.

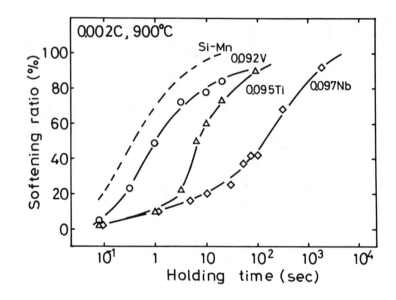

Fig.9   The comparison of Nb, V, and Ti effect on the softening behavior in 0.002C steels.

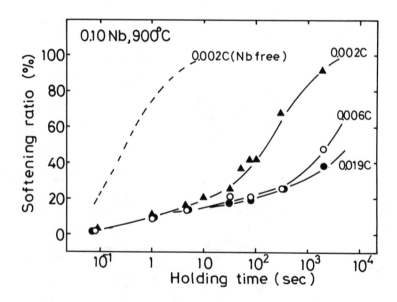

Fig.10   The effect of C content on the softening behavior in 0.10 Nb steels at 900°C.

taken from the plate which was hot rolled with the same reduction and held
at the same temperatures as in the double compression tests and rapidly
quenched after several holding intervals. Nb(C,N) precipitates were observed
in extracted replica by transmission electron microscope and analyzed chemically. Fig. 12 shows the change of the amount of Nb as Nb(C,N) with holding
time at 900°C for three Nb steels with varied C content. No precipitation
is observed to take place for the whole range of holding time in 0.002C-
0.097Nb steel. On the contrary, increasing amount of Nb(C,N) was observed
after 8 sec. in 0.006C-0.097Nb and 0.019C-0.095Nb steels. Comparing this
result in Fig. 12 with the one in Fig. 10, it is found that softening
begins to be retarded in higher C steels almost at the same holding time
when strain induced precipitation starts. It indicates that strain induced
precipitation retards softening including both the onset and the progress of
recrystallization in higher C steels. Photo.2 shows the extracted strain
induced precipitates Nb(C,N) at 900°C in 0.019C-0.095Nb steel. No precipitation is observed for holding time less than 8 sec., but in longer holding
times, precipitates are observed at the grain boundaries and in the matrix
of elongated prior austenite. The average diameter of Nb(C,N) at the grain
boundaries with holding time of 30 minutes is about 250Å. Precipitates in
the matrix are finer than those at the grain boundaries and considered to be
precipitated along subboundaries. The average diameter of these precipitates
is about 100Å. From these results, it is considered that precipitation
takes place first at the grain boundaries and afterward along subboundaries.
Fig. 13 shows the change of the amount of strain induced precipitated Nb(C,N)
with holding time at several temperatures in 0.019C-0.095Nb steels. Precipitation proceeds fastest at 900°C and 850°C, and more slowly at higher temperatures. The amount of precipitates tends to saturate with holding time of
about 15 minutes.

Fig. 14 shows softening behavior at 900°C in higher C or higher N steels
with 0.10V. Increase of C or N content does not influence softening behavior
in V steels. Consideration of solubility of VC(14) and VN(16) shows the
possibility of VN precipitation but not of VC precipitation at this temperature. Therefore, the results in Fig. 14 suggest that recrystallization seems
to be complete before the start of strain induced precipitation of VN.
Fig. 15 shows softening behavior at 900°C in higher C steels with 0.10Ti.
Softening in longer holding times than 1 sec is retarded by increased C
content, and it is probably due to the start of strain induced precipitation
of TiC. The comparison between four types of steels, C-Mn, C-Mn-Nb, C-Mn-V,
and C-Mn-Ti, is made in Fig. 16 on the basis of softening behavior at 900°C.
While Nb or Ti has a great retarding effect on the softening behavior, V has
only a slight effect which should be considered to result from the solute V
alone.

Discussion

Comparison between the retarding effects of static recrystallization by
solute atom and precipitate

It was shown that solute atom of microalloying elements retards the
onset of recrystallization in deformed austenite, while strain induced
precipitate of those retards both the onset and the progress of recrystallization. In this section, comparison was made between the retarding effects
by solute atom and precipitate. Fig. 17 shows the change of $t_{x=20\%}$ which
corresponds to the onset of recrystallization with Nb content in 0.002C and 0.019C

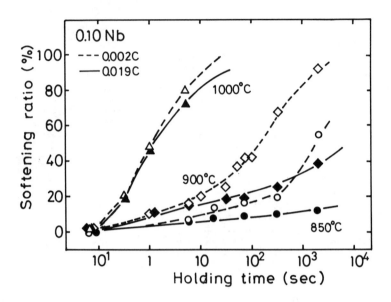

Fig.11 The effect of deformation temperature on the softening
behavior in 0.002C-0.097Nb and 0.019C- 0.095Nb steels.

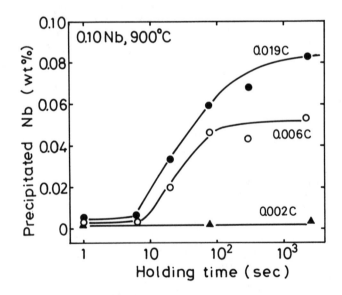

Fig.12 The progress of strain induced precipitation of
Nb(C,N) in 0.002C-0.097Nb, 0.006C-0.097Nb, and
0.019C-0.095Nb steels at 900°C.

1 sec.   **8 sec.**

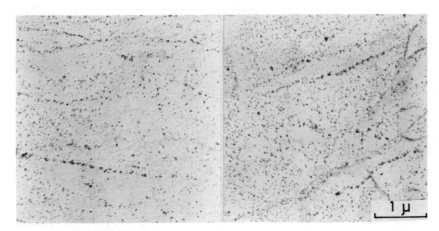

300 sec.   **1800 sec.**

Photo.2   Extracted Nb(C,N) observation by electron microscope.
(0.019C-0.095Nb steel held at 900°C for 1-1800 second after
deformation)

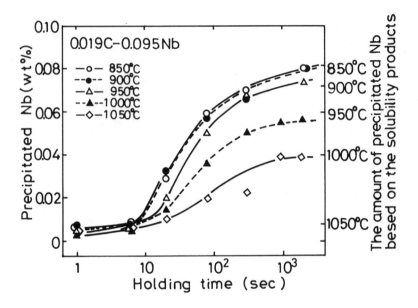

Fig.13   The progress of strain induced precipitation of Nb(C,N)
in 0.019C-0.095Nb steel.

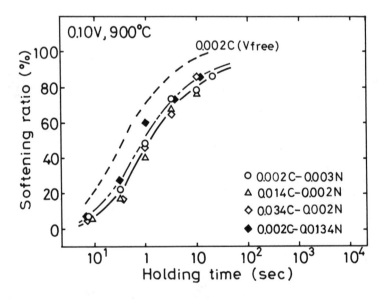

Fig.14   The effect of C and N content on the softening
behavior in 0.10V steels.

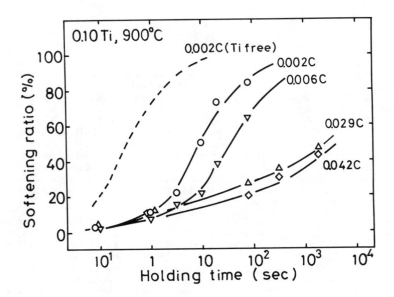

Fig.15 The effect of C content on the softening behavior
in 0.10Ti steels.

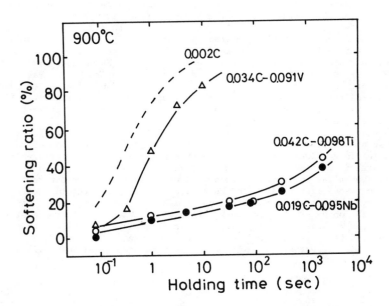

Fig.16 The comparison of Nb, V, and Ti effect on the softening
behavior in 0.02–0.04C steels.

steels. As mentioned before, almost all Nb atoms remain dissolved in 0.002C steel and the amount of solute Nb increases with total Nb content. On the contrary, all Nb up to 0.16% can precipitate stoichiometrically in 0.019C steel and the stoichiometric amount of Nb as Nb(C,N) increases with Nb content. Therefore, the change of t x=20% with Nb content in 0.002C steel and 0.019C steel gives the retarding rate of solute Nb and Nb(C,N) respectively for the onset of recrystallization. The retarding rates with Nb content in these two steels are almost same at 1000°C but different at 900°C and 850°C. The onset of recrystallization is delayed significantly due to the strain induced precipitation in 0.019C steel. Therefore it can be said that strain induced precipitated Nb(C,N) has much greater retarding effect on the onset of recrystallization than solute Nb.

A similar result was obtained for t x=50% which roughly corresponds to about 30% recrystallization, as shown in Fig. 18. t x=50% increases with the increase of solute Nb in 0.002C steels. This is essentially due to the increase of tx=20% but not resulting from the retardation of the progress of recrystallization. Comparing the two steels, a considerable difference appears at the lower temperatures. The retarding rate due to Nb(C,N) is much higher than solute Nb and the progress of recrystallization is delayed very much.

The interaction between strain induced precipitation and the progress of recrystallization in 0.019C-0.095Nb steel was shown in Fig. 19 in comparison with the recrystallization behavior in 0.002C and 0.002C-0.097Nb steels. Precipitation-temperature-time diagram for 0.019C-0.095Nb steel was produced with data in Fig. 13. In the region where the amount of Nb as Nb(C,N) is below 20% of total Nb, tx=20% and tx=50% which correspond to the onset of recrystallization and 30% recrystallization respectively increase monotonously with the temperature decrease. The recrystallization behavior is not influenced by C content in the two Nb steels and retarded only by solute atom. In the region where the amount of Nb as Nb(C,N) is above 20%, tx=20%, and tx=50% increases abruptly for the higher C steel and recrystallization behavior in two steels of 0.019C-0.095Nb and 0.002C-0.097Nb begins to differ from each other. The recrystallization is significantly retarded by strain induced precipitates, even though the onset of recrystallization is delayed by solute Nb at all temperatures. All these results suggests that firstly the onset of recrystallization is retarded by solute Nb because of retarded recovery and then the onset and the progress of recrystallization is conciderably retarded by precipitated Nb(C,N) after precipitation starts in the longer period of time at low temperatures. In other words, solute Nb has an important role at lower temperatures to retard the onset of recrystallization so that the strain induced precipitation starts before recrystallization is complete and to make the retarding effect of recrystallization by precipitate work effectively. When the retarding effect by solute atom is weak as in V steel, recrystallization is complete before strain induced precipitation and significant retarding effect cannot be expected. This difference is well demonstrated between V and Nb steels in Fig. 16.

Strain induced precipitation kinetics was studied in Fig. 13, on the right horizontal axis of which the amount of precipitates calculated for each temperature is also indicated on the basis of the solubility products obtained by Irvine et.al (14). The measured amount of Nb as Nb(C,N) with holding time of 30 minutes is 3 times larger than the calculated amount at the maximum. Fig. 17 shows solubility product calculated using the results for holding time of 30 minutes, comparing with the results by Irvine et.al.

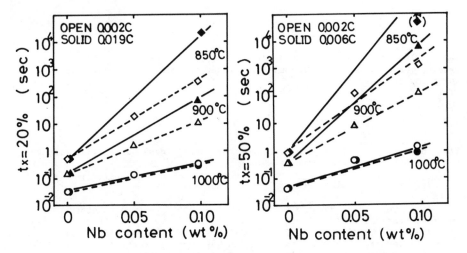

Fig.17 The effect of Nb content on tx=20% in 0.002C and 0.019C steels.

Fig.18 The effect of Nb content on tx=50% in 0.002C and 0.019C steels.

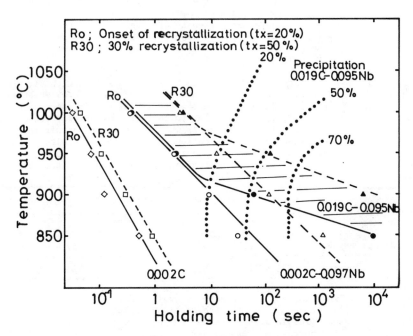

Fig.19 Recrystallization-precipitation diagram for C-Mn and C-Mn-Nb steels.

The results for 0.036C-0.096Nb steel without strain agree with Irvine's result. On the other hand, solubility product in strained austenite is found to follow the equation (4).

$$\log \ [\text{Nb}][\text{C}+\frac{12}{14}\text{N}] = - \ \frac{6570}{T} + 1.93 \qquad (4)$$

From those results, it may be concluded that strain does not only accelerate but also enhance Nb(C,N) precipitation.

The effect of solute atom on static recrystallization

As shown in Fig. 9, solute atom of microalloying elements delays the onset of static recrystallization by delaying recovery stage. Fig. 21 shows the change of $t_{x=20\%}$ at 900°C and 1000°C with Nb, V, and Ti content. Fig. 22 shows the change of $t_{x=50\%}$ with Nb, V and Ti content. $t_{x=20\%}$ increases with the increase of Nb, V, and Ti at both temperatures. $t_{x=50\%}$ also increases with the increase of microalloying content,but, comparing Fig. 21 and Fig. 22, this is found to be simply due to the increase of $t_{x=20\%}$ on a basis of log t kinetics. Solute drag effect has been proposed as a mechanism for the retarding effect of static recrystallization by solute atom. And it is presumably resulting from lattice distortion by solute atom in the matrix. Therefore the degree of the retarding effect by solute atom may be explained by the degree of lattice distortion. From this point of view, softening behavior in deformed austenite of steels with other alloying elements which show various atomic radius was investigated and compared with that of steels with solute Nb, V, and Ti using similar experimental conditions.

Fig. 23 shows the softening behavior in 0.09C-1.2Mn steels with Ni, Cr, Mn and Mo. For comparison, the data for base steel and 0.002C-0.097Nb steel were included. Softening behavior of the steels with Ni and Cr which have similar atomic radius as Fe is very similar to that of base steel. Softening process of the steels with alloying elements which have larger atomic radius is somewhat retarded, although the retardation is small compared with that of Nb. The retarding rate, $t_{x=20\%}$ and $t_{x=50\%}$ per atomic % of solute atom at 900°C, was listed in Table II. The retarding rate increases in the following ascending order: Ni<Cr<Mn<V<Mo<Ti<Nb.

Table II The retarding rate by solute atoms							
Solute atom	Ni	Cr	Mn	V	Mo	Ti	Nb
$t_{x=20\%}$/at%(sec/%)	0.13	0.17	0.36	2.3	8.5	39	210
$t_{x=50\%}$/at%(sec/%)	0.75	0.93	3.2	10	30	120	3000

The effect of solute atom has been generally evaluated by misfit factor calculated with the atomic radius of base metal and solute atom. The half of interatomic distance (16) is used to calculate the atomic radius. As demonstrated in Fig. 24, the result does not seem to give a consistent correlation.

The degree of lattice distortion can be indirectly detected by the change of lattice constant. Therefore, the correlation between the retarding

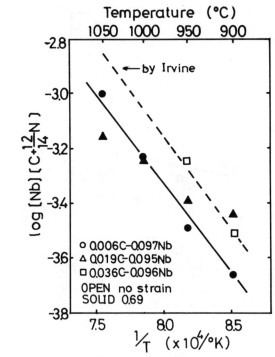

Fig.20 The temperature dependance of the
        solubility product [Nb][C+12/14N]
        in deformed austenite.

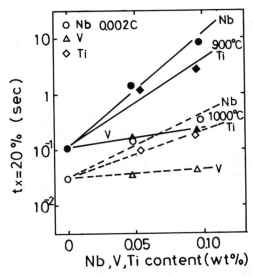

Fig.21 The effect of Nb, V, and Ti content
        on tx=20% in 0.002C steels.

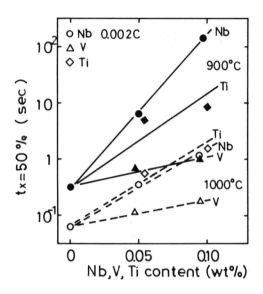

Fig.22 The effect of Nb, V, and Ti content
on tx=50% in 0.002C steels.

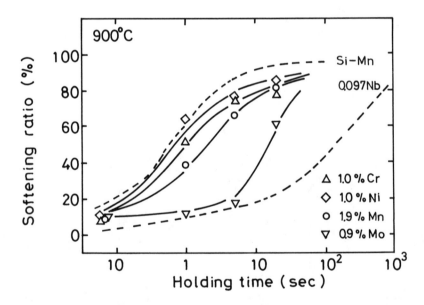

Fig.23 The effect of Cr, Ni, Mn, and Mo on the softening behavior.

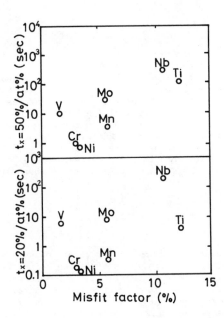

Fig.24 The correlation between tx=20%,
        tx=50% and misfit factor.

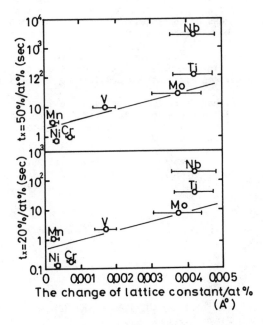

Fig.25 The correlation between tx=20%,
        tx=50% and the change of lattice
        constant.

rate for solute atom and the lattice constant change in austenite phase was examined by using high temperature X-ray diffractometer. Tungsten powder which was coated to the sample surface was used for calibration for the diffracted angle. The result shown in Fig. 25 seems to give a rather consistent correlation between the retarding rate and the lattice constant change per atomic %. Alloying elements yielding large change in lattice constant lead to high retarding rate at least in terms of expected orders. However, a closer analysis reveals that Nb and Ti show some extra deviation from the expected value for the retarding rate. So Nb and Ti may form cluster-type pre-precipitation, and this can be responsible for the deviation through the interaction with the dislocations, subboundaries and grain boundaries.

Conclusion

The effect of microalloying elements on recovery and recrystallization in deformed austenite was systematically investigated using the stress-strain curve in the double compression test. Particular emphasis was placed on the differentiation of the role of solute atom from that of strain induced precipitate. To meet this purpose, 0.002C steels with Nb, V, and Ti were prepared by decarburizing method in addition to higher C steels with Nb, V, and Ti. The results obtained are summarized as follows:

(1) Nb, V, and Ti in solution retard recovery in deformed austenite and consequently delay the onset of recrystallization, but do not much influence the progress of recrystallization. Solute Nb has the strongest effect among the elements studied.

(2) The retarding effect of solute atom is well correlated with the change of lattice constant in austenite. This effect seems to be caused by the interaction between dislocations and lattice distorsion.

(3) Both onset and progress of recrystallization are substantially delayed as soon as strain induced precipitation of Nb(C,N) and TiC begins. The retarding effect of VN is very small, because recrystallization is completed before precipitation of VN starts.

(4) Quantitative analysis indicates that the amount of Nb(C,N) precipitates is larger in strained austenite than in strain-free austenite at a given temperature. The solubility products of strain induced precipitate Nb(C.N) is given as follows:

$$\log [Nb][C + \frac{12}{14}N] = -\frac{6570}{T} + 1.93$$

Acknowledgements

The authors are grateful to Dr. I. Kozasu and Mr. M. Niikura for helpful discussions and reading manuscript.

References

(1) R.A.P. Djaic, and J.J. Jonas, "Static Recrystallization of Austenite between Intervals of Hot Working", Journal of The Iron and Steel Institute, vol.210, 1972, P256.

(2) R.A. Petkovic, M.J. Luton, and J.J. Jonas, "Flow Curves and Softening Kinetics in High Strength Low Alloy Steels", The Hot Deformation of Austenite, J.B. Ballance, ed.; AIME, New York, 1977, P68.

(3) M.J. Stewart, "The Effect of Niobium and Vanadium on The Softening of Austenite during Hot Working", ibid, P233.

(4) J.N. Cordea, and R.E. Hook, "Recrystallization Behavior in Deformed Austenite of High Strength Low Alloy Steels", Metallurgical Transactions, vol.1, 1970, P111.

(5) T.L. Capeletti, L.A. Jackman, and W.J. Childs, "Recrystallization Following Hot-Working of a High-Strength Low-Alloy Steel and a 304 Stainless Steel at the Temperature of Deformation", ibid, vol.3, 1972, P789.

(6) H. Weiss, A. Gittins, G.G. Brown, W.J. McG. Tegart, "Recrystallization of a Niobium-Titanium in The Austenite Range", Journal of The Iron and Steel Institute, vol.211, 1973, P703.

(7) A. LeBon, J. Rofes-Vernis, and C. Rossand, "Recrystallization and Precipitation during Hot Working of a Nb-Bearing HSLA Steel", Metal Science, vol.9, 1975, P36.

(8) K.J. Irvine, T. Gladman, J. Orr, and F.B. Pickering, "Controlled Rolling of Structural Steels", Journal of The Iron and Steel Institute, vol.208, 1970, P717.

(9) R. Priestner, C.C. Earley, and J. H. Rendall, "Observations on the Behaviour of Austenite during the Hot Working of Some Low-Carbon Steels", Journal of The Iron and Steel Institute, vol.206, 1968, P1252.

(10) I. Kozasu, T. Shimizu, and H. Kubota, "Recrystallization of Austenite of Si-Mn Steels with Minor Alloying Elements after Hot Rolling", Transactions of The Iron and Steel Institute of Japan, vol.11, 1971, P367.

(11) B.L. Phillipo, and F.A.A. Crane, "Structure and Strength of C-Mn-Nb steels during Hot Rolling", Journal of The Iron and Steel Institute, vol.211, 1973, P653.

(12) H. Sekine, and T. Maruyama, "Retardation of Recrystallization of Austenite during Hot-rolling in Nb-containing Low-carbon steels", Transactions of The Iron and Steel Institute of Japan, vol.16, 1976, P427.

(13) L. Darken, and R. Gurry, Physical Chemistry of Metals, McGraw-Hill, Inc., New York, 1953, P440.

(14) K.J. Irvine, F.B. Pickering, and T. Gladman, "Grain-Refined C-Mn Steels", Journal of the Iron and Steel Institute, vol.205, 1967, P161.

(15) R. Coladas, J. Masounave, and J. P. Bailon, "The Influence of Niobium on the Austenite Processing of Medium and High Carbon Steels", The Hot Deformation of Austenite, J.B. Ballance, ed.; AIME, New York, 1977, P341.

(16) K. Bungardt, K. Kind, and W. Olsen, "Die Loslichkeit des Vanadinkarbids im Austenit", Archiv für das Eisenhüttenwesen, vol.27, 1956, P61.

(17) W.B. Pearson, A Handbook of Lattice Spacing and Structures of Metals and Alloys, Pergamon Press, London, 1958, P124.

DISCUSSION

Comment: I want to congratulate the authors for such a clear exposition of the solute effects and the relative solute effects of these elements. I think there are two important techniques that they have used which should bear investigation by other people. One is that they did use very low-carbon and low-nitrogen steels similar to those used by Luton and Petkovic some years ago. The other is that they used very short unloading and reloading times in their apparatus. This is the kind of testing you couldn't do long ago with simpler kinds of testing machines. By going down to fractions of a second, it is possible to look at effects that occur well before precipitation starts, and this is a very interesting type of investigation to do. I know that at Sheffield, tests of unloading times on the order of 50 milliseconds or less are used. At McGill, we are also doing tests with unloading times of about 50 milliseconds, and we have observed things similar to those that Dr. Yamamoto described. The other comment I would like to make is that many people do try and explain their results in the solute range on the basis of atomic size differences, and we have seen some plots of that type. Dr. Yamamoto pointed out, some of the points don't fall purely on an atomic-sized difference scale, and these do correlate very well with an electron difference. Of course, the ordering events do depend on electron differences, and I do think you have to look at both atomic size differences and electron differences in order to get a good explanation of the relative effects of these elements.

Comment: If I understood the author correctly, at the point where there was a deviation between the softening and the lattice misfit, I think you had said something about chemistry. I'd like to support Tony DeArdo's remarks here and suggest there are precipitates long before they are seen, especially when utilizing coarse extraction or replication techniques. Sid Brenner (of USS) is doing some studies with the field-ion probe and he is seeing precipitates on the order of 5-10 angstroms, where they "shouldn't" be.

Comment: With regard to the conclusions that Nb(C,N) and TiC precipitates
retard recrystallization of austenite very strongly whereas V(C,N) does
not, it is important to point out that this is merely a question of which
temperature the comparison is made at. In the region above 850°C examined
by Yamamoto et. al the effect of V is very limited because the solubility
of V(C,N) is too large to cause any significant precipitation. If, however,
the comparison is made at a lower temperature, as done by W. Roberts (1),
where the supersaturation with respect to V(C,N) is large enough to cause
abundant precipitation, it is found that the impedance of recrystallization
may even be stronger than for Nb(C,N) (1).

1. W. Roberts, Scand. J. Metallurgy g (1980) 13-20.

Comment: The conclusion of the effect of microalloying elements on recrystallization obtained here is based on the investigation at the temperatures
above 850°C, because the γ-α transformation in the ultra-low carbon-base
steels started at around 820°C. With regard to the result by W. Roberts,
two comments will be made; first it is more important in controlled rolling
to raise the stop-temperature of static recrystallization of austenite.
Secondly, the effect of the solute or precipitates among the various
microalloying elements should be compared in the base of atomic (or weight)
% addition of solute element or unit volume fraction of the precipitates,
respectively. The fact accepted generally in both cases is that Nb is more
effective element than V.

HOT ROLLING OF C-Mn-Ti STEEL

L.A. Leduc[*] and C.M. Sellars
Department of Metallurgy,
University of Sheffield, England

Experimental cast slabs of plain carbon steel and of a similar steel
containing nearly stoichiometric additions of titanium and nitrogen have
been hot rolled and tested in plane strain compression at hot working
temperatures and strain rates. In the titanium steel, fine particles
($\sim$11nm) of TiN prevent grain coarsening on initial reheating to temperatures
of 1150°C but do not retard either dynamic or static recrystallisation.
Statically recrystallised grain sizes are similar to those in C-Mn steel
for the same initial grain size and deformation conditions, but no grain
growth takes place in the titanium steel. Austenite grain sizes of
20-30μm can therefore be produced after relatively small total rolling
reductions. The grain size depends critically on the reduction in the
last pass but is insensitive to finishing temperatures of up to 1050°C. On
cooling, finer ferrite grain sizes are obtained than in the C-Mn steel, with
the expected beneficial effects on strength and toughness.

[*] Now at Departamento de Metallurgia, Universidad Nacional Autonoma
Mexico, Mexico, D.F.

Introduction

In the rolling of thick plates and heavy sections, finishing temperatures tend to be high and, in the case of sections, they are also non-uniform. Conventional controlled rolling, which relies on retarding recrystallisation of austenite at low finish rolling temperatures is therefore not feasible for grain refining these products. However, the alternative of allowing full recrystallisation of the austenite, but preventing grain growth (1) appears a viable possibility for obtaining finer structures than are attainable in plain C-Mn steels. This would be particularly effective if grain coarsening during reheating could also be avoided. The progressive refining of grain size by repeated recrystallisation between passes would then not be essential and total reductions sufficient for consolidation (∿70% for plate from continuously cast slab (2)) should result in satisfactory structures.

The presence of TiN particles has been shown by several authors (3-5) to result in high grain coarsening temperatures, particularly with Ti:N ratios around the stoichiometric value. The purpose of the present work is to examine the influence of these particles on the recrystallisation and subsequent grain growth of austenite and on the ferrite grain structure and properties obtained after cooling.

Experimental Materials and Procedure

Steels of composition shown in Table 1 were air melted and sand cast as slabs 25 x 75 x 375 mm and as billets 76 mm diameter x 200 mm long. As shown previously (2), the sand cast slabs had structures of similar scale to those in commercial continuously cast slabs. The slabs were machined to 20 x 50 mm and lengths of about 100 mm were then used for experimental hot rolling. The cylindrical billets were initially hot rolled* to 20 x 55 mm slabs, shot blasted and then sectioned into 120 mm lengths for experimental hot rolling, which was carried out on an instrumented Hille 50 mill with 140 mm diameter rolls and speed adjusted to give mean equivalent strain rates during rolling in the range 2 to 6 sec^{-1}. Temperature was continuously monitored by a 1.5 mm diameter "Pyrotenax" inconel sheathed thermocouple inserted into a transverse hole drilled to the centreline of the slabs. The output was coupled to a high speed amplifier, with an automatic back-off system, and recorded on a U.V. recorder.

Table 1. Analyses of Experimental Steels

Steel	Analysis, wt %					
	C	Mn	Si	Al	N	Ti
C-Mn	0.12	1.46	0.31	0.032	0.005	-
Low Ti	0.12	1.47	0.30	0.036	0.005	0.02
Ti	0.12	1.30	0.14	0.028	0.015	0.06

A series of single pass rolling experiments was carried out on the cast slabs, which were reheated for 20 min to 1150°C to obtain a constant grain size and were then transferred for 10 min to a second furnace set at a temperature slightly higher than the rolling temperature to minimise temperature gradients before rolling. After rolling either the whole slab or a

* By British Steel Corporation, Hoyle Street Laboratories.

piece sheared from it was quenched for metallographic study. When cold, slabs were shot blasted and measured and then reheated as before for a further rolling pass. The overall sequence of pass reductions employed for these experiments was 18%, 29%, 11%, 20%.

The slabs hot rolled from the round billet were reheated to 1200°C or above for 30 min and were used for multipass rolling experiments with various combinations of pass reductions for direct comparison between the C-Mn steel and the Ti steel. Each schedule was carried out in duplicate for both steels and all schedules finished with slabs about 10 mm thick, from which a piece was quenched for metallographic study. The slabs were air cooled, a further sample was taken and the remaining material was shot blasted and used for plane strain compression specimens. These were machined to 50 x 70 mm and were then given a commercial dull 5µm chromium plating to protect them from oxidation and to provide a surface that is wetted by the glass lubricant used for the tests.

Plane strain compression tests were carried out on a computer controlled servohydraulic machine provided with superalloy tools of 15 mm working face width. Tests were performed at constant equivalent strain rates of 0.5, 5 and 50 sec^{-1} at temperatures of 900°, 975° and 1050°C after reheating the specimens to 1050°C for 15 min. For the C-Mn steel the tests at 900°C were also repeated on specimens reheated to 900°C to obtain an equivalent grain size to that in the Ti steel. Data on load and displacement collected by the computer were converted to equivalent true stress-strain curves and were corrected for spread and friction effects. In addition to tests carried out to equivalent true strains of 2.5, others were terminated at different strains and the specimens were held at temperature for times up to 100 sec after recrystallisation before quenching to determine recrystallised grain size and grain growth. Some tests were also interrupted for various times after a strain of 0.1 to determine the static restoration kinetics.

For metallographic study, plane strain compression specimens were sectioned longitudinally at mid-width. The rolled slabs were also sectioned longitudinally, but in this case at 0.21 of the width, where the thermal history during rolling is most nearly equivalent to the mean for the transverse section. For determination of austenite grain structures, specimens were tempered for 3 hours at 300°C before mechanical polishing and etching in a solution of 100 ml saturated aqueous picric acid, 10 ml wetting agent (Teepol) and 250 mg Na_2O_2 for 3 min at 60°C. Ferrite-pearlite structures were etched in 2% nital. Mean linear intercept grain sizes were determined in both the longitudinal and through-thickness directions by counting a minimum of 600 grains.

After two of the multipass rolling schedules, the slabs were cooled in vermiculite to simulate air cooling of thicker material and longitudinal tensile specimens (5.64 mm gauge diameter and 28.2 gauge length) and Charpy specimens, with the notch in the through thickness direction were machined from them.

Results

Reheating the cast steels for 30 min at different temperatures showed the stabilising effect of TiN particles on austenite grain size. This is illustrated in fig 1, which compares the C-Mn and low Ti steels. The higher Ti steel gave essentially the same results as the latter, with a grain coarsening temperature of 1150-1175°C. In both Ti steels optical microscopy showed a more or less uniform distribution of cuboidal TiN particles of size 1-5µm, which accounted for most of the volume fraction of TiN calculated from

the chemical composition. Electron metallography of carbon extraction replicas also showed the presence of a fine dispersion of TiN particles of mean
size about 11 nm, which appeared to be identical in both steels.

During the single pass rolling experiments it was found that on a second
reheating to 1150°C the fine austenite grain size was again stabilised in
the Ti steel, but after the third and fourth reheating exaggerated grain
coarsening to mean grain sizes of 1400 and 940μm had taken place. The coarse
TiN particles appeared to be unchanged by repeated reheating, but after the
third reheat the fine particles had coarsened to a mean size of about 20 nm.

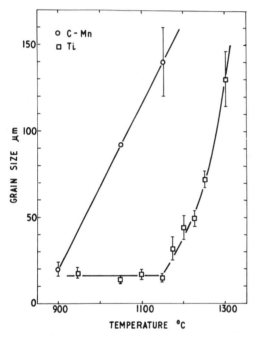

Fig. 1 -
Austenite grain size
after reheating the
C-Mn and low Ti steels
in the as-cast
condition.

Single Pass Rolling

In all cases after single pass rolling followed by air cooling to ∿750°C
and quenching the microstructures were fully martensitic. This is consistent with the observations from temperature changes in air cooled slabs that
the Ar_3 temperature was in the range 705° to 750°C with the value depending
on the austenite grain size, the retained strain and the thickness of the
slab, i.e. the cooling rate. The austenite structures observed are shown in
fig 2 in terms of the aspect ratio (AR) and mean grain size ($\bar{d}$) as a function
of the rolling temperature, where

$$AR = d_L/d_T \qquad (1)$$

and

$$\bar{d} = (d_L d_T)^{\frac{1}{2}} \qquad (2)$$

d_L and d_T are the mean linear intercept grain sizes measured in the longitudinal and through thickness directions respectively. The lines in the
figure are calculated, as shown in the Appendix, from the recrystallisation
kinetics, the original and recrystallised grain sizes and the observed

cooling rates, and give the temperatures for 50% recrystallisation marked by the vertical lines. Within the scatter of the experimental points for the C-Mn steel, the aspect ratio and grain size changes lead to reasonably consistent recrystallisation temperatures. Because of the relatively small elongation of the grains after 11% reduction, satisfactory measurements of aspect ratio could not be made in this case. In the case of the Ti steel, the results after the first two reheatings only are shown. In these cases the structure was fully recrystallised for all rolling conditions, although the recrystallised grains tended to be slightly elongated in the rolling direction, particularly after the first 18% reduction. After the third and fourth reheatings, the coarse irregular grain structures made it impossible to obtain satisfactory quantitative measurements, but in neither case was there evidence of significant recrystallisation up to the highest rolling temperatures of about 1050°C.

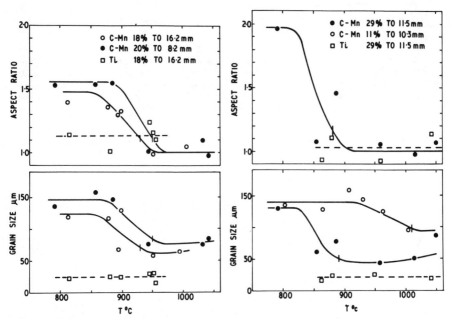

Fig. 2 - Austenite grain size and aspect ratio after single pass rolling reductions (a) following first and fourth reheats and (b) following second and third reheats to 1150°C.

Multipass Rolling

These schedules were carried out on previously hot rolled material with reheating temperatures of 1200°C or higher, which are well above the grain coarsening temperature for the Ti steel. Some results of schedules which gave nearly similar austenite grain sizes after rolling are shown in Table 2. These illustrate the typical results for all the rolling schedules, namely that after reheating the austenite grain size of the Ti steel is smaller than that of the C-Mn steel and that this difference is maintained after rolling. In the schedules illustrated, full recrystallisation is expected (6) after each pass and the rolling loads for the C-Mn and Ti steels were always the same.

Table 2. Multipass Rolling Data

Steel	Reheat $T^{o}C$	Reheat $\bar{d}\gamma$ μm	Rolling Schedule Pass $T^{o}C$/Reduction %	$\bar{d}\gamma$ μm	$\bar{d}\alpha$ μm
C-Mn	1250	320	1120/32 + 980/28	44	18.5
Ti		200		34	12.2
C-Mn	1265	415	1135/20 + 995/38	49	-
Ti		340		32	12.5
C-Mn	1200	170	1130/20 + 1050/15+	42	18.2
Ti		120	970/15 + 900/15	31	12.3
C-Mn	1210	180	1135/20 + 1055/16+	40	18.8
Ti		100	970/22 + 895/9	36	12.6

Plane Strain Compression Testing

Typical stress-strain curves obtained at 900°C are shown in fig 3.
These have the characteristic form associated with the occurrence of dynamic
recrystallisation and show that when the grain sizes of the C-Mn and Ti
steels are the same, the stress-strain curves are the same, but with a coarse
grain size in the C-Mn steel the occurrence of dynamic recrystallisation is
considerably retarded.

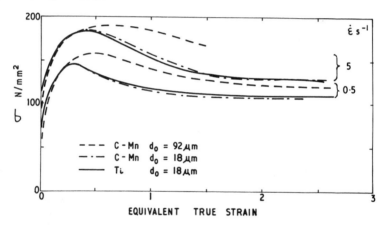

Fig. 3 - Equivalent true stress-equivalent true strain
curves derived from plane strain compression
tests at 900°C.

From the tests in which deformation was interrupted at a strain of 0.1
(i.e. before the onset of dynamic recrystallisation) the fractional restoration (S) during the delay time was determined from the stress-strain curves
as

$$S = \frac{\sigma_1 - \sigma_2}{\sigma_1 - \sigma_2^*} \qquad (3)$$

where σ_1 is the flow stress at the end of the first deformation, σ_2 is flow stress at 0.02 offset strain in the second deformation and $\sigma_2{}^*$ is the equivalent value when full restoration has occurred. This definition was chosen because the change in grain size produced by recrystallisation led to different offset flow stresses in the first and second deformations in fully recrystallised material. The restoration curves as a function of time of interruption are shown in fig 4.

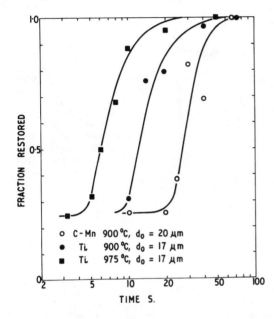

Fig. 4 –
Static restoration curves after plane strain deformation to an equivalent true strain of 0.1 at 5 sec^{-1}.

The recrystallised austenite grain sizes obtained on holding at temperature after deformation to different strains are given in fig 5 together with the results from single pass rolling. This shows that plane strain compression and rolling give similar recrystallised grain sizes for the C-Mn and Ti steels when the initial grain sizes are similar, but that rolling of the coarser grain sized C-Mn steel results in significantly larger recrystallised grain size.

Holding the Ti steel at 975°C for times up to 100 sec after complete recrystallisation gave no detectable change in grain size, whereas in the same times C-Mn steel shows significant grain growth at 950°C even with a recrystallised grain size of 50µm (7).

Ferrite Structure and Properties

After air cooling, the ferrite structure in the Ti steel was always entirely polygonal whereas in the C-Mn steel there was a significant fraction of Widmanstatten structure even from austenite grain sizes of 40-50µm. After slower cooling in vermiculite, both steels were entirely polygonal and no evidence of carbide films at ferrite grain boundaries was found in either steel.

The ferrite grain size obtained after all rolling schedules was smaller in the Ti steel than in the C-Mn steel, see for example Table 2. The relationship between austenite grain size and ferrite grain size obtained

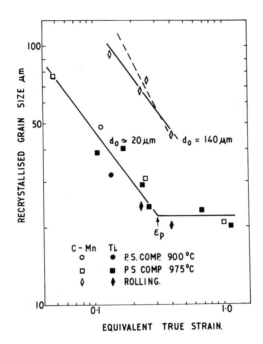

Fig. 5 -
Dependence of recryst-
allised austenite grain
size on the equivalent
true strain of the prior
deformation in plane
strain compression or
rolling. ε_p marks the
strain to the peak flow
stress at 900°C and
5 sec^{-1}.

from multipass rolling (finishing thickness 10 mm), from single pass rolling
to a finishing thickness of 11.5 mm and, for the C-Mn steel, also to a
finishing thickness of 8.3 mm is shown in fig. 6. Within the scatter of
results, no effect of the different cooling rates at 750°C, from about 2.25°C
/sec for 11.5 mm to about 3.0°C/sec for 8.3 mm could be detected. However,
the slower cooling rates in vermiculite of ∿1.4°C/sec and ∿0.9°C/sec at
750°C for material rolled to 10 mm and to 13 mm tended to give larger grains,
e.g. 14.0 and 16.2μm in the Ti steel and 18.5 and 22.1μm in the C-Mn compared
with ∿12.5 and ∿18μm expected after air cooling the two steels at 10 mm
thickness.

Mechanical properties measured on vermiculite cooled steels are shown in
Table 3.

Table 3. Mechanical Properties

Steel	$\bar{d}\alpha$ μm	Y.S. N/mm^2	T.S. N/mm^2	Elong. %	RA %	Charpy 20J ITT, °C
C-Mn	18.5	295	435	35	76	- 30
Ti	14.0	328	465	30	74	- 43
C-Mn	22.1	286	430	37	75	- 23
Ti	16.2	318	461	31	75	- 35

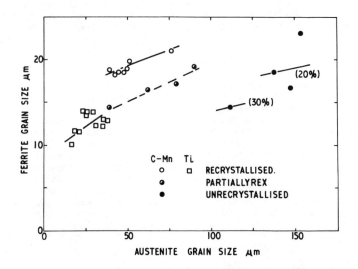

Fig. 6 - Dependence of ferrite grain size on austenite
grain size in air cooled slabs of about 10 mm
thickness.

Discussion

The observations on the grain coarsening behaviour of the Ti steels and
on the distributions of TiN particles are in agreement with previous work
(3-5). The coarse particles precipitated during solidification account for
the majority of the Ti in both steels, leading to a higher volume fraction in
the steel of higher Ti and N content. These particles are too widely spaced
to have significant effect on grain coarsening, although they probably
contribute to the somewhat smaller grain size in the Ti steel than in C-Mn
steel on reheating to temperatures well above the grain coarsening temperature (c.f. Table 2).

The fine particles appear similar in both the Ti steels and lead to
similar grain coarsening temperatures. George and Irani (3) estimated the
volume fraction of small particles to be $\sim 7.5 \times 10^{-5}$. Taking this figure
with an initial mean austenite grain size of 20μm and substituting them into
the Gladman equation (8) gives a critical particle size for grain coarsening
of 13.5 nm in good agreement with the observed size. The decrease in grain
coarsening temperature with repeated reheating through the ferrite austenite
transformation is also in agreement with previous work and is associated both
with coarsening of the particle distribution and with refinement of the
austenite grain size produced by transformation, e.g. ~ 20μm on first reheating and ~ 15μm on second reheating in the present work on the higher Ti steel.

The high temperature stress-strain curves on C-Mn steel (Fig 3) show the
major effect of initial grain size on dynamic recrystallisation kinetics
reported previously for nickel (9), C-Mn steel (10) and stainless steel (11).
They also show that the presence of TiN particles has no significant effect.
Similarly the measurements of static restoration kinetics indicate that the
TiN particles have little effect on the kinetics of static recrystallisation.
In fact, if the observations at 900°C, fig 4, are corrected for the small

difference in grain size between the C-Mn and Ti steel by assuming a
dependence of recrystallisation time on d^2 (6), it still appears that static
recrystallisation may be marginally more rapid in the Ti steel. This con-
trasts with the observations of Ouchi et al (5) who found that the strain
to give the start of recrystallisation in a constant holding time was
slightly larger for a 0.019% Ti steel than for C-Mn steel.

The single pass rolling results on C-Mn steel, Fig. 2, show a decreasing temperature range for recrystallisation with increasing rolling reduction. From the observed cooling rates and Appendix equation (A8) the temperatures for 50% recrystallisation have been converted to equivalent times
for isothermal holding at 900°C, Table 4. These times are also corrected
to a constant grain size of 140μm and show a strong dependence on rolling
reduction, in general agreement with observations on the effect of strain
in laboratory tests (6). In fact the strain dependence is somewhat less
than expected from laboratory tests and if a grain size correction of the
form used is valid down to 20μm the rolling results also indicate a time for
50% recrystallisation after a strain of 0.1 about an order of magnitude
less than observed by the restoration measurements in plane strain compression. The reason for these differences requires further study.

Table 4. Static Recrystallisation Kinetics

Steel	R	$\bar{d}_o$	T_{50}	$t_{50}(900°C)$	$t_{50}(140/\bar{d}_o)^2$
	%	μm	°C	sec	sec
C-Mn	18	124	930	31	40
	29	131	890	9	10
	11	140	1010	134	130
	20	147	950	30	27
Ti	18	20	(775)	(0.8)	(40)
	29	15	(714)	(0.1)	(10)
	11	1400	(1190)	(13,000)	(130)
	20	940	(1060)	(1,200)	(27)

In table 4 it has also been assumed that recrystallisation times for
the Ti steel are the same as for the C-Mn steel, and the recrystallisation
temperatures for the observed grain sizes have been estimated. Although
using grain size ratios over such a wide range for correction is open to
considerable doubt, the results do show that full recrystallisation after
the first two reheats and no recrystallisation after the second two reheats
would be expected over the experimental rolling temperature range as a
result of grain size differences alone. Again the experimental observations
on the Ti steel therefore show no direct effect of Ti on the recrystallisation kinetics. It should be noted that in the present steel, although the
Ti:N ratio of 4 slightly exceeds the stoichiometric value of 3.42, it would
be expected from the solubility data for TiC (12), that any carbide formed
from excess Ti on cooling would redissolve at ∿950°C, which is considerably
lower than the reheating temperatures employed. Also reprecipitation would
not be expected during or after rolling over the experimental temperature
range.

Recrystallised grain sizes in the Ti and C-Mn steels for given deformation conditions also appear to be identical, when the initial grain size is
the same, fig 5, showing no constraining effect of TiN particles on grain
growth under the higher driving force during recrystallisation, even though
grain growth after recrystallisation is prevented. The dependence of

recrystallised grain size on strain at strains below that at which dynamic recrystallisation occurs, followed by a constant grain size for higher strains in the steels of initial grain size 20μm is in general agreement with previous results on C-Mn steels of coarser initial grain size (6). The slope of the dependence is, however, somewhat lower than that found previously and shown by the broken line through the points for an initial grain size of 140μm. This may indicate that the previously proposed functional relationships are an oversimplification for a wide range of grain sizes.

In single pass rolling, the prevention of grain growth after recrystallisation leads to a constant austenite grain size at temperatures up to 1050°C, fig 2. In multipass rolling this effect and the restraint on grain growth on reheating to well above the grain coarsening temperature both contribute to a smaller grain size in the Ti steel than in C-Mn steel (Table 2). With a larger number of passes only the former effect would be expected to be significant (13). On transformation to ferrite, the Ti steel also gives a smaller grain size than C-Mn steel, even for the same austenite grain size, possibly because pinning effects of TiN particles lead to irregularities in the austenite boundaries, which give an effectively larger grain boundary area.

The room temperature strength and toughness of the Ti steel are consistently somewhat better than for the C-Mn steel given the same rolling treatment. The differences are consistent with the differences in ferrite grain size (14) and show no effects attributable to the presence of TiN. Elongation to failure tends to be lower for the Ti steel, probably because of the presence of the coarse particles, as no evidence was found of grain boundary carbide films, which may occur on slow cooling Ti steels of Ti:N ratio significantly higher than stoichiometric.

Conclusions

It is concluded that the addition of titanium in stoichiometric ratio with nitrogen in C-Mn steel produces a small volume fraction of fine TiN particles, which restrain grain growth of austenite, but have no direct effect on the kinetics of dynamic or static recrystallisation or on the grain size produced by static recrystallisation. Maximum benefit from these effects could be realised in rolling of heavy plate or sections from continuously cast products, when only a single reheat is required and advantage can be taken of the high grain coarsening temperature of Ti steel to obtain a fine initial grain size before rolling. Full recrystallisation, but no grain growth then takes place after each pass, leading to a final austenite grain size which does not depend on the total rolling reduction or on the finishing temperature, at least up to 1050°C. The austenite grain size does, however, depend critically on the reduction in the final pass, decreasing with increasing reduction, so that relatively coarse grain sizes would be obtained on recrystallisation after a small finishing pass.

On cooling, a finer ferrite grain size is obtained in the Ti steel than in C-Mn steel of the same austenite grain size, leading to the expected benefits in strength and toughness.

Acknowledgements

The authors are grateful for the financial support provided to Luis Leduc by the Bank of Mexico and the National University of Mexico.

651

References

1. C. M. Sellars and J. A. Whiteman, "A Look at the Metallurgy of the Hot Rolling of Steel", Metallurgist and Materials Technologist, 6 (1974) pp. 441-447.
2. L. Leduc, T. Nadarajah and C.M. Sellars, "Density Changes During Hot Rolling of Cast Steel Slabs", Metals Technology, 7 (1980) pp. 269-273.
3. T. J. George and J. J. Irani, "Control of Austenite Grain Size by Additions of Titanium", J. Aust. Inst. Metals, 13 (1968) pp. 94-106.
4. T. J. George, G. Bashford and J. K. MacDonald, "Grain Size Control in Structural Steels", J. Aust. Inst. Metals, 16 (1971) pp. 36-48.
5. C. Ouchi, T. Sanpei, T. Okita and I. Kozasu, "Microstructural Changes of Austenite during Hot Rolling and their Effect on Transformation Kinetics", pp. 316-340 in The Hot Deformation of Austenite, John B. Ballance, ed; AIME, New York, N.Y., 1977.
6. C. M. Sellars, "The Physical Metallurgy of Hot Working", pp. 3-15 in Hot Working and Forming Processes, C. M. Sellars and G. J. Davies eds; Metals Society, London, 1980.
7. S. R. Foster, "Simulation of Hot Rolling of Low Carbon Steels", Ph.D. research, University of Sheffield.
8. T. Gladman, "On the Theory of the Effect of Precipitate Particles on Grain Growth in Metals", Proc. Roy. Soc., 294A (1966) 298-309.
9. J. P. Sah, G. J. Richardson and C. M. Sellars, "Grain Size Effects during Dynamic Recrystallisation of Nickel", Metal Sci., 8 (1974) pp. 325-331.
10. S. Sakui, T. Sakai and K. Takeishi, "Hot Deformation of Austenite in a Plain Carbon Steel", Trans. Iron Steel Inst. Japan, 17 (1977) pp. 718-725.
11. W. Roberts, H. Boden and B. Ahlblom, "Dynamic Recrystallisation Kinetics" Metal Sci., 13 (1979) pp. 195-205.
12. K. J. Irvine, F. B. Pickering and T. Gladman, "Grain-Refined C-Mn Steels" J. Iron Steel Inst., 205 (1967) pp. 161-182.
13. C. M. Sellars and J. A. Whiteman, "Recrystallisation and Grain Growth in Hot Rolling", Metal Sci., 13 (1979) pp. 187-194.
14. F. B. Pickering, "High-Strength, Low-Alloy Steels - A Decade of Progress" pp. 9-30 in Microalloying 75, Union Carbide Corp., New York, N.Y. 1977.
15. L. A. Leduc, "Hot Rolling of Titanium-Bearing Steels", Ph.D. Thesis, University of Sheffield, 1980.
16. D. R. Barraclough, "Hot Working of Stainless Steel and Low Alloy Steel", Ph.D. Thesis, University of Sheffield, 1974.

Appendix

If during recrystallisation it is assumed that N_v nuclei per unit volume form instantaneously and then grow uniformly with a constant shape, e.g. as tetrakaidecahedra, but without increasing their number, then the volume fraction recrystallised (X) is related to their mean linear intercept size $(\bar{d})$ and to the mean linear intercept grain size of the fully recrystallised material $(\bar{d}_{rex})$ by the relationship

$$X = (\bar{d}/\bar{d}_{rex})^3 \tag{A1}$$

The number of recrystallising grains per unit length $(N_{L,r})$ is related to $\bar{d}$ as

$$N_{L,r} = X/\bar{d} \tag{A2}$$

Hence, substituting from equation (A1),

$$N_{L,r} = x^{2/3}/\bar{d}_{rex} \tag{A3}$$

The number of unrecrystallised grains per unit length ($N_{L,u}$) is related to the number of grains per unit length in the initial deformed structure (N_L^*) as

$$N_{L,u} = N_L^* \ (1 - X) \tag{A4}$$

In a partially recrystallised structure the total number of grains per unit length is then

$$N_L^M = \frac{1}{\bar{d}_L} = N_L^* \ (1 - X) + x^{2/3}/\bar{d}_{rex} \tag{A5}$$

If the original undeformed structure is equiaxed grains of size $\bar{d}_o$, then on plane strain deformation without grain boundary sliding, the values of N_L^* in the longitudinal and through thickness directions are related to the fractional reduction in thickness (R) as

$$N_{L,L}^* = (1 - R)/\bar{d}_o \tag{A6}$$

and

$$N_{L,T}^* = 1/(1 - R)\bar{d}_o \tag{A7}$$

Then, by substitution of equations (A6) and (A7) into equation (A5) the mean grain size in the two directions in the recrystallising structure is obtained and from equations (1) and (2) the aspect ratio and overall mean grain size can be calculated as a function of X and compared with experimental observations. Good agreement between the predicted form of curve and observed grain sizes has been found (15) for data on 304 stainless steel (16), for which the metallography is more reliable than for transformed austenite.

During rolling, temperature changes continuously during recrystallisation after a pass. When the cooling rate is known, an equivalent temperature compensated time for recrystallisation can be obtained using the relationship (13)

$$W = \Sigma \ (\exp - Q_{rex}/R \ T_i) \ \delta t_i \tag{A8}$$

where δt_i is a short time interval for which the temperature is T_i, and Q_{rex} is the activation energy for recrystallisation, taken to be 300 kJ/mol for C-Mn steel (6). Taking the kinetics of recrystallisation to follow an Avrami equation with an exponent of 2 gives

$$X = 1 - \exp - 0.693 \ (W/W_{0.5})^2 \tag{A9}$$

where $W_{0.5}$ is the temperature compensated time for 50% recrystallisation after a specific deformation. As it has been found previously (6) that recrystallisation times for C-Mn steels are not sensitive to temperature of deformation, equations (A8) and (A9) have been used to calculate X as a function of rolling temperature to obtain the curves in fig 2.

DISCUSSION

Q: You passed over very quickly that the austenite grain size might more or less explode if the austenite to ferrite transformation is passed several times. You also quickly said something that had to do with the solution effects on the precipitate structure. Therefore, I wonder if you could possibly elaborate a bit on what you actually think is happening.

A: If you hold these steels isothermally at temperatures in the austenite range, the coarsening rate of the Ti nitrides (even with a little bit of excess Ti as we had) is very, very slow. But, when one goes through the transformation repeatedly, for reasons we don't really understand, you do seem to accelerate the coarsening. But more importantly, every time you cycle through the transformation, you tend to refine the austenite grain size that you get when you reheat again. If you look at the Gladman equation you will note that, with the same distribution of particles, your grain-coarsening temperature will decrease drastically after the initial grain size decreases. I agree with that and it is the same sort of explanation we would like to propose.

Q: In your last figure, you showed two Ti-steels (0.02 and 0.06%) and both have a stoichiometric ratio of Ti to nitrogen. In the recrystallized Ti-steels, is this the result for both Ti steels or only one?

A: These are the results of the high Ti-steel. However, the addition of more nitrogen and more Ti is not a very sensible thing to do, because you produce a larger volume fraction of coarse precipitates and they don't do you any good. I didn't show the mechanical property results, but they may in fact slightly decrease the ductility and you hardly have any effect at all on this very small volume fraction of fine particles. After we completed the experiment by adding Ni and Ti, we realized it was the wrong thing to do, and that the minimum amount of Ti to tie-up the nitrogen would be just as effective and less dangerous if you could control it.

Q: I have just seen your analysis for measuring grain size and getting mean grain size and aspect ratios. However, the grains have three dimensions, and I believe you should have obtained your mean grain size from the cube root to the product of the three dimensions, rather than the square root of the product of two of the dimensions. Although, perhaps it won't make much difference in the final analysis.

A: At the plane strain conditions, it makes no difference, but if the grain size changes, the long transverse dimension changes as well. It is a small point, but it might be worth looking at the derivation again to see if it makes any difference. The same thing occurs when aluminum nitride behaves in much the same way as the Ti nitride steel. In some work we did in 1967, we found that the killed steel and nonkilled steel behave in just the same way.

HSLA Ti - CONTAINING STEELS[*]

J-C. Herman   P. Messien   T. Gréday

Centre de Recherches Métallurgiques

11,rue E. Solvay - B-4000 Liège (Belgium)

Titanium containing low carbon (.08%) steels are studied to perform: austenite grain control, refinement of the ferritic microstructure, ferritic substructure strengthening and inclusion shape control. Within the solidification range, titanium combines with nitrogen and sulfur ; conditions for the globularization of the sulfides are computed according to the steel chemistry. The hot-working process governs the TiC microprecipitation and the resulting steel properties. TiC formed in austenite promotes grain refinement and delays the recrystallization. Carbon content is the leading parameter for the mechanical properties of Ti-steels owing to its action on size and distribution of the TiC particles and on the formation of filamental Fe_3C. Yield strengths up to 600 MPa are obtained when an ultra fine precipitation stabilizes the ferritic substructure. Refinement of the ferrite grain size occurs with an increase of the Ti and Mn contents ; elongated grains are also obtained as a result of the uncomplete recrystallization of the austenite. When TiC precipitates in the upper part of the γ range further grain refinement and improved impact strength are obtained.

[*] This research has been carried out with a financial support of the IRSIA (Belgium Institute for Scientific Research in Industry and Agriculture)

Introduction

In HSLA steels, the austenitic grain size is controlled by two possible mechanisms : the carbide -nitride microprecipitation and the hot-deformation process generally performed by controlled rolling.

The effects of such a thermomechanical treatment have been widely studied in the last decade, more precisely with respect to dynamic or static recrystallization process of the austenite (1-3). The extended use of HSLA steels produced by controlled rolling has also promoted the study of the influence of such carbide or nitride forming elements as Nb, V and Ti on these recrystallization processes (4-8).

Recently, we have reviewed the potential capabilities of titanium additions which can be used in low carbon steels to reach at least one of the following objectives :

- Austenite grain size control during hot-rolling by an appropriate TiC precipitation which finally promotes a fine ferritic microstructure associated to high yield and impact strengths.

- Formation of a substructure in the ferrite phase as a result of the same TiC microprecipitation.

- Inclusion shape control (sulfide spheroidization) to improve the impact strength in the transverse direction.

Materials and Experimental Procedure

The study was carried out using laboratory steels prepared by vacuum induction melting of 20 kgs heats. These low carbon (.05-.14%) Al-killed steels are characterized by various manganese (.3-1.2%) and titanium (.03%-.18%) additions. The levels of sulfur and phosphorus are constant and two levels of nitrogen .004-5% and .010% have also been used (Table I).

Table I. Chemical Composition Ranges of the Steels (Wt %)

C	Mn	Si	Ti	N_2
.05	.6	.01	.09	.005
.07-.08	.3-1.2	.01	.02-.17	.004-5-.010
.10	.6	.20	.09	.006
.11-.12	.6-1.0	.20	.09-.16	.006
.14	1.0	.20	.16	.006

★ Al_{tot}: .03-.04, S : .015, P : .01%

The square base ingots were forged to 20mm (.79 in.) thick slabs before hot-rolling. The processing conditions for hot-rolling are shown in Table II. Variations of the finishing rolling temperature from 920°C to 820°C (1688°F to 1508°F) have been realized on a limited number of ingots in view to assess the influence of this parameter on both the austenitic and ferritic grains sizes.

Tensile specimens, 31.6mm (1.24in.) long and 4mm (0.16in.) in diameter, were machined in both the rolling and transverse directions to rolling.

```
Table II - Rolling Conditions

1. Reheating temperature : 1200°C (2191°F) - 2 h ;
2. Starting rolling temperature : 1100°C (2012°F) ;
3. Finishing temperature : 920°C (1688°F) to 820°C (1508°F)_;

4. Final thickness : 8 mm (.315 in.).
5. Number of passes : 3 ; reduction at the last pass : 25%.
6. Coiling temperatures (strips) : 620°C (1148°F) - 680°C (1256°F)
                                    720°C (1328°F).
   Cooling rates (plates) : 2°C/sec (3.6°F/sec) - 12°C/sec (21.6°F/sec) -
                            30°C/sec (54°F/sec).
```

The mechanical properties were measured at a strain rate of 2mm/min (0.079 in./min). Yield strength, ultimate tensile strength and total elongation were determined on triplicate specimens using a gage length of 20mm (0.70 in).

Impact tests using reduced (5x10x55mm) Charpy V-notch specimens have been performed both in rolling and transverse directions and testing temperatures have been varied from 40°C to -20°C (104°F to -4°F).

Optical microstructural analysis was performed in view to characterize both the inclusion content of the steels (TiN, TiS, (Ti,Mn)S, MnS) and the grain sizes and anisotropy, as measured by the ratio longitudinal/transverse of the mean intercept lengths. Moreover, microstructural examinations using transmission electron microscopy were also used to characterize the microprecipitation of the carbonitrides leading to its quantification by quantitative metallography techniques.

Results and Discussion

Thermodynamic Aspects of the Titanium Precipitation

The precipitation behavior of the titanium carbides and its resulting precipitation hardening is hardly influenced by the carbon and titanium contents which determine the starting precipitation temperature of the carbides formed in the austenitic range as well as during the $\gamma-\alpha$ transformation.

We have established a thermodynamical model based on the partition of the elements (9) to calculate the contents in carbon, manganese, titanium sulfur and nitrogen of the interdendritic liquid during the solidification, assuming that the equilibrium is reached at any temperature during the cooling ; the solidus curve has being varied according to the enrichments of the liquid in the above mentioned elements, consequently we were able to precise the composition of the precipitates appearing until the end of the solidification as well as the temperature of precipitation. As a result of this model, the titanium fraction combined with nitrogen and sulfur has been computed in function of the steels analysis. Three mains conclusions are derived from these computations :

- According to the temperature range of the precipitation (homogeneous liquid, interdendritic liquid or slid),the mean size of the TiN precipitates is changed. To control the austenic grain size during the slab reheating treatment,a maximum of fine precipitates (area $\leq 1\mu^2$) is needed ; they are only formed when the steel chemistry is such that a majority of the TiN precipitates in the solid phases when the titanium content is well adapted to the nitrogen content ; as an example for .005% nitrogen, titanium must be .03% or less.

657

The mean size of titanium nitrides increases when the precipitation starts at a higher temperature (homogeneous or interdendritic liquid) thus when titanium-or nitrogen content increases. Nitrides of larger size ($\leq 15\mu^2$) are observed when the steel chemistry is such that a precipitation in the homogeneous liquid is avoided (Fig. 1).

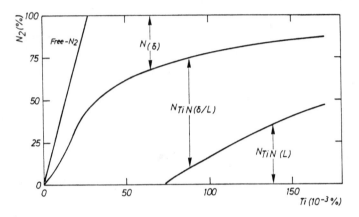

Fig. 1 - Influence of the titanium content on the nitrogen distribution between titanium nitrides precipitated in solid ($N_{(\delta)}$, in inter-dendritic liquid ($N_{TiN(\delta/L)}$ and in homogeneous liquid ($N_{TiN(L)}$); steel composition : C=.08%, Si=.3%, Mn=.9%, N_2=.008%, S=.015%.

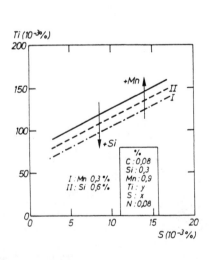

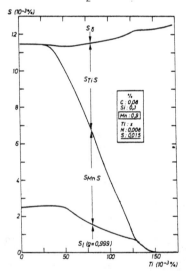

Fig. 2 - Minimum titanium content required to change the typical elon-gated morphology of the sulfide in-clusions into a globular form ; in-fluence of titanium, sulfur, manga-nese, and silicon contents.

Fig. 3.- Evolution of the sulfur dis-tribution between ferrite, TiS, MnS and liquid fraction at a solidifica-tion rate g=.999,with the titanium content.

- Higher is the sulfur content of the steel, higher will be (Fig. 2) the titanium addition needed to realize a complete globularization of the sulfides. When the steel contains .015% S this is only obtained for a minimum addition of 0.150% Ti ; then the titanium fraction combined with the sulfur (TiS) is .022%. At lower titanium content, pure manganese and titanium sulfides and complex (Ti,Mn)S sulfides are formed ; in any case, the titanium fraction combined with the sulfur can be computed (Fig. 3).

- The titanium fractions combined to nitrogen and sulfur being so calculated the free titanium content Ti_F is derived. This Ti_F content is then available for the precipitation hardening.

Dissolution of the Titanium Carbides during the Slab Reheating

The slab reheating temperature is the first parameter to be considered in order to determine the importance of the precipitation hardening after hot-rolling. A higher carbon content of the steels, but the free titanium content being constant, will increase the dissolution temperature of the titanium carbides (Fig. 4) ; as a result it can be concluded that, for an usual free titanium content of .12%, a maximum carbon content of .08% is required for a complete dissolution of the carbides during a slab reheating at 1200°C (2192°F). All the slabs having been reheated at this temperature, different starting conditions as regard the initial (undissolved) precipitation are thus realized before hot-rolling according to the carbon and titanium contents of the steels.

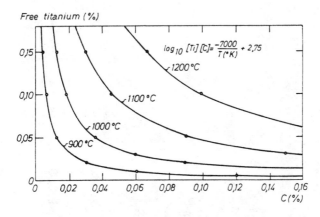

Fig. 4 - Solubility isotherms of the titanium carbide (7).

The undissolved carbides can control the austenic grain size during rolling and they act also as preferential sites for the precipitation of the carbides (precipitate-induced precipitation) (8) which are formed at the higher temperatures of the austenitic range;such precipitates reduce the precipitation hardening as it has been confirmed by microstructural examinations : the lower is the carbon content of the steel, the finest is the TiC precipitation (Fig. 5 - Table III) and the larger is the precipitation hardening. This results from the lower starting temperature for the carbides precipitation during hot-rolling. Moreover a finer precipitation is then obtained during the subsequent γ-α transformation,with a maximum hardenability.

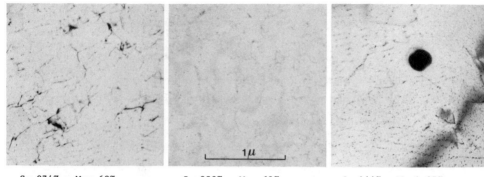

C=.074% - Mn=.60%　　　　C=.052% - Mn=.63%　　　　C=.144% - Mn=1.07%
　Ti$_F$ = .085%　　　　　　　Ti$_F$ = .090%　　　　　　　Ti$_F$ = .115%

Fig. 5 - Influence of the titanium and carbon contents on the TiC precipitation ; coiling temperature : 680°C (1256°F).

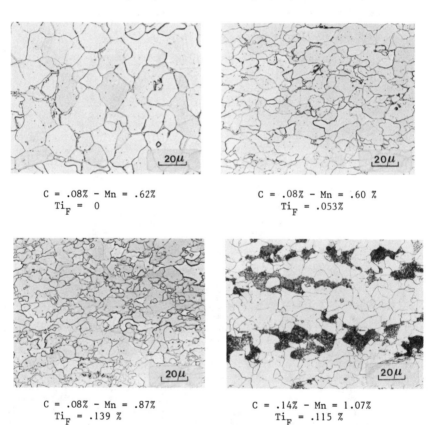

C = .08% - Mn = .62%　　　　　C = .08% - Mn = .60 %
　　Ti$_F$ = 0　　　　　　　　　　Ti$_F$ = .053%

C = .08% - Mn = .87%　　　　　C = .14% - Mn = 1.07%
　　Ti$_F$ = .139 %　　　　　　　Ti$_F$ = .115 %

Fig. 6 - Ferrite grains refinement ; influence of the titanium and carbon
contents ; coiling temperature : 680°C (1256°F).

Table III. TiC Granulometry as Determined by Quantitative Metallography

Free titanium %	Carbon %	TiC Mean Diameter (Å)	Percentage of carbides			
			d < 125 Å	125< d < 250 Å	250< d <500 Å	d > 500 Å
.130	.120	193	39	30	16	15
.083	.085	95	60	21	13	6
.140	.075	89	63	24	9	4
.085	.050	60	100	–	–	–

As a result of our observations, an increasing control of the austenitic grain size during rolling is obtained, even without undissolved initial
carbides, by an increase of the free titanium content of the steel which
induces precipitation in the higher temperature range during the rolling
process. As a consequence, a finer ferritic grain size is obtained at larger free titanium contents. Moreover, it has been observed that, in these
conditions, there is a limitation (Fig. 6) in the supplementary grain refinement resulting from an increase of the carbon content which leads to an
incomplete dissolution of the titanium carbides at 1200°C (2192°F). Consequently and as far as the mechanical strength is concerned, these limited
grain refinement cannot justify the choice of an increased carbon content
of the steel which induces the carbides precipitation at higher temperature
during rolling but reduces markedly the precipitation hardening (Fig. 7).

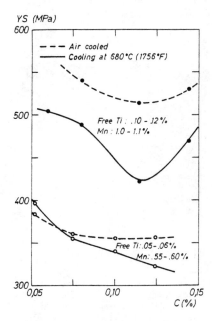

Fig. 7. Influence of the carbon content on the yield strength of titanium
steels.

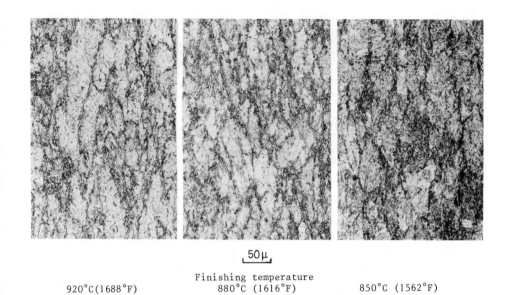

$\lfloor 50\,\mu \rfloor$

Finishing temperature

920°C(1688°F)   880°C (1616°F)   850°C (1562°F)

Fig. 8.- Austenitic grain size vs the finishing temperature ; steel composition:
C:.085%, Mn:.81%, Ti:.178% ; magnification : 200 x.

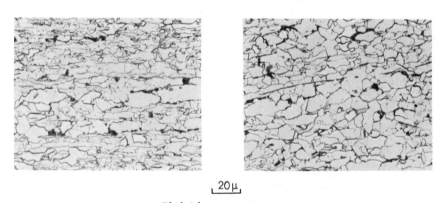

$\lfloor 20\,\mu \rfloor$

Finishing temperature

820°C(1508°F)   880°C (1616°F)

Fig. 9 - Influence of the finishing temperature on the homogeneity of the ferritic grains ; steel composition : C:.085%, Mn:.81%, Ti:.178%; coiling
temperature : 680°C (1256°F) ; magnification : 500 x.

Influence of the Thermomechanical Treatment

A minimum austenitic grain size (Fig. 8) is obtained when use is made of a finishing temperature of 880°C (1616°F). It promotes an optimal homogeneity of the ferritic microstructure at room temperature (Fig. 9).

Final accelerated cooling of plates increases not only the heterogeneity of the ferritic grain size but also promotes a substructure in the ferrite and refines the titanium precipitation (Fig. 10). As a result important modifications of the mechanical properties have been measured (Table V) : their anisotropy increases at low finishing temperatures owing to the elongation of the substructure cells into the rolling direction.

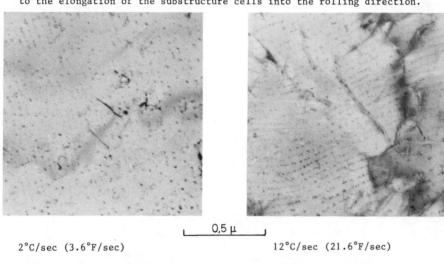

| 0,5 μ |

2°C/sec (3.6°F/sec)   12°C/sec (21.6°F/sec)

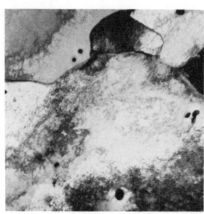

| 5 μ |

2°C/sec (3.6°F/sec)   12°C/sec (21.6°F/sec)

Fig. 10 - Influence of the cooling rate on the titanium carbides size and the substructure ; steel composition : C:.085%, Mn:.81%, Ti:.178%; finishing temperature : 880°C (1616°F).

Table IV. Influence of the Cooling Rate (C.R.) and the Finishing Temperature (F.T.) on the Anisotropy of the Mechanical Properties ; Steel Composition : C=.085%, Mn=.81%, Ti=.178%

	F.T. ＼ C.R.	2°C/sec (3.6°F/sec)	12°C/sec (21.6°F/sec)	30°C/sec (54°F/sec)
R.D.* T.D.*	820°C (1508°F)	530 516	575 557	656 641
R.D. T.D.	850°C (1562°F)	525 510	563 556	601 683
R.D. T.D.	880°C (1616°F)	512 505	563 556	613 661

* R.D. and T.D.: measured respectively in rolling and in transverse directions.

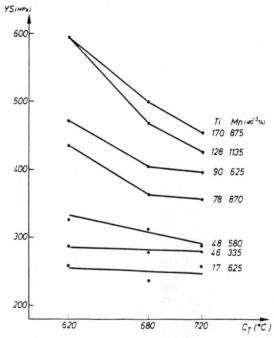

Fig. 11 - Effect of the coiling temperature (C_T) on the yield strength.

A similar behavior is observed for coiled strips : reducing the coiling temperature modifies and refines the precipitation mainly when the titanium content is high and ,consequently, the mechanical strength is increased.

Nevertheless, as far as a maximum stability of the mechanical properties is required with respect to variations in the rolling process, a coiling temperature of 680°C (1256°F) is to be used (Fig. 11).

The heterogeneity, more particularly the anisotropy, of the ferritic microstructure depends also on the steel chemistry and especially on the Ti/$_C$ ratio (Fig. 12.a). This anisotropy of the ferritic grains, enhanced at higher manganese contents, induces an anisotropy of the Charpy V-notch toughness, especially for high titanium steels although their sulfides globularization is complete (Fig. 12.b).

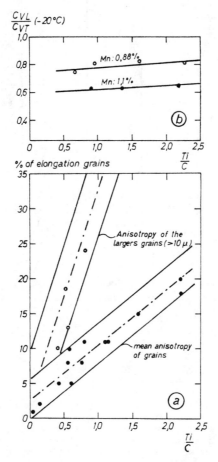

Fig. 12 - Influence of the Ti/C ratio on the anisotropy of the ferritic grains (a) and Charpy V-notch toughness anisotropy (b).

High Strength Titanium Steels

The range of the mechanical strengths covered by titanium steels depends on their carbon, manganese and titanium contents (Table V). Statistical analysis has been used to express the influence of the steel chemistry on the mechanical properties, at a constant carbon level (.08%).

Table V. Influence of the Chemical Analysis on the Mechanical Properties
in the Rolling Direction (coiling : 680°C - 1256°F)

Carbon	Mn = .6%			Mn = 1-1.2%		
content	.08%	.10%	.125%	.08%	.125%	.144%
Ti=.09% Y.S. MPa	358	343	324	374		
T.S. MPa	457	431	418	477		
A %	26.6	32.3	29.2	27.1		
C_v(J) -20°C	17	26	25	212		
Ti=.16%	500				422	470
	610				552	610
	24.0				25.2	23.6
	20				76	46

Table VI. Influence of the Chemical analysis on the Yield Strength of Low
Carbon (.08%) Steels (S=.015%) coiled at different temperatures(C.T)

$$YS \ (MPa) = A_1 \ [\% \ Ti \ ] + A_2 \ [\% \ Mn \ ] + A_3 \ [\% \ N \ ] + A_0$$

A_i C.T.	720°C (1328°F)	680°C (1256°F)	620°C (1148°F)	Air cooling 2°C/sec 3.6°F/sec
A_o	284	283	178	203
A_1	1430	1650	2480	1450
A_2	9	28	119	130
A_3	11300	11300	11300	7300

Four regression equations have been calculated. They correspond to three
coiling temperatures (strips) and one cooling rate (plate), the finishing
rolling temperature being 880°C (Table VI). The Ti-hardening coefficient
decreases with an increase in the coiling temperature. Moreover, higher
is the carbon content of the steel, lower will be the titanium hardening
coefficient (Fig. 13). The Mn-hardening coefficient is higher than the
one derived from a classical Petch analysis (5). Its dependence in respect to the coiling temperature and the cooling rate results from a refinement of the TiC precipitates formed during the allotropic transformation
which takes place at lower temperature when the manganese content increases (7).

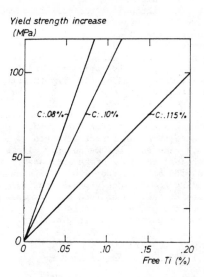

Fig. 13 – Influence of the carbon content on the hardening coefficient of titanium ; air cooling.

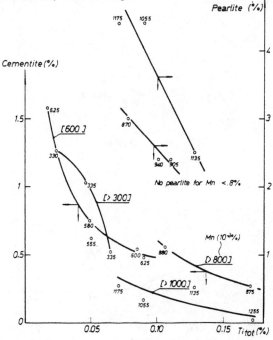

Fig. 14 – Effect of the manganese and titanium contents on pearlite and filamental cementite at grain boundaries. Volume fractions ; coiling temperature : 680°C (1250°F).

Large modifications in the percentage and the size of the pearlite are also observed as a result of variations of the coiling temperature, the cooling rate and the chemistry of the steel (C, Mn, Si contents). The most important parameters to be considered as far as the impact strength of the steels is concerned,are the percentage and the thickness of filamental cementite in the ferritic grains boundaries (Fig. 14). Both the percentage and the thickness of these cementite particles are reduced by increasing the Ti and Mn contents, the former acting by lowering the steel free carbon content during the γ–α transformation and the latter by diminishing the eutectoid carbon content. In high carbon (.1 – .25 %) titanium steels, appearance of cementite,the percentage of which increases with the steel carbon content, is the essential factor which explains that no significant improvement of the Charpy V-notch impact strength is obtained while the mechanical strength is reduced. One means only allows to enhance the notch impact strength : it consists of an increase of manganese content to 1-1.2% associated with a low carbon (.05-.08%) content and a sufficient titanium content leading to sulfides globularization (Fig. 15). Nevertheless, as a consequence of the use of such higher manganese contents, the anisotropy of the ferritic grains increases (Fig. 16) and larger titanium additions are required to perform a given globularization level of the sulfides (Fig. 2).

Higher Impact Strength in HSLA-Titanium Steels.

Higher impact strength is obtained in HSLA titanium hot-rolled plates or strips (YS ≤ 450 MPa) when use is made of a heat treatment which refines the microstructure and reduces the cementite precipitation in the ferritic grains boundaries.

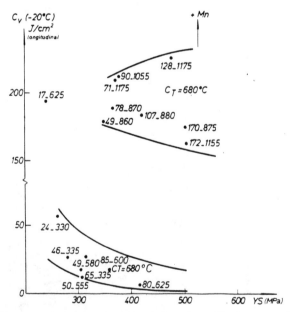

Fig. 15 - Influence of the steel composition (Mn, Ti) in the relation between yield strength and Charpy V impact strength at -20°C (-4°F).

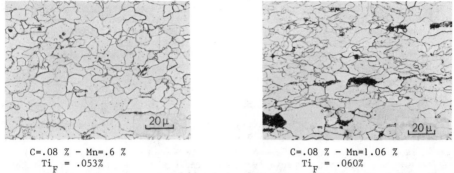

C=.08 % – Mn=.6 %
Ti_F = .053%

C=.08 % – Mn=1.06 %
Ti_F = .060%

Fig. 16 – Influence of the manganese content on the anisotropy of the ferritic
grains; coiling temperature:680°C (1256°F); magnification : 500 x.

As hot-rolled   Heat treated

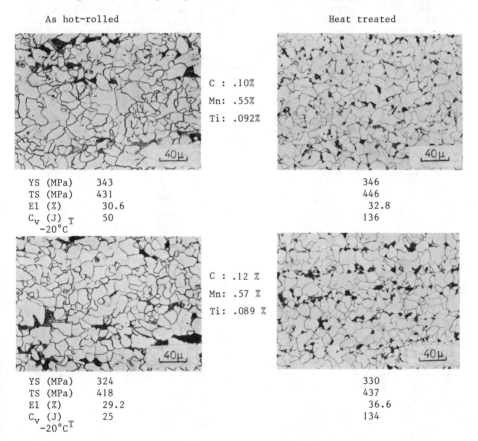

C : .10%
Mn: .55%
Ti: .092%

YS (MPa)	343	346
TS (MPa)	431	446
El (%)	30.6	32.8
C_V (J) $_T$ −20°C	50	136

C : .12 %
Mn: .57 %
Ti: .089 %

YS (MPa)	324	330
TS (MPa)	418	437
El (%)	29.2	36.6
C_V (J) $_{-20°C}$ ^{T}	25	134

Fig. 17 – Comparison of the microstructures and mechanical properties of as
hot-rolled (finishing : 880°C – 1616°F), coiling : 680°C – 1256°F)
and heat-treated (900°C – 1652°F) titanium steels.

In low carbon titanium steels, these improvments are achieved simultaneously with an important drop of the tensile strength, as a result of the coalescence of both the TiC fine precipitates and the substructure. In opposition for higher carbon contents (.12-.15%), the carbides precipitation which has occured in the austenitic range, is not affected by the heat treatment so that the mechanical strength can be improved owing to the microstructure refinement, simultaneously with an important increase of the impact strength (Fig. 17).

Conclusions

The present work has been carried out to precise the effects of a titanium addition to HSLA steels. As a result, we conclude :

- Yield strengths within the range 300-600 MPa are obtained together with good impact strength, especially in the transverse direction, owing to the sulfides globularization when sufficiently high Ti contents are used.

- The slab reheating temperature is an essential parameter which determines together with the steel carbon content, the features of the subsequent TiC precipitation and thus its hardenability ; ultra fine precipitates occuring simultaneously with the phase transformation induce not only an important precipitation hardening but also stabilize the substructure in low carbon (.05-.08%) steels ; as a consequence lower is the carbon content higher will be the yield strength for given titanium and manganese contents.

- Reduced strength levels are measured for steels with higher carbon contents (.1-.15). This results from a precipitation of TiC in the austenite.It is not linked to any marked improvment of the impact strength owing to an increase of the percentage of cementite particles in the grain boundaries ; this effect can be limited by an increase of the titanium and manganese contents in which case the notch toughness is enhanced.

- Austenite grain size control during hot-rolling is achieved through an important precipitation in the upper part of the austenitic range ; ferrite grains refinement increases with the titanium content.

- A heat treatment of the hot-rolled steel can achieved a still better combination of yield and impact strengths as a result of a supplementary grain refinement and a reduction of the percentage of intergranular cementite; for steel carbon contents in the range .12-.15%, this toughness improvment is obtained without any change in the tensile properties.

References

1. M. Lamberigts and T. Gréday, "Synthesis Report on Research into the Thermomechanical Treatments of Steels", Steel Research Reports ; Commission of the European Communities - EUR 5828, 1977, pp. 5-14-43-64.

2. C.M. Sellars, "The Influence of Particles on Recrystallization during Thermomechanical Processing", paper presented at the 1st Risø International Symposium, Sept. 1980, p. 297-301.

3. W.J. Mc G. Tegart and A. Gittins "The Hot Deformation of Austenite", in the Hot Deformation of Austenite, John Ballance, ed. AIME, New-York, 1977, p. 7-46.

4. R.A. Petkovic, M.J. Luton and J.J. Jonas, "Flow Curves and Softening Kinetics in H.S.L.A. Steels", in The Hot Deformation Austenite, John Ballance, ed., A.I.M.E., New York, N.Y., 1977.

5. M. Lamberigts and T. Gréday, "Precipitation and Recrystallization in Dispersion-Hardened Steels", C.R.M. Metall. Rep., 38, Liège (1974) pp. 23-38.

6. P. Messien and T. Gréday, "Réchauffage à basse température et laminage contrôlé", Revue de Métallurgie, Paris (1979), pp. 172-181.

7. L. Meyer, F. Heisterkamp and W. Mueschenborn, "Columbium, Titanium and Vanadium in Normalized, Thermo-Mechanically Treated and Cold-Rolled Steels", in Microalloying 75, John Crane, ed., Union Carbide Corporation, New York, N.Y. 1977, p. 153-167.

8. I. Takahashi, T. Kato, T. Tanaka and T. Mori, "Development of a High-Strength, Hot-Rolled Steel with 100 Ksi Yield Strength", paper presented at 106th AIME Annual Meeting, Atlanta, Ga, March, 1977.

9. E.T. Turkdogan and R.A. Grange, "Microsegregation in Steel", Journal of the Iron and Steel Institute, (1970), pp. 482-494.

CONFERENCE SUMMATION

Michael Korchynsky
UNION CARBIDE CORPORATION
Robinson Plaza II, Rt. 60
Pittsburgh, Pennsylvania 15205

We are at the end of the three long, tiring but certainly very exciting days, and the time has come to answer the question of what we have accompli-shed. That is a subjective task, and I can assure you I didn't volunteer for it. By the infallible decision of the Organizing Committee, I have been told that I am the appointed man, so please bear with me. Since an evaluation of our three days of discussion is a very subjective project, and I presume every one of us has to do it on his own, perhaps I can suggest the criteria by which we assess what we have accomplished in these sessions.

First of all, it's important to know what progress has been made in the science of the deformation of microalloyed austenite in the last six years. As a benchmark we could consider the state of the art as it existed five or six years ago, say, at the time of the Washington symposium on microalloyed steels. As you probably recall, in the proceedings of Microalloying 75, we had some classical papers, like LeBon's, Tanaka's, Kozasu's, and Fukuda's. This was at a time when both Professor Sellars and Professor Jonas had established the methodology of evaluating the hot defor-mation behavior of austenite and had already created a rather large group of followers. Whether we made any progress or not in this rather well established research area would be one of the criteria.

The second way of assessing our progress could be related to the following: The reason why we are interested in the deformation of microalloyed austenite is not strictly a scientific problem. The reason for us joining the bandwagon was the demands of technology. The high-strength, low-alloy steels had been developed in the last 15 years and there was a need to answer many technological questions, such as, how to improve the processing of these steels. And for that reason, the second criterion should be, "How much has our effort helped to advance the technology?"

And finally, the third, and perhaps most important, aspect is "What now? What are the future trends, and where are we going to go from here?" There will be other conferences on similar subjects in the near future, and we would like to know whether there are any directions that we should follow more intensively than others.

When we look at the whole gamut of topics covered in these three days, one is relating reheating of slabs or billets to grain coarsening. Somehow we have rediscovered the phenomenon of grain coarsening, and get excited about the three stages of coarsening, with Gladman's theory being invoked. Gradually we realize that the stability, and the dispersion of particles is very important phenomenon. Work by Pickering, Fitzsimons, and Tiitto referred to this phenomenon, and they pointed out that when we add either titanium or niobium or vanadium, the nature of the precipitated phase is very important. There are great differences between vanadium or niobium carbides and their nitrides. In studies of grain coarsening, the type of the precipitated phase becomes a very important factor. This introduces the concept of using the most stable compound, titanium nitride. Its use opens a new field of creating fine dispersions of stable particles to prevent grain coarsening during reheating and retain fine austenitic grain structure prior to the deformation.

Another area, touched on in several papers, deals with the question of how the microalloy additions, either as solutes or as precipitates, affect the dynamic or static recrystallization. Whether the retardation or even the suppression of recrystallization is due to precipitation or solute atoms, was a subject of arguments and controversy. Even today this question is not completely resolved in everybody's mind. Pickering in his work indicated that solute atoms seem to influence primarily the nucleation step of recrystallization. Fitzsimons and Tiitto have shown that the dynamic recrystallization is very strongly influenced by nitrogen in vanadium steels. Santella felt that the degree of supersaturation of austenite is one of the important factors controlling the inhibition of recrystallization. Cuddy had a model developed which associated the suppression of recrystallization to the increase of the solute atoms in austenite. A very large quantity of work on this subject was done by Tamura, who in a whole gamut of steels, studied the phenomenon of dynamic recrystallization. Similarly, the work by Bacroix indicated that there are some subtle interactions which influence the activity of interstitial atoms. For instance, the effect of molybdenum on both carbon and nitrogen may strongly influence the kinetics of retardation of austenite. We heard a very interesting discussion on this subject by Professor Sakai. Professor Gorczyca pointed out that whenever we talk about recrystallization, we have to be careful to realize how nonhomogenious this phenomenon can be. Perhaps the most significant amount of light on this subject was thrown by the paper by Yamamoto. This is a very elegant piece of work which demonstrates under what conditions precipitates retard recrystallization of austenite.

We all are fully aware that the boundary where solutes become precipitates or where a precipitate becomes a cluster is a gray area. Therefore, many of us will not be fully satisfied with the answer obtained thus far. But considering what we knew before, Yamamoto's paper provides a much more solid experimental evidence to assess this problem.

A very large number of papers dealt with the technological aspects and the simulation of rolling processes. In DiMicco's work, a very large number of rolling schedules was evaluated in a systematic way. Katsumata did some work to assess, among other things, the effect of initial grain size. The effect seems to be diminishing or of minor importance when the total deformation exceeds 70%. Very interesting work showing the effect of titanium nitride was given by Boyd. Work presented by Li Shuchung has indicated that under certain conditions low-temperature deformation may lead to grain coarsening. Extension of this work to the low-temperature field, to rolling in the $\alpha + \gamma$ field, was shown by Hashimoto. The improvements in transition temperature obtained by low-temperature rolling were quite impressive. Whether the presence of texture and associated splitting is or is not a tolerable engineering characteristic of these materials, remains to be seen.

Important from the standpoint of processing was the study reported by DiCello and Aichbhoumik on the increase in the resistance to deformation in niobium steels. The primary factor is the stiffness of the nonrecrystallized, work-hardened austenite, aggravated by concurrent precipitation. Process simulation dealing with specific materials was reported by Messien: titanium carbide precipitation was used as the major strengthening mechanism, associated with some grain refinement and sulfide shape control. Coldren gave a review on some Mo-Nb-V steels. Because of apparent suppression of precipitation of both niobium and vanadium in austenite, the molybdenum addition contributed to higher precipitation strengthening during and after γ to α transformation.

A new trend compared to what we knew before, has been reported by Sekine. He emphasized the importance of utilizing the recrystallization controlled rolling, and to capitalize on the refining of grains by recrystallization process. I believe that in the future recrystallization rolling may provide an alternative to contemporary low temperature controlled rolling. If the metallurgy of grain refinement through recrystallization rolling can be developed, then we will be able to take advantage of the associated increased productivity and economy.

Whenever we talk about controlled rolling, one problem which appears in work reported by DeMicco, Katsumata, and Tanaka is the occurance of mixed grains. Their occurance seems to be aggravated when we have strain-induced grain growth due to low deformations. It seems that in this instance, a high recrystallization temperature might be a curse rather than a blessing. Ways of combatting this problem have been reported in great detail by Tanaka. His contribution is a very valuable guideline on how to cope with this serious problem of mixed grains.

When discussing the effect of microalloying on the behavior of austenite, we fully appreciate that there is some effect of these elements in solution. But in HSLA steels, the major effect is associated with precipitation. First, a strain-induced precipitation in austenite, and then, precipitation taking place during and after transformation from austenite to ferrite. It is somewhat surprising that while we so often refer to the effects of vanadium, niobium, or titanium on properties of microalloyed austenite, only rarely we ask the question, "What is the precipitated phase?". In the past, we used the term "carbonitrides". Whenever we added one of these microalloying elements, we would use the formula C + 12/14 N. This suggested that with respect to the microalloying elements: V, Nb, or Ti, carbon and nitrogen are equivalent and we can treat them together. Recently, more and more data are being generated that show this is not the case, that nitrides and carbides are different species. The thermodynamic studies at the Swedish Research Institute by Roberts have shown convincingly that, depending on the nitrogen content of the steel, we have predominantly either nitrides, or, when nitrogen is exhausted, carbides. The region in which we have carbonitrides, at least in low-carbon vanadium steels, is very restricted. Interest in defining whether we have carbides or nitrides represents a new trend, and is evident in the work by Lagneborg and by Houghton at McMaster.

Since the solubility of nitrides is significantly lower and - consequently - the stability of their dispersion greater compared to carbides, the role of nitrogen might be more important in microalloyed steels than we appreciated in the past.

The approaches to how to identify the precipitating species are
different. The relatively simple model proposed by Wilson uses the equilibrium solubility data. This method may upset some people since we are
dealing with a kinetic situation. Nevertheless this is an attempt to show
the sequence of precipitation. Promising are both the thermodynamic approach,
reported by Lagneborg, and refined experimental techniques reported by
Houghton. We assume that there is a series of C-curves for precipitation
of various species. Perhaps the next step in development of microalloyed
steels would be an attempt to capitalize on these various C-curves spaced
over temperatures, so that in each temperature region of hot working use
is being made of different species.

. When thinking about precipitates: carbides versus nitrides, then
titanium nitride, the most stable compound, attracts special attention.
Work reported by Boyd and by Sellars falls in this category. This is
different from the paper by Messien, where the role of titanium was primarily to provide precipitation strengthening by titanium carbide precipitation
and inclusion shape control. Here we are talking about the very small
additions of titanium, 0.01-0.015%, in combination with nitrogen. This
work started initially primarily to control the grain coarsening tendency
in the heat-affected zone under conditions of very intensive heat input.
Dispersion of TiN is also a very effective way in controlling grain size
in slabs and ingots, as was discussed by Sellars. The technology of
adding very finely controlled amounts of titanium (0.01-0.015%) is available.
Through injection, we can control the amount of this addition with accuracy
of $\pm$ 20 ppm. At the same time, we are not compelled to control with equal
accuracy nitrogen. We are interested to achieve a certain degree of titanium nitride distribution, and an excess of nitrogen can be taken care of
effectively by some other means. Here one important point should be noted:
In several papers, the characteristic grain coarsening temperature of
titanium-nitrogen steels was correlated with the amount of titanium and
nitrogen. Nothing could be more misleading than that, since the grain
coarsening temperature of these steels can vary by $300^{\circ}C$ for identical
chemical compositions, depending on the degree of dispersion of titanium
nitride. In these steels, the control of microstructure, i.e., of the
degree of dispersion, is critical, and, to make use of these steels, continuous casting is a necessity.

The theme of this conference revolves around means of "conditioning"
the austenite, either by grain refinement, by introducing deformation bands,
or by having some precipitates, to achieve the main objective: to assure
the finest ferritic grain. The transformation of γ to α from a variety of
"conditioned" microstructures was reported by Pickering, Baker, Priestner,
and others. One of the important questions is the ratio of austenite to
ferrite grain size the $d_{\gamma}/d\alpha$ ratio. It can be very small, perhaps 1.0,
when we refine the austenitic grain size to 5-7 microns, or as high as 8-12,
if the austenite is coarser. A new parameter useful in maximizing ferritic
grain refinement from a given austenitic grain structure is the rate of
cooling. Whether this is primarily the undercooling effect, lowering the
A_3 temperature, or some other phenomenon is not clear. What we know is that
by accelerated cooling we can apparently increase N (the nucleation frequency) and decrease G (the growth), and the net result is grain refinement.

In our efforts to maximize the grain refinement of ferrite, of
importance is the question:, "Is there any particular shape of austenitic
grains that gives us maximum ferrite grain refinement?" In other words,
if we use the term introduced years ago by Kozasu, S_v (the effective austenitic grain boundary area per unit volume), are all austenitic grain
boundaries equivalent, and all that we have to achieve is to maximize S_v?

In the past, there was the prevailing opinion that this is not the case, that the ragged grain boundaries of deformed "pancaked" austenite provide more nucleation sites for ferrite. For that reason, for a given S_v, the transformation from nonrecrystallized austenite was expected to give greater grain refinement than transformation from fully recrystallized grains. Results reported by Lagneborg suggest that for high S_v values ($S_v \gtrsim 100$ mm^{-1}), the ferritic grain size depends only on S_v, and not on the shape of austenite grains: recrystallized (equiaxed) or deformed (pancaked). This suggests that for an effective refinement of ferrite, deformation of austenite below its recrystallization temperature may not be essential. In attempting to achieve maximum grain refinement of ferrite, perhaps we do not have to produce pancaked austenite grains, provided the recrystallized grains are sufficiently fine.

What does this mean to us for the future; where do we go from here? The only paper that tried to project the future was that by Dr. Sekine, where some of the ideas of what the future may hold for us have been outlined. Using some of his ideas and some of mine, I would like to submit for your consideration the following. First of all, it appears that the trend to start with the smallest austenitic grains in billets and slabs will continue. We may use for this purpose titanium nitride or exploit lower reheating temperatures, provided that the lower slab reheating temperature is compatible with the microalloy system being used, or something that was just touched upon by Professor Sellars. He has shown that because of the rapid heating employed in his experiments, the austenitic grains were much smaller than would be obtained under industrial conditions. The concept of rapid heating is an engineering reality and there are mills in this country which use induction heating for heating slabs. Perhaps this new processing parameter, the rate of heating, might be an important new development to assure finest initial austenitic grain size by preventing coarsening of dispersed phases.

The second new trend, which has already been started, is a study of the sequence of precipitation when there are more than one microalloy element in a steel. In particular, alloy systems which have C-curves of precipitation well separated over a temperature range might be particularly useful. It appears to me that primary interest should be devoted to the nucleation step, and for that reason, the analysis of the detailed final composition of the particles precipitated may be of secondary importance. Years ago, in one of the AIME memorial lectures, Professor Morris Cohen suggested that one who can control the nucleation step can develop new properties in old materials. In the HSLA steels alloy systems, the nucleation step may be more important than the subsequent growth step and accumulation on the nucleus. For example, comparing carbides and nitrides, nitrides are one which probably precipitate first, and they should be utilized to greater extent.

The third aspect would be to see what processing conditions are both technically and also economically attractive. Economically attractive are those which can be applied to existing mills, since the probability that we will rebuild our mills to be able to have 30-40% reduction in the final pass is somewhat slim. For that reason, some of our intellectual effort should be aimed at how to maximize S_v, the grain boundary area of austenite, under the most economic condition. The interplay between finishing temperature and accelerated cooling rate, may be important to get grain refinement by combining somewhat higher finishing temperature with accelerated cooling. For that reason, the concept of accelerated cooling, which has been adopted all over the world for hot strip mill products, is being adapted to plate products as well. At the Pittsburgh Works of J & L this concept was successfully used 15 years ago.

Last year, Nippon Kokan announced that they have a much larger installation for accelerated cooling of plates for linepipe. This gives them the opportunity either to increase the strength level without loss of toughness or to reduce the alloy content for a given strength level.

The majority of papers presented in these three days, were devoted to plate products. In plate products, the strength is a more trivial property than the strength/toughness combination. We know that to get good shelf energy and to meet the drop-weight tear test, heavy reduction at low temperatures below 900° are essential. But the improvement in toughness cannot be correlated with grain refinement. There are some other factors which are not completely understood. Whether this is related to texture or to the amount and distribution of carbides should be studied very intensively, because this might be one of the bottlenecks for products of both high strength and good toughness. At Nippon Kokan, accelerated cooling is applied after controlled rolling, i.e., heavy deformation in the low temperature region. Controlled rolling is believed to be essential to assure high toughness. Whether or not the necessity for controlled rolling could be partially or fully be eliminated by lowering the carbon content, is an open question.

These are the few ideas which come to my mind where the future may lead us, and when we meet three or four years from now at a similar meeting, you can challenge me in how wrong I was!

SUBJECT INDEX

AUTHOR INDEX